THE WILEY BICENTENNIAL—KNOWLEDGE FOR GENERATIONS

*E*ach generation has its unique needs and aspirations. When Charles Wiley first opened his small printing shop in lower Manhattan in 1807, it was a generation of boundless potential searching for an identity. And we were there, helping to define a new American literary tradition. Over half a century later, in the midst of the Second Industrial Revolution, it was a generation focused on building the future. Once again, we were there, supplying the critical scientific, technical, and engineering knowledge that helped frame the world. Throughout the 20th Century, and into the new millennium, nations began to reach out beyond their own borders and a new international community was born. Wiley was there, expanding its operations around the world to enable a global exchange of ideas, opinions, and know-how.

For 200 years, Wiley has been an integral part of each generation's journey, enabling the flow of information and understanding necessary to meet their needs and fulfill their aspirations. Today, bold new technologies are changing the way we live and learn. Wiley will be there, providing you the must-have knowledge you need to imagine new worlds, new possibilities, and new opportunities.

Generations come and go, but you can always count on Wiley to provide you the knowledge you need, when and where you need it!

WILLIAM J. PESCE
PRESIDENT AND CHIEF EXECUTIVE OFFICER

PETER BOOTH WILEY
CHAIRMAN OF THE BOARD

Visualizing
PSYCHOLOGY

Siri Carpenter, Ph. D.

Karen Huffman

Palomar College

In collaboration with
THE NATIONAL GEOGRAPHIC SOCIETY

CREDITS

VICE PRESIDENT AND PUBLISHER Jay O'Callaghan

MANAGING DIRECTOR Helen McInnis

DIRECTOR OF DEVELOPMENT Barbara Heaney

DEVELOPMENT EDITOR Johnna Barto

EXECUTIVE EDITOR Christopher Johnson

EDITORIAL ASSISTANT Eileen McKeever

EXECUTIVE MARKETING MANAGER Jeffrey Rucker

PRODUCTION MANAGER Kelly Tavares

SENIOR PRODUCTION EDITOR Sujin Hong

PRODUCTION ASSISTANT Courtney Leshko

CREATIVE DIRECTOR Harry Nolan

COVER DESIGNER Harry Nolan

INTERIOR DESIGN Vertigo Design

PHOTO RESEARCHER Tara Sanford/Mary Ann Price/
Stacy Gold, National Geographic Society

ILLUSTRATION COORDINATOR Anna Melhorn

MEDIA EDITOR Lynn Pearlman

COVER IMAGE

Main image: George Logan/Taxi/Getty

Smaller images (left to right): Anne Keiser/NG Image
Collection, Stacy Gold/NG Image Collection, Cary
Wolinsky/NG Image Collection, Jimmy Chin/NG Image
Collection, Howard Sochurek/NG Image Collection

This book was set in Times New Roman by PrePress Company, Inc., printed and bound by Quebecor World. The cover was printed by Phoenix Color.

To order books or for customer service please, call
1-800-CALL WILEY (225-5945).

ISBN 978-0-471-76796-1

Printed in the United States of America

10 9 8 7 6 5 4 3 2

Visualizing Psychology is designed to help your students learn effectively. Created in collaboration with the National Geographic Society and our Wiley Visualizing Consulting Editor, Professor Jan Plass of New York University, *Visualizing Psychology* integrates rich visuals and media with text to direct student's attention to important information. This approach represents complex processes, organizes related pieces of information, and integrates information into clear representations. Beautifully illustrated, *Visualizing Psychology* shows your students what the discipline is all about, its main concepts and applications, while also instilling an appreciation and excitement about the richness of the subject.

Visuals, as used throughout this text, are instructional components that display facts, concepts, processes, or principles. They create the foundation for the text and do more than simply support the written or spoken word. The visuals include diagrams, graphs, maps, photographs, illustrations, schematics, animations, and videos.

Why should a textbook based on visuals be effective? Research shows that we learn better from integrated text and visuals than from either medium separately. Beginners in a subject benefit most from reading about the topic, attending class, and studying well-designed and integrated visuals. A visual, with good accompanying discussion, really can be worth a thousand words!

Well-designed visuals can also improve the efficiency with which information is processed by a learner. The more effectively we process information, the more likely it is that we will learn. This processing of information takes place in our working memory. As we learn we integrate new information in our working memory with existing knowledge in our long-term memory.

Have you ever read a paragraph or a page in a book, stopped, and said to yourself:

"I don't remember one thing I just read?" This may happen when your working memory has been overloaded, and the text you read was not successfully integrated into long-term memory. Visuals don't automatically solve the problem of overload, but well-designed visuals can reduce the number of elements that working memory must process, thus aiding learning.

You, as the instructor, facilitate your student's learning. Well-designed visuals, used in class, can help you in that effort. Here are six methods for using the visuals in *Visualizing Psychology* in classroom instruction.

1. **Assign students to study visuals in addition to reading the text.**

   Instead of assigning only one medium of presentation, it is important to make sure your students know that the visuals are just as essential as the written material.

2. **Use visuals during class discussions or presentations.**

   By pointing out important information as the students look at the visuals during class discussions, you can help focus students' attention on key elements of the visuals and help them begin to organize the information and develop an integrated model of understanding. The verbal explanation of important information combined with the visual representation can be highly effective.

3. **Use visuals to review content knowledge.**

Students can review key concepts, principles, processes, vocabulary, and relationships displayed visually. Better understanding results when new information in working memory is linked to prior knowledge.

4. **Use visuals for assignments or when assessing learning.**

Visuals can be used for comprehension activities or assessments. For example, students could be asked to identify examples of concepts portrayed in visuals. Higher-level thinking activities that require critical thinking, deductive and inductive reasoning, and prediction can also be based on visuals. Visuals can be very useful for drawing inferences, for predicting, and for problem solving.

5. **Use visuals to situate learning in authentic contexts.**

Learning is made more meaningful when a learner can apply facts, concepts, and principles to realistic situations or examples. Visuals can provide that realistic context.

6. **Use visuals to encourage collaboration.**

Collaborative groups often are required to practice interactive processes such as giving explanations, asking questions, clarifying ideas, and arguing points of view. These interactive, face-to-face processes provide the information needed to build a verbal mental model. Learners also benefit from collaboration in many instances such as decision making or problem solving.

Visualizing Psychology not only aids student learning with extraordinary use of visuals, but it also offers an array of remarkable photos, media, and film from the National Geographic Society collections. Students using *Visualizing Psychology* also benefit from the long history and rich, fascinating resources of National Geographic.

National Geographic has also performed an invaluable service in fact-checking *Visualizing Psychology*: they have verified every fact in the book with two outside sources, ensuring the accuracy and currency of the text.

Given all of its strengths and resources, *Visualizing Psychology* will immerse your students in the discipline, its main concepts and applications, while also instilling an appreciation and excitement about the richness of the subject area.

Additional information on learning and instructional design is provided in a special guide to using this book, *Learning from Visuals: How and Why Visuals Can Help Students Learn*, prepared by Matthew Leavitt of Arizona State University. This article is available at the Wiley web site: www.wiley.com/college/visualizing. The online *Instructor's Manual* also provides guidelines and suggestions on using the text and visuals most effectively.

Visualizing Psychology also offers a rich selection of visuals in the supplementary materials that accompany the book. To complete this robust package the following materials are available: Test Bank with visuals used in assessment, PowerPoints, Image Gallery to provide you with the same visuals used in the text, web-based learning materials for homework and assessment including images, video, and media resources from National Geographic.

PREFACE

Do I contradict myself?
Very well then I contradict myself,
(I am large, I contain multitudes).
WALT WHITMAN, 1819-1892

Our own senses tell us – and psychological science confirms – that human beings are extraordinarily diverse. Bold, shy, cheerful, gloomy, analytical, artistic, clever, foolish, conscientious, careless, energetic, lazy, tense, relaxed, trusting, cautious – each person is a unique collection of qualities. We all see our world and respond to it just a little differently from any other person. Yet we also can observe remarkable constants in human thought, feeling, and behavior. Even in our endless variety, we are part of an integrated whole. To paraphrase the poet Walt Whitman, we contain multitudes. This debut edition of *Visualizing Psychology* invites readers to explore the complexities and nuances of behavior – both human and nonhuman – that make the study of psychology so compelling.

As you might expect, the compelling (and rewarding) nature of psychology has attracted the attention and devotion of literally millions of readers, along with a multitude of psychology books. Why do we need another text? What makes *Visualizing Psychology* unique? Your authors and the editors and publisher of this text all believe that *active learning* and *critical thinking* (two synonymous and inseparable terms) are key ingredients to true understanding and lifelong learning. Therefore, we have developed and incorporated a large set of active learning and critical thinking pedagogical tools that will help you, the reader, personally unlock the fascinating mysteries and excitement of psychology. These tools will also teach you how to apply the wealth of insights and knowledge from psychological science to your everyday life. Best of all, active learning and critical thinking can make your study and mastery of psychology easier and more rewarding.

As the name implies, *Visualizing Psychology* is also unique in its focus on visuals. Based in part on the old saying that a "picture is worth a thousand words," this text covers the basic content of a standard psychology text enhanced by an educationally sound and carefully designed visual art program. For example, each chapter of this text contains a unique *Visualizing* feature and *Process Diagram* that present a key concept or topic and then explore it in detail using a combination of illustrative photos or figures. While reading through the text, be sure to take full advantage of these and other special study tools, including *What a Scientist Sees*, *With Your Own Eyes*, *Concept Checks*, *Visual Summaries*, and so on. On pages x–xiii of this Preface, we provide a brief description and selected samples of these special features.

Through this premier art program, combined with our strong emphasis on active learning and critical thinking, *Visualizing Psychology* provides readers with a new and innovative approach to the understanding of psychology's major issues, from stereotyping to stem cells. In the context of an engaging visual presentation, we offer solid discussions of critical psychological concepts, ranging from stress's impact on health to the psychological foundations of prejudice.

This book is intended to serve as a broad overview of the entire field of psychology. Despite its shortened and condensed nature, *Visualizing Psychology*, like most other large survey and general education texts, contains a large number of unfamiliar terms and complex concepts. Do not be dismayed. The language of psychology is new to all but the most seasoned scholars. With a little hard work and concentrated study, you can master this material, and your work will pay off with immediate and unforeseen rewards that can last a lifetime.

As you can see, your authors feel passionate about psychology, and we believe that the study of psychology offers all of us an incomparable window into not only ourselves, but also the world that sustains us. We're eager to share our passion for psychology with you. We also welcome feedback from our readers. Please feel free to contact us at siri@nasw.org and khuffman@palomar.edu.

ORGANIZATION

Poet Walt Whitman describes himself as containing "multitudes," and the field of psychology is similarly multitudinous. However, if you're like most beginning introductory psychology students, you probably think of psychology as primarily the study of abnormal behavior and therapy. You'll be surprised, and hopefully delighted,

to discover that our field is much larger and richer than just these two areas. To organize such a diverse and complex field, our book is divided into 15 chapters that are arranged in a somewhat "microscopic/telescoping" fashion. We tend to move from the smallest element of behavior (the neuron and neuroscience) all the way out to the largest (the group, culture, and social psychology). In this section, we'll provide a brief summary of the major topics explored in each chapter:

- **CHAPTER 1** describes psychology's history, its different theoretical perspectives and fundamental questions, and how psychologists go about answering those questions.

- **CHAPTER 2** explains the neural and other biological bases of behavior, and lays the groundwork for further discussions of biological foundations that appear in later chapters.

- **CHAPTER 3** examines interactions among stress, health, and behavior.

- **CHAPTERS 4 THROUGH 8** present aspects of cognition, including sensation, perception, consciousness, learning, memory, thinking, language, and intelligence. These chapters examine both cognition under healthy circumstances and cases where cognition goes awry. Throughout these discus-

Self-actualization needs: to find self-fulfillment and realize one's potential

Esteem needs: to achieve, be competent, gain approval, and excel

Belonging and love needs: to affiliate with others, be accepted, and give and receive attention

Safety needs: to feel secure and safe, to seek pleasure and avoid pain

Physiological needs: hunger, thirst, and maintenance of internal state of the body sions, we provide examples and exercises that connect basic research on cognition to real-world situations.

■ **CHAPTERS 9 AND 10** of *Visualizing Psychology* explore human development across the life span, including physical, cognitive, social, moral, and personality development. We have organized these chapters topically, rather than chronologically, to help students appreciate the trajectory that each facet of development takes over the course of a lifetime.

■ **CHAPTERS 11 AND 12** discuss processes and qualities that are integral to our most basic experiences and interactions with one another: motivation, emotion, and personality.

■ **CHAPTER 13** addresses six major categories of psychological disorders. But first, we begin by discussing what constitutes "abnormal behavior," how psychological disorders are identified and classified. We also explore how psychological disorders vary across cultures.

■ **CHAPTER 14** describes and evaluates major forms of therapy, organizing the most widely used treatments into three groups: insight therapies, behavior therapies, and biomedical therapies.

■ **CHAPTER 15**, social psychology, is in some ways the culmination of all the previous chapters, as there is no aspect of psychology that is irrelevant to how we think about, feel about, and act toward oth-

ers. In this final chapter, we explore a range of social psychological phenomena, ranging from perceptions of others' intentions, to romantic attractions, to prejudice and discrimination.

ILLUSTRATED BOOK TOUR

A number of pedagogical features using visuals have been developed specifically for *Visualizing Psychology*. Presenting the highly varied and often technical concepts woven throughout psychological science raises challenges for reader and instructor alike. The Illustrated Book Tour on the following pages provides a guide to the diverse features contributing to *Visualizing Psychology's* pedagogical plan.

CHAPTER INTRODUCTIONS illustrate certain concepts in the chapter with concise stories that underscore some of psychology's most compelling questions. These narratives are featured alongside striking accompanying photographs. The chapter openers also include illustrated **CHAPTER OUTLINES** that use thumbnails of illustrations from the chapter to refer visually to the content.

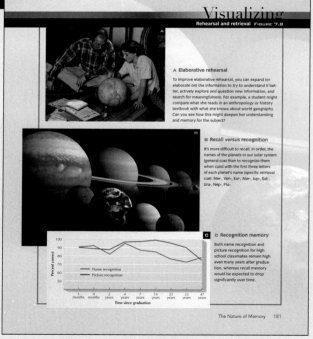

PROCESS DIAGRAMS present a series of figures or a combination of figures and photos that describe and depict a complex process, helping students to observe, follow, and understand the process.

VISUALIZING features are specially designed multi-part visual spreads that focus on a key concept or topic in the chapter, exploring it in detail or in broader context using a combination of photos and figures.

WHAT A SCIENTIST SEES features highlight a concept or phenomenon, using photos and figures that would stand out to a professional in the field, and helping students to develop observational skills.

WITH YOUR OWN EYES features provide an opportunity for students to observe how particular psychological phenomena unfold.

WHAT IS HAPPENING IN THIS PICTURE? is an end-of-chapter feature that presents students with a photograph relevant to chapter topics but that illustrates a situation students are not likely to have encountered previously. The photograph is paired with questions designed to stimulate creative thinking.

OTHER PEDAGOGICAL FEATURES

LEARNING OBJECTIVES at the beginning of each section head indicate in behavioral terms what the student must be able to do to demonstrate mastery of the material in the chapter.

CONCEPT CHECK questions at the end of each section give students the opportunity to test their comprehension of the learning objectives.

cognition Mental activities involved in acquiring, storing, retrieving, and using knowledge.

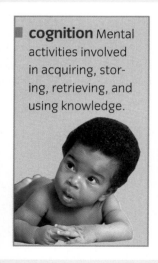

MARGINAL GLOSSARY TERMS (IN GREEN BOLDFACE) introduce each chapter's most important terms, often reinforced with a thumbnail photograph. Other terms appear in black boldface and are defined in the text or in italics.

ILLUSTRATIONS AND PHOTOS support concepts covered in the text, elaborate on relevant issues, and add visual detail. Many of the photos originate from National Geographic's rich sources.

Tools for studying the brain TABLE 1.2
Each biological method has strengths and weaknesses, but all provide invaluable insights. Findings from these tools are discussed in a number of chapters in this text.

Method	Description	Sample Results
Brain dissection	Careful cutting and study of a cadaver brain to reveal structural details.	Brain dissections of Alzheimer's disease victims often show identifiable changes in various parts of the brain (Chapter 7).
	Brain dissection. Structures of the brain can be examined by dissecting the... donated their...	
Ablation/lesions	Su... (a... of... ob... me...	
Clinical observations/ case studies	Ot... pe... ass...	
Electrical recordings	Us... pe... bra... an...	

TABLES AND GRAPHS, with data sources cited at the end of the text, summarize and organize important information.

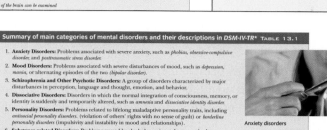

Summary of main categories of mental disorders and their descriptions in *DSM-IV-TR* TABLE 13.1

1. **Anxiety Disorders:** Problems associated with severe anxiety, such as *phobias, obsessive-compulsive disorder,* and *posttraumatic stress disorder.*
2. **Mood Disorders:** Problems associated with severe disturbances of mood, such as *depression, mania,* or alternating episodes of the two (*bipolar disorder*).
3. **Schizophrenia and Other Psychotic Disorders:** A group of disorders characterized by major disturbances in perception, language and thought, emotion, and behavior.
4. **Dissociative Disorders:** Disorders in which the normal integration of consciousness, memory, or identity is suddenly and temporarily altered, such as *amnesia* and *dissociative identity disorder.*
5. **Personality Disorders:** Problems related to lifelong maladaptive personality traits, including *antisocial personality disorders.* (violation of others' rights with no sense of guilt) or *borderline personality disorders* (impulsivity and instability in mood and relationships).
6. **Substance-related Disorders:** Problems caused by alcohol, cocaine, tobacco, and other drugs.
7. **Somatoform Disorders:** Problems related to unusual preoccupation with physical health or physical symptoms with no physical cause.
8. **Factitious Disorders:** Disorders that the individual adopts to satisfy some economic or psychological need.
9. **Sexual and Gender Identity Disorders:** Problems related to unsatisfactory sexual activity, finding unusual objects or situations arousing, or gender identity problems.
10. **Eating Disorders:** Problems related to food, such as *anorexia nervosa* and *bulimia.*
11. **Sleep Disorders:** Serious disturbances of sleep, such as *insomnia* (too little sleep), *sleep terrors,* or *hypersomnia* (too much sleep).
12. **Impulse Control Disorders (not elsewhere classified):** Problems related to *kleptomania* (impulsive stealing), *pyromania* (setting of fires), and *pathological gambling.*
13. **Adjustment Disorders:** Problems involving excessive emotional reaction to specific stressors such as divorce, family discord, or economic concerns.
14. **Disorders usually first diagnosed in infancy, childhood, or early adolescence:** Problems that appear before adulthood, including mental retardation and language development disorders.
15. **Delirium, Dementia, Amnestic, and Other Cognitive Disorders:** Problems caused by known damage to the brain, including Alzheimer's disease, strokes, and physical trauma to the brain.

Anxiety disorders

Mood disorders

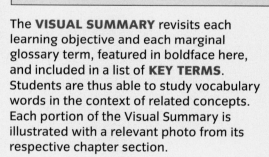

VISUAL SUMMARY

1 Introducing Psychology
1. **Psychology** is the scientific study of behavior and mental processes. The discipline places high value on **empirical evidence** and **critical thinking.** One of the most enduring debates in science has been the **nature-nurture** controversy.
2. Psychology's four basic goals are to describe, explain, predict, and change behavior and mental processes through the use of scientific methods.
3. Psychologists work as therapists, researchers, teachers, and consultants in a wide range of settings.

2 Origins of Psychology
1. Wilhelm Wundt, the father of psychology, and his followers, including Edward Titchener, were interested in studying conscious experience. Their approach, **structuralism,** sought to identify the basic structures of mental life through introspection.
2. The functionalist school, led by William James, studied how the mind functions to adapt humans and other animals to their environment.
3. Contemporary psychology reflects seven major perspectives: psychoanalytic or psychodynamic, behavioral, humanistic, cognitive, neuroscientific or biopsychological, evolutionary, and sociocultural. Most contemporary psychologists embrace a unifying perspective known as the **biopsychosocial model.** This model views biological processes, psychological factors, and social forces as interrelated influences on behavior.
4. Despite early limitations, women and minorities have made important contributions to psychology. Pioneers include Margaret Floy Washburn, Mary Calkins, Francis Cecil Sumner, and Kenneth B. Clark.

3 The Science of Psychology
1. Basic research is aimed at advancing general scientific understanding, whereas applied research works to address real-world problems.
2. Most scientific investigations involve six basic steps, collectively known as the scientific method. Scientific progress comes from repeatedly challenging and revising existing theories and building new ones.
3. Psychologists must maintain high ethical standards. This includes respecting the rights of both human and nonhuman research participants and of psychotherapy clients. Informed consent and debriefing are critical elements of any research that involves human participants. Researchers and clinicians are held professionally responsible for their actions by the APA, by their institutions, and by local and state agencies.

28 CHAPTER 1 Introduction and Research Methods

The **VISUAL SUMMARY** revisits each learning objective and each marginal glossary term, featured in boldface here, and included in a list of **KEY TERMS.** Students are thus able to study vocabulary words in the context of related concepts. Each portion of the Visual Summary is illustrated with a relevant photo from its respective chapter section.

4 The Biology of Learning
1. Learning creates structural changes in the brain. Early evidence for such changes came from research on animals raised in enriched versus deprived environments.

5 Conditioning and Learning in Everyday Life
1. Prejudice is a negative instance of classical conditioning. Influential research by Clark and Clark indicated that children learn to associate inferior qualities with darker skin and more positive qualities with lighter skin. Classical conditioning is a primary tool for advertisers and others who want to manipulate beliefs, emotions, or behavior. Classical conditioning is also used in medicine and in the development and treatment of phobias.

2. Learning is an evolutionary adaptation that enables organisms to survive and prosper in a constantly changing world. Researchers have identified biological constraints that limit the generality of conditioning principles: biological preparedness and instinctive drift.

2. Operant conditioning plays a role in the development of prejudice. Operant conditioning also underlies biofeedback. Even accidental reinforcement can cause superstitious behavior.

3. Cognitive-social learning plays a role in many areas of everyday life, including the development of prejudice and media influences on our behavior.

KEY TERMS
- learning p. 144
- classical conditioning p. 144
- conditioning p. 144
- unconditioned stimulus (UCS) p. 144
- unconditioned response (UCR) p. 144
- neutral stimulus (NS) p. 144
- conditioned stimulus (CS) p. 144
- conditioned response (CR) p. 144
- conditioned emotional responses p. 146
- stimulus generalization p. 146
- stimulus discrimination p. 147
- extinction p. 147
- spontaneous recovery p. 147
- reconditioning p. 148
- higher-order conditioning p. 148
- operant conditioning p. 149
- reinforcement p. 149
- punishment p. 149
- law of effect p. 149
- positive reinforcement p. 150
- negative reinforcement p. 150
- Premack principle p. 150
- schedules of reinforcement p. 150
- continuous reinforcement p. 150
- partial (or intermittent schedule of reinforcement p. 150
- fixed ratio p. 151
- variable ratio p. 151
- fixed interval p. 151
- variable interval p. 151
- shaping p. 151
- positive punishment p. 153
- negative punishment p. 153
- cognitive-social theory p. 156
- insight p. 157
- insight learning p. 157
- cognitive map p. 157
- latent learning p. 157
- observational learning p. 158
- taste aversion p. 161
- biological preparedness p. 162
- instinctive drift p. 162
- biofeedback p. 164

CRITICAL AND CREATIVE THINKING QUESTIONS

1. How might Watson and Rayner, who conducted the famous "Little Albert" study, have designed a more ethical study of conditioned emotional responses?
2. What are some examples of ways in which observational learning has benefited you in your life? Are there instances in which observational learning has worked to your disadvantage?
3. Do you have any taste aversions? How would you use information in this chapter to remove them?
4. Most classical conditioning is involuntary. Considering this, is it ethical for politicians and advertisers to use classical conditioning to influence our thoughts and behavior? Why or why not?

Critical and Creative Thinking Questions 169

CRITICAL AND CREATIVE THINKING QUESTIONS encourage critical thinking and highlight each chapter's important concepts and applications

SELF-TEST: SENSATION AND PERCEPTION

(Check your answers in Appendix B.)

1. Transduction is the process of converting _____.
   a. neural impulses into mental representations of the world
   b. receptors into transmitters
   c. a physical stimulus into neural impulses
   d. receptors into neural impulses

2. Sensory reduction refers to the process of _____
   a. reducing your dependence on a single sensory system
   b. decreasing the number of sensory receptors that are stimulated
   c. filtering and analyzing incoming sensations before sending a neural message to the cortex
   d. reducing environmental sensations by physically preventing your sensory organs from seeing, hearing, etc.

3. HOW THE EYE SEES: Identify the parts of the eye, placing the appropriate label on the figure below:
   cornea rod
   iris cone
   pupil fovea
   lens blind spot
   retina

4. A visual acuity problem that occurs when the cornea and lens focus an image in front of the retina is called
   a. farsightedness
   b. hyperopia
   c. myopia
   d. presbyopia

5. HOW THE EAR HEARS: Identify the parts of the ear, placing the appropriate label on the figure below:
   pinna stapes
   tympanic membrane oval window
   malleus cochlea
   incus

6. Chronic exposure to loud noise can cause permanent
   a. auditory illusions
   b. auditory hallucinations
   c. nerve deafness
   d. conduction deafness

7. Most information related to smell is processed in the
   a. nasal cavity
   b. temporal lobe
   c. olfactory bulb
   d. parietal lobe

8. Most of our taste receptors are found on the
   a. papillae
   b. gustatory cells
   c. taste buds
   d. frenulum

112 CHAPTER 4 Sensation and Perception

SELF-TESTS at the end of each chapter provide a series of multiple choice questions, many of them incorporating visuals from the chapter, that review the major concepts.

MEDIA AND SUPPLEMENTS

INSTRUCTOR SUPPLEMENTS
VIDEOS

A rich collection of video material, drawn from the award-winning National Geographic Film Collection, has been selected to accompany and enrich the text. Each chapter includes video clips, available online as digitized streaming video, that illustrate and expand on a concept or topic to aid student understanding.

 This logo in the text indicates that video from the archives of the National Geographic Society is available.

Accompanying each of the videos is contextualized commentary and questions that can further develop student understanding.

The videos are available on the main website: www.wiley.com/college/carpenter

POWERPOINT PRESENTATIONS AND IMAGE GALLERY

A complete set of highly visual PowerPoint presentations by Karen Huffman and Megan Benoit Ratcliff is available online to enhance classroom presentations. Tailored to the text's topical coverage and learning objectives, these presentations are designed to convey key text concepts, illustrated by embedded text art.

All photographs, figures, maps, and other visuals from the text are online and can be used as you wish in the classroom. These online electronic files allow you to easily incorporate them into your PowerPoint presentations as you choose, or to create your own overhead transparencies and handouts.

As noted previously, by using visuals in class discussions, you can help focus students' attention on key elements of the visuals and help them begin to organize the information and develop greater understanding. The classroom explanation of important information combined with the visual representation can be highly effective.

TEST BANK (AVAILABLE IN WILEYPLUS AND ELECTRONIC FORMAT)

The visuals from the textbook are also included in the Test Bank by Lynda M. Federoff, Indiana University of Pennsylvania. The Test Bank has approximately 1200 test items, with at least 25 percent incorporating visuals from the book. The test items include multiple choice and essay questions testing a variety of comprehension levels. The test bank is available in two formats: online in MS Word files and a Computerized Test Bank, a multi-platform CD-ROM. The easy-to-use test-generation program fully supports graphics, print tests, student answer sheets, and answer keys. The software's advanced features allow you to create an exam to your exact specifications.

INSTRUCTOR'S MANUAL (AVAILABLE IN ELECTRONIC FORMAT)

The Manual begins with a special introduction on Using Visuals in the Classroom, prepared by Matthew Leavitt of Arizona State University, in which he provides guidelines and suggestions on how to use the visuals in teaching the course. For each chapter, materials by Kathy Weatherford of Trident Technical College and Megan Benoit Ratcliff include suggestions and directions for using web-based learning modules in the classroom and for homework assignment and other course activities.

WEB-BASED LEARNING MODULES

www.wiley.com/college/carpenter

A robust suite of multi-media learning resources have been designed for *Visualizing Psychology*, again focusing on and using the visuals from the book. Delivered via the web, the content is organized into tutorial animations.

Tutorial Animations: Animations visually support the learning of a difficult concept, process, or theory, many of them built around a specific feature such as a Process Diagram, Visualizing piece, or key visual in the chapter. The animations go beyond the content and visuals presented in the book, providing additional visual examples and descriptive narration.

BOOK COMPANION SITE (WWW.WILEY.COM/COLLEGE/CARPENTER)

Instructor Resources on the book companion site include the Test Bank, Instructor's Manual, all illustrations and photos in the textbook in jpeg format, as well as select flash animations for use in classroom presentations.

WILEYPLUS

Visualizing Psychology is available with **WileyPLUS**, a powerful online tool that provides instructors and students with an integrated suite of teaching and learning resources in one easy-to-use Web site.

WileyPLUS Helping Teachers Teach and Students Learn

www.wiley.com/college/carpenter

This title is available with *WileyPLUS*, a powerful online tool that provides instructors and students with an integrated suite of teaching and learning resources in one easy to use Web site. *WileyPLUS* is organized around the activities you and your students perform in class.

For Instructors

Prepare & Present: Create class presentations using a wealth of Wiley-provided resources—such as an online version of the textbook, PowerPoint slides, animations, overviews, and visuals from the Wiley Image Gallery—making your preparation time more efficient. You may easily adapt, customize, and add to this content to meet the needs of your course.

Create Assignments: Automate the assigning and grading of homework or quizzes by using Wiley-provided question banks, or by writing your own. Student results will be automatically graded and recorded in your gradebook. *WileyPLUS* can link the pre-lecture quizzes and test bank questions to the relevant section of the online text.

Track Student Progress: Keep track of your students' progress via an instructor's gradebook, which allows you to analyze individual and overall class results to determine students' progress and level of understanding.

Administer Your Course: *WileyPLUS* can easily be integrated with another course management system, gradebook, or other resources you are using in your class, providing you with the flexibility to build your course, your way.

For Students

WileyPLUS provides immediate feedback on student assignments and a wealth of support materials. This powerful study tool will help your students develop their conceptual understanding of the class material and increase their ability to answer questions.

A "**Study and Practice**" area links directly to text content, allowing students to review the text while they study and answer. Resources include National Geographic videos, concept animations and tutorials, visual learning interactive exercises, and links to websites that offer opportunities for further exploration.

An "**Assignment**" area keeps all the work you want your students to complete in one location, making it easy for them to stay "on task." Students will have access to a variety of interactive self- assessment tools, as well as other resources for building their confidence and understanding. In addition, all of the pre-lecture quizzes contain a link to the relevant section of the multimedia book, providing students with context-sensitive help that allows them to conquer problem-solving obstacles as they arise.

A **Personal Gradebook** for each student will allow students to view their results from past assignments at any time.

Please view our online demo at **www.wiley.com/college/wileyplus.** Here you will find additional information about the features and benefits of *WileyPLUS*, how to request a "test drive" of *WileyPLUS* for this title, and how to adopt it for class use.

ACKNOWLEDGMENTS

CLASS TESTING AND STUDENT FEEDBACK

To make certain that *Visualizing Psychology* meets the needs of today's students, we asked several instructors to class test chapters. The feedback that we received from students and instructors confirmed our belief that the visualizing approach taken in this book is highly effective in helping students to learn. We wish to thank the following instructors and their students who provided us with helpful feedback and suggestions:

Sheree Barron
*Georgia College &
State University*

Dale V. Doty
*Monroe Community
College*

William Rick Fry
*Youngstown State
University*

Andy Gauler
*Florida Community
College at Jacksonville*

Bonnie A. Green
*East Stroudsburg
University*

Janice A. Grskovic
*Indiana University
Northwest*

Richard Keen
Converse College

Robin K. Morgan
*Indiana University
Southwest*

Jack A. Palmer
*University of
Louisiana at Monroe*

Marianna Rader
*Florida Community
College at Jacksonville*

Melissa Terlecki
Cabrini College

PROFESSIONAL FEEDBACK

Throughout the process of writing and developing this text and the visual pedagogy, we benefited from the comments and constructive criticism provided by the instructors listed below. We offer our sincere appreciation to these individuals for their helpful review:

John Broida
University of Southern Maine

Tracie Burt
Southeast Arkansas College

Barbara Canaday

Richard Cavasina
*California University of
Pennsylvania*

Michelle Caya
*Community College of Rhode
Island*

Diane Cook
Gainesville State College

Dale V. Doty
Monroe Community College

Steve Ellyson
Youngstown State University

Nolen Embry-Bailey
*Bluegrass Community and
Technical College*

Melanie Evans
*Eastern Connecticut State
University*

William Rick Fry
Youngstown State University

Betsy Goldenberg
*University of Massachusetts,
Lowell*

Jeffrey Henriques
University of Wisconsin

Richard Keen
Converse College

Jean Mandernach
*University of Nebraska at
Kearney*

Jan Mendoza
*Brooks College/Golden West
College*

Tara Mitchell
Lock Haven University

Robin K. Morgan
Indiana University Southeast

Ronald Mulson
*Hudson Valley Community
College*

Larry Peck
*Erie Community College –
North*

Lori Perez
*Community College of Baltimore
County*

Marianna Rader
*Florida Community College at
Jacksonville*

Clayton Teem
Gainesville State College

FOCUS GROUPS AND TELESESSION PARTICIPANTS

A number of professors and students participated in focus groups and telesessions, providing feedback on the text, visuals, and pedagogy. Our thanks to the following instructors for their helpful comments and suggestions.

American Psychological Association Focus Group participants:

Sheree Barron
Georgia College & State University

Joni Caldwell
Union College

Stephen Ray Flora
Youngstown State University

Regan Gurung
University of Wisconsin, Green Bay

Visualizing format/ pedagogy Focus Group and Telesession participants:

Sylvester Allred
Northern Arizona University

David Bastedo
San Bernardino Valley College

Ann Brandt-Williams
Glendale Community College

Natalie Bursztyn
Bakersfield College

Stan Celestian
Glendale Community College

O. Pauline Chow
Harrisburg Area Community College

Diana Clemens-Knott
California State University, Fullerton

Mitchell Colgan
College of Charleston

Linda Crow
Montgomery College

Smruti Desai
Cy-Fair College

Brett Heintz
Delgado Community College

J. Kris Leppien-Christensen
Saddleback College

Mike Majors
Delgado Community College

Debra Murray
Viterbo University

Jack A. Palmer
University of Louisiana at Monroe

Melissa S. Terlecki
Cabrini College

Charles Dick
Pasco-Hernando Community College

Donald Glassman
Des Moines Area Community College

Mark Grobner
California State University, Stanislaus

Michael Hackett
Westchester Community College

Gale Haigh
McNeese State University

Roger Hangarter
Indiana University

Michael Harman
North Harris College

Terry Harrison
Arapahoe Community College

Javier Hasbun
University of West Georgia

Hasiotis, Stephen
University of Kansas

Adam Hayashi
Central Florida Community College

American Psychological Society Focus Group participants:

Jonathan Bates
Hunter College

Michell E. Berman
University of Southern Mississippi

Will Canu
University of Missouri - Rolla

Patricia C. Ellerson
Miami University of Ohio

Laura Hubbard
University of California, Berkeley

James Hutcheon
Georgia Southern University

Scott Jeffrey
Community College of Baltimore County, Catonsville Campus

Matthew Kapell
Wayne State University

Arnold Karpoff
University of Louisville

Dale Lambert
Tarrant County College NE

Arthur Lee
Roane State Community College

Harvey Liftin
Broward Community College

Walter Little
University at Albany, SUNY

Mary Meiners
San Diego Miramar College

Scott Miller
Penn State University

Jane Murphy
Virginia College Online

Renee Engeln-Maddox
Loyola University Chicago

Julie Evey
University of Southern Indiana

Bonnie A. Green
East Stroudsburg University

Janice A. Grskovic
Indiana University Northwest

Keith Happaney
Lehman College

Hector L. Torres
Medial College of Wisconsin

Bethany Myers
Wichita State University

Terri Oltman
Westwood College

Keith Prufer
Wichita State University

Ann Somers
University of North Carolina, Greensboro

Donald Thieme
Georgia Perimeter College

Kip Thompson
Ozarks Technical Community College

Judy Voelker
Northern Kentucky University

Arthur Washington
Florida A&M University

Stephen Williams
Glendale Community College

Feranda Williamson
Capella University

SPECIAL THANKS

Special thanks also go to the superb editorial and production teams at John Wiley and Sons who guided us through the challenging steps of developing this book. We thank in particular: Executive Editor Chris Johnson, who expertly launched and directed our process; Helen McInnis, Managing Director, Wiley Visualizing, who oversaw the concept of the book; Johnna Barto, Development Editor, who supplied her usual invaluable expertise in developing the book; Barbara Heaney, Director of Product and Market Development, who provided valuable suggestions for page layout; Kelly Tavares, Production Manager, who stepped in whenever we needed expert advice; Sujin Hong, Production Editor, who guided the book through production; Jay O'Callaghan, Vice President and Publisher, who oversaw the entire project; and Jeffrey Rucker, Executive Marketing Manager for Wiley Visualizing, who adeptly represents the Visualizing imprint. Like any cooperative effort, writing a book requires an immense support team, and we are deeply grateful to this remarkable group of people.

We wish also to acknowledge the contributions of Vertigo Design for the interior design concept, and Harry Nolan, Wiley's Creative Director who gave art direction, refined the design and other elements and the cover. We appreciate the efforts of Hilary Newman and Tara Sanford in obtaining some of our text photos, and Sandra Rigby and Anna Melhorn for their expertise in managing the illustration program.

We also wish to thank those who worked on the media and ancillary materials: Tom Kulesa, Senior Media Editor, and Lynn Pearlman, Media Editor, for their expert work in developing the video and electronic components, and a host of others—all of whom contributed to the development of the wide assortment of ancillaries.

We thank Mary Ann Price for her unflagging, always swift work in researching and obtaining many of our text images, and Stacy Gold, Research Editor and Account Executive at the National Geographic Image Collection, for her valuable expertise in selecting NGS photos.

Many other individuals at National Geographic offered their expertise and assistance in developing this book: Francis Downey, Vice President and Publisher, and Richard Easby, Supervising Editor, National Geographic School Division; Mimi Dornack, Sales Manager, and Lori Franklin, Assistant Account Executive, National Geographic Image Collection; and Devika Levy, Jim Burch, and Michael Garrity of the National Geographic Film Library. We appreciate their contributions and support.

Next, we would like to offer thanks to individuals at Pre-Press Company, Inc.—particularly Alyssa Morin and Stephanie Kling, Project Managers, and Gordon Laws, Development Manager, and to Ann Greenberger, all of whom provided unfailing dedication to shaping drafts of each chapter of our book. Their careful and professional approach was critical to the successful production of this book.

From Siri Carpenter: thank you to my husband, Joe Carpenter, for his thoughtful advice and steadfast support throughout the production of this book. My appreciation also to colleagues who provided helpful feedback in one way or another: Tracy Banaszynski, Jennifer Randall Crosby, Brian Detweiler-Bedell, Jerusha Detweiler-Bedell, Meghan Dunn, and Kristi Lemm.

From Karen Huffman: continuing appreciation to my students. They taught me what students want to know and inspired me to write the book. In addition, these individuals deserve special recognition: John Bryant, Sandy Harvey, Michelle Hernandes, Jordan Hertzog-Merino, Richard Hosey, Christine McLean, and Kandis Mutter. They provided careful editing of this text, library research, and a unique sense of what should and should not go into an introduction to psychology text. I sincerely appreciate their contributions. Last, and definitely not least, I thank my beloved husband, Bill Barnard, who has provided countless hours of careful editing, advice, and unwavering support throughout the work on this book.

CONTENTS *in Brief*

CONTENTS

VISUALIZING FEATURES

Visualizing Features: Multi-part visual presentations that focus on a key concept or topic in the chapter

PROCESS DIAGRAMS

Process Diagrams: A series or combination of figures and photos that describe and depict a complex process

FOR INSTRUCTORS

WileyPLUS is built around the activities you perform in your class each day. With WileyPLUS you can:

Prepare & Present
Create outstanding class presentations using a wealth of resources such as PowerPoint™ slides, image galleries, interactive simulations videos, and more. You can even add materials you have created yourself.

Create Assignments
Automate the assigning and grading of homework or quizzes by using the provided question banks, or by writing your own.

Track Student Progress
Keep track of your students' progress and analyze individual and overall class results.

Now Available with WebCT and Blackboard!

"It has been a great help, and I believe it has helped me to achieve a better grade."

Michael Morris,
Columbia Basin College

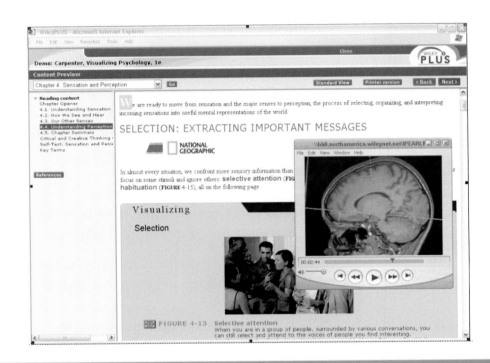

FOR STUDENTS

You have the potential to make a difference!
WileyPLUS is a powerful online system packed with features to help you make the most of your potential and get the best grade you can!

With WileyPLUS you get:

• A complete online version of your text and other study resources.

• Problem-solving help, instant grading, and feedback on your homework and quizzes.

• The ability to track your progress and grades throughout the term.

For more information on what *WileyPLUS* can do to help you and your students reach their potential, please visit www.wiley.com/college/*wileyplus*.

76% of students surveyed said it made them better prepared for tests. *

*Based on a survey of 972 student users of *WileyPLUS*

Visualizing
PSYCHOLOGY

Introduction and Research Methods

What might compel a person to willingly dangle hundreds of feet above the ground? What binds mothers—human and nonhuman alike—to their young? How can chronic stress contribute to serious health problems, such as cancer and heart disease? What happens when ancient cultural practices collide with the instruments and demands of modern living? All of these questions, and countless more, are the province of psychology.

Psychology is a dynamic field that affects every part of our lives. Our innermost thoughts, our relationships, our politics, our "gut" feelings, and our deliberate decisions are all shaped by a complex psychology that affects us at every level, from the cellular to the cultural. Psychology encompasses not only humankind but our nonhuman compatriots as well—from rats and pigeons to cats and chimps.

Psychologists work in an incredible range of areas, perhaps more than you realize. In addition to studying and treating abnormal behavior, psychologists study sleep, dreaming, stress, health, drugs, personality, sexuality, motivation, emotion, learning, memory, childhood, aging, death, love, conformity, intelligence, creativity, and much more.

We invite you to let us know how your study of psychology (and this text) affects you and your life. What questions have captured your imagination? What have you learned about yourself and your world? Where will you go next in your journeys? You can reach us at siri@nasw.org and khuffman@palomar.edu.

We look forward to hearing from you.

Warmest Regards,
Siri Carpenter and Karen Huffman

Introducing Psychology

LEARNING OBJECTIVES

Describe how scientific psychology differs from pseudopsychologies.

Review psychology's four main goals.

Identify some of the diverse professional roles that psychologists fill.

WHAT IS PSYCHOLOGY?

psychology The scientific study of behavior and mental processes.

The term **psychology** derives from the roots *psyche*, meaning "mind," and *logos*, meaning "word." Modern psychology is the scientific study of **behavior** and **mental processes**. Behavior is anything we do—from sleeping to rock climbing. Mental processes are our private, internal experiences—thoughts, perceptions, feelings, memories, and dreams.

For many psychologists, the most important part of the definition of psychology is the word *scientific*. Psychology places high value on **empirical evidence** and **critical thinking** (FIGURE 1.1).

Be careful not to confuse scientific psychology with **pseudopsychologies**, which only give the appearance of science. (*Pseudo* means "false.") Pseudopsychologies include claims made by psychics (who purport to be able to read thoughts and foretell the future), palmistry (reading people's character from the markings on their palms), psychometry (determining facts about an object by handling it), psychokinesis (moving objects by purely mental means), and astrology (the study of how the positions of the stars and planets influence people's personalities and affairs) (FIGURE 1.2).

empirical evidence
Information acquired by direct observation and measurement using systematic scientific methods.

critical thinking
The process of objectively evaluating, comparing, analyzing, and synthesizing information.

Test your knowledge of psychology FIGURE 1.1

Answer true or false to the following statements:

1. The best way to learn and remember information is to "cram," or study it intensively during one concentrated period.
2. Most brain activity stops during sleep.
3. Advertisers and politicians often use subliminal persuasion to influence our behavior.
4. Punishment is the most effective way to permanently change behavior.
5. Eyewitness testimony is often unreliable.
6. Polygraph ("lie detector") tests can accurately and reliably reveal whether a person is lying.
7. Behaviors that are unusual or that violate social norms indicate a psychological disorder.
8. People with schizophrenia have two or more distinct personalities.
9. Similarity is one of the best predictors of long-term relationships.
10. In an emergency, as the number of bystanders increases, your chance of getting help decreases.

Answers: 1. False (Chapter 1). **2.** False (Chapter 5). **3.** False (Chapter 4). **4.** False (Chapter 6). **5.** True (Chapter 7). **6.** False (Chapter 11). **7.** False (Chapter 13). **8.** False (Chapter 13). **9.** True (Chapter 15). **10.** True (Chapter 15).

"The Amazing Randi" FIGURE 1.2

The magician James Randi has dedicated his life to educating the public about fraudulent pseudopsychologists. Along with the prestigious MacArthur Foundation, Randi has offered $1 million to "anyone who proves a genuine psychic power under proper observing conditions" (*About James Randi*, 2002; Randi, 1997). After many years, the money has never been collected. If you would like more information, visit Randi's website at www.randi.org.

The Goals of Psychology

Scientific psychology has four basic goals: to describe, explain, predict, and change behavior and mental processes through the use of scientific methods. Let's consider each within the context of aggressive behavior.

Psychologists usually attempt to describe, or name and classify, particular behaviors by making careful scientific observations. For example, if someone says, "Males are more aggressive than females," does "aggressive" mean angry? Prone to yelling? Likely to throw the first punch? Scientific description requires specificity.

To explain a behavior or mental process, we need to discover and understand its causes. One of the most enduring debates in science has been the **nature-nurture controversy** (Gardiner & Kosmitski, 2005; McCrae, 2004). To what extent are we controlled by biological and genetic factors (the nature side) or by environment and learning (the nurture side)? Today, almost all scientists agree that most psychological and even physical traits reflect an interaction between nature and nurture. For example, research indicates that there are numerous interacting causes or explanations for aggression, including culture, learning, genes, brain damage, and high levels of testosterone (Anderson, 2004; Burns & Katovich, 2003; Trainor, Bird, & Marler, 2004; Uhlmann & Swanson, 2004).

After describing and explaining a behavior or event, psychologists try to predict the conditions under which that behavior or event is likely to occur. For instance, knowing that alcohol leads to increased aggression (Buddie & Parks, 2003), we might predict that more fights will erupt in places where alcohol is consumed than in those where alcohol isn't consumed.

To psychologists, change means applying psychological knowledge to prevent unwanted outcomes or to bring about desired goals. In almost all cases, change as a goal of psychology is positive. For example, psychologists help people stop addictive behaviors, improve their work environments, become less depressed, improve their family relationships, and so on.

nature-nurture controversy

Ongoing dispute over the relative contributors of nature (heredity) and nurture (environment) to the development of behavior and mental processes.

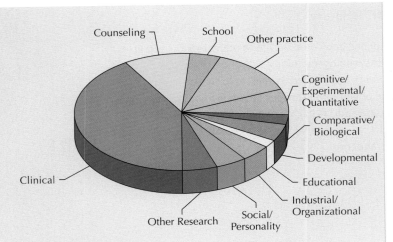

Counseling — School — Other practice

Cognitive/
Experimental/
Quantitative

Comparative/
Biological

Developmental

Educational

Industrial/
Organizational

Social/
Personality

Other Research

Clinical

Percentage of psychology degrees awarded by subfield FIGURE 1.3

This is a small sampling of the numerous specialty areas in psychology.

Source: American Psychological Association, 2004.

Many people think of psychologists only as therapists, but there are many psychologists who have no connection with therapy. Instead, they work as researchers, teachers, and consultants in academic, business, industry, and government settings, or in a combination of settings (FIGURE 1.3). For more information about what psychologists do—or how to pursue a career in psychology—check out the websites of the American Psychological Association (APA; www.apa.org) and the Association for Psychological Science (APS; www.psychologicalscience.org).

CONCEPT CHECK **STOP**

Why are critical thinking and empirical evidence so important to scientific psychology?

What professional fields might psychologists work in?

Origins of Psychology

LEARNING OBJECTIVES

Describe the different perspectives offered by early psychologists.

Identify a fundamental difference between the psychoanalytic and behavioristic perspectives.

Explore the important role of women and minorities in psychology's history.

Explain the central idea underlying the biopsychosocial model.

Although people have always been interested in human nature, it was not until the first psychological laboratory was founded in 1879 that psychology as a science officially began. As interest in the new field grew, psychologists adopted various perspectives on the "appropriate" topics for psychological research and the "proper" research methods. These diverse viewpoints and subsequent debates molded and shaped modern psychological science.

A BRIEF HISTORY: PSYCHOLOGY'S INTELLECTUAL ROOTS

Wilhelm Wundt (VILL-helm Voont), generally acknowledged as the "father of psychology," established the first psychological laboratory in Germany in 1879, helped train the first scientific psychologists, and wrote one of psychology's most important books, *Principles of Physiological Psychology.*

Wundt and his followers were primarily interested in how we form sensations, images, and feelings. Their chief methodology was termed "introspection," which involved monitoring and reporting on conscious experiences. Edward Titchener brought Wundt's ideas to the United States. Wundt and Titchener's approach, now known as **structuralism**, sought to identify the basic building blocks, or structures, of mental life through introspection and then to determine how these elements combine to form the whole of experience. Because introspection could not be used to study animals, children, or more complex mental disorders, structuralism failed as a working psychological approach. Although short-lived, structuralism established a model for studying mental processes scientifically.

Structuralism's intellectual successor, **functionalism**, studied how the mind functions to adapt humans and other animals to their environment. Functionalism was strongly influenced by Charles Darwin's theory of evolution (Segerstrale, 2000). William James was the leading force in the functionalist school (**FIGURE 1.4**). Although functionalism also eventually declined, it expanded the scope of psychology to include research on emotions and observable behaviors, initiated the psychological testing movement, and influenced modern education and industry.

During the late 1800s and early 1900s, while functionalism was prominent in the United States, the **psychoanalytic school** was forming in Europe (Gay, 2000). Its founder, Austrian physician Sigmund Freud, believed that many psychological problems are caused by conflicts between "acceptable" behavior and "unacceptable" unconscious sexual or aggressive motives (Chapter 12). The theory provided a basis for a system of therapy known as *psychoanalysis* (Chapter 14).

Freud's nonscientific approach and emphasis on sexual and aggressive impulses have long been controversial, and today there are few strictly Freudian psychoanalysts left. But the broad features of his theory have profoundly affected psychotherapy, psychiatry, and modern psychodynamic psychologists, who focus on the importance of unconscious processes and unresolved past conflicts.

In the early 1900s, another major perspective appeared that dramatically shaped the course of psychology. Unlike earlier approaches, the **behaviorist perspective** emphasizes objective, observable environmental influences on overt behavior. Behaviorism's founder, John B. Watson (1913), rejected the practice of introspection and the influence of unconscious forces. Instead, Watson adopted Russian physiologist Ivan Pavlov's concept of *conditioning* (Chapter 6) to explain behavior as a result of observable stimuli (in the environment) and observable responses (behavioral actions).

Most early behaviorist research was focused on learning; nonhuman animals were ideal subjects for this research. One of the most well-known behaviorists, B. F. Skinner, was convinced that behaviorist approaches could be used to "shape" human behavior. Therapeutic techniques rooted in the behavioristic perspective

William James (1842–1910) FIGURE 1.4

William James broadened psychology to include animal behavior, biological processes, and behaviors. His book *Principles of Psychology* (1890) became the leading psychology text.

have been most successful in treating observable behavioral problems, such as phobias and alcoholism (FIGURE 1.5).

Although the psychoanalytic and behaviorist perspectives dominated American psychology for some time, in the 1950s a new approach emerged—the **humanist perspective**, which stressed *free will* (voluntarily chosen behavior) and *self-actualization* (a state of self-fulfillment). According to Carl Rogers and Abraham Maslow (two central humanist figures), all individuals naturally strive to develop and move toward self-actualization. Like psychoanalysis, humanist psychology developed an influential theory of personality and a form of psychotherapy (Chapters 12 and 14).

One of the most influential modern approaches, the **cognitive perspective**, recalls psychology's earliest days, in that it emphasizes thoughts, perception, and information processing. Modern cognitive psychologists, however, study how we gather, encode, and store information using a vast array of mental processes. These include perception, memory, imagery, concept formation, problem solving, reasoning, decision making, and language. Many cognitive psychologists use what is called an *information-processing approach*, likening the mind to a computer that sequentially takes in information, processes it, and then produces a response.

During the last few decades, scientists have explored the role of biological factors in almost every area of psychology. Using sophisticated tools and technologies, scientists who take this **neuroscientific/biopsychological perspective** examine behavior through the lens of genetics and biological processes in the brain and other parts of the nervous system.

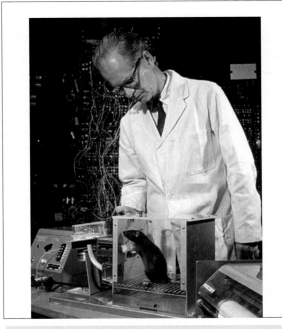

B. F. Skinner (1904–1990) FIGURE 1.5

B. F. Skinner was one of the most influential psychologists of the twentieth century. He believed that by using basic learning principles to shape human behavior, we could change what he perceived as the negative course of humankind.

The **evolutionary perspective** stresses natural selection, adaptation, and evolution of behavior and mental processes. Its proponents argue that natural selection favors behaviors that enhance an organism's reproductive success.

Finally, the **sociocultural perspective** emphasizes social interactions and cultural determinants of behavior and mental processes. Although we are often unaware of their influence, factors such as ethnicity, religion, occupation, and socioeconomic class have an enormous psychological impact on our mental processes and behavior.

Early schools of psychological thought, like structuralism and functionalism, have almost entirely disappeared or have been blended into newer, broader perspectives. Contemporary psychology reflects seven major perspectives: psychoanalytic/psychodynamic, behaviorist, humanist, cognitive, neuroscientific/biopsychological, evolutionary, and sociocultural (TABLE 1.1). Yet the complex behaviors and mental processes we humans and other animals exhibit require complex explanations. That is why most contemporary psychologists do not adhere to one single intellectual perspective. Instead, a more integrative, unifying theme—the **biopsychosocial model**—has gained wide acceptance. This model views biological processes (e.g., genetics, brain functions, neurotransmitters, and evolution), psychological factors (e.g., learning, thinking, emotion, personality, and motivation), and social forces (e.g., family, school, culture, ethnicity, social class, and politics) as interrelated, inseparable influences.

biopsychosocial model A unifying theme of modern psychology that considers biological, psychological, and social processes.

Seven major perspectives in modern psychology TABLE 1.1

Perspectives	Prominent Figures	Major Emphases
Psychoanalytic/ psychodynamic (1895–present)	Sigmund Freud Carl Jung Alfred Adler Karen Horney	Unconscious processes and unresolved past conflicts; the "mind" is like an iceberg with the largest, "unconscious," section hidden beneath the surface
Behaviorist (1906–present)	Ivan Pavlov Edward Thorndike John B. Watson B. F. Skinner	Objective, observable environmental influences on overt behavior; the "mind" is like a "black box" that is unobservable and unmeasurable
Humanist (1950s–present)	Carl Rogers Abraham Maslow	Free-will, self-actualization, and human nature as naturally positive and growth-seeking
Cognitive (1950s–present)	Jean Piaget Albert Ellis Albert Bandura Robert Sternberg Howard Gardner Clark Hull	Thoughts, perception, and information processing
Neuroscientific/ biopsychological (1950s–present)	Johannes Müller Karl Lashley David Hubel James Olds Roger Sperry Candace Pert Torsten Wiesel	Genetics and biological processes in the brain and other parts of the nervous system
Evolutionary (1980s–present)	Charles Darwin Konrad Lorenz E. O. Wilson David Buss Margo Wilson	Natural selection, adaptation, and evolution of behavior and mental processes
Sociocultural (1980s–present)	John Berry Patricia Greenfield Richard Brislin	Social interaction and the cultural determinants of behavior and mental processes

BIOPSYCHOSOCIAL MODEL

The biopsychosocial model combines and interacts with the seven major perspectives.

Biopsychosocial model diagram with segments: Behaviorist, Humanist, Psychoanalytic/Psychodynamic, Cognitive, Bio-psycho-social, Sociocultural, Neuroscientific/Biopsychological, Evolutionary

www.wiley.com/college/carpenter

The Biopsychosocial Model

Look at the photo of the child to the right. What might be the cause of her emotional arousal? Now look at the photo below, which shows the child within a broader context. With this "bigger picture" (the child's immediate surroundings, her parents' guiding influence, and her culture's enthusiasm for exciting sporting events) in mind, can you better understand why she might be feeling and behaving as she is? The biopsychosocial model recognizes that there is usually no single cause for our behavior or our mental states. For example, our moods and feelings are often influenced by genetics and neurotransmitters (biology), our learned responses and patterns of thinking (psychology), and our socioeconomic status and cultural views of emotion (social).

WOMEN AND MINORITIES IN PSYCHOLOGY

During the late 1800s and early 1900s, most colleges and universities provided little opportunity for women and minorities, either as students or as faculty members.

Mary Calkins (1863–1930)
FIGURE 1.6

Despite these early limitations, both women and minorities have made important contributions to psychology.

One of the first women to be recognized in the field was Mary Calkins (FIGURE 1.6). Calkins performed valuable research on memory, and in 1905 she served as the first female president of the APA. Her achievements are particularly noteworthy, considering the significant discrimination that she overcame. Even after she completed all the requirements for a Ph.D. at Harvard University and was described by William James as his brightest student, the university refused to grant the degree to a woman. The first woman to receive a Ph.D. in psychology was Margaret Floy Washburn (in 1894), who wrote several influential books and served as the second female president of the APA.

Francis Cecil Sumner became the first African American to earn a Ph.D. in psychology. He earned it from Clark University in 1920 and later chaired one of the country's leading psychology departments, at Howard University. In 1971, one of Sumner's students, Kenneth B. Clark, became the first African American to be elected APA president. Clark's research with his wife, Mamie Clark, documented the harmful effects of prejudice and directly influenced the Supreme Court's ultimate ruling against racial segregation in schools (FIGURE 1.7).

Sumner and Clark, Calkins and Washburn, along with other important minorities and women, made

Kenneth Clark (1914–1975) and Mamie Clark
FIGURE 1.7

significant and lasting contributions to the developing science of psychology. In recent years, people of color and women are being actively encouraged to pursue graduate degrees in psychology. However, white (non-Hispanic) people still make up the majority of new doctorate recipients in psychology.

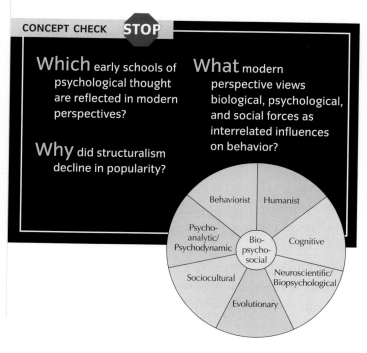

CONCEPT CHECK **STOP**

Which early schools of psychological thought are reflected in modern perspectives?

What modern perspective views biological, psychological, and social forces as interrelated influences on behavior?

Why did structuralism decline in popularity?

The Science of Psychology

In science, research strategies are generally categorized as either basic or applied. **Basic research** is typically conducted in universities or research laboratories by researchers who are interested in advancing general scientific understanding. Basic research meets the first three goals of psychology (description, explanation, and prediction). In contrast, **applied research** is generally conducted outside the laboratory, and it meets the fourth goal of psychology—to change existing real-world problems.

applied research
Research designed to solve practical problems.

basic research
Research conducted to advance scientific knowledge rather than for practical application.

Basic and applied research frequently interact, with one building on the other. For example, after basic research documented a strong relationship between alcohol consumption and increased aggression, applied research led some sports stadium owners to limit the sale of alcohol during the final quarter of football games and the last two innings of baseball games.

THE SCIENTIFIC METHOD: A WAY OF DISCOVERING

Like scientists in any other scientific field, psychologists follow strict, standardized scientific procedures so that others can understand, interpret, and repeat or test their findings. Most scientific investigations involve six basic steps (**FIGURE 1.8**). The **scientific method** is cyclical and cumulative, and scientific progress comes from repeatedly challenging and revising existing theories and building new ones. If numerous scientists, using different participants in varied settings, can repeat, or *replicate*, a study's findings, there is much greater scientific confidence in the findings. If the findings cannot be replicated, researchers look for explanations and conduct further studies. When different studies report contradictory findings, researchers often average or combine the results of all such studies and reach conclusions about the overall weight of the evidence, a popular statistical technique called *meta-analysis*.

ETHICAL GUIDELINES: PROTECTING THE RIGHTS OF OTHERS

The two largest professional organizations of psychologists, the American Psychological Association (APA) and the Association for Psychological Science (APS), both recognize the importance of maintaining high ethical standards in research, therapy, and all other areas of professional psychology. The preamble to the APA's publication *Ethical Principles of Psychologists and Code of Conduct* (1992) admonishes psychologists to maintain their competence, to retain objectivity in applying their skills, and to preserve the dignity and best interests of their clients, colleagues, students, research participants, and society.

The scientific method FIGURE 1.8

www.wiley.com/
college/carpenter

Cycle continues

Cycle begins

Step 1
Literature review
The scientist conducts a *literature review*, reading what has been published in major professional, scientific journals on her subject of interest.

Step 6
Theory
After one or more studies on a topic, researchers generally advance a *theory* to explain their results. This new theory then leads to new (possibly different) hypotheses and new methods of inquiry.

Step 2
Testable hypothesis, operationally defined
The scientist makes a *testable hypothesis*, or a specific prediction about how one factor, or *variable*, is related to another. To be scientifically testable, the variables must be *operationally defined* —that is, stated precisely and in measurable terms.

SCIENTIFIC METHOD

Step 5
Peer-reviewed scientific journal
The scientist writes up the study and its results and submits it to a *peer-reviewed scientific journal.* (Peer-reviewed journals ask other scientists to critically evaluate submitted material.) On the basis of these peer reviews, the study may then be accepted for publication.

Step 3
Research design
The scientist chooses the best *research design* to test the hypothesis and collect the data. She might choose naturalistic observation, case studies, surveys, experiments, or other methods.

Step 4
Statistical analysis
The scientist performs *statistical analyses* on the raw data to determine whether the findings support or reject her hypothesis. This allows her to organize, summarize, and interpret numerical data.

Respecting the rights of human participants

The APA has developed guidelines regulating research with human participants. One of the chief principles is that an investigator should obtain the participant's **informed consent** before initiating an experiment. The researcher should fully inform the participant as to the nature of the study, including physical risks, discomfort, or unpleasant emotional experiences. The researcher must also explain that participants are free to decline to participate or to withdraw from the research at any time.

■ **informed consent** A participant's agreement to take part in a study after being told what to expect.

■ **debriefing** Informing participants after a study about the purpose of the study, the nature of the anticipated results, and any deception used.

Participants in a given research study may not respond naturally if they know the true purpose behind the study. Therefore, the APA acknowledges the need for some deception in certain research areas. But when deception is used, important guidelines and restrictions apply, including **debriefing** participants at the end of the experiment.

APA guidelines also stipulate that all information acquired about people during a study must be held confidential and not published in such a way that individuals' privacy is compromised. If research participation is a course requirement or an opportunity for extra credit, the student must be given the choice of an alternative activity of equal value.

Finally, a human subjects committee or institutional review board must first approve all research using human participants conducted at a college, university, or any other reputable institution. Participating in psychological research can be a fascinating way to learn more about the nuts and bolts of the process. Many university psychology departments offer opportunities for students and community members to participate in research. In addition, the Association for Psychological Science (APS) has a website (http://psych.hanover.edu/research/exponnet.html) with links to ongoing studies that need participants.

Respecting the rights of nonhuman animals

Although they are involved in only 7 to 8 percent of psychological research (American Psychological Association, 1984), nonhuman animals—mostly rats and mice—have made significant contributions to almost every area of psychology, including the brain and nervous system, health and stress, sensation and perception, sleep, learning, memory, and emotion. Nonhuman animal research has also produced significant gains for animals themselves—for example, by suggesting more natural environments for zoo animals and more successful breeding techniques for endangered species.

Despite the advantages, using nonhuman animals in psychological research remains controversial (Guidelines for the Treatment, 2005). While debate continues over ethical questions surrounding such research, psychologists take great care in handling research animals. Researchers also actively search for new and better ways to protect them (Atkins, Panicker, & Cunningham, 2005; Guidelines for the Treatment, 2005; Sherwin et al., 2003). In all institutions where nonhuman animal research is conducted, animal care committees ensure proper treatment of research animals, review projects, and set guidelines in accordance with the APA standards for the care and treatment of nonhuman research animals.

Respecting the rights of psychotherapy clients

Like psychological scientists, therapists must maintain the highest of ethical standards; they must also uphold their clients' trust. All personal information and therapy records must be kept confidential, with records being available only to authorized persons and with the client's permission. However, the public's right to safety ethically outweighs the client's right to privacy. Therapists are legally required to break confidentiality if a client threatens violence to him- or herself or to others, if a client is suspected of abusing a child or elderly person, and in other limited situations. In general, however, a counselor's primary obligation is to protect client disclosures (Corey, 2005).

Any member of the APA who disregards the association's principles for the ethical treatment of human or nonhuman research participants or therapy clients may be censured or expelled from the organization. Clinicians who violate the ethical guidelines for working with clients risk severe sanctions and can permanently lose their licenses to practice. In addition, both researchers and clinicians are held professionally and legally responsible by their institutions as well as by local and state agencies.

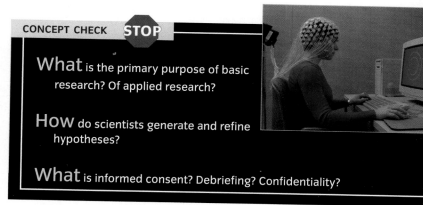

CONCEPT CHECK **STOP**

What is the primary purpose of basic research? Of applied research?

How do scientists generate and refine hypotheses?

What is informed consent? Debriefing? Confidentiality?

Research Methods

LEARNING OBJECTIVES

Explain why only experiments can identify the cause and effect underlying particular patterns of behavior.

Describe the three key types of descriptive research.

Explain what is meant by the axiom "Correlation is not causation."

Identify some important methods used in biological research.

Psychologists draw on four major types of psychological research—experimental, descriptive, correlational, and biological (FIGURE 1.9). All have advantages and disadvantages, and most psychologists use several methods to study a single problem. In fact, when multiple methods lead to similar conclusions, scientists have an especially strong foundation for concluding that one variable does affect another in a particular way.

Experimental Research: Manipulation and control of variables	Descriptive Research: Naturalistic observation, surveys, case studies	Correlational Research: Statistical analyses of relationships between variables	Biological Research: Studies the brain and other parts of the nervous system
Purpose: Identify cause and effect (Meets the **explanation** goal of psychology)	**Purpose:** Observe, collect, and record data (Meets the **descriptive** goal of psychology)	**Purpose:** Identify relationships and how well one variable predicts another (Meets the **predictive** goal of psychology)	**Purpose:** Identify causation, as well as, description, and prediction (Meets one or more of the **four goals of psychology**)

| **Advantages:** Allows researchers precise control over variables and to identify cause and effect | **Disadvantages:** Ethical concerns, practical limitations, artificiality of lab conditions, uncontrolled variables may confound results, researcher and participant biases | **Advantages:** Minimizes artificiality, easier to collect data, allows description of behavior and mental processes as they occur | **Disadvantages:** Little or no control over variables, researcher and participant biases, cannot explain cause and effect | **Advantages:** Helps clarify relationships between variables that cannot be examined by other methods and allows prediction | **Disadvantages:** Researchers cannot identify cause and effect | **Advantages:** Shares many or all of the advantages of experimental, descriptive, and correlational research | **Disadvantages:** Shares many or all of the disadvantages of experimental, descriptive, and correlational research |

The four major methods of psychological research FIGURE 1.9

EXPERIMENTAL RESEARCH: A SEARCH FOR CAUSE AND EFFECT

experiment A carefully controlled scientific procedure that determines whether variables manipulated by the experimenter have a causal effect on other variables.

independent variable (IV) Variable that is manipulated to determine its causal effect on the dependent variable.

dependent variable (DV) Variable that is measured; it is affected by (or dependent on) the independent variable.

The most powerful research method is the **experiment**, in which an experimenter manipulates and controls the variables to determine cause and effect. Only through an experiment can researchers examine a single factor's effect on a particular behavior (Ray, 2003). That's because the only way to discover which of many factors has an effect is to experimentally isolate each one. As illustrated in FIGURE 1.10, an experiment has two critical components an **independent variable (IV)** and a **dependent variable (DV)**, and experimental and control groups.

The goal of any experiment is to learn how the dependent variable is affected by (depends on) the independent variable. Experiments can also have different *levels* of an independent variable. For example, consider the experiment testing the effect of TV violence on aggression described in Figure 1.10. Two experimental groups could be created, with one group watching two hours of violent programming and the other watching six hours; the control group would watch only nonviolent programming. Then a researcher could relate differences in aggressive behavior (DV) to the *amount* of violent programming viewed (IV).

In experiments, all extraneous variables (such as time of day, lighting conditions, and participants' age and gender) must be held constant across experimental and control groups so that they do not affect the different groups' results.

In addition to the scientific controls mentioned (e.g., operational definitions, a control group, and consistent extraneous variables), a good scientific experiment also protects against potential sources of error from both the researcher and the participants.

Experimenters can unintentionally let their beliefs and expectations affect participants' responses, producing flawed results. For example, imagine what might happen if an experimenter breathed a sigh of relief when a participant gave a response that supported the researcher's hypothesis. One way to prevent such **experimenter bias** from destroying the validity of participants' responses is to establish objective methods for collecting and recording data. For example, an experimenter might use audiotape recordings to present stimuli and computers to record responses.

Another option is to use "blind observers" (neutral helpers) to collect and record the data without knowing what the researcher has predicted. In addition, it is sometimes possible to arrange a **double-blind study**, where neither the observer nor the participant knows which group received the experimental treatment.

Experimenters can also skew their results when they assume that behaviors that are typical in their own culture are typical in all cultures—a bias known as **ethnocentrism**. One way to avoid this problem is to have researchers from two cultures conduct the same study twice, once with their own culture and once with at least one other culture. This kind of **cross-cultural sampling** isolates group differences in behavior that stem from researchers' ethnocentrism.

In addition to potential problems from the researcher, the participants themselves can also introduce error or bias into an experiment. First, **sample bias** can occur if the sample of participants does not accurately reflect the composition of the larger population that they are drawn from. For example, critics suggest that much psychological literature is biased because it primarily uses white participants. One way to minimize sample bias is to randomly select participants who constitute a *representative sample* of the entire population of interest. Assigning participants to experimental groups using a chance, or random system, such as a coin toss, also helps prevent sample bias. This procedure of *random assignment* ensures that each participant is equally likely to be assigned to any particular group.

Bias can also occur when participants are influenced by the experimenter or the experimental conditions. For example, participants may try to present themselves in a good light (the **social desirability response**) or deliberately attempt to mislead the researcher.

1. To test the hypothesis that watching violent television increases aggression, experimenters might randomly assign children to one of two groups: **experimental group** participants, who watch a prearranged number of violent television programs, and **control group** participants, who watch the same amount of television, except the programs that they watch are nonviolent.

2. Experimenters then observe the children and count how many times— say, within one hour—each child hits, kicks, or punches a large, plastic "Bobo doll" (an operational definition of aggression).

    Having at least two groups—a control group and an experimental group—allows the performance of one group to be compared with that of another.

Hypothesis
"Watching violence on TV increases aggression."

Random Assignment
to experimental conditions

Experimental Group

Control Group

Independent Variable (IV)
(Violent or nonviolent program)

Dependent Variable (DV)
(Number of times child hits Bobo doll)

Groups Compared

www.wiley.com/college/carpenter

A double-blind experimental design FIGURE 1.11

In a double-blind experiment to test a new drug, both the experimenters administering the drug and the participants are unaware (or "blind") as to who is receiving a *placebo* (a fake pill or injection) and who is receiving the drug itself. Researchers use placebos (thus treating experimental and control group participants exactly the same) because they know that the mere act of taking a pill or receiving an injection can change a participant's condition or responses.

Researchers attempt to control for this type of participant bias by offering anonymous participation and other guarantees of privacy and confidentiality. Single- and double-blind studies and placebos, which ensure that participants are unaware of which group they're in, offer additional safeguards (FIGURE 1.11). Finally, one of the most effective (but controversial) ways of preventing participant bias is by deceiving them (temporarily) about the nature of the research project.

"Just pretend we're not here, Ms. Robinson..."

DESCRIPTIVE RESEARCH: NATURALISTIC OBSERVATION, SURVEYS, AND CASE STUDIES

Almost everyone observes and describes others in an attempt to understand them, but in conducting **descriptive research**, psychologists do it systematically and scientifically. The key types of descriptive research are naturalistic observation, surveys, and case studies. Most of the problems and safeguards discussed with regard to the experimental method also apply to the nonexperimental methods.

When conducting **naturalistic observation**, researchers systematically measure and record participants' behavior, without interfering. Many settings lend themselves to naturalistic observation, from supermarkets to airports to outdoor settings (FIGURE 1.12).

descriptive research Research methods used to observe and record behavior (without producing causal explanations).

Naturalistic observation FIGURE 1.12

Studying behavior in its natural environment allows behavior to unfold naturally (without interference).

The chief advantage of naturalistic observation is that researchers can obtain data about natural behavior, rather than about behavior that is a reaction to an artificial experimental situation. But naturalistic observation can be difficult and time-consuming, and the lack of control by the researcher makes it difficult to conduct observations for behavior that occurs infrequently.

Psychologists use **surveys** (FIGURE 1.13) to measure a variety of psychological behaviors and attitudes. (The survey technique includes tests, questionnaires, polls, and interviews.) One key advantage of surveys is that researchers can gather data from many more people than is possible with other research methods. Unfortunately, most surveys rely on self-reported data, and not all participants are completely honest. Although they can help predict behavior, survey techniques cannot explain causes of behavior.

What if a researcher wants to investigate photophobia (fear of light)? In such a case, it would be difficult to find enough participants to conduct an experiment or to use surveys or naturalistic observation. For rare disorders or phenomena, researchers try to find someone who has the problem and study him or her intensively. Such in-depth studies of a single research participant are called **case studies** (FIGURE 1.14).

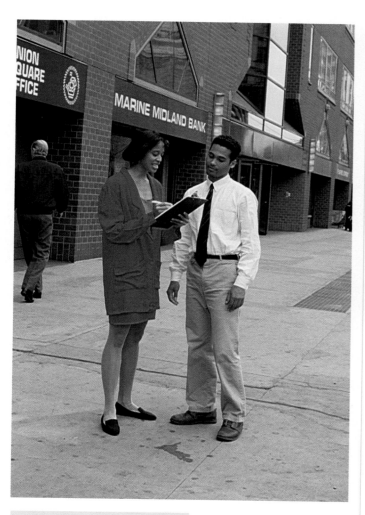

Surveys FIGURE 1.13

In conducting surveys, researchers often use questionnaires to gather data from a wide selection of people.

An early case study FIGURE 1.14

In 1848, a railroad foreman named Phineas Gage had a metal rod (13 pounds, 1¼ inches in diameter, and 3½ feet long) blown through the front of his face and brain. Amazingly, Gage was soon up and moving around; he didn't receive medical treatment until 1½ hours later. However, Gage suffered a serious personality transformation. Before the accident, he had been capable, energetic, and well-liked. Afterward, he was described as "fitful, capricious, impatient of advice, obstinate, and lacking in deference to his fellows" (Macmillan, 2000, p. 13). Gage's injury and "recovery" were carefully documented and recorded by his attending physician, Dr. J. M. Harlow. As his story illustrates, the case study method offers unique advantages. However, this method also poses serious research limits, including lack of generalizability, recorder bias, and inaccurate or biased recall among participants.

CORRELATIONAL RESEARCH: LOOKING FOR RELATIONSHIPS

> **■ correlational research** Scientific study in which the researcher observes or measures (without directly manipulating) two or more variables to find relationships between them.

When nonexperimental researchers want to determine the degree of relationship (correlation) between two variables, they turn to **correlational research**. As the name implies, when any two variables are "co-related," a change in one is accompanied by a change in the other.

Using the correlational method, researchers measure participants' responses on two or more variables of interest. Next, the researchers analyze their results using a statistical formula that results in a **correlation coefficient**, a numerical value that indicates the degree and direction of the relationship between the two variables. Correlation coefficients are expressed as a number ranging from +1.00 to −1.00. The sign (+ or −) indicates the direction of the correlation, positive or negative (**FIGURE 1.15A** on the following page).

The number (0 to +1.00 or −1.00) associated with a correlation indicates the strength of the relationship. Both +1.00 and −1.00 are the strongest possible relationships. Thus, if you had a correlation of +0.92 or −0.92, you would have a *strong correlation*. By the same token, a correlation of +0.15 or −0.15 would represent a *weak correlation*.

Correlational research is an important research method for psychologists, and understanding correlations can also help us live safer and more productive lives. For example, correlational studies have repeatedly found high correlations between birth defects and a pregnant mother's use of alcohol (Bearer et al., 2004–2005; Gunzerath et al., 2004). This kind of information enables us to reliably predict our relative risks and to make informed decisions (**FIGURE 1.15B**).

However, people sometimes do not understand that a correlation between two variables does not mean that one variable causes another (**FIGURE 1.15C**). People sometimes read media reports about correlations between stress and cancer, for example, or between family dynamics and homosexuality, and infer that "stress causes cancer" or that "withdrawn fathers and overly protective mothers cause their sons to become homosexuals." They fail to realize that a third factor, perhaps genetics, may cause greater susceptibility to both of the correlated phenomena. Although correlational studies do sometimes point to possible causes, only the experimental method manipulates the independent variable under controlled conditions and, therefore, can support conclusions about cause and effect.

BIOLOGICAL RESEARCH: TOOLS FOR EXPLORING THE BRAIN AND NERVOUS SYSTEM

> **■ biological research** Scientific studies of the brain and other parts of the nervous system.

Biological research examines the biological processes that are involved in our feelings, thoughts, and behavior.

The earliest explorers of the brain dissected the brains of deceased humans and conducted experiments on other animals using *lesioning* techniques (systematically destroying brain tissue to study the effects on behavior and mental processes). By the mid-1800s, this research had produced a basic map of the nervous system, including some areas of the brain. Early researchers also relied on clinical observations and case studies of living people who had experienced injuries, diseases, and disorders that affected brain functioning.

Modern researchers still use such methods, but they also employ other techniques to examine biological processes that underlie our behavior (**TABLE 1.2**). Recent advances in brain science have led to various types of brain-imaging scans, which can be used in both clinical and laboratory settings (Benazzi, 2004; Haller et al., 2005). Most of these methods are relatively *noninvasive*—that is, their use does not involve breaking the skin or entering the body.

Three types of correlation

 Each dot on these graphs (called *scatterplots*) represents one participant's score on two factors, or variables.

 In a positive correlation, the two factors move (or vary) in the same direction.

In a negative correlation, the two factors vary in opposite directions— that is, as one factor increases, the other factor decreases.

Sometimes there is no relationship between two variables—a zero correlation.

Pregnancy and smoking

B Research shows that cigarette smoking is strongly linked to serious fetal damage. The more the mother smokes the more the fetus is damaged. Is this a positive or negative correlation?

Correlation is not causation

C Does a high correlation between young children's foot size and their reading speed mean that having small feet causes slower reading? Or that reading quickly causes feet to grow? Obviously not! Instead, both are caused by a third variable—an increase in children's ages.

Tools for studying the brain TABLE 1.2

Each biological method has strengths and weaknesses, but all provide invaluable insights. Findings from these tools are discussed in a number of chapters in this text.

Method	Description	Sample Results
Brain dissection *Brain dissection.* Structures of the brain can be examined by dissecting the brains of deceased people who donated their bodies for scientific study.	Careful cutting and study of a cadaver brain to reveal structural details.	Brain dissections of Alzheimer's disease victims often show identifiable changes in various parts of the brain (Chapter 7).
Ablation/lesions	Surgically removing parts of the brain (ablation), or destroying specific areas of the brain (lesioning), is followed by observation for changes in behavior or mental processes.	Lesioning specific parts of the rat's hypothalamus greatly affects its eating behavior (Chapter 11).
Clinical observations/ case studies	Observing and recording changes in personality, behavior, or sensory capacity associated with brain disease or injuries.	Damage to one side of the brain often causes numbness or paralysis on the body's opposite side.
Electrical recordings *Electroencephalogram (EEG).* Electrodes are attached to the patient's scalp, and the brain's electrical activity is displayed on a computer monitor or recorded on a paper chart.	Using electrodes attached to a person's or animal's skin or scalp, brain activity is recorded to produce an electroencephalogram.	Reveals areas of the brain most active during a particular task or changes in mental states, like sleeping and meditation (Chapter 5); also traces abnormal brain waves caused by brain malfunctions, like epilepsy or tumors.
Electrical stimulation of the brain (ESB)	Using an electrode, a weak electric current stimulates specific areas or structures of the brain.	Penfield (1958) mapped the surface of the brain and found that different areas have different functions.

Tools for studying the brain (continued) TABLE 1.2

Method	Description	Sample Results
CT (computed tomography) scan *CT scans* This CT scan used X-rays to locate a brain tumor. The tumor is the deep purple mass at the top left.	Computer created cross-sectional X-rays of the brain; least expensive type of imaging and widely used in research.	Reveals the effects of strokes, injuries, tumors, and other brain disorders.
PET (positron emission tomography) scan *PET scans and brain functions.* In these two scans, the left one shows brain activity when the eyes are open, whereas the one on the right is with the eyes closed. Note the increased activity, red and yellow, in the top of the photo when the eyes are open.	Radioactive form of glucose is injected into the bloodstream; scanner records amount of glucose used in particularly active areas of the brain and produces computer-constructed picture of the brain.	Originally designed to detect abnormalities, also used to identify brain areas active during ordinary activities (reading, singing, etc.)
MRI (magnetic resonance imaging) scan *Magnetic resonance imaging (MRI).* Note the fissures and internal structures of the brain. The throat, nasal airways, and fluid surrounding the brain are dark.	A high-frequency magnetic field is passed through the brain by means of electromagnets.	Produces high-resolution three-dimensional pictures of the brain useful for identifying abnormalities and mapping brain structures and function.
fMRI (functional magnetic resonance imaging) scan	A newer, faster version of the MRI that detects blood flow by picking up magnetic signals from blood that has given up its oxygen to activate brain cells.	Indicates which areas of the brain are active or inactive during ordinary activities or responses (like reading or talking); also shows changes associated with disorders.

CONCEPT CHECK **STOP**

How do psychologists guard against bias?

Which type of psychological research involves observing participants' behavior in the real world?

What is the difference between a positive and a negative correlation?

What does it mean to call a procedure "noninvasive"?

Getting the Most from Your Study of Psychology

LEARNING OBJECTIVES

Describe the steps you can take to read more accurately.

Explain how visual features can enhance learning.

Examine your current time-management habits, and identify how you might improve them.

Explain the benefits of distributed study and overlearning.

Summarize the grade-improvement and test-taking strategies that students can use to ensure success in their courses.

In this section, we offer several well-documented techniques that will help you "work smarter"—not just longer or harder (Dickinson, O'Connell, & Dunn, 1996), so that you can get the most from your study of psychology (or any other subject). Mastering these skills will initially take some time, but you'll save hundreds of hours later on.

FAMILIARIZATION

Have you ever noticed that you can read a paragraph many times and still remember nothing from it? Often you must make a conscious effort to learn. There are a number of ways to actively read (and remember) information in a text. The first step is to familiarize yourself with the general text so that you can take full advantage of its contents. In *Visualizing Psychology*, the Preface, Table of Contents, Glossary (both running through the text of each chapter and at the end of each chapter), References, Name Index, and Subject Index will help give you a bird's-eye view of the rest of the text. In addition, as you scan the book to familiarize yourself with its contents, you should also take note of the many tables, figures, photographs, and special feature boxes, all of which will enhance your understanding of the subject.

ACTIVE READING

The most important tool for college sucess is the ability to read and master the assigned class text. One of the best ways to read actively is to use the **SQ4R method**, which was developed by Francis Robinson (1970). The initials stand for six steps in effective reading: **S**urvey, **Q**uestion, **R**ead, **R**ecite, **R**eview, and w**R**ite. As you might have guessed, *Visualizing Psychology* was designed to incorporate each of these steps (**FIGURE 1.16**).

VISUAL LEARNING

Our brains are highly tuned to visual cues as well as verbal cues. Photographs, drawings, and other graphical information help us solidify our understanding, organize and internalize new material, recognize patterns and interrelationships, and think creatively.

In some books, photographs and illustrations merely repeat, visually, concepts that are also stated in words. *Visualizing Psychology* is different. We have explicitly designed the book to take advantage of readers' capacity to process information through both visual and verbal channels. The text is tightly integrated with a rich array of visual features that will help you solidify your understanding of the concepts contained in the text. (The Illustrated Book Tour in the Preface describes these features in detail.) The photographs, drawings, diagrams, and graphs in this book carry their own weight—that is, they serve a specific instructional purpose, above and beyond what is stated in words. They are as essential as the text itself; be sure to pay attention to them.

Survey
Each chapter of the text opens with an illustrated Chapter Outline, and each section includes Learning Objectives. Together they provide an overview, or survey, of the chapter. Knowing what to expect from a chapter can help you focus on the main points.

Question
To maintain your attention and increase comprehension, each chapter includes a number of Learning Objectives. Use them to form questions, then keep the questions in mind while reading each of the chapter's sections.

Read
Try to answer the questions you formed in the previous step as you read the chapter. Read carefully—in short, concentrated time periods.

Recite
After you have read one small section, stop and silently recite what you've just read or make a written summary. To better retain the material, think of personal examples of major concepts and how you might apply a certain concept to real-life situations.

Review
Review the Visual Summary at the end of each chapter. Answer the Concept Check questions that conclude each major section and the Critical and Creative Thinking Questions at the end of each chapter. Write down your responses before checking your answers. When you finish the chapter, look back over your answers, the chapter headings, and the Learning Objectives for each section. Repeat this review process before each class quiz or exam.

wRite
In addition to the writing you do in the previous steps, take brief notes in the text margins or on a separate sheet of paper. This will help keep you focused during your reading.

Improving Your Grade

If you are a student, the general learning tools outlined here will make you more efficient and successful in your courses. In addition, here are several specific strategies for grade improvement and test taking that can further improve your performance:

- *Take good notes.* Effective note taking depends on active

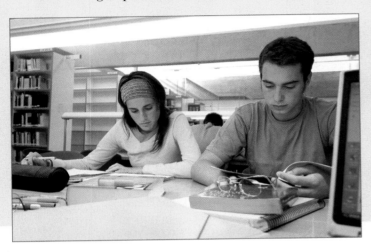

listening. Ask yourself, "What is the main idea?" Write down key ideas and supporting details and examples.

- *Understand your professor.* The amount of lecture time spent on various topics is generally a good indication of what the instructor considers important.

- *General test taking.* On multiple-choice exams, carefully read each question and all of the alternative answers before responding. Be careful to respond to all questions, and make sure that you

have recorded your answers correctly. Finally, practice your test taking by responding to the Concept Check questions, the Critical and Creative Thinking Questions, and the Self-Test in each chapter.

- *Skills courses.* Improve your reading speed and comprehension and your word-processing/typing skills by taking additional courses designed to develop these specific abilities.

- *Additional resources.* Don't overlook important human resources. Your instructors can provide useful tips. Enlist roommates, classmates, friends, family members as your conscience-coaches.

TIME MANAGEMENT

If you find that you can't always strike a good balance between work, study, and social activities or that you aren't always good at budgeting your time, here are four basic time-management strategies:

- *Establish a baseline.* Before attempting any changes, simply record your day-to-day activities for one to two weeks. You may be surprised by how you spend your time.

- *Set up a realistic schedule.* Make a daily and weekly "to do" list, including all required activities, basic maintenance tasks (like laundry, cooking, child care, and eating), and a reasonable amount of "down time." Then create a daily schedule of activities that includes time

for each of these. To make permanent time-management changes, shape your behavior, starting with small changes and building on them.

- *Reward yourself.* Give yourself immediate, tangible rewards for sticking with your daily schedule.

- *Maximize your time.* Time-management experts, such as Alan Lakein (1998), suggest that you should try to minimize the amount of time you spend worrying and complaining and fiddling around getting ready to study ("fretting and prepping"). Also be alert for hidden "time opportunities"—spare moments that normally go to waste, which you might instead use productively.

DISTRIBUTED STUDY

Spaced practice is a much more efficient way to study and learn than massed practice (Chapter 7). That is, you will learn material more thoroughly if you distribute your study over time, rather than trying to cram all the information in at once.

OVERLEARNING

Many people tend to study new material just to the point where they can recite the information, but they do not attempt to understand it more deeply. For best results, you should know how key terms and concepts are related to one another. You should also be able to generate examples other than the ones in the text. You should repeatedly review the material (by visualizing the phenomena that are described and explained in the text and by rehearsing what you have learned) until the information is firmly locked in place. You will find this confidence-building exercise particularly important if you suffer from test anxiety.

CONCEPT CHECK **STOP**

What are the elements of active reading?

How can a textbook take advantage of readers' visual learning abilities?

Why is distributed practice more effective than massed practice?

What steps will you take to get the most from your study of psychology?

S R Q R R R

What is happening in this picture ?

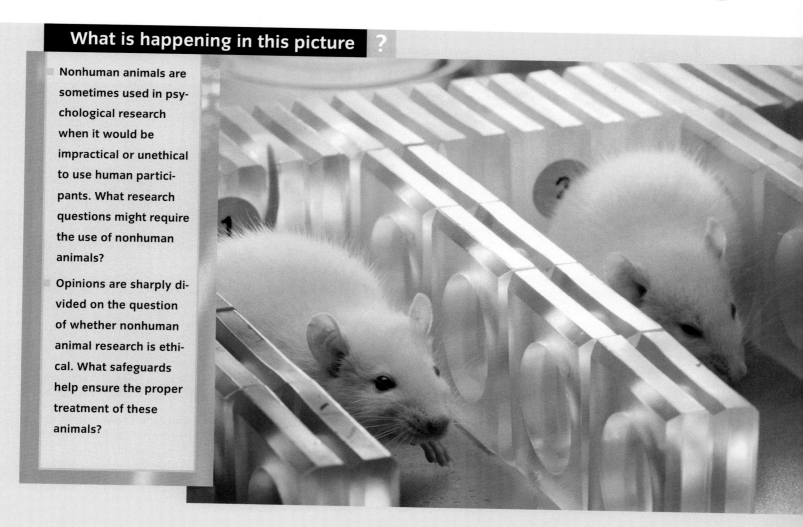

Nonhuman animals are sometimes used in psychological research when it would be impractical or unethical to use human participants. What research questions might require the use of nonhuman animals?

Opinions are sharply divided on the question of whether nonhuman animal research is ethical. What safeguards help ensure the proper treatment of these animals?

VISUAL SUMMARY

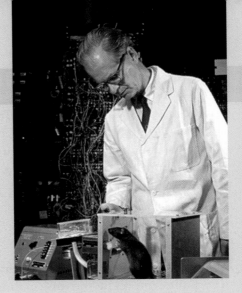

1 Introducing Psychology

1. **Psychology** is the scientific study of behavior and mental processes. The discipline places high value on **empirical evidence** and **critical thinking**. One of the most enduring debates in science has been the **nature-nurture controversy**.

2. Psychology's four basic goals are to describe, explain, predict, and change behavior and mental processes through the use of scientific methods.

3. Psychologists work as therapists, researchers, teachers, and consultants in a wide range of settings.

2 Origins of Psychology

1. Wilhelm Wundt, the father of psychology, and his followers, including Edward Titchener, were interested in studying conscious experience. Their approach, structuralism, sought to identify the basic structures of mental life through introspection.

2. The functionalist school, led by William James, studied how the mind functions to adapt humans and other animals to their environment.

3. Contemporary psychology reflects seven major perspectives: psychoanalytic or psychodynamic, behavioral, humanistic, cognitive, neuroscientific or biopsychological, evolutionary, and sociocultural. Most contemporary psychologists embrace a unifying perspective known as the **biopsychosocial model**. This model views biological processes, psychological factors, and social forces as interrelated influences on behavior.

4. Despite early limitations, women and minorities have made important contributions to psychology. Pioneers include Margaret Floy Washburn, Mary Calkins, Francis Cecil Sumner, and Kenneth B. Clark.

3 The Science of Psychology

1. **Basic research** is aimed at advancing general scientific understanding, whereas **applied research** works to address real-world problems.

2. Most scientific investigations involve six basic steps, collectively known as the scientific method. Scientific progress comes from repeatedly challenging and revising existing theories and building new ones.

3. Psychologists must maintain high ethical standards. This includes respecting the rights of both human and nonhuman research participants and of psychotherapy clients. **Informed consent** and **debriefing** are critical elements of any research that involves human participants. Researchers and clinicians are held professionally responsible for their actions by the APA, by their institutions, and by local and state agencies.

4 Research Methods

1. Experimental research manipulates and controls variables to determine cause and effect. An **experiment** has two critical components: **independent** and **dependent variables**, and experimental and control groups. A good scientific experiment protects against potential sources of error from both the researcher and the participants.

2. **Descriptive research** involves systematically observing and describing behavior without manipulating variables. The three major types of descriptive research are naturalistic observation, surveys, and case studies.

3. **Correlational research** allows researchers to observe the relationship between two variables. Researchers analyze their results using a correlation coefficient. Correlations can be positive or negative. A correlation between two variables does not necessarily mean that one causes the other.

4. **Biological research** focuses on internal, biological processes that are involved in our feelings, thoughts, and behavior. Recent advances in brain imaging have improved scientists' ability to examine these processes and to do so non-invasively.

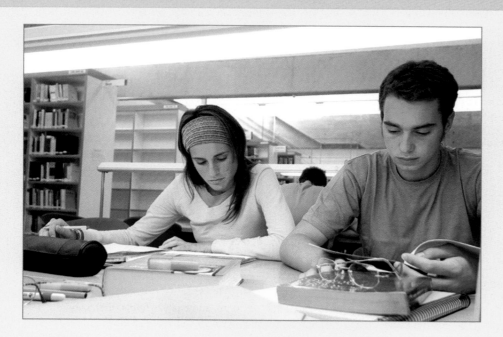

5 Getting the Most from Your Study of Psychology

1. Several well-documented techniques will help readers understand and absorb the material in this book most completely. These include familiarization, active reading, visual learning, time management, distributed study, and overlearning.

2. For students, several additional strategies for grade improvement and test taking can further improve course performance. These include more effective note taking, understanding the professor, general test-taking strategies, study skills courses, and other helpful resources.

KEY TERMS

- psychology p. 4
- behavior p. 4
- mental processes p. 4
- empirical evidence p. 4
- critical thinking p. 4
- pseudopsychologies p. 4
- nature-nurture controversy p. 5
- structuralism p. 7
- functionalism p. 7
- psychoanalytic school p. 7
- behaviorist perspective p. 7
- humanist perspective p. 8
- cognitive perspective p. 8
- neuroscientific/biopsychological perspective p. 8

- evolutionary perspective p. 8
- sociocultural perspective p. 8
- biopsychosocial model p. 8
- basic research p. 12
- applied research p. 12
- scientific method p. 12
- informed consent p. 14
- debriefing p. 14
- experiment p. 16
- independent variable (IV) p. 16
- dependent variable (DV) p. 16
- experimental group p. 16
- control group p. 16
- experimenter bias p. 16

- double-blind study p. 16
- ethnocentrism p. 16
- cross-cultural sampling p. 16
- sample bias p. 16
- social-desirability response p. 16
- descriptive research p. 18
- naturalistic observation p. 18
- surveys p. 19
- case studies p. 19
- correlational research p. 20
- correlation coefficient p. 20
- biological research p. 20
- SQ4R method p. 24

CRITICAL AND CREATIVE THINKING QUESTIONS

1. Scientific psychologists are among the least likely to believe in psychics, palmistry, astrology, and other paranormal phenomena. Why might that be?

2. Which psychological perspective would most likely be used to study and explain why some animals, such as newly hatched ducks or geese, follow and become attached to (or imprinted on) the first large moving object they see or hear?

3. Why is the scientific method described as a cycle, rather than as a simple six-step process?

4. Imagine that a researcher recruited research participants from among her friends, then assigned them to experimental or control groups based on their gender. Why might this be a problem?

5. Which modern methods of examining how the brain influences behavior are noninvasive?

6. What do you think keeps most people from fully employing the active learning strategies and study skills presented in this chapter?

SELF-TEST

(Check your answers in Appendix B.)

1. In your text, psychology is defined as the _____.
   a. science of conscious and unconscious forces on behavior
   b. empirical study of the mind
   c. scientific study of the mind
   d. scientific study of behavior and mental processes

2. According to your textbook, the goals of psychology are to _____.
   a. explore the conscious and unconscious functions of the human mind
   b. understand, compare, and analyze human behavior
   c. improve psychological well-being in all individuals from conception to death
   d. describe, explain, predict, and change behavior and mental processes

3. The father of psychology is _____ .

a. Sigmund Freud
b. B. F. Skinner
c. Wilhelm Wundt
d. William James

4. The biopsychosocial model is known as a(n)

_____ .

a. integrative, unifying model
b. concept formation
c. consolidation model
d. eclectic conceptualization

5. The term *basic research* is **BEST** defined as research that

_____ .

a. is basic to one field only
b. is intended to advance scientific knowledge rather than for practical application
c. is done to get a grade or a tenured teaching position
d. solves basic problems encountered by humans and animals in a complex world

6. Identify and label the 6 steps in the scientific method: choose the best research design to test the hypothesis; after more studies, advance a theory to explain the results; perform statistical analysis; conduct a review of the literature; make a testable hypothesis; write up study and its results and submit to scientific journal

Step 1
Step 6
Step 2
SCIENTIFIC METHOD
Step 5
Step 3
Step 4

7. According to your text, debriefing is _____ .

a. interviewing subjects after a study to find out what they were thinking during their participation
b. explaining the purpose of the study, anticipated results, and deception used when the study is over
c. disclosing potential physical and emotional risks, and the nature of the study prior to its beginning
d. interviewing subjects after a study to determine whether any deception was effective in preventing them from learning the true purpose of the study

8. _____ are manipulated; _____ are measured.

a. Dependent variables; independent variables
b. Surveys; experiments
c. Statistics; correlations
d. IVs; DVs

9. The **BEST** definition of a double-blind study is research in which

_____ .

a. nobody knows what they are doing
b. neither the participants in the treatment group nor the control group knows which treatment is being given to which group
c. both the researcher and the participants are unaware of who is in the experimental and control groups
d. two control groups (or placebo conditions) must be used

10. In a case study, a researcher is most likely to

_____ .

a. interview many research subjects who have a single problem or disorder
b. conduct an in-depth study of a single research participant
c. choose and investigate a single topic
d. use any of these options, which describe different types of case studies

11. In _____research, a researcher observes or measures (without directly manipulating) two or more variables to find relationships between them, without inferring a causal relationship.

a. experimental   c. basic
b. correlational   d. applied

12. Label SQ4R on the diagram.
S_____
Q_____
R_____
R_____
R_____
R_____

S
R   Q
R   R
R

Neuroscience and Biological Foundations

In 1996, when world-class bicyclist Lance Armstrong was diagnosed with testicular cancer, it seemed uncertain whether he would survive, let alone mount his bike again. By the time the cancer was discovered, it had already spread to his lungs and his brain. Armstrong's doctors told him that he had only about a forty percent chance of survival. (Later one of his doctors revealed that the real estimate was three percent, but they had given him the higher figure to keep him from becoming discouraged!)

After major surgeries and intensive chemotherapy, however, Armstrong overcame the cancer, and by 1998 he was racing again—with remarkable success. He won his first Tour de France race in 1999, and then went on to win it an unprecedented six more consecutive times before retiring from professional cycling in 2005. The Tour de France is by far the most prestigious of all cycling competitions in the world. It is also one of the most physically demanding; the race runs over three weeks and covers 3000 to 4000 kilometers (1800 to 2500 miles), over the steep French Pyrennes mountains and across the flat coastal plains.

What made Armstrong's extraordinary achievements possible? Is he "genetically gifted"? What part of his brain allowed him to precisely monitor and meter out limited energy supplies over long days on his bike? Are there special chemicals in his brain—and those of other top athletes—that make it possible to cope with the pain and exhaustion from hours, days, and years of grueling practices and races? In this chapter, we'll look at the biological processes that make it possible for great achievers like Lance Armstrong—as well as the rest of us—to absorb and organize the massive influx of sights, sounds, thoughts, emotions, and memories that compete, every minute, for the brain's attention.

Our Genetic Inheritance

behavioral genetics The study of the relative effects of heredity and environment on behavior and mental processes.

evolutionary psychology A branch of psychology that studies the ways in which natural selection and adaptation can explain behavior and mental processes.

Thousands of years of evolution have contributed to what we are today. Our ancestors foraged for food, fought for survival, and passed on traits that were selected and transmitted down through the generations. How do these transmitted traits affect us today? For answers, psychologists often turn to **behavioral genetics** (how heredity and environment affect us) and **evolutionary psychology** (how the natural process of adapting to our environment affects us).

BEHAVIORAL GENETICS: IS IT NATURE OR NURTURE?

Ancient cultures, including the Egyptian, Indian, and Chinese, believed the heart was the center of all thoughts and emotions. But we now know that the brain and the rest of the nervous system are the power behind our psychological life and much of our physical being. This chapter introduces you to the field of **neuroscience** and *biopsychology*, the scientific study of the biology of behavior and mental processes. We will discuss genetics and heredity, the nervous system, and the functions of each part of the brain.

At the moment of your conception, your mother and father each contributed to you 23 *chromosomes*. Thousands of *genes* make up each chromosome (**FIGURE 2.1**). For some traits, such as blood type, a single pair of genes (one from each parent) determines what characteristics you will possess. But most traits are determined by a combination of many genes.

When the two genes for a given trait conflict, the outcome depends on whether the gene is *dominant* or *recessive*. A dominant gene reveals its trait whenever the gene is present. In contrast, the gene for a reces-

neuroscience An interdisciplinary field studying how biological processes relate to behavioral and mental processes.

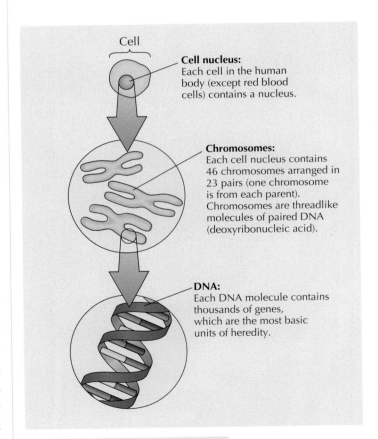

Cell nucleus: Each cell in the human body (except red blood cells) contains a nucleus.

Chromosomes: Each cell nucleus contains 46 chromosomes arranged in 23 pairs (one chromosome is from each parent). Chromosomes are threadlike molecules of paired DNA (deoxyribonucleic acid).

DNA: Each DNA molecule contains thousands of genes, which are the most basic units of heredity.

Hereditary code FIGURE 2.1

Gene–environment interaction FIGURE 2.2

Children who are malnourished may not reach their full potential genetic height or maximum intelligence. Can you see how environmental factors interact with genetic factors to influence many traits?

sive trait will normally be expressed only if the other gene in the pair is also recessive.

It was once assumed that characteristics such as eye color, hair color, or height were the result of either one dominant gene or two paired recessive genes. But modern geneticists believe that each of these characteristics is *polygenic*, meaning they are controlled by multiple genes. Many polygenic traits like height or intelligence are also affected by environmental and social factors (FIGURE 2.2). Fortunately, most serious genetic disorders are not transmitted through a dominant gene. Can you understand why?

How do scientists research human inheritance? To determine the influences of heredity or environment on complex traits like aggressiveness, intelligence, or sociability, scientists rely on these indirect methods: twin, family, and adoption studies, and studies of genetic abnormalities.

Psychologists are especially interested in the study of twins because they have a uniquely high proportion of shared genes. Identical (*monozygotic*—one egg) twins share 100 percent of the same genes, whereas fraternal (*dizygotic*—two egg) twins share, on average, 50 percent of their genes, just like any other pair of siblings (FIGURE 2.3).

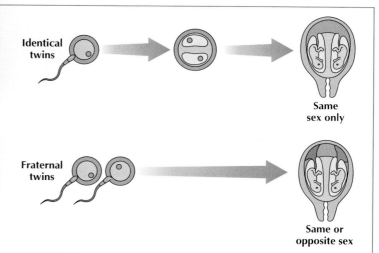

Identical and fraternal twins FIGURE 2.3

Monozygotic, or identical, twins develop from a single egg fertilized by a single sperm. They share the same placenta and have the same sex and same genetic make-up.
Fraternal twins are formed when two separate sperm fertilize two separate eggs.

Because both identical and fraternal twins share the same parents and develop in relatively the same environment, they provide a valuable "natural experiment." If heredity influences a trait or behavior to some degree, identical twins should be more alike than fraternal twins. Twin studies have provided a wealth of information on the relative effects of heredity on behavior. For example, studies of intelligence show that identical twins have almost identical IQ scores, whereas fraternal twins are only slightly more similar in their IQ scores than are non-twin siblings (Bouchard, 2004; Plomin, 1999). The difference suggests a genetic influence on intelligence.

Psychologists interested in behavioral genetics can also study entire families. If a specific trait is inherited, blood relatives should show increased trait similarity, compared with unrelated people. Also, closer relatives, like siblings, should be more similar than distant relatives. Family studies have shown that many traits and mental disorders, such as intelligence, sociability, and depression, do indeed run in families.

Studying families with children who have been adopted provides valuable information for researchers. If adopted children are more like their biological family in some trait, then genetic factors probably had the greater influence. Conversely, if adopted children resemble their adopted family even though they do not share similar genes, then environmental factors may predominate.

Finally, research in behavioral genetics explores disorders and diseases that result when genes malfunction. For example, researchers believe that genetic or chromosomal abnormalities are important factors in Alzheimer's disease and schizophrenia.

Findings from these four methods have allowed behavioral geneticists to estimate the *heritability* of various traits. That is, to what degree are individual differences a result of genetic, inherited factors rather than differences in

the environment? If genetics contributed nothing to the trait, it would have a heritability estimate of 0 percent. If a trait was completely due to genetics, it would have a heritability estimate of 100 percent. Keep in mind, however, that heritability estimates apply to groups, not individuals (**FIGURE 2.4**).

As we've seen, behavioral genetics studies help explain the role of heredity (nature) and the environment (nurture) in our individual behavior. To increase our understanding of genetic dispositions, we also need to look at universal behaviors transmitted from our evolutionary past.

EVOLUTIONARY PSYCHOLOGY: DARWIN EXPLAINS BEHAVIOR AND MENTAL PROCESSES

Evolutionary psychology suggests that many behavioral commonalities, from eating to fighting with our enemies, emerged and remain in human populations because they helped our ancestors (and ourselves)

Height and heritability FIGURE 2.4

Height has one of the highest heritability estimates—around 90 percent (Plomin, 1990). However, it's impossible to predict with certainty an individual's height from a heritability estimate. The same is true regardless of whether the trait in question is athletic ability, intelligence, depression, or cancer.

survive. This perspective stems from the writings of Charles Darwin (1859), who suggested that natural forces select traits that are adaptive to the organism's survival. This process of **natural selection** occurs when a particular genetic trait gives a person a reproductive advantage over others. Some people mistakenly believe that natural selection means "survival of the fittest." But what really matters is *reproduction*—the survival of the genome. Because of natural selection, the fastest or otherwise most fit organisms will be most likely to live long enough to pass on their genes to the next generation (Figure 2.4).

    Genetic mutations also help explain behavior. Everyone likely carries at least one gene that has mutated, or changed from the original. Very rarely, a mutated gene will be significant enough to change an individual's behavior. It might cause someone to be more social, more risk taking, more shy, more careful. If the gene then gives the person reproductive advantage, he or she will be more likely to pass on the gene to future

generations. However, this mutation doesn't guarantee long-term survival. A well-adapted population can perish if its environment changes.

CONCEPT CHECK STOP

What is the difference between a dominant and a recessive trait?

Why doesn't knowing the heritability of a trait predict what characteristics a given individual will have?

What is natural selection?

How are heredity and evolution linked to human behavior?

Neural Bases of Behavior

LEARNING OBJECTIVES

Describe how neurons communicate throughout the body.

Explain the role that neurotransmitters play.

Compare and contrast the functions of neurotransmitters and hormones.

Your brain and the rest of your nervous system essentially consist of **neurons**. Each one is a tiny information-processing system with thousands of connections for receiving and sending electrochemical signals to other neurons. Each human body may have as many as one *trillion* neurons. (Be careful not to confuse the term *neuron* with the term *nerve*. Nerves are large bundles of axons outside the brain and spinal cord.)

    Neurons are held in place and supported by **glial cells**. Glial cells surround neurons, perform cleanup tasks, and insulate one neuron from another

so their neural messages do not get scrambled. They also play a direct role in nervous system communication (Rose & Konnerth, 2002; Wieseler-Frank, Maier, & Watkins, 2005). However, the "star" of the communication show is still the neuron.

    No two neurons are alike, but most share three basic features: **dendrites**, the

neuron A nerve cell that receives and conducts electrical impulses from the brain.

Dendrites receive information from other neurons and sensory receptors.

Cell body receives information from dendrites, and if enough stimulation is received the message is passed on to the axon.

Axon carries neuron's message to other body cells.

Myelin sheath covers the axon of some neurons to insulate and help speed neural impulses.

Terminal buttons of axon form junctions with other cells and release chemicals called neurotransmitters.

The structure of a neuron FIGURE 2.5

Arrows indicate direction of information flow: dendrites → cell body → axon → terminal buttons of axon.

cell body, and an **axon** (FIGURE 2.5). To remember how information travels through the neuron, think of these three in reverse alphabetical order: *Dendrite* → *Cell Body* → *Axon*.

HOW DO NEURONS COMMUNICATE?

A neuron's basic function is to transmit information throughout the nervous system. Neurons "speak" in a type of electrical and chemical language. The process of neural communication begins within the neuron itself, when the dendrites and cell body receive electrical "messages." These messages move along the axon in the form of a neural impulse, or **action potential** (FIGURE 2.6).

Nerve impulses move much more slowly than electricity through a wire. A neural impulse travels along a bare axon at only about 10 meters per second. (Electricity moves at 36 million meters per second.)

Some axons, however, are enveloped in fatty insulation, the **myelin sheath**. This sheath blankets the axon, with the exception of periodic *nodes*, points at which the myelin is very thin or absent. In a myelinated axon, the nerve impulse moves about 10 times faster than in a bare axon because the action potential jumps from node to node rather than traveling along the entire axon.

Communication *within* the neuron (FIGURE 2.6) is not the same as communication *between* neurons (FIGURE 2.7 on page 40). Within the neuron, messages travel electrically. But messages are transmitted chemically from one neuron to the next. The chemicals that transmit these messages are called **neurotransmitters**.

neurotransmitters Chemicals that neurons release, which affect other neurons.

Communication within the neuron – the action potential FIGURE 2.6

www.wiley.com/college/carpenter

Resting, Polarized Membrane

Depolarization (sodium ions flow in)

Step 1: When the neuron is active, or *resting*, it is in a *polarized* state. The fluid outside the axon has more positively charged ions than the fluid inside, yielding a net negative charge inside. This is somewhat analogous to a car or flashlight battery. Batteries have positive and negative poles that are also "polarized." And, just as a battery has an electrical potential difference between the poles (1.5 volts for flashlight battery), the axon membrane has a potential difference of about –70mV across it. (The inside is more negative by 70/1000 of a volt.)

Step 2: When the cell body of the neuron receives sufficient stimulation via the dendrites from adjacent neurons, the electrical potential of the axon membrane near the cell body changes. When this potential change reaches a specific level, special voltage-controlled channels open, allowing a rapid inflow of positive sodium (Na^+) ions. This changes the previously negative charge inside the axon to positive—*depolarizing* the membrane.

(potassium ions flow out)

Depolarization

Step 3: This depolarization produces an imbalance of ions in the adjacent section on the axon membrane. Pores in this neighboring area now open, and more positively charged sodium ions flow in. Meanwhile, potassium channels in the previously depolarized section open, and the positive potassium ions (K^+) move out to balance the electrical charge. Thus, the resting potential is restored.

Flow of depolarization

Action potential

Action potential

Action potential

Summary: Can you see how imbalances on one section of the membrane cause nearby ion channels to open and the next section of the axon depolarizes, and the next, and the next, in a chain reaction? Through the sequential process of depolarization, followed by repolarization, the action potential moves continuously down the axon. This is like a line of dominoes falling, each one toppling the next. This neural message is called the *action potential*. According to the "all-or-none law," there is no such thing as a partial action potential.

Neural Bases of Behavior   39

Communication between neurons – neurotransmitter action at the synapse FIGURE 2.7

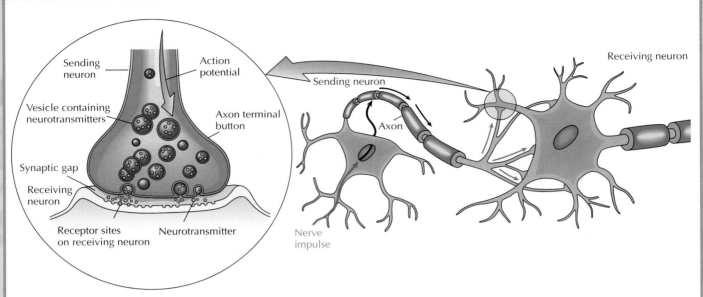

1. Sending a chemical signal   When action potentials reach the branching axon terminals, they trigger the terminal buttons at the axon's end to open and release thousands of neurotransmitters into the *synaptic gap*, the tiny opening between the sending and receiving neuron. These chemicals then move across the synaptic gap and attach to the membranes of the receiving neuron. In this way, they carry the message from the sending neuron to the receiving neuron.

2. Receiving a chemical signal   After a chemical message flows across the synaptic gap, it attaches to specific receiving neurons. It's important to know that each receiving neuron gets multiple neurotransmitter messages. As you can see in this close-up photo, the axon terminals from thousands of other nearby neurons almost completely cover the cell body of the receiving neuron. It's also important to understand that neurotransmitters deliver either excitatory or inhibitory messages, and that the receiving neuron will only produce an action potential and pass along the message if the number of excitatory messages outweigh the inhibitory messages. Why do we need competing messages? Like an accelerator and brake on a car, your body needs similar "on" (excitatory) and "off" (inhibitory) mechanisms. Using these two "switches," your body manages an amazing balance between overexcitation, leading to seizures, and underexcitation, leading to coma and death.

How Poisons and Drugs Affect Our Brain

Neurotransmitters help explain how poisons and mind-altering drugs affect the brain.

A Like a key fitting into a lock, receptor sites on receiving neurons' dendrites recognize neurotransmitters by their particular shape.

B Molecules without the correct shape don't fit the receptors, so they cannot stimulate the dendrite.

C Some *agonist drugs*, like the poison in the black widow spider or the nicotine in cigarettes, are similar enough in structure to a certain neurotransmitter (in this case, *acetylcholine*) that they mimic its effects on the receiving neuron.

D Some *antagonist drugs* block neurotransmitters like acetylcholine. Because acetylcholine is vital in muscle action, blocking it paralyzes muscles, including those involved in breathing, which can be fatal.

Examples of antagonists to acetylcholine include most snake venom and some poisons, like *botulinum* toxin (Botox®), which is used to treat some medical conditions involving abnormal muscle contraction—as well as for some cosmetic purposes.

Normal neurotransmitter activation

Postsynaptic receptor site

Nerve impulse

A

Blocked neurotransmitter activation

B

Agonist drug "mimics" neurotransmitter

C

Antagonist drug fills receptor space and blocks neurotransmitter

D

Researchers have discovered hundreds of substances that function as neurotransmitters. These substances regulate a wide variety of physiological processes.

One of the benefits of studying your brain and its neurotransmitters is that it will help you understand some common medical problems (**TABLE 2.1** on the next page discusses some of the better understood neurotransmitters). For example, we know that decreased levels of the neurotransmitter dopamine are as-

sociated with Parkinson's disease (PD), whereas excessively high levels of dopamine appear to contribute to some forms of schizophrenia.

Perhaps the best-known neurotransmitters are the endogenous opioid peptides, commonly known as **endorphins** (a contraction of *endogenous* [self-produced] and *morphine*). These chemicals mimic the effects of opium-based drugs such as morphine—they elevate mood and reduce pain. They also affect memory, learning, blood pressure, appetite, and sexual activity.

How neurotransmitters affect us TABLE 2.1

Neurotransmitter	Known or suspected effects
Serotonin	Mood, sleep, appetite, sensory perception, temperature regulation, pain suppression, and impulsivity. Low levels associated with depression.
Acetylcholine (ACh)	Muscle action, cognitive functioning, memory, rapid-eye-movement (REM) sleep, emotion. Suspected role in Alzheimer's disease.
Dopamine (DA)	Movement, attention, memory, learning, and emotion. Excess DA associated with schizophrenia, too little with Parkinson's disease. Also plays a role in addiction and the reward system.
Norepinephrine (NE) (or noradrenaline)	Learning, memory, dreaming, emotion, waking from sleep, eating, alertness, wakefulness, reactions to stress. Low levels of NE associated with depression, high levels with agitated, manic states.
Epinephrine (or adrenaline)	Emotional arousal, memory storage, and metabolism of glucose necessary for energy release.
Gamma aminobutyric acid (GABA)	Neural inhibition in the central nervous system. Tranquilizing drugs, like Valium, increase GABA's inhibitory effects and thereby decrease anxiety.
Endorphins	Mood, pain, memory, and learning.

HORMONES: A GLOBAL COMMUNICATION SYSTEM

We've just seen how the nervous system uses neurotransmitters to transmit messages throughout the body. A second type of communication system also exists. This second system is made up of a network of glands, called the **endocrine system** (FIGURE 2.8). Rather than neurotransmitters, this

The endocrine system
FIGURE 2.8

This figure shows the major endocrine glands, along with some internal organs to help you locate the glands.

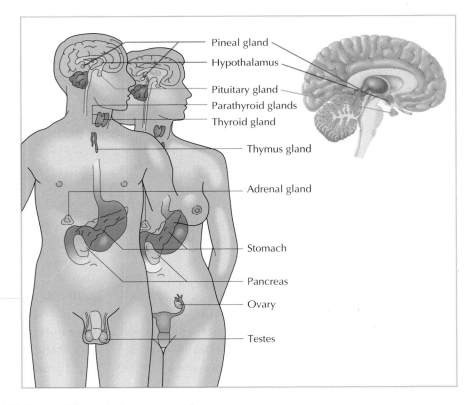

- Pineal gland
- Hypothalamus
- Pituitary gland
- Parathyroid glands
- Thyroid gland
- Thymus gland
- Adrenal gland
- Stomach
- Pancreas
- Ovary
- Testes

Neurotransmitters
send individual messages

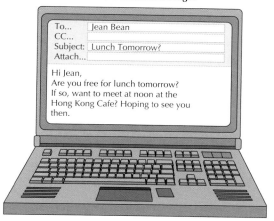

To...	Jean Bean
CC...	
Subject:	Lunch Tomorrow?
Attach...	

Hi Jean,
Are you free for lunch tomorrow?
If so, want to meet at noon at the
Hong Kong Cafe? Hoping to see you
then.

Hormones
send global messages

To...	Friends; family; co-workers
CC...	
Subject:	Party!
Attach...	

Hi Everybody,
Jean Bean and I are hosting a party on
Saturday night at 9 p.m. Please come,
and tell your friends!

Why do we need two communication systems? FIGURE 2.9

You can think of neurotransmitters as individual e-mails that you send to particular people. Neurotransmitters deliver messages to specific receptors, which other neurons nearby probably don't "overhear." Hormones, in contrast, are like a global e-mail message that you send to everyone in your address book. Endocrine glands release hormones directly into the bloodstream and travel throughout the body, carrying messages to any cell that will listen. Hormones also function like your global e-mail recipients forwarding your message to yet more people. For example, a small part of the brain called the hypothalamus releases hormones that signal the pituitary (another small brain structure), which stimulates or inhibits the release of other hormones.

hormones Chemicals manufactured by endocrine glands and circulated in the bloodstream to produce bodily changes or maintain normal bodily function.

system uses **hormones** to carry its messages.

Your endocrine system has several important functions. It helps regulate long-term bodily processes, such as growth and sexual characteristics. It also maintains ongoing bodily processes (such as digestion and elimination). Finally, hormones control the body's response to emergencies. In times of crisis, the hypothalamus sends messages through two pathways—the neural system and the endocrine system (primarily the pituitary). The pituitary sends hormonal messages to the adrenal glands (located right above the kidneys). The adrenal glands then release **cortisol**, a "stress hormone" that boosts energy and blood sugar levels, *epinephrine* (commonly called adrenaline), and *norepinephrine*. (Remember from Table 2.1 that these same chemicals also serve as neurotransmitters.)

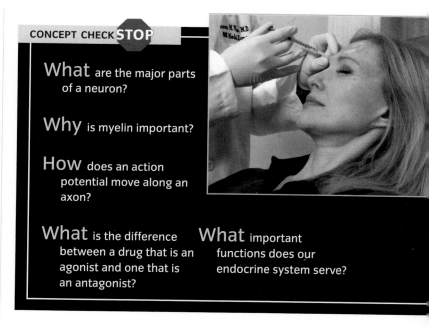

CONCEPT CHECK STOP

What are the major parts of a neuron?

Why is myelin important?

How does an action potential move along an axon?

What is the difference between a drug that is an agonist and one that is an antagonist?

What important functions does our endocrine system serve?

Nervous System Organization

Have you heard the expression "Information is power?" Nowhere is this truer than in the human body. Without information, we could not survive. Neurons within our nervous system must take in sensory information from the outside world and then decide what to do with the information. Just as the circulatory system handles blood, which conveys chemicals and oxygen, our nervous system uses chemicals and electrical processes that convey information.

The nervous system is divided and subdivided into several branches (FIGURE 2.10). One main part of our nervous system includes the brain and a bundle of nerves that form the *spinal cord*. Because this system is located in the center of your body (within your skull and spine), it is called the **central nervous system (CNS)**. The CNS is primarily responsible for processing and organizing information.

The second major part of your nervous system includes all the nerves outside the brain and spinal cord. This **peripheral nervous system (PNS)** carries messages (action potentials) to and from the central nervous system to the periphery of the body. Now, let's take a closer look at the CNS and the PNS.

central nervous system (CNS) The brain and spinal cord.

peripheral nervous system (PNS) All nerves and neurons connecting the CNS to the rest of the body.

CENTRAL NERVOUS SYSTEM (CNS): THE BRAIN AND SPINAL CORD

The central nervous system (CNS) is the branch of the nervous system that makes us unique. Most other animals can smell, run, see, and hear far better than we can. But thanks to our CNS, we can process information and adapt to our environment in ways that no other animal can. Unfortunately, our CNS is also incredibly fragile. Unlike neurons in the PNS that can regenerate and require less protection, serious damage to neurons in the CNS is usually permanent. However, the brain may not be as "hard wired" as we once believed.

Scientists long believed that after the first two or three years of life, humans and most animals are unable to repair or replace damaged neurons in the brain or spinal cord. We now know that the brain is capable of lifelong **neuroplasticity** and **neurogenesis**.

Neuroplasticity Rather than being a fixed, solid organ, the brain is capable of changing its structure and function as a result of usage and experience (Abbott, 2004; Neumann et al., 2005). This "rewiring" is what makes our brains so wonderfully adaptive. For example, it makes it possible for us to learn a new sport or a foreign language.

Remarkably, this rewiring has even helped "remodel" the brain following strokes. For example, psychologist Edward Taub and his colleagues (2002, 2004) have had success working with stroke patients (FIGURE 2.11).

Neurogenesis Our brains continually replace lost cells with new cells that originate deep within the brain and migrate to become part of its circuitry. The source of these newly created

neuroplasticity The brain's ability to reorganize and change its structure and function through the life span.

neurogenesis The division and differentiation of nonneuronal cells to produce neurons.

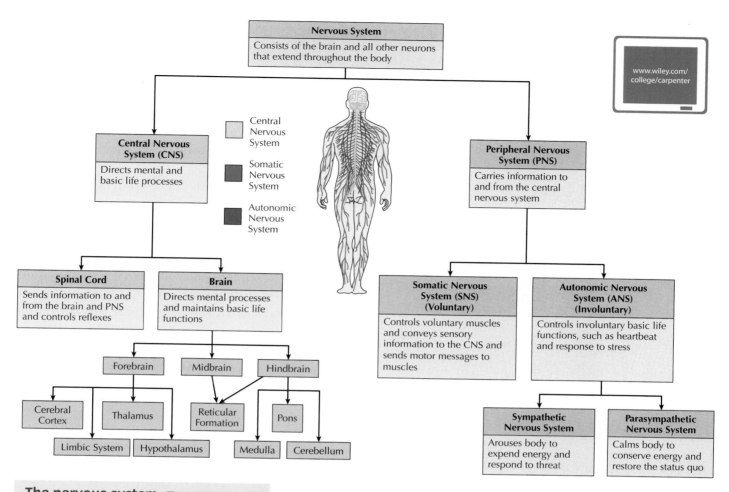

Nervous System

Consists of the brain and all other neurons that extend throughout the body

Central Nervous System

Somatic Nervous System

Autonomic Nervous System

Central Nervous System (CNS)

Directs mental and basic life processes

Peripheral Nervous System (PNS)

Carries information to and from the central nervous system

Spinal Cord

Sends information to and from the brain and PNS and controls reflexes

Brain

Directs mental processes and maintains basic life functions

Forebrain | Midbrain | Hindbrain

Cerebral Cortex | Thalamus | Reticular Formation | Pons

Limbic System | Hypothalamus | Medulla | Cerebellum

Somatic Nervous System (SNS) (Voluntary)

Controls voluntary muscles and conveys sensory information to the CNS and sends motor messages to muscles

Autonomic Nervous System (ANS) (Involuntary)

Controls involuntary basic life functions, such as heartbeat and response to stress

Sympathetic Nervous System

Arouses body to expend energy and respond to threat

Parasympathetic Nervous System

Calms body to conserve energy and restore the status quo

The nervous system FIGURE 2.10

cells is neural **stem cells**—rare, immature cells that can grow and develop into any type of cell. Their fate depends on the chemical signals they receive (Abbott, 2005; Kim, 2004). Stem cells have been used for bone marrow transplants, and clinical trials using stem cells to repopulate or replace cells devastated by injury or disease have helped patients suffering from strokes, Alzheimer's, Parkinson's, epilepsy, stress, and depression (Chang et al., 2005; Gruber et al., 2003; Kim et al., 2002; Wickelgren, 2002).

■ **stem cells**
Precursor (immature) cells that give birth to new specialized cells; a stem cell holds all the information it needs to make bone, blood, brain—any part of a human body—and can also copy itself to maintain a stock of stem cells.

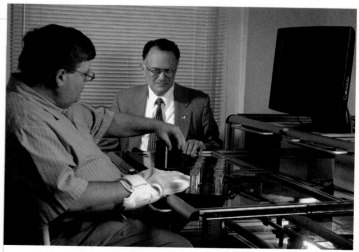

A breakthrough in neuroscience FIGURE 2.11

By immobilizing the unaffected arm or leg and requiring rigorous and repetitive exercise of the affected limb, psychologist Edward Taub and colleagues "recruit" stroke patients' intact brain cells to take over for damaged cells. The therapy has restored function in some patients as long as 21 years after their strokes.

Nervous System Organization   45

Does this mean that people paralyzed from spinal cord injuries might be able to walk again? At this point, neurogenesis in the brain and spinal cord is minimal. However, one possible bridge might be to transplant embryonic stem cells in the damaged area of the spinal cord. Researchers have transplanted mouse embryonic stem cells into a damaged rat spinal cord (Jones, Anderson, & Galvin, 2003; McDonald et al., 1999). When the damaged spinal cord was viewed several weeks later, the implanted cells had survived and spread throughout the injured spinal cord area. More important, the transplant rats also showed some movement in previously paralyzed parts of their bodies. Medical researchers have also begun human trials using nerve grafts to repair damaged spinal cords (Lopez, 2002; Saltus, 2000).

Now that we have discussed the remarkable adaptability of the central nervous system, let's take a closer look at its components: the brain and the spinal cord.

Because of its central importance for psychology and behavior, we'll discuss the brain in detail in the next major section. But the spinal cord is also impor-

tant. Beginning at the base of the brain and continuing down the back, the spinal cord carries vital information from the rest of the body into and out of the brain.

But the spinal cord doesn't simply relay messages. It can also initiate some automatic behaviors on its own. We call these involuntary, automatic behaviors **reflexes** or **reflex arcs** because the response to the incoming stimuli is automatically "reflected" back (**FIGURE 2.12**).

We're all born with numerous reflexes, many of which fade over time. But even as adults, we still blink in response to a puff of air in our eyes, gag when something touches the back of the throat, and urinate and defecate in response to pressure in the bladder and rectum. Reflexes even influence our sexual responses. Certain stimuli, such as the stroking of the genitals, can lead to arousal and the reflexive muscle contractions of orgasm in both men and women. However, in order to have the passion, thoughts, and emotion we normally associate with sex, the sensory information from the stroking and orgasm must be carried to the brain.

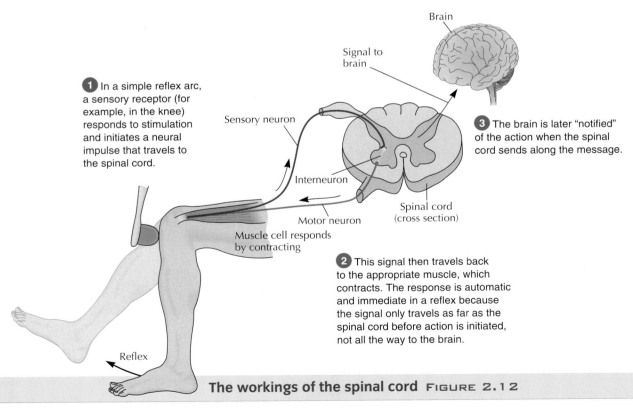

1 In a simple reflex arc, a sensory receptor (for example, in the knee) responds to stimulation and initiates a neural impulse that travels to the spinal cord.

Sensory neuron

Signal to brain

Brain

3 The brain is later "notified" of the action when the spinal cord sends along the message.

Interneuron

Motor neuron

Spinal cord (cross section)

Muscle cell responds by contracting

2 This signal then travels back to the appropriate muscle, which contracts. The response is automatic and immediate in a reflex because the signal only travels as far as the spinal cord before action is initiated, not all the way to the brain.

Reflex

The workings of the spinal cord FIGURE 2.12

Reflexes provide an evolutionary advantage. If messages had to travel all the way to the brain before they could be acted upon, an animal (human or nonhuman) might be fatally wounded in the meantime.

Infant Reflexes

If you have an infant in your home, you can easily check his or her reflexes. First, try brushing the side of the baby's cheek. Note how he or she will automatically (reflexively) turn his or her head toward the stimulation and be-gin to suck. This is called the *rooting reflex*. Now place your finger or a spoon in the infant's hand, and note how he or she grasps it—the *grasp reflex*. To test for the *Babinski reflex*, stroke the sole of the foot. The infant will automatically spread his or her toes. These infant reflexes fade as we age.

PERIPHERAL NERVOUS SYSTEM (PNS): CONNECTING THE CNS TO THE REST OF THE BODY

The chief function of the peripheral nervous system (PNS) is to carry information to and from the central nervous system. It links the brain and spinal cord to the body's sense receptors, muscles, and glands.

The PNS is subdivided into the somatic nervous system and the autonomic nervous system. In a kind of "two-way street," the **somatic nervous system (SNS)** (also called the skeletal nervous system) first carries sensory information to the CNS, and then carries messages from the CNS to skeletal muscles (**FIGURE 2.13**).

The other subdivision of the PNS is the **autonomic nervous system (ANS)**. The ANS is responsible for involuntary tasks, such as heart

> **somatic nervous system (SNS)** Subdivision of the peripheral nervous system (PNS). The SNS connects the sensory receptors and controls the skeletal muscles.

> **autonomic nervous system (ANS)** Subdivision of the peripheral nervous system (PNS) that controls involuntary functions. It includes the *sympathetic* nervous system and the *parasympathetic* nervous system.

Spinal Cord (CNS)
Sensory Neuron
Sensory receptors in hand
Interneuron
Muscles used to purse lips and blow on coffee
Motor Neuron

Sensory and motor neurons FIGURE 2.13

In order for you to be able to function, your brain must communicate with your body. This is the job of the *somatic nervous system,* which receives sensory information, sends it to the brain, and allows the brain to direct the body to act. Messages (action potentials) within the nervous system can cross the synapse in only one direction. *Sensory neurons* carry messages to the CNS. *Motor neurons* carry messages *away* from the CNS. *Interneurons* internally communicate and intervene between the sensory inpts and the motor outputs. Most of the neurons in the brain are interneurons.

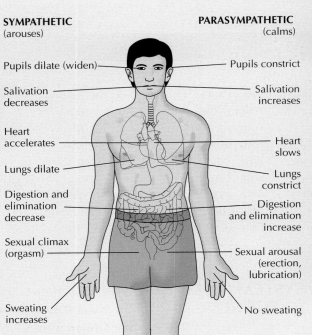

SYMPATHETIC (arouses)		PARASYMPATHETIC (calms)
Pupils dilate (widen)		Pupils constrict
Salivation decreases		Salivation increases
Heart accelerates		Heart slows
Lungs dilate		Lungs constrict
Digestion and elimination decrease		Digestion and elimination increase
Sexual climax (orgasm)		Sexual arousal (erection, lubrication)
Sweating increases		No sweating

Stress, high activity, fight-or-flight

Relaxation, low stress, rest-and-digest

Actions of the autonomic nervous system (ANS) FIGURE 2.14

rate, digestion, pupil dilation, and breathing. Like an automatic pilot, the ANS can sometimes be consciously overridden. But as its name implies, the autonomic system normally operates on its own (autonomously).

The autonomic nervous system is further divided into two branches, the sympathetic and parasympathetic, which tend to work in opposition to each other to regulate the functioning of such target organs as the heart, the intestines, and the lungs (**FIGURE 2.14**). Like two children on a teeter-totter, one will be up while the other is down, but they essentially balance each other out.

During stressful times, either mental or physical, the **sympathetic nervous system** mobilizes bodily resources to respond to the stressor. This emergency response is often called the "fight-or-flight" response. If you noticed a dangerous snake coiled and ready to strike, your sympathetic nervous system would increase your heart rate, respiration, and blood pressure, stop your digestive and eliminative processes, and release hormones, such as cortisol, into the bloodstream. The net result of sympathetic activation is to get more oxygenated blood and energy to the skeletal muscles, thus allowing you to cope with the stress—to "fight or flee."

In contrast to the sympathetic nervous system, the **parasympathetic nervous system** is responsible for returning your body to its normal functioning by slow-

ing your heart rate, lowering your blood pressure, and increasing your digestive and eliminative processes.

The sympathetic nervous system provides an adaptive, evolutionary advantage. At the beginning of human evolution, when we faced a dangerous bear or an aggressive human intruder, there were only two reasonable responses—fight or flight. This automatic mobilization of bodily resources can still be critical even in modern times. However, less life-threatening events, such as traffic jams, also activate our sympathetic nervous system. Our bodies still respond to these sources of stress with sympathetic arousal. As the next chapter discusses, ongoing sympathetic system response to such chronic, daily stress can become detrimental to our health.

CONCEPT CHECK STOP

What important roles does the spinal cord play?

How does neurogenesis begin?

What are the tasks of each of the subdivisions of the peripheral nervous system?

Sexual Arousal

The complexities of sexual interaction—and in particular, the difficulties that couples sometimes have in achieving sexual arousal or orgasm—illustrate the balancing act between the sympathetic and parasympathetic nervous systems.

Parasympathetic dominance: Sexual arousal and excitement require that the body be relaxed enough to allow increased blood flow to the genitals—in other words, the nervous system must be in *parasympathetic dominance*. Parasympathetic nerves carry messages from the central nervous system directly to the sexual organs, allowing for a localized response (increased blood flow and genital arousal).

Sympathetic dominance: When a person experiences strong emotions, such as anger, anxiety, or fear, the body shifts to *sympathetic dominance*, which can disrupt the sexual response patterns that occur under parasympathetic dominance. During sympathetic dominance, impulses from the central nervous system are sent to transfer stations outside their end destinations; the transfer stations quickly relay the messages all around the body. Blood flow to the genitals and organs decreases as the body readies for a "fight or flight." As a result, the person is unable (or less likely) to become sexually aroused. Any number of circumstances—for example, performance anxiety, fear of unwanted pregnancy or disease, or tensions between partners—can trigger sympathetic dominance.

A Tour Through the Brain

LEARNING OBJECTIVES

Identify the major structures of the hindbrain, midbrain, and forebrain, and of the cerebral cortex.

Summarize the major roles of the lobes of the cerebral cortex.

Describe what scientists have learned from split-brain research.

Explain why it's a mistake to believe that the right brain is usually "neglected."

Describe some examples of localization of function in the brain.

We begin our exploration of the brain at the lower end, where the spinal cord joins the base of the brain, and move upward toward the skull. As we move from bottom to top, "lower," basic processes, like breathing, generally give way to more complex mental processes (**FIGURE 2.15** and **FIGURE 2.16**).

LOWER-LEVEL BRAIN STRUCTURES: THE HINDBRAIN, MIDBRAIN, AND PARTS OF THE FOREBRAIN

Brain size and complexity vary significantly from species to species. For example, fish and reptiles have smaller, less complex brains than do cats and dogs. The most complex brains belong to whales, dolphins, and higher primates such as chimps, gorillas, and humans. The billions of neurons that make up the human brain control much of what we think, feel, and do. Certain brain structures are specialized to perform certain tasks, a process known as **localization of function**. However, most parts of the brain perform integrating, overlapping functions.

The hindbrain You are deep in REM sleep as you begin your last dream of the night. Your vital signs increase as the dream gets more exciting. But then your sleep—and your dream—is shattered by a buzzing alarm clock. All your automatic behaviors and survival responses in this scenario are controlled or influenced by parts of the hindbrain. The **hindbrain** includes the medulla, pons, and cerebellum.

The **medulla** is essentially an extension of the spinal cord, with many nerve fibers passing through it carrying information to and from the brain. It also controls many essential automatic bodily functions, such as respiration and heart rate.

The **pons** is involved in respiration, movement, sleeping, waking, and dreaming (among other things). It also contains axons that cross from one side of the brain to the other (*pons* is Latin for "bridge").

localization of function Specialization of various parts of the brain for particular functions.

Damage to the brain FIGURE 2.15

The 2005 debate over Terri Schiavo was largely about whether her husband should be allowed to remove her feeding tube because she was still able to move and breathe on her own and showed some reflexive responses. Terri's parents and others believed that these lower-level brain functions were sufficient proof of life. Advocates on the other side felt that once the cerebral cortex ceases functioning, the "person" is dead and there is no ethical reason to keep the body alive. What do you think?

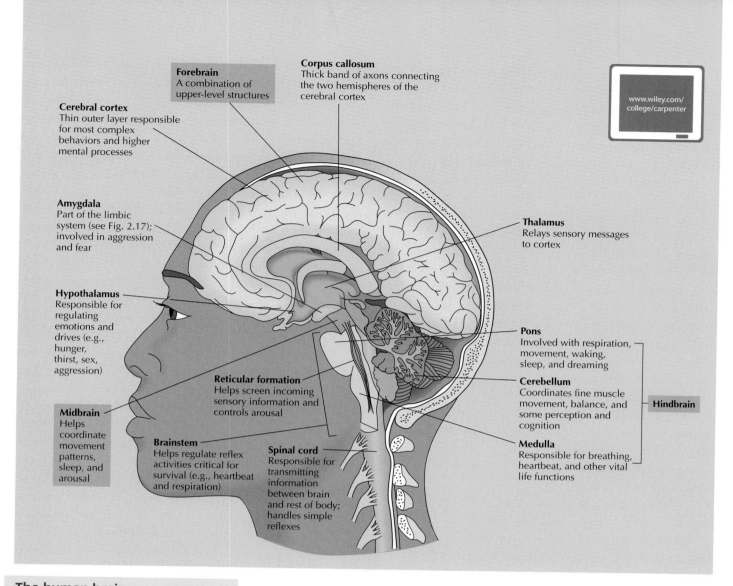

Forebrain
A combination of upper-level structures

Corpus callosum
Thick band of axons connecting the two hemispheres of the cerebral cortex

Cerebral cortex
Thin outer layer responsible for most complex behaviors and higher mental processes

Amygdala
Part of the limbic system (see Fig. 2.17); involved in aggression and fear

Thalamus
Relays sensory messages to cortex

Hypothalamus
Responsible for regulating emotions and drives (e.g., hunger, thirst, sex, aggression)

Pons
Involved with respiration, movement, waking, sleep, and dreaming

Reticular formation
Helps screen incoming sensory information and controls arousal

Cerebellum
Coordinates fine muscle movement, balance, and some perception and cognition

Hindbrain

Midbrain
Helps coordinate movement patterns, sleep, and arousal

Brainstem
Helps regulate reflex activities critical for survival (e.g., heartbeat and respiration)

Spinal cord
Responsible for transmitting information between brain and rest of body; handles simple reflexes

Medulla
Responsible for breathing, heartbeat, and other vital life functions

The human brain FIGURE 2.16

This drawing summarizes key functions of some of the brain's major structures. The brainstem, which includes parts of the hindbrain, midbrain, and forebrain, provides a handy geographical landmark.

The cauliflower-shaped **cerebellum** ("little brain" in Latin) is, evolutionarily, a very old structure. It coordinates fine muscle movement and balance. Researchers using functional magnetic resonance imaging (fMRI) have also shown that parts of the cerebellum are active during perceptual, cognitive, and language tasks (Paquier & Mariën, 2005; Rönnberg, Rudner & Ingvar, 2004; Thompson, 2005; Wild et al., 2003).

The midbrain

The **midbrain** helps us orient our eye and body movements to visual and auditory stimuli, and works with the pons to help control sleep and level of arousal. It also contains a small structure involved with the neurotransmitter dopamine, which deteriorates in Parkinson's disease.

Running through the core of the hindbrain, midbrain, and brainstem is the **reticular formation** (RF). This diffuse, finger-shaped network of neurons filters incoming sensory information and alerts the higher brain centers to important events. Without your reticular formation, you would not be alert or perhaps even conscious.

The forebrain   The **forebrain** is the largest and most prominent part of the human brain. It includes the thalamus, hypothalamus, limbic system, and cerebral cortex (**FIGURE 2.17**). The first three structures are located near the top of the brainstem. The cerebral cortex (discussed separately in the next section) is wrapped above and around them. (*Cerebrum* is Latin for "brain" and *cortex* is Latin for "covering" or "bark.")

The **thalamus** integrates input from the senses, and it may also be involved in learning and memory (Bailey & Mair, 2005; Ridley et al., 2005). Think of the thalamus as an air traffic control center that receives information from all aircraft and directs them to landing or takeoff areas. The thalamus receives input from nearly all sensory systems and directs the information to the appropriate cortical areas. Because the thalamus is the brain's major sensory relay center to the cerebral cortex, damage or abnormalities might cause the cortex to misinterpret or not receive vital sensory information. Interestingly, brain-imaging research links thalamus abnormalities to schizophrenia, a serious psychological disorder involving problems with sensory filtering and perception (Clinton & Meador-Woodruff, 2004; Preuss et al., 2005).

Beneath the thalamus lies the kidney bean–sized **hypothalamus** (*hypo-* means "under"). It has been called the "master control center" for emotions and many basic motives such as hunger, thirst, sex, and aggression (Hinton et al., 2004; Williams et al., 2004). It regulates the body's internal environment, including temperature control, which it accomplishes by regulating the endocrine system. Hanging down from the hypothalamus, the *pituitary gland* is usually considered the master endocrine gland because it releases hormones that activate the other endocrine glands. The hypothalamus influences the pituitary through direct neural connections and by releasing its own hormones into the blood supply of the pituitary. The hypothalamus also directly influences some important aspects of behavior, such as eating and drinking patterns.

An interconnected group of forebrain structures, known as the **limbic system**, is located roughly along the border between the cerebral cortex and the lower-level brain structures.

The limbic system is generally responsible for emotions, drive reduction, and memory. However, the major focus of interest in the limbic system, and particularly the *amygdala*, has been its production and regulation of aggression and fear (Blair, 2004; Pontius, 2005; Rumpel et al., 2005). Another well-known function of the limbic system is its role in pleasure or reward (Dackis & O'Brien, 2001; Olds & Milner, 1954). Even though limbic system structures and neurotransmitters are instrumental in emotional behavior, the cerebral cortex also tempers emotion in humans.

Limbic system: Regulates fear and other emotions

Thalamus: Integrates input from the senses

Hypothalamus: Controls emotions and basic motivations, such as hunger

Cerebral cortex: Governs higher mental processes

Structures of the forebrain FIGURE 2.17

THE CEREBRAL CORTEX: THE CENTER OF "HIGHER" PROCESSING

cerebral cortex
Thin surface layer on the cerebral hemispheres that regulates most complex behavior, including receiving sensations, motor control, and higher mental processes.

The gray, wrinkled **cerebral cortex** is responsible for most complex behaviors and higher mental processes. It plays such a vital role that many consider it the essence of life. In fact, physicians may declare a person legally dead when the cortex dies, even when the lower-level brain structures and the rest of the body are fully functioning.

Although the cerebral cortex is only about one-eighth of an inch thick, it's made up of approximately 30 billion neurons and nine times as many glial cells. It contains numerous "wrinkles" called *convolutions* (think of a crumpled-up newspaper), which allow it to fit in the restricted space of the skull.

The full cerebral cortex and the two cerebral hemispheres beneath it closely resemble an oversized walnut. The division, or *fissure*, down the center marks the left and right *hemispheres* of the brain, which make up about 80 percent of the brain's weight. They are mostly filled with axon connections between the cortex and the other brain structures. Each hemisphere controls the opposite side of the body.

The cerebral hemispheres are divided into eight distinct areas or lobes—four in each hemisphere (FIGURE 2.18). Like the lower-level brain structures, each lobe specializes in somewhat different tasks—another example of localization of function. However, some functions overlap between lobes.

The frontal lobes The large **frontal lobes** coordinate messages received from the other three lobes. An area at the very back of the frontal lobes, known as the *motor cortex*, instigates all voluntary movement. In the lower left frontal lobe lies *Broca's area*. In 1865, French physician Paul Broca discovered that damage to this area causes difficulty in speech, but not language comprehension. This type of impaired language ability is known as *Broca's aphasia*. Finally, the frontal lobes control most higher functions that distinguish humans from other animals, such as thinking, personality, emotion, and memory. Abnormalities in the frontal lobes are often observed in patients with schizophrenia (Chapter 13). And as the case of Phineas Gage (Chapter 1) and other research indicate, damage to the frontal lobe affects

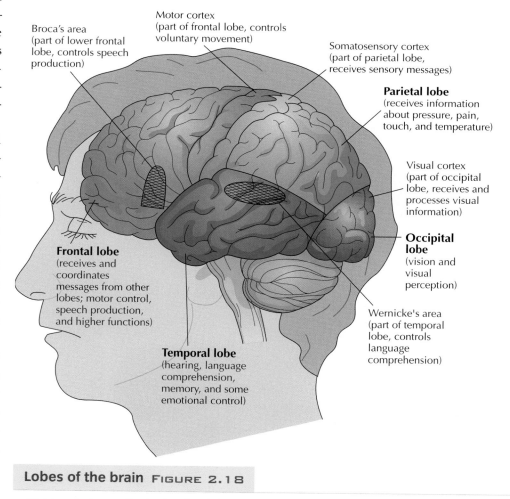

Broca's area
(part of lower frontal lobe, controls speech production)

Motor cortex
(part of frontal lobe, controls voluntary movement)

Somatosensory cortex
(part of parietal lobe, receives sensory messages)

Parietal lobe
(receives information about pressure, pain, touch, and temperature)

Visual cortex
(part of occipital lobe, receives and processes visual information)

Occipital lobe
(vision and visual perception)

Wernicke's area
(part of temporal lobe, controls language comprehension)

Frontal lobe
(receives and coordinates messages from other lobes; motor control, speech production, and higher functions)

Temporal lobe
(hearing, language comprehension, memory, and some emotional control)

Lobes of the brain FIGURE 2.18

Divisions between the brain's four lobes—frontal, parietal, temporal, and occipital—are marked by visibly prominent folds, which provide convenient landmarks.

Accidental proof of specialized brain functions? FIGURE 2.19

In 1998, construction worker Travis Bogumill was accidentally shot with a nail gun near the rear of his right frontal lobe. Remarkably, Bogumill experienced only an impaired ability to perform complex mathematical problems. This case supports experimental research showing that the frontal lobes and short-term memory are responsible for reasoning, problem solving, mathematical calculation, and thinking about future rewards or actions (Evans, 2003; Hill, 2004; Neubauer et al., 2004).

Left Motor Cortex
(on fontal lobes)

Right Somatosensory Cortex
(on parietal lobes)

Messages sent from motor cortex to various body regions

Messages sent to somatosensory cortex from various body regions

Frontal lobes

Motor cortex

Somatosensory cortex

Parietal lobes

Body representation of the motor cortex and somatosensory cortex FIGURE 2.20

This drawing represents a vertical cross section taken from the left hemisphere's motor cortex and right hemisphere's somatosensory cortex. If body areas were truly proportional to the amount of tissue on the motor and somatosensory cortices, our bodies would look like the oddly shaped human figures draped around the outside edge of the cortex.

motivation, drives, creativity, self-awareness, initiative, reasoning, and emotional behavior (FIGURE 2.19).

The parietal lobes, the temporal lobes, and the occipital lobes

The **parietal lobes** interpret bodily sensations including pressure, pain, touch, temperature, and location of body parts. A band of tissue on the front of the parietal lobe, called the *somatosensory cortex*, receives information about touch in different body areas. Areas of the body with more somatosensory and motor cortex devoted to them (such as the hands and face) are most sensitive to touch and have the most precise motor control (FIGURE 2.20).

The **temporal lobes** are responsible for hearing, language comprehension, memory, and some emotional control. The *auditory cortex* (which processes sound) is located at the top front of each temporal lobe. This area processes incoming sensory information and sends it to the parietal lobes, where it is combined with other sensory information.

An area of the left temporal lobe, *Wernicke's area*, is involved in language comprehension. About a decade after Broca's discovery, German neurologist Carl Wernicke noted that patients with damage in this area could not understand what they read or heard, but they could speak quickly and easily. However, their speech was often unintelligible because it contained made-up words, sound substitutions, and word substitutions. This syndrome is now referred to as *Wernicke's aphasia*.

The **occipital lobes** are responsible, among other things, for vision and visual perception. Damage to the occipital lobe can produce blindness, even though the eyes and their neural connection to the brain are perfectly healthy.

The association areas

One of the most popular myths in psychology is that we use only 10 percent of our brain. This myth might have begun with early research showing that approximately three-fourths of the cortex is "uncommitted" (with no precise, specific function responsive to electrical brain stimulation). These areas are not dormant, however. They are clearly involved in interpreting, integrating, and acting on information processed by other parts of the brain. They are called **association areas** because they associate, or connect, various areas and functions of the brain. The association areas in the frontal lobe, for example, help in decision making and planning. Similarly, the association area right in front of the motor cortex is involved in the planning of voluntary movement.

TWO BRAINS IN ONE? A HOUSE DIVIDED

We mentioned earlier that the brain's left and right cerebral hemispheres control opposite sides of the body. Each hemisphere also has separate areas of specialization. (This is another example of *localization* of function, yet it is technically referred to as *lateralization*.)

Early researchers believed that the right hemisphere was "subordinate" or "nondominant" to the left, with few special functions or abilities. In the 1960s, landmark research with *split-brain* patients began to change this view.

The primary connection between the two cerebral hemispheres is a thick, ribbon-like band of nerve fibers under the cortex called the **corpus callosum** (FIGURE 2.21). In some rare cases of severe epilepsy, when other forms of treatment have failed,

Crisscrossing of visual information FIGURE 2.21

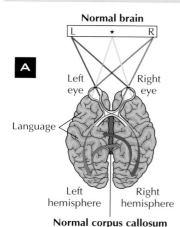

A Visual images from the left half of each of your two eyes connect only to the right half of your brain, and vice versa. Both hemispheres transmit information across your corpus callosum to the other side.

Normal brain
L ★ R
Left eye — Right eye
Language
Left hemisphere — Right hemisphere
Normal corpus callosum

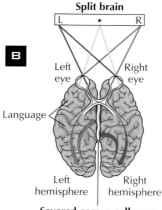

B When the corpus callosum is cut, the "split-brain" patient does not receive the visual information that is normally shared across the two hemispheres.

Split brain
L ★ R
Left eye — Right eye
Language
Left hemisphere — Right hemisphere
Severed corpus callosum

surgeons cut the corpus callosum to stop the spread of epileptic seizures from one hemisphere to the other. Because this operation cuts the only direct communication link between the two hemispheres, it reveals what each half of the brain can do in isolation from the other. The resulting research has profoundly improved our understanding of how the two halves of the brain function.

If you met and talked with a split-brain patient, you probably wouldn't even know he or she had had the operation. The subtle changes in split-brain patients normally appear only with specialized testing (**FIGURE 2.22**).

Dozens of studies on split-brain patients, and newer research on people whose brains are intact, have documented several differences between the two brain hemispheres (**FIGURE 2.23**). In one study, researchers reported that different aspects of personality appear in the different hemispheres. For example, in one patient, the right hemisphere seemed more disturbed by childhood memories of being bullied than did the left (Schiffer et al., 1998).

Interestingly, left and right brain specialization is not usually reversed in left-handed people. About 68 percent of left-handers and 97 percent of right-handers have their major language areas on the left hemisphere. This suggests that even though the right side of the brain is dominant for movement in left-handers, other skills are often localized in the same brain areas as for right-handers.

What about the popular conception of the "neglected right brain?" Courses and books directed at "right-brain thinking" often promise to increase your intuition, creativity, and artistic abilities by waking up your

Split-brain research FIGURE 2.22

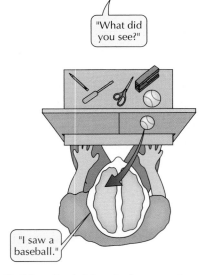

A When a split-brain patient stares straight ahead and a picture of a screwdriver is flashed only to the left visual field, the information goes only to the nonverbal right hemisphere and he cannot name what he saw.

B When asked to "pick up what you saw," his left hand can touch the items hidden behind the screen and easily identify the screwdriver. This shows that the right hemisphere received the photo image of the screwdriver, but the patient could not name it because the information did not reach the left hemisphere, where language is normally processed.

C When the left hemisphere receives an image of a baseball (from the right visual field), the patient easily names it.

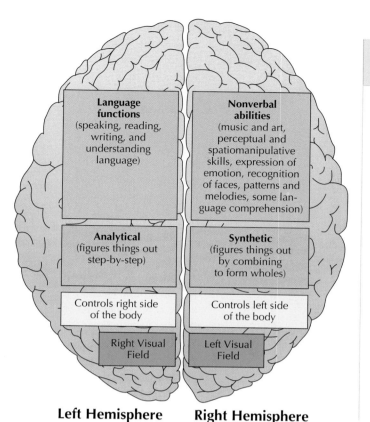

Language functions (speaking, reading, writing, and understanding language)

Nonverbal abilities (music and art, perceptual and spatiomanipulative skills, expression of emotion, recognition of faces, patterns and melodies, some language comprehension)

Analytical (figures things out step-by-step)

Synthetic (figures things out by combining to form wholes)

Controls right side of the body

Controls left side of the body

Right Visual Field

Left Visual Field

Left Hemisphere   **Right Hemisphere**

Functions of the left and right hemispheres
FIGURE 2.23

The left hemisphere specializes in verbal and analytical functions. The right hemisphere focuses on nonverbal abilities, such as spatiomanipulative skills, art and musical abilities, and visual recognition tasks. Keep in mind that both hemispheres are activated when we perform almost any task or respond to any stimuli.

"neglected" and "underused" right brain (e.g., Bragdon & Gamon, 1999; Edwards, 1999). This myth of the neglected right brain arose from popularized accounts of split-brain patients and exaggerated claims and unwarranted conclusions about differences between the left and right hemispheres. The fact is that the two hemispheres work together in a coordinated, integrated way, with each making important contributions.

CONCEPT CHECK   **STOP**

What are the main parts of the hindbrain, the midbrain, and the forebrain?

How does each part of the cerebral cortex function?

How do Broca's aphasia and Wernicke's aphasia differ?

Why is the amygdala important?

What are the findings of split brain research?

What is happening in this picture ?

The ability to curl one's tongue lengthwise is one of the few traits that depend on only one dominant gene. What does it mean if both of your parents are "noncurlers"?

Unlike tongue-curling ability, most traits are polygenic, meaning that they are controlled by more than one gene. Can you think of why humans and other animals have evolved to possess such "complex traits"?

1 Our Genetic Inheritance

1. **Neuroscience** studies how biological processes relate to behavioral and mental processes.

2. Genes (dominant or recessive) hold the code for inherited traits. Scientists use **behavioral genetics** methods to determine the relative influences of heredity and environment (heritability) on complex traits.

3. **Evolutionary psychology** suggests that many behavioral commonalities emerged and remain in human populations through natural selection, because they helped ensure our "genetic survival."

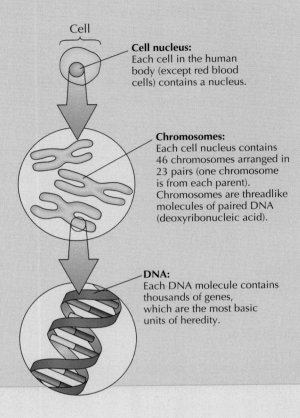

Cell

Cell nucleus:
Each cell in the human body (except red blood cells) contains a nucleus.

Chromosomes:
Each cell nucleus contains 46 chromosomes arranged in 23 pairs (one chromosome is from each parent). Chromosomes are threadlike molecules of paired DNA (deoxyribonucleic acid).

DNA:
Each DNA molecule contains thousands of genes, which are the most basic units of heredity.

2 Neural Bases of Behavior

1. **Neurons**, supported by glial cells, receive and send electrochemical signals to other neurons and to the rest of the body. Their major components are dendrites, a cell body, and an axon.

2. Within a neuron, a neural impulse, or action potential, moves along the axon. An action potential is an all-or-none event.

3. Neurons communicate with each other using **neurotransmitters**, which are released at the synapse and attach to the receiving neuron. Neurons receive input from many synapses, some excitatory and some inhibitory. Hundreds of different neurotransmitters regulate a wide variety of physiological processes. Many poisons and drugs act by mimicking or interfering with neurotransmitters.

4. The endocrine system uses **hormones** to broadcast messages throughout the body. The system regulates long-term bodily processes, maintains ongoing bodily processes, and controls the body's response to emergencies.

3 Nervous System Organization

1. The **central nervous system (CNS)** includes the brain and spinal cord. The CNS allows us to process information and adapt to our environment in ways that no other animal can. The spinal cord transmits information between the brain and the rest of the body, and initiates involuntary reflexes. Although the CNS is very fragile, recent research shows that the brain is capable of lifelong **neuroplasticity** and **neurogenesis**. Neurogenesis is made possible by **stem cells**.

2. The **peripheral nervous system (PNS)** includes all the nerves outside the brain and spinal cord. It links the brain and spinal cord to the body's sense receptors, muscles, and glands. The PNS is subdivided into the **somatic nervous system (SNS)** and the **autonomic nervous system (ANS)**.

3. The ANS includes the sympathetic nervous system and the parasympathetic nervous system. The sympathetic nervous system mobilizes the body's "fight-or-flight" response. The parasympathetic nervous system returns the body to its normal functioning.

4 A Tour Through the Brain

1. The brain is divided into the hindbrain, the midbrain, and the forebrain. The brainstem includes parts of each of these. Certain brain structures are specialized to perform certain tasks (**localization of function**).

2. The hindbrain (including the medulla, pons, and cerebellum) controls automatic behaviors and survival responses.

3. The midbrain helps us orient our eye and body movements, helps control sleep and arousal, and is involved with the neurotransmitter dopamine. The reticular formation runs through the core of the hindbrain, midbrain, and brainstem.

4. Forebrain structures (including the thalamus, hypothalamus, and limbic system) integrate input from the senses, control basic motives, regulate the body's internal environment, and regulate emotions, learning, and memory.

5. The **cerebral cortex**, part of the forebrain, governs most higher processing and complex behaviors. It is divided into two hemispheres, each controlling the opposite side of the body. The corpus callosum links the hemispheres. Each hemisphere is divided into frontal, parietal, temporal, and occipital lobes. Each lobe specializes in somewhat different tasks, but a large part of the cortex is devoted to integrating actions performed by different brain regions.

6. Split-brain research shows that each hemisphere performs somewhat different functions, although they work in close communication.

CRITICAL AND CREATIVE THINKING QUESTIONS

1. Imagine that scientists were able to identify specific genes linked to serious criminal behavior, and it was possible to remove or redesign these genes. Would you be in favor of this type of gene manipulation? Why or why not?

2. Why is it valuable for scientists to understand how neurotransmitters work at a molecular level?

3. What are some everyday examples of neuroplasticity—that is, of how the brain is changed and shaped by experience?

(Check your answers in Appendix B.)

1. Behavioral genetics is the study of _____ .

   a. the relative effects of behavior and genetics on survival
   b. the relative effects of heredity and environment on behavior and mental processes
   c. the relative effects of genetics on natural selection
   d. how genetics affects correct behavior

2. Evolutionary psychology studies _____ .

   a. the ways in which humans adapted their behavior to survive and evolve
   b. the ways in which humankind's behavior has changed over the millennia
   c. the ways in which humans can evolve to change behavior
   d. the ways in which natural selection and adaptation can explain behavior and mental processes

3. This is a measure of the degree to which a characteristic is related to genetic, inherited factors.

   a. heritability
   b. inheritance
   c. the biological ratio
   d. the genome statistic

4. The term _____ refers to the evolutionary concept that those with adaptive genetic traits will live and reproduce.

   a. natural selection
   b. evolution
   c. survival of the fittest
   d. all of these options

5. Label the following parts of a neuron, the cell of the nervous system responsible for receiving and transmitting electrochemical information:

   a. dendrites
   b. cell body
   c. axon
   d. myelin sheath
   e. terminal buttons of axon

6. Your textbook's definition of an **action potential** is _____ .

   a. the likelihood that a neuron will take action when stimulated
   b. the tendency for a neuron to be potentiated by neurotransmitters
   c. a neural impulse that carries information along the axon of a neuron
   d. the firing of a nerve, either toward or away from the brain

7. Too much of this neurotransmitter may be related to schizophrenia whereas too little of this neurotransmitter may be related to Parkinson's disease.

   a. acetylcholine
   b. dopamine
   c. norepinephrine
   d. serotonin

8. Chemicals that are manufactured by endocrine glands and circulated in the bloodstream to change or maintain bodily functions are called _____ .

   a. vasopressors
   b. gonadotropins
   c. hormones
   d. steroids

9. Label the main glands of the endocrine system:

   a. pineal
   b. pituitary
   c. adrenal
   d. thyroid
   e. thymus

10. The central nervous system _____ .

    a. consists of the brain and spinal cord
    b. is the most important nervous system
    c. includes the automatic and other nervous systems
    d. all of these options

11. The peripheral nervous system _____ .

    a. is composed of the spinal cord and peripheral nerves
    b. is less important than the central nervous system
    c. is contained within the skull and spinal column
    d. includes all the nerves and neurons outside the brain and spinal cord

12. The _____ nervous system is responsible for fight or flight whereas the _____ nervous system is responsible for maintaining calm.

    a. central; peripheral
    b. parasympathetic; sympathetic
    c. sympathetic; parasympathetic
    d. autonomic; somatic

13. Label the following structures/areas of the brain:

    a. forebrain
    b. midbrain
    c. hindbrain
    d. thalamus
    e. hypothalamus
    f. cerebral cortex

14. Label the four lobes of the brain:

    a. frontal lobe
    b. parietal lobe
    c. temporal lobe
    d. occipital lobe

15. Identify the following functions of the brain as either left or right hemisphere.

    a. language function, _____ hemisphere
    b. nonverbal abilities, _____ hemisphere
    c. control of left side of body and right visual field, _____ hemisphere
    d. control of right side of body and left visual field, _____ hemisphere

Stress and Health Psychology 3

Where were you on September 11, 2001? Even now, years later, you can probably recall in crystalline detail the events of that Tuesday morning—the suddenness of the terrorist attacks, the buildings burning and tumbling to the ground, even the incongruously cheerful skies above the destruction. Many people who were directly affected by the September 11 attacks—and even some who were not—remember the day and its aftermath as the most stressful time in their lives. How does such extreme stress affect people's health and well-being, both immediately and down the road? Is there any way that people can cope with stress of such magnitude? What about life's more mundane aggravations—rude drivers, long lines at the grocery store, nagging decisions about which job to take or which college to attend? Do these also take a toll on our well-being? If so, what can we do about it?

Throughout most of history, people have understood that emotions and thoughts affect physical health. However, in the late 1800s, after discovering physiological causes for infectious diseases such as typhoid and syphilis, scientists began to focus primarily on the physiological causes of disease. Today, the major causes of death have shifted from contagious diseases (such as pneumonia, influenza, tuberculosis, and measles) to noncontagious diseases (such as cancer, cardiovascular disease, and chronic lung disease), and the focus has returned to psychological behaviors and lifestyles (National Center for Health Statistics, 2004). In this chapter, we explore how biological, psychological, and social factors (the *biopsychosocial model*) affect illness as well as health and well-being.

Understanding Stress

Anything that places a demand on the body can cause **stress**. The trigger that prompts the stressful reaction is called a **stressor**. Stress reactions can occur in response to either internal cognitive stimuli, external stimuli, or environmental stimuli (Sanderson, 2004; Sarafino, 2005).

Pleasant or beneficial stress, such as moderate exercise, is called **eustress**. Stress that is unpleasant or objectionable, as from chronic illness, is called **distress** (Selye, 1974). The total absence of stress would mean the total absence of stimulation, which would eventually lead to death. Because health psychology has been chiefly concerned with the negative effects of stress, we will adhere to convention and use the word "stress" to refer primarily to harmful or unpleasant stress.

stress The body's nonspecific response to any demand made on it; physical and mental arousal to situations or events that we perceive as threatening or challenging.

SOURCES OF STRESS

Early stress researchers Thomas Holmes and Richard Rahe (1967) believed that any **life change** that required some adjustment in behavior or lifestyle could cause some degree of stress (**FIGURE 3.1**). They also believed that exposure to numerous stressful events within a short period could have a direct detrimental effect on health.

To investigate the relationship between change and stress, Holmes and Rahe created a Social Readjustment Rating Scale (SRRS) that asked people to check off all the life events they had experienced in the previous year (**TABLE 3.1**).

The SRRS scale is an easy and popular way to measure stress. Cross-cultural studies have shown that most people rank the magnitude of stressful events in similar ways (De Coteau, Hope, & Anderson, 2003; Scully, Tosi, & Banning, 2000). But the SRRS is not foolproof. For example, it only shows a correlation between stress and illness; it does not prove that stress actually causes illnesses.

Not all stressful situations are single events such as a terrorist attack, a death, or a birth. **Chronic stressors**, such as a bad marriage, poor working conditions, or an intolerable political climate, can be significant too. Even the stress of low-frequency noise is associated with measurable hormonal and cardiac changes (Waye et al., 2002). Our social lives can also be chronically stressful because making and maintaining friendships involves considerable thought and energy (Sias et al., 2004).

Perhaps the largest source of chronic stress is work. People often experience stress associated with keeping or changing jobs or with job performance (Moore, Grunberg, & Greenberg, 2004). The most stressful jobs are those that make great demands on performance and concentration but allow little creativity or

Stressful ordeal or exciting opportunity?

FIGURE 3.1

Different people may perceive any given event differently, depending on how they interpret and appraise the event (Holt & Dunn, 2004). Also, some people are better able to deal with change than others, perhaps because of good coping skills, general physical health, healthier lifestyles, or even genetic predisposition.

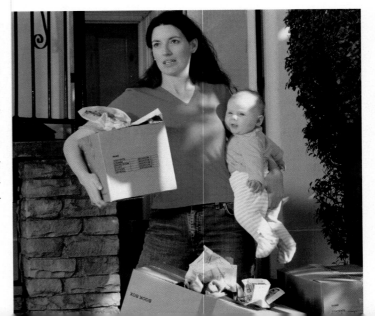

Measuring life changes TABLE 3.1

Social Readjustment Rating Scale

To score yourself on this scale, add up the "life change units" for all life events you have experienced during the last year and compare your score with the following standards: 0–150 = No significant problems; 150–199 = Mild life crisis (33 percent chance of illness); 200–299 = Moderate life crisis (50 percent chance of illness); 300 and above = Major life crisis (80 percent chance of illness).

Life events	Life change units	Life events	Life change units
Death of spouse	100	Son or daughter leaving home	29
Divorce	73	Trouble with in-laws	29
Marital separation	65	Outstanding personal achievement	28
Jail term	63	Spouse begins or stops work	26
Death of a close family member	63	Begin or end school	26
Personal injury or illness	53	Change in living conditions	25
Marriage	50	Revision of personal habits	24
Fired at work	47	Trouble with boss	23
Marital reconciliation	45	Change in work hours or conditions	20
Retirement	45	Change in residence	20
Change in health of family member	44	Change in schools	20
Pregnancy	40	Change in recreation	19
Sex difficulties	39	Change in church activities	19
Gain of a new family member	39	Change in social activities	18
Business readjustment	39	Mortgage or loan for lesser purchase (car, major appliance)	17
Change in financial state	38	Change in sleeping habits	16
Death of a close friend	37	Change in number of family get-togethers	15
Change to different line of work	36	Change in eating habits	15
Change in number of arguments with spouse	35	Vacation	13
Mortgage or loan for major purchase	31	Christmas	12
Foreclosure on mortgage or loan	30	Minor violations of the law	11
Change in responsibilities at work	29		

Source: Reprinted from *Journal of Psychosomatic Research*, Vol. III; Holmes and Rahe: "The Social Readjustment Rating Scale," 213–218, 1967, with permission from Elsevier.

opportunity for advancement (Angenendt, 2003; Lewig & Dollard, 2003).

Stress at work can also cause serious stress at home, not only for the worker but for other family members as well. In our private lives, divorce, child and spousal abuse, alcoholism, and money problems can place severe stress on all members of a family (DiLauro, 2004; Luecken & Lemery, 2004).

In addition to chronic stressors, the minor **hassles** of daily living can pile up and become a major source of stress. We all share many hassles, such as time pressures and financial concerns. But our reactions to these hassles vary. Persistent hassles can lead to a form of physical, mental, and emotional exhaustion known as **burnout** (Sarafino, 2005; Linzer et al., 2002) (**FIGURE 3.2**).

Some authorities believe that hassles can be more significant than major life events in creating stress (Kraaij, Arensman, & Spinhoven, 2002; Lazarus, 1999). Divorce is extremely stressful, but it may be so because of the increased number of hassles—a change in finances, child-care arrangements, longer working hours, and so on.

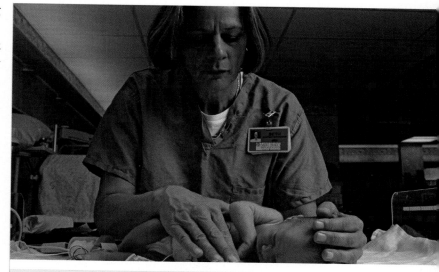

Is nursing a stressful career? FIGURE 3.2

Over time, some people in chronically stressful professions who think of their job as a "calling" become emotionally drained and disillusioned and feel a loss of personal accomplishment—they "burn out." Burnout can cause more work absences, less productivity, and increased risk for physical problems. What other occupations might pose an especially high risk for burnout?

Like hassles, **frustration** can cause stress. And the more motivated we are, the more frustrated we are when our goals are blocked.

Finally, stress can arise when we experience **conflict**—that is, when we are forced to make a choice between at least two incompatible alternatives. There are three basic types of conflict: approach–approach, avoidance–avoidance, and approach–avoidance.

In an **approach–approach conflict**, a person must choose between two or more favorable alternatives. For example, imagine that you must choose between two jobs: one that will be inherently interesting and one that will look impressive on your résumé. Either choice will benefit you in some way. The requirement to choose is the source of stress.

An **avoidance–avoidance conflict** involves making a choice between two or more unpleasant alternatives that will lead to negative results no matter which choice is made. Avoidance–avoidance conflicts can lead to intense stress.

An **approach–avoidance conflict** occurs when a person must choose between alternatives that will have both desirable and undesirable results. We have all been faced with such decisions as "I want to spend more time in a close relationship, but that means I won't be able to see my old friends as much." This conflict leads to a great deal of ambivalence.

The longer any conflict exists or the more important the decision, the more stress a person will experience (**FIGURE 3.3**). Generally, approach–approach conflicts are the easiest to resolve and produce the least stress. Avoidance–avoidance conflicts, on the other hand, are usually the most difficult because all choices lead to unpleasant results.

NATIONAL GEOGRAPHIC

■ **approach–approach conflict** Forced choice between two or more desirable alternatives.

■ **avoidance–avoidance conflict** Forced choice between two or more undesirable alternatives.

■ **approach–avoidance conflict** Forced choice between two or more alternatives that have both desirable and undesirable results.

Sophie's Choice FIGURE 3.3

In the book (and film) *Sophie's Choice*, Sophie and her two children are sent to a German concentration camp. A soldier demands that Sophie give up either her daughter or her son, or else both children will be killed. Obviously, both alternatives will have tragic results. What kind of conflict does this example illustrate?

HOW STRESS AFFECTS THE BODY

When mentally or physically stressed, your body undergoes several physiological changes. The **sympathetic nervous system** and the **HPA Axis** control the most significant of these changes (**FIGURE 3.4**).

Cortisol, a key element of the HPA Axis, plays a critical role in the long-term effects of stress. Prolonged elevation of cortisol has been linked to increased levels of depression, post-traumatic stress disorder (PTSD), memory problems, unemployment, and even drug and alcohol abuse (Bremner et al., 2004; Cowen, 2002; Sinha et al., 2003; Wüst et al., 2005). Perhaps most important, increased cortisol is directly related to impairment of immune system functioning.

The discovery of the relationship between stress and the immune system is very important. When the immune system is impaired, we are at greatly increased risk of suffering from a number of diseases, including bursitis, colitis, Alzheimer's disease, rheumatoid arthritis, periodontal disease, and even the common cold (Cohen et al., 2002; Hawkley & Cacioppo, 2004; Segerstrom & Miller, 2004; Theoharides & Cochrane, 2004).

Stress—an interrelated system FIGURE 3.4

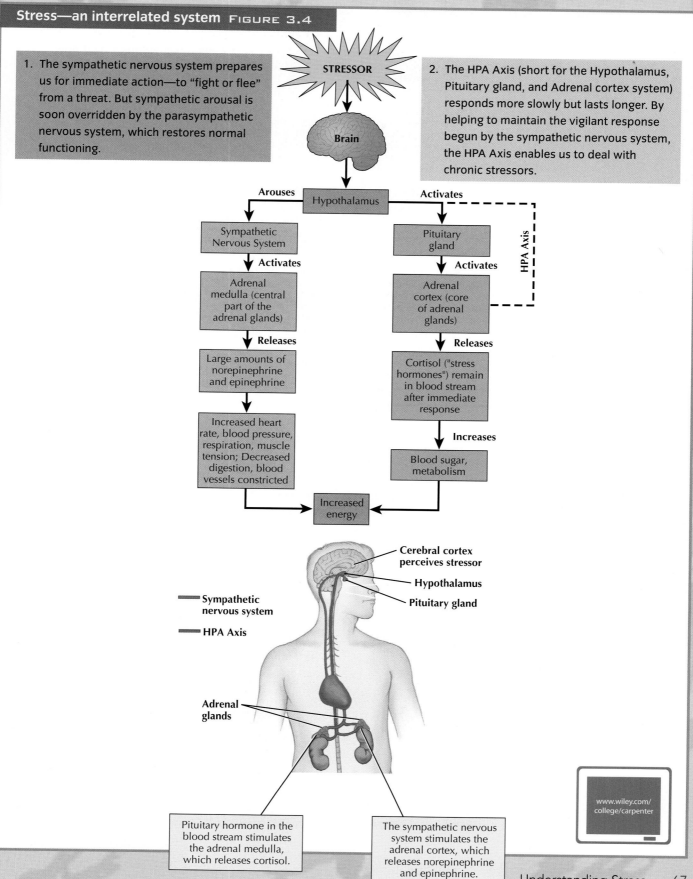

1. The sympathetic nervous system prepares us for immediate action—to "fight or flee" from a threat. But sympathetic arousal is soon overridden by the parasympathetic nervous system, which restores normal functioning.

2. The HPA Axis (short for the Hypothalamus, Pituitary gland, and Adrenal cortex system) responds more slowly but lasts longer. By helping to maintain the vigilant response begun by the sympathetic nervous system, the HPA Axis enables us to deal with chronic stressors.

STRESSOR

Brain

Arouses ← Hypothalamus → **Activates**

HPA Axis

Sympathetic Nervous System

Activates

Adrenal medulla (central part of the adrenal glands)

Releases

Large amounts of norepinephrine and epinephrine

Increased heart rate, blood pressure, respiration, muscle tension; Decreased digestion, blood vessels constricted

Pituitary gland

Activates

Adrenal cortex (core of adrenal glands)

Releases

Cortisol ("stress hormones") remain in blood stream after immediate response

Increases

Blood sugar, metabolism

Increased energy

Cerebral cortex perceives stressor
Hypothalamus
Pituitary gland

— Sympathetic nervous system

— HPA Axis

Adrenal glands

Pituitary hormone in the blood stream stimulates the adrenal medulla, which releases cortisol.

The sympathetic nervous system stimulates the adrenal cortex, which releases norepinephrine and epinephrine.

www.wiley.com/college/carpenter

Understanding Stress   67

Stress causes a number of detrimental physiological changes. Hans Selye [SELL-yay], a physiologist renowned for his work on stress since the 1930s, described a generalized, three-part physiological reaction to severe stressors. He termed it the **general adaptation syndrome (GAS)** (Selye, 1936).

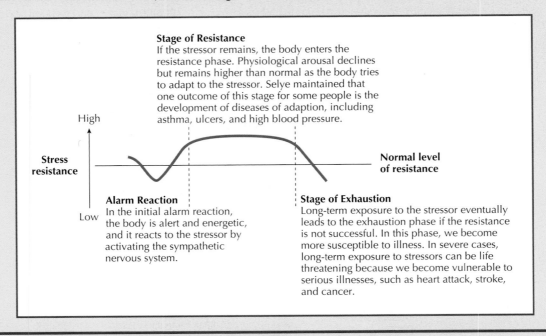

Stage of Resistance
If the stressor remains, the body enters the resistance phase. Physiological arousal declines but remains higher than normal as the body tries to adapt to the stressor. Selye maintained that one outcome of this stage for some people is the development of diseases of adaption, including asthma, ulcers, and high blood pressure.

High

Stress resistance

Normal level of resistance

Low

Alarm Reaction
In the initial alarm reaction, the body is alert and energetic, and it reacts to the stressor by activating the sympathetic nervous system.

Stage of Exhaustion
Long-term exposure to the stressor eventually leads to the exhaustion phase if the resistance is not successful. In this phase, we become more susceptible to illness. In severe cases, long-term exposure to stressors can be life threatening because we become vulnerable to serious illnesses, such as heart attack, stroke, and cancer.

■ **psychoneuro-
immunology**
[sye-koh-NEW-roh-IM-you-NOLL-oh-gee] The interdisciplinary field that studies the effects of psychological factors on the immune system.

Knowledge that psychological factors have considerable control over infectious diseases has upset the long-held assumption in biology and medicine that these diseases are "strictly physical." The clinical and theoretical implications are so important that a new field of biopsychology has emerged: **psychoneuroimmunology**.

CONCEPT CHECK STOP

What are the three types of conflict that can cause stress?

Who is most susceptible to burnout?

What is the HPA Axis?

Why does chronic stress threaten the immune system?

What occurs during the exhaustion phase of the GAS?

Stress and Illness

LEARNING OBJECTIVES

Explain why an immune system compromised by stress might be more vulnerable to cancer growth.

Define the personality patterns that can influence how we respond to stress.

Describe the key symptoms of post-traumatic stress disorder (PTSD).

Explain how biological and psychological factors can jointly influence the development of gastric ulcers.

s we've just seen, stress has dramatic effects on our bodies. This section explores how stress is related to four serious illnesses—cancer, coronary heart disease, post-traumatic stress disorder, and gastric ulcers. Next we'll briefly discuss each of these.

CANCER: A VARIETY OF CAUSES—EVEN STRESS

Cancer is among the leading causes of death for adults in the United States. It occurs when a particular type of primitive body cell begins rapidly dividing and then forms a tumor that invades healthy tissue. Unless destroyed or removed, the tumor eventually damages organs and causes death. More than 100 types of cancer have been identified. They appear to be caused by an interaction between environmental factors and inherited predispositions.

In a healthy person, whenever cancer cells start to multiply, the immune system checks the uncontrolled growth by attacking the abnormal cells (**FIGURE 3.5**).

As mentioned earlier, stress causes the adrenal glands to release hormones that suppress the immune system. The compromised immune system is then less able to resist infection and cancer development (Goebel & Mills, 2000; Shi et al., 2003).

The good news is that we can substantially reduce our risk of developing cancer by making changes that reduce our stress levels and enhance our immune

The healthy immune system in action FIGURE 3.5

Stress can compromise the immune system, but the actions of a healthy immune system are shown here. The round structure near the left center of this photomicrograph is a T-lymphocyte, a type of white blood cell produced by the immune system. It has just killed a cancer cell, the sweet potato–shaped structure.

systems. For example, researchers have found that interrupting people's sleep decreased the number of *natural killer cells* (a type of immune system cell) by 28 percent (Irwin et al., 1994). Fortunately, these researchers also found that a normal night's sleep after the deprivation returned the killer cells to their normal levels.

CARDIOVASCULAR DISORDERS: THE LEADING CAUSE OF DEATH IN THE UNITED STATES

Cardiovascular disorders cause over half of all deaths in the United States (American Heart Association, 2004). Understandably, health psychologists are concerned because stress is a major contributor to these deaths. **Heart disease** is a general term for all disorders that eventually affect the heart muscle and lead to heart failure. *Coronary heart disease* occurs when the walls of the coronary arteries thicken, reducing or blocking the blood supply to the heart. Symptoms of such disease include *angina* (chest pain due to insufficient blood supply to the heart) and *heart attack* (death of heart muscle tissue). Controllable factors that contribute to heart disease include stress, smoking, certain personality characteristics, obesity, a high-fat diet, and lack of exercise (Brummett et al., 2004; David et al., 2004; Hirao-Try, 2003).

When the body is stressed, the autonomic nervous system releases epinephrine and cortisol into the bloodstream. These hormones increase heart rate and release fat and glucose from the body's stores to give muscles a readily available source of energy.

If no physical "fight or flight" action is taken (and this is most likely the case in our modern lives), the fat that was released into the bloodstream is not burned as fuel. Instead, it may adhere to the walls of blood vessels. These fatty deposits are a major cause of blood supply blockage, which causes heart attacks.

The effects of stress on heart disease may be amplified if a person tends to be hard driving, competitive, ambitious, impatient, and hostile—a **Type A personality**. The antithesis of the Type A personality is the **Type B personality**, which is characterized by a laid-back, relaxed attitude toward life.

Have you ever wondered why some people survive in the face of great stress (personal tragedies, demanding jobs, and even a poor home life) while others do not? Suzanne Kobasa was among the first to study this question (Kobasa, 1979; Maddi, 2004; Turnipseed, 2003). Examining male executives with high levels of stress, she found that some people are more resistant to stress than others because of a personality factor called **hardiness**, a resilient type of optimism that comes from three distinctive attitudes.

First, hardy people feel a strong sense of commitment to both their work and their personal life. They also make intentional commitments to purposeful

> **Type A personality**
> Behavior characteristics that include intense ambition, competition, exaggerated time urgency, and a cynical, hostile outlook.

> **Type B personality**
> Behavior characteristics consistent with a calm, patient, relaxed attitude.

activity and problem solving. Second, these people see themselves as being in control of their lives, rather than as victims of their circumstances. Finally, hardy people look at change as a challenge, an opportunity for growth and improvement—not as a threat.

The important lesson from this research is that hardiness is a learned behavior, not something based on luck or genetics. If you're not one of the hardy souls, you can develop the trait.

Of course, Type A personality and lack of hardiness are not the only controllable risk factors associated with heart disease. *Smoking, obesity,* and *lack of exercise* are also very important factors. Smoking restricts blood circulation, and obesity stresses the heart by causing it to pump more blood to the excess body tissue. A high-fat diet, especially one that is high in cholesterol, contributes to the fatty deposits that clog blood vessels. Lack of exercise contributes to weight gain and prevents the body from obtaining important exercise benefits, including strengthened heart muscle, increased heart efficiency, and the release of neurotransmitters, such as serotonin, that alleviate stress and promote well-being (**FIGURE 3.6**).

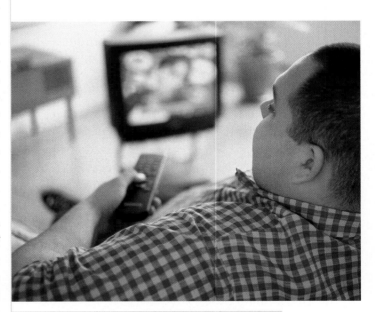

Controllable risk factors FIGURE 3.6

Obesity and lack of exercise are important risk factors for heart disease. Can you see how the two phenomena also tend to feed into each other?

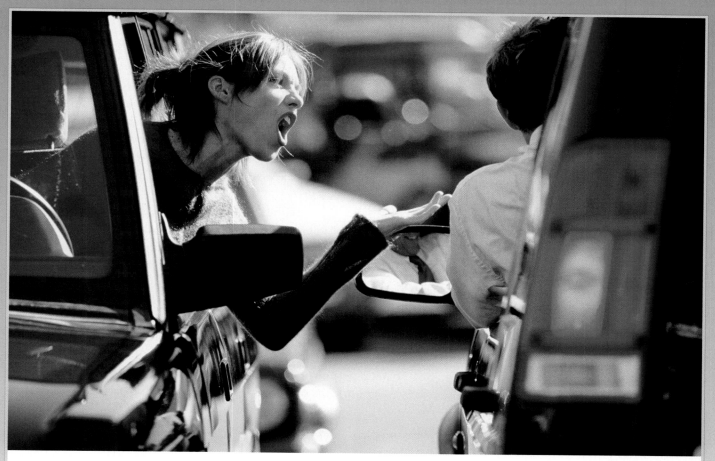

Type A Personality and Hostility

Among Type A characteristics, *hostility* is the strongest predictor of heart disease (Krantz & McCeney, 2002; Mittag & Maurischat, 2004). In particular, the constant stress associated with cynical hostility—constantly being "on watch" for problems—translates physiologically into higher blood pressure and heart rate and production of stress-related hormones. In addition, people who are hostile, suspicious, argumentative, and competitive tend to have more interpersonal conflicts. This can heighten autonomic activation, leading to increased risk of cardiovascular disease (Boyle et al., 2004; Bruck & Allen, 2003; Niaura et al., 2002; Vanderwerker & Prigerson, 2004).

   Health psychologists have developed two types of behavior modification for people with Type A personalities. The *shotgun approach* aims to eliminate or modify all Type A behaviors. The major criticism of this approach is that it eliminates not only undesirable Type A behaviors but also desirable traits, such as ambition. In contrast, the *target behavior approach* focuses on only those Type A behaviors that are likely to cause heart disease—namely, cynical hostility.

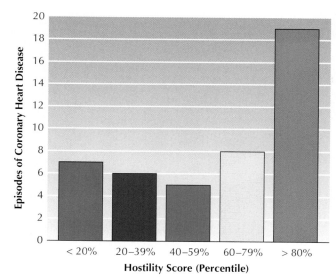

Source: Niaura et al., *Health Psychology,* 2002.

POST-TRAUMATIC STRESS DISORDER (PTSD): A DISEASE OF MODERN TIMES?

One of the most powerful examples of the effects of severe stress is **post-traumatic stress disorder (PTSD)**. Children as well as adults can experience the symptoms of PTSD, which include feelings of terror and helplessness during the trauma and recurrent flashbacks, nightmares, impaired concentration, and emotional numbing afterward. These symptoms may continue for months or years after the event. Some victims of PTSD turn to alcohol and other drugs, which often compounds the problem (Schnurr & Green, 2004).

> **post-traumatic stress disorder (PTSD)** An anxiety disorder following exposure to a life-threatening or other extreme event that evoked great horror or helplessness. It is characterized by flashbacks, nightmares, and impaired functioning.

During the Industrial Revolution, workers who survived horrific railroad accidents sometimes developed a condition very similar to PTSD. It was called "railway spine" because experts thought the problem resulted from a twisting or concussion of the spine. Later, doctors working with combat veterans referred to the disorder as "shell shock" because they believed it was a response to the physical concussion caused by exploding artillery. Today, we know that PTSD is caused by any exposure to extraordinary stress (FIGURE 3.7).

PTSD's essential feature is *severe anxiety* (a state of constant or recurring alarm and fearfulness) that develops after experiencing a traumatic event, learning about a violent or unexpected death of a family member, or even being a witness or bystander to violence (American Psychiatric Association, 2002).

According to the Facts for Health website (www.factsforhealth.org), approximately 10 percent of Americans have had or will have PTSD at some point in their lives. TABLE 3.2 summarizes the primary symptoms of PTSD and offers five important tips for coping with traumatic events.

Coping with extreme trauma FIGURE 3.7

Can you see how finding sources of help for those who have experienced trauma is important for both practical and psychological reasons?

GASTRIC ULCERS: ARE THEY CAUSED BY STRESS?

Beginning in the 1950s, psychologists reported strong evidence that stress can lead to **ulcers**—painful lesions to the lining of the stomach and upper part of the small intestine. Correlational studies have found that people who live in stressful situations develop ulcers more often than people who don't. And numerous experiments with laboratory animals have shown that stressors, such as shock or confinement to a very small space for a few hours, can produce ulcers in some laboratory animals (Andrade & Graeff, 2001; Bhattacharya & Muruganandam, 2003; Gabry et al., 2002; Landeira-Fernandez, 2004).

The relationship between stress and ulcers seemed well established until researchers reported a bacterium (*Helicobacter pylori* or *H. pylori*) associated with ulcers. Because many people prefer medical explanations (like bacteria or viruses) to psychological ones, the idea that stress causes ulcers has been largely abandoned.

Identifying PTSD and coping with crisis TABLE 3.2

Primary symptoms of post-traumatic stress disorder (PTSD)

- Re-experiencing the event through vivid memories or flashbacks
- Feeling "emotionally numb"
- Feeling overwhelmed by what would normally be considered everyday situations
- Diminished interest in performing normal tasks or pursuing usual interests
- Crying uncontrollably
- Isolating oneself from family and friends and avoiding social situations
- Relying increasingly on alcohol or drugs to get through the day
- Feeling extremely moody, irritable, angry, suspicious, or frightened
- Having difficulty falling or staying asleep, sleeping too much, and experiencing nightmares
- Feeling guilty about surviving the event or being unable to solve the problem, change the event, or prevent the disaster
- Feeling fear and sense of doom about the future

Five important tips for coping with crisis

1. Recognize your feelings about the situation and talk to others about your fears. Know that these feelings are a normal response to an abnormal situation.
2. Be willing to listen to family and friends who have been affected and encourage them to seek counseling if necessary.
3. Be patient with people. Tempers are short in times of crisis, and others may be feeling as much stress as you.
4. Recognize normal crisis reactions, such as sleep disturbances and nightmares, withdrawal, reverting to childhood behaviors, and trouble focusing on work or school.
5. Take time with your children, spouse, life partner, friends, and coworkers to do something you enjoy.

Source: American Counseling Association and adapted from Pomponio, 2002.

Most ulcer patients have the *H. pylori* bacterium in their stomachs, and it clearly damages the stomach wall. Antibiotic treatment helps many patients. However, approximately 75 percent of normal control subjects' stomachs also have the bacterium. This suggests that the bacterium may cause the ulcer, but only in people whose systems are compromised by stress. Behavior modification and other psychological treatments, working alongside antibiotics, can help ease ulcers. Finally, studies of the amygdala (a part of the brain involved in emotional response) show that it plays an important role in gastric ulcer formation (Henke, 1992; Tanaka et al., 1998). Apparently, stressful situations and direct stimulation of the amygdala cause an increase in stress hormones and hydrochloric acid and a decrease in blood flow in the stomach walls. This combination leaves the stomach more vulnerable to attack by the *H. pylori* bacteria.

In sum, it appears that *H. pylori*, increased hydrochloric acid, increased stress hormones, and decreased blood flow lead to the formation of gastric ulcers. Once again, we see how biological, psychological, and social forces influence one another (the biopsychosocial model). For now, the psychosomatic explanation for ulcers is back in business (Overmier & Murison, 2000).

CONCEPT CHECK STOP

How does stress suppress the immune system?

Which aspect of the Type A personality is most strongly related to heart disease?

What are the components of psychological hardiness?

What is the primary feature of PTSD?

Health Psychology in Action

Health psychology is the study of how biological, psychological, and social factors affect health and illness. In this section, we consider the psychological components of two major health risks—tobacco and alcohol. We also explore the psychological factors that affect chronic pain.

> **health psychology**
> The study of how biological, psychological, and social factors affect health and illness.

TOBACCO: HAZARDOUS TO YOUR HEALTH

Most people know that smoking is bad for their health and that the more they smoke, the more at risk they are. *Tobacco* use endangers both smokers and those who breathe secondhand smoke, so it is not surprising that most health psychologists and medical professionals are concerned with preventing smoking and getting those who already smoke to stop.

The first puff on a cigarette is rarely pleasant. So why do people start smoking? The answer is complex. First, smoking usually starts when people are young. A survey of U.S. middle schools, grades 6 to 8, found that one in eight students were experimenting with some form of tobacco, such as cigarettes, cigars, and chewing tobacco. Peer pressure and imitation of role models (such as celebrities) are particularly strong factors in young people's decision to begin smoking.

What Does a Health Psychologist Do?

Health psychologists are interested in how people's lifestyles and activities, emotional reactions, ways of interpreting events, and personality characteristics influence their physical health and well-being.

As researchers, they are particularly interested in the relationship between stress and the immune system.

As practitioners, health psychologists can work as independent clinicians or as consultants with physicians, physical and occupational therapists, and other health care workers. The goal of the health psychologist is to reduce psychological distress and unhealthy behaviors. They also help patients and families make critical decisions and prepare psychologically for surgery or other treatment.

Health psychologists also educate the public about health maintenance. They provide information about the effects of stress, smoking, alcohol, and lack of exercise, as well as other health issues. In addition, health psychologists help people cope with chronic problems, such as pain, diabetes, and high blood pressure, as well as unhealthful behaviors, such as anger expression and lack of assertiveness.

How might the health psychologist in this photo help a patient coping with a serious illness? Consider how the psychologist and patient would need to address the three factors (biological, psychological, and social) associated with health and illness.

Second, nicotine is addictive. So once a person begins to smoke, there is a biological need to continue. Nicotine addiction appears to be very similar to heroin, cocaine, and alcohol addiction (Brody et al., 2004). When we inhale tobacco smoke, the nicotine quickly increases the release of acetylcholine and norepinephrine in our brain. These *neurotransmitters* (see Chapter 2) increase alertness, concentration, memory, and feelings of pleasure. Nicotine also stimulates the release of dopamine, the neurotransmitter that is most closely related to reward centers in the brain (Brody et al., 2004; Noble, 2000).

Finally, smokers learn to associate smoking with pleasant things, such as good food, friends, and sex. (Ironically, as shown in **FIGURE 3.8**, anti-smoking laws may actually create stronger bonds among smokers and enhance their resolve to keep smoking.)

In addition to these social and psychological associations, smokers also learn to associate the act of smoking with the "high" that nicotine gives them. When smokers are deprived of cigarettes, they go through an unpleasant *physical withdrawal*. Nicotine relieves the withdrawal symptoms, so smoking is rewarded. Is it any surprise that many scientists believe that the best way to reduce the number of smokers is to stop people from ever taking that first puff?

Some people find that the easiest way for them to stop smoking is to suddenly and completely stop. However, the success rate for this "cold turkey" approach is extremely low. Even with medical aids, such as patches, gum, or pills, it is still very difficult to quit. Any

Smokers unite? FIGURE 3.8

Anti-smoking laws may have made quitting smoking even more difficult for some. Being forced to gather together outside to smoke may forge stronger social bonds among smokers and strengthen their resolve to "suffer the tyranny" of anti-smoking laws. In addition, having to wait longer for their next nicotine dose increases the severity of withdrawal symptoms (Palfai et al., 2000). In effect, the smoker gets repeated previews of just how unpleasant quitting would be.

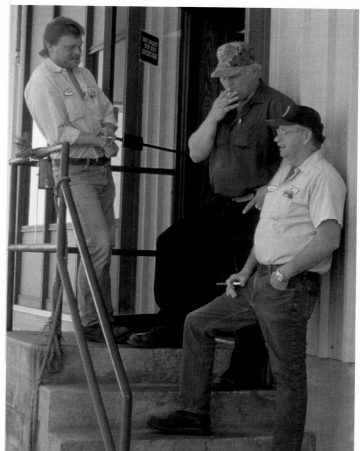

program designed to help smokers break their habit must combat the social rewards of smoking as well as the physical addiction to nicotine (Koop, Richmond, & Steinfeld, 2004).

Some **smoking cessation programs** combine nicotine replacement therapy with cognitive and behavioral techniques. This helps smokers identify stimuli or situations that make them feel like smoking and then change or avoid them (Brandon et al., 2000). Smokers can also refocus their attention on something other than smoking or remind themselves of the benefits of not smoking (Taylor et al., 2000). Behaviorally, they might cope with the urge to smoke by chewing gum, exercising, or chewing on a toothpick after a meal instead of lighting a cigarette.

ALCOHOL: A PERSONAL AND SOCIAL HEALTH PROBLEM

The American Medical Association considers *alcohol* to be the most dangerous and physically damaging of all drugs (American Medical Association, 2003). After tobacco, it is the leading cause of premature death in America and most European countries (Cohen, Mason, & Farley, 2004; Grilly, 2006; John & Hanke, 2003). Drinking alcohol may also cause serious brain damage (Crews et al., 2004). Alcohol also seems to increase aggression, which helps explain why it's a major factor in most murders, suicides, spousal assaults, incidents of child abuse, and accidental deaths in the United States (Pappas, McKenry, & Catlett, 2004; Sebre et al., 2004; Sher, Grekin, & Williams, 2005).

> ■ **binge drinking**
> Occurs when a man consumes five or more drinks in a row, or a woman consumes four or more drinks at a time, on at least three occasions during the previous two weeks.

Most people are now aware of the major risks of drinking alcohol and driving—heavy fines, loss of driver's license, serious injuries, jail time, and even death. But the effects of drinking alcohol itself can also be fatal.

Because alcohol depresses neural activity throughout the brain, if blood levels of alcohol rise to a certain level, the brain's respiratory center stops functioning and the person dies. This is why **binge drinking** is so dangerous (FIGURE 3.9).

Binge drinking is of particular concern on college campuses. A national survey of college students found that between 1993 and 2001, approximately 44 percent of college students were binge drinkers (Wechsler et al., 2002). This survey also found that white students are significantly more likely to drink heavily (50.2 percent) than are students of other ethnicities.

Alcohol is a serious health problem for all college-age students. Research shows that every year:

- 1,400 college students die from alcohol-related causes,
- 500,000 students suffer nonfatal, alcohol-related injuries,
- 1.2–1.5 percent of students attempt suicide because of alcohol or other drug use,
- 400,000 students have unprotected sex, and more than 100,000 are too intoxicated to know whether they consented to sexual intercourse (Task Force of the National Advisory, 2002).

College administrators, however, are increasingly aware of the problems of binge drinking and other types of alcohol abuse. They are developing policies and programs that go beyond traditional educa-

Binge drinking FIGURE 3.9

Unfortunately, many college students believe that heavy drinking is harmless fun and a natural part of college life. But excessive alcohol consumption, including binge drinking, carries serious health consequences and can even be fatal.

Do You Have an Alcohol Problem?

In our society, drinking alcohol (within limits) is generally considered normal and appropriate. However, many people abuse alcohol. If you'd like to make a quick check of your own drinking behavior, place a mark next to each of the symptoms that describes your current drinking behavior.

Seven Signs of Alcohol Dependence Syndrome

_____ • Drinking increases, sometimes to the point of almost continuous daily consumption.

_____ • Drinking is given higher priority than other activities, in spite of its negative consequences.

_____ • More and more alcohol is required to produce behavioral, subjective, and metabolic changes; large amounts of alcohol can be tolerated.

_____ • Even short periods of abstinence bring on withdrawal symptoms, such as sweatiness, trembling, and nausea.

_____ • Withdrawal symptoms are relieved or avoided by further drinking, especially in the morning.

_____ • The individual is subjectively aware of a craving for alcohol and has little control over the quantity and frequency of intake.

_____ • If the person begins drinking again after a period of abstinence, he or she rapidly returns to the previous high level of consumption and other behavioral patterns.

Source: World Health Organization, 2004.

tional programs to also include the physical, social, legal, and economic environment on college campuses and the surrounding communities (Kapner, 2004).

CHRONIC PAIN: AN ONGOING THREAT TO HEALTH

Normally, pain is essential to our survival and well-being. It alerts us to dangerous or harmful situations and forces us to rest and recover from injury (Watkins & Maier, 2000). In contrast, **chronic pain**—the kind that comes with a chronic disease or continues long past the healing of a wound—does not serve a useful function.

Chronic pain is a serious problem with no simple solution. Although exercise increases the body's supply of *endorphins*, which help block the perception of pain, patients with chronic pain tend to decrease their activity and exercise.

Psychological factors such as anxiety frequently intensify pain and increase the related anguish and disability (Currie & Wang, 2004; Eriksen et al., 2003;

Keefe, Abernethy, & Campbell, 2005). Even well-meaning family members can exacerbate pain just by asking about it, because talking about pain focuses attention on it and increases its intensity (Sansone, Levengood, & Sellbom, 2004).

Health psychologists often use psychologically oriented treatments to treat chronic pain. For example, a health psychologist may begin a **behavior modification** program for both the patient with chronic pain and his or her family. Such programs often involve establishing an individualized pain management plan that incorporates daily exercise and relaxation techniques. Health psychologists also monitor each patient's adherence to the plan and provide rewards for following through with the treatment program.

chronic pain
Continuous or recurrent pain lasting six months or longer.

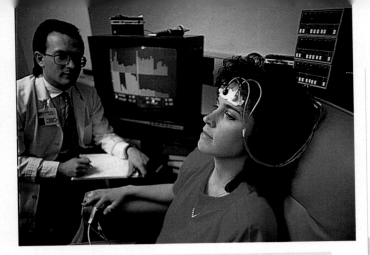

Biofeedback for chronic pain FIGURE 3.10

Biofeedback using an EMG is most helpful when the pain involves extreme muscle tension, such as with tension headache and lower back pain.

In **biofeedback** for chronic pain, an electromyograph (EMG) measures muscle tension by recording electrical activity in the skin. When a patient is sufficiently relaxed, the machine signals with a tone or light. This feedback teaches patients self-regulation skills that help control their pain (**FIGURE 3.10**). Biofeedback is sometimes as effective as more expensive and length-

ier forms of pain treatment (Engel, Jensen, & Schwartz, 2004; Hammond, 2005; Hawkins & Hart, 2003).

Finally, special **relaxation techniques**—like the breathing and muscle relaxation exercises that are taught in childbirth classes—can help divert attention from pain, reducing discomfort (Astin, 2004).

CONCEPT CHECK STOP

Why is it so difficult for people to quit smoking?

What are some key indicators of problem drinking?

How do behavior modification programs for chronic pain treatment work?

Why might relaxation techniques ease chronic pain, even without eliminating the pain itself?

Health and Stress Management

LEARNING OBJECTIVES

Compare emotion-focused and problem-focused forms of coping.

Explain the role that interpretation plays in shaping our responses to stressors.

Review some major resources for combating stress.

COPING WITH STRESS

Because we can't escape stress, we need to learn how to effectively cope with it. Simply defined, **coping** is an attempt to manage stress in some effective way. It is not one single act but a process that allows us to deal with various stressors (FIGURE 3.11).

Our level of stress generally depends on both our interpretation of and our reaction to stressors. **Emotion-focused forms of coping** are emotional or cognitive strategies that change how we view a stressful situation.

For example, suppose you were refused a highly desirable job. You might reappraise the situation and decide that the job wasn't the right match for you or that you weren't really qualified or ready for it.

When faced with unavoidable stress, people commonly use **defense mechanisms**. That is, they unconsciously distort reality to protect their egos and to avoid anxiety. For instance, fantasizing about what you will do on your next vacation can help relieve stress. Sometimes, however, defense mechanisms can be destructive. For example, people often fabricate excuses for not attaining particular goals. Such *rationalization* keeps us from seeing a situation more clearly and realistically and can prevent us from developing valuable skills or qualities.

■ emotion-focused forms of coping Coping strategies based on changing one's perceptions of stressful situations.

People often combine *problem-focused* and *emotion-focused* coping strategies to resolve complex stressors or to respond to a stressful situation that is in flux. In some situations, an emotion-focused strategy can allow people to step back from an especially overwhelming problem. Then they can reappraise the situation and use the problem-solving approach to look for solutions. Can you see how each form of coping is represented in these two photos?

Some research (Gaab et al., 2005; Lazarus, 1999) suggests that our emotional response to a stressful event depends largely on how we interpret the event. We can learn to cope with stress by using cognitive, emotional, and behavioral strategies that reduce our feeling of threat.

problem-focused forms of coping Coping strategies that use problem-solving strategies to decrease or eliminate the source of stress.

Emotion-focused forms of coping that are accurate reappraisals of stressful situations and that do not distort reality may alleviate stress in some situations (Giacobbi, Foore, & Weinberg, 2004; Patterson, Holm, & Gurney, 2004). Many times, however, it is necessary and more effective to use **problem-focused forms of coping**, which deal directly with the situation or the stressor to eventually decrease or eliminate it (Bond & Bunce, 2000). These direct coping strategies include:

- identifying the stressful problem,

- generating possible solutions,

- selecting the appropriate solution,

- applying the solution to the problem, thus eliminating the stress.

RESOURCES FOR HEALTHY LIVING: FROM GOOD HEALTH TO MONEY

A person's ability to cope effectively depends on the stressor itself—its complexity, intensity, and duration—and on the type of coping strategy used. It also depends on available resources. Seven important resources for healthy living and stress management are health and exercise, control, positive beliefs, social skills, social support, material resources, and relaxation.

Health and exercise   All stressors cause physiological changes. Therefore, an individual's health significantly affects his or her ability to cope. The stronger and healthier people are, the better they cope (and the longer they stay in the resistance phase of the General Adaptation Syndrome without entering the exhaustion phase). Exercising and keeping fit will probably help you minimize anxiety, depression, and tension, which are associated with stress. Researchers have found that people who engaged in strenuous exercise programs experienced greater reductions in anxiety than did those in moderate programs (Broman-Fulks et al., 2004; Hong, 2000).

external locus of control Belief that chance or outside forces beyond one's control determine one's fate.

internal locus of control Belief that one controls one's own fate.

Control   Believing that you are the "master of your own destiny" is an important resource for effective coping. People with an **external locus of control** feel powerless to change their circumstances and are less likely to make healthy changes, follow treat-

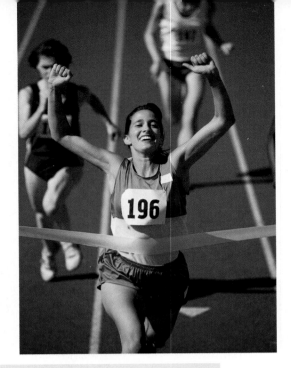

More than just fun FIGURE 3.12

Exercise—especially aerobic exercise—uses up the hormones that are secreted into the bloodstream during stress, giving the immune system a boost. Exercise also helps relieve muscle tension; improves cardiovascular efficiency; and increases strength, flexibility, and stamina for encountering future stressors. Likewise, a person's locus of control affects how he or she responds to stress. Does this photo seem to illustrate an internal or external locus of control?

ment programs, or positively cope with a situation. Conversely, people with an **internal locus of control** believe that they are in charge of their own destinies and are therefore able to adopt more positive coping strategies. They tend to recover more quickly from illness (Ewart & Fitzgerald, 1994) and to experience less psychological stress than those with an external locus of control (Hamid & Chan, 1998; Lu et al., 2000) (FIGURE 3.12).

Positive beliefs   A positive self-image and attitude can be especially significant coping resources. Even temporarily raising self-esteem reduces the amount of anxiety caused by stressful events (Greenberg et al., 1993). Also, hope can sustain a person in the face of severe odds, as is often documented in news reports of people who have triumphed over seemingly unbeatable circumstances.

Social skills   Social situations can prompt both pleasure and stress. For some people, merely meeting someone new and trying to find something to talk about can be very stressful. People who acquire social skills (such as knowing appropriate behaviors for certain situations, having conversation starters up their sleeves, and expressing themselves well) suffer less anxiety than people who do not. In fact, people who lack social skills are more at risk for developing illness than those who have them (Cohen & Williamson, 1991). Social skills not only help us interact with others but also communicate our needs and desires, enlist help when we need it, and decrease hostility in tense situations.

Social support

Having the support of others helps offset the stressful effects of divorce, the loss of a loved one, chronic illness, pregnancy, physical abuse, job loss, and work overload (Greeff & Van Der Merwe, 2004; Shen, McCreary, & Myers, 2004; Southwick, Vythilingam, & Charney, 2005). When we are faced with stressful circumstances, our friends and family often help us take care of our health, listen, hold our hands, make us feel important, and provide stability to offset the changes in our lives.

For people with specific problems, support and self-help groups can be very helpful. There are groups for alcoholics and families of alcoholics, for former drug addicts, for divorced people, for single parents, for cancer patients, for parents who have had a young child die, and so on. Support groups help people cope not only because they provide other people to lean on but also because people can learn techniques for coping from others with similar problems.

Material resources

We've all heard the saying "Money isn't everything," but money and the things that money can buy can be very real resources. Money increases the number of options available for eliminating sources of stress or reducing the effects of stress. When faced with the minor hassles of everyday living, or when faced with chronic stressors or major catastrophes, people with money and the skills to effectively use it generally fare better and experience less stress than people without money (Chi & Chou, 1999; Ennis, Hobfoll, & Schroeder, 2000).

Relaxation

One of the most effective ways of dealing with physical stress reactions is to make a conscious decision to relax. There are a variety of relaxation techniques. As we discovered in an earlier section, biofeedback is often used in the treatment of chronic pain, but it is also useful in teaching people to relax and manage their stress (Hammond, 2005). **Progressive relaxation** helps reduce or relieve the muscular tension commonly associated with stress (Khandai, 2004; Vocks et al., 2004). Using this technique, patients first tense and then relax specific muscles, such as those in the neck, shoulders, and arms. This technique teaches people to recognize the difference between tense and relaxed muscles.

CONCEPT CHECK **STOP**

What is a defense mechanism?

Why would it sometimes be useful for people to combine coping strategies?

Why is exercise important for counteracting stress?

How can social skills protect us against stress?

What is happening in this picture ?

Anti-smoking laws, which have become increasingly common since the 1990s, have forced smokers to go outside to smoke. What psychological effect of these laws do cigarette companies exploit in this advertisement?

VISUAL SUMMARY

1 Understanding Stress

1. **Stress** is the body's nonspecific response to any demand placed on it. Pleasant or beneficial stress is called eustress. Stress that is unpleasant or objectionable is called distress. Life changes, chronic stressors, minor hassles, frustrations, and conflict all can cause stress. There are three basic types of conflict: **approach–approach conflict, avoidance–avoidance conflict**, and **approach–avoidance conflict**.

2. The sympathetic nervous system and the HPA Axis control significant physiological responses to stress. The sympathetic nervous system prepares us for immediate action; the HPA Axis responds more slowly but lasts longer. Researchers' understanding that psychological factors influence many diseases has spawned a new field of biopsychology called **psychoneuroimmunology**.

3. The general adaptation syndrome (GAS) describes the body's three-stage reaction to stress. The phases are the initial alarm reaction, the resistance phase, and the exhaustion phase (if resistance to stress is not successful).

2 Stress and Illness

1. Stress makes the immune system less able to resist infection and cancer development.

2. Increased stress hormones can cause fat to adhere to blood vessel walls, increasing the risk of heart attack. Having a **Type A personality** amplifies the risk of stress-related heart disease. The **Type B personality** is its antithesis.

3. Exposure to extraordinary stress can cause **post-traumatic stress disorder (PTSD)**, a type of severe anxiety that increases the likelihood of substance abuse problems.

4. Stress increases the risk of developing gastric ulcers among people who have the *H. pylori* bacterium in their stomachs.

3 Health Psychology in Action

1. **Health psychology**, the study of how biological, psychological, and social factors affect health and illness, is a growing field in psychology.

2. Most approaches to helping people quit smoking include cognitive and behavioral techniques to aid smokers in their withdrawal from nicotine, along with nicotine replacement therapy.

3. Alcohol is a leading cause of premature death and may cause serious brain damage. Alcohol also seems to increase aggression. Alcohol, particularly **binge drinking**, is of major concern on college campuses. Although many college students believe that heavy drinking is a harmless part of college life, college administrators are becoming increasingly aware of the problems of binge drinking and other types of alcohol abuse.

4. **Chronic pain** is a serious problem with no simple solution. Psychological factors frequently intensify pain and increase related problems. To treat chronic pain, health psychologists often focus their efforts on psychologically oriented treatments, including behavior modification, biofeedback, and relaxation techniques.

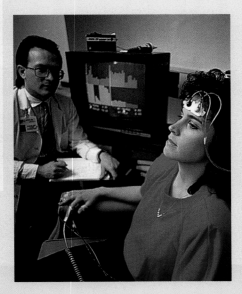

4 Health and Stress Management

1. Our level of stress generally depends on both our interpretation of and our reaction to stressors. **Emotion-focused forms of coping** are emotional or cognitive strategies that change how we view a stressful situation. Some emotion-focused defense mechanisms, such as rationalization, can be destructive because they prevent us from developing valuable skills or qualities. It is often more effective to use **problem-focused forms of coping** strategies. People often combine problem-focused and emotion-focused coping strategies to resolve complex stressors or to respond to a stressful situation that is in flux.

2. Seven important resources for healthy living and stress management are health and exercise, positive beliefs, social skills, social support, material resources, control, and relaxation. A more specific look at control tells us that believing that one controls one's own fate—**an internal locus of control**—is a healthy way to manage stress. Having an **external locus of control** means believing that chance or outside forces beyond one's control determine one's fate.

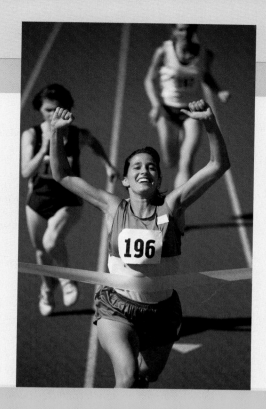

KEY TERMS

CRITICAL AND CREATIVE THINKING QUESTIONS

1. What are the major sources of stress in your life?

2. If you were experiencing a period of high stress, what would you do to avoid illness?

3. Why are smoking-prevention efforts often aimed at adolescents?

4. What are the benefits of involving a person's family in behavior modification programs to treat chronic pain?

5. Which forms of coping do you most often draw on?

6. Which of the resources for stress management seem most important to you?

(Check your answers in Appendix B.)

1. In your text, the physical and mental arousal to situations that we perceive as threatening or challenging is called

   _____.

   a. distress
   b. eustress
   c. stress
   d. all of these options

2. _____ is one of the largest sources of **CHRONIC** stress for adults.

   a. Birth
   b. Work
   c. Christmas
   d. Moving

3. A state of physical, emotional, and mental exhaustion attributable to long-term involvement in emotionally demanding situations is called _____.

   a. primary conflict
   b. technostress
   c. burnout
   d. secondary conflict

4. In an *approach-approach conflict,* a person must choose between two or more goals that will lead to _____, whereas in an *avoidance-avoidance conflict,* a person must choose between two or more goals that will lead to

   _____.

   a. less conflict; more conflict
   b. frustration; hostility
   c. a desirable result; an undesirable result
   d. effective coping; ineffective coping

5. Label the *HPA Axis,* which enables us to deal with chronic stressors:

   a. hypothalamus
   b. pituitary
   c. adrenal cortex

6. _____ is the field that studies the effects of psychological factors on the immune system.

   a. Psychosomatology
   b. Neurobiology
   c. Psychoneuroimmunology
   d. Biopsychology

7. Research suggests that one particular *Type A* characteristic, illustrated in this photo, is **MOST** associated with heart disease. What characteristic is this?

   a. intense ambition
   b. impatience
   c. cynical hostility
   d. time urgency

8. People who experience flashbacks, nightmares, and impaired functioning following a life-threatening or other horrifying event are _____.

   a. suffering from a psychosomatic illness
   b. experiencing post-traumatic stress disorder
   c. having a nervous breakdown
   d. weaker than people who take such events in stride

9. Which of the following is **TRUE** of *health psychology*?

   a. It studies the relationship between psychological behavior and physical health.
   b. It studies the relationship between psychological behavior and illness.
   c. It emphasizes wellness and the prevention of illness.
   d. All of these options

10. The effects of nicotine are related to the release of _____ in the brain.

  a. acetylcholine
  b. norepinephrine
  c. dopamine
  d. all of these options

11. At several parties in the past couple of weeks, Arash consumed 5 drinks in a row and his girlfriend consumed 4 drinks in a row. This means that _____ met the definition for binge drinking.

  a. Arash
  b. his girlfriend
  c. both Arash and his girlfriend
  d. neither Arash nor his girlfriend

12. Any *pain* that continues or recurs for six months or more is considered to be _____ .

  a. life-threatening
  b. chronic
  c. psychosomatic
  d. caused by an incompetent physician

13. Biofeedback using _____ measures muscle tension and provides feedback regarding a patient's level of relaxation.

  a. EMG   c. EGG
  b. EKG   d. EEG

14. *Emotion-focused forms of coping* are based on changing your _____ when faced with stressful situations.

  a. feelings
  b. perceptions
  c. strategies
  d. all of these options

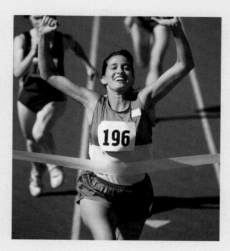

15. Research suggests that people with higher _____ have less psychological stress than those with higher _____ .

  a. external locus of control; internal locus of control
  b. internal locus of control; external locus of control
  c. emotion-focused coping styles; problem-focused coping styles
  d. problem-focused coping styles; emotion-focused coping styles

Sensation and Perception

Imagine that your visual field were suddenly inverted and reversed, so that things you expected to be on your right would be on your left, and things you expected to be above your head would be below your head. You would certainly have trouble getting around. Do you think you could ever adapt to this distorted world?

To answer that question, psychologist George Stratton (1896) wore special lenses for eight days. For the first few days, Stratton had a great deal of difficulty navigating in this environment and coping with everyday tasks. But by the third day, his experience had begun to change. He noted:

> *Walking through the narrow spaces between pieces of furniture required much less care than hitherto. I could watch my hands as they wrote, without hesitating or becoming embarrassed thereby.*

By the fifth day, Stratton had almost completely adjusted to his strange perceptual environment. His expectations of how the world should be arranged had changed. As Stratton's experiment shows, we are able to adapt even our most basic perceptions by retraining our brains to adapt to unfamiliar physical sensations, creating a newly coherent world. This chapter focuses on two separate, but inseparable, aspects of how we experience the world: sensation and perception. The boundary between these two processes is not precise. **Sensation** is the process of receiving, translating, and transmitting raw sensory data from the external and internal environments to the brain. **Perception** is the "higher level" process of selecting, organizing, and interpreting sensory data into useful mental representations of the world.

Understanding Sensation

LEARNING OBJECTIVES

Describe how raw sensory stimuli are converted to signals in the brain.

Explain how the study of thresholds helps to explain sensation.

Describe why adapting to sensory stimuli provides an evolutionary advantage.

Identify the factors that govern pain perception.

When presented with a high-pitched tone, a musician reported, "It looks like fireworks tinged with a pink-red hue. The color feels rough and unpleasant, and it has an ugly taste—rather like that of a briny pickle" (Luria, 1968). This musician was describing a rare condition known as **synesthesia**, which means "mixing of the senses." People with synesthesia routinely blend their sensory experiences. They may "see" temperatures, "hear" colors, or "taste" shapes. To appreciate how extraordinary synesthesia is, we must first understand the basic processes of normal, nonblended sensations. For example, how do we turn light and sound waves from the environment into something our brain can comprehend? To do this, we must have both a means of detecting stimuli and a means of converting them into a language the brain can understand.

PROCESSING: DETECTION AND CONVERSION

Our eyes, ears, skin, and other sense organs all contain special cells called receptors, which receive and process sensory information from the environment. For each sense, these specialized cells respond to a distinct stimulus, such as sound waves or odor molecules. During the process of **transduction**, the receptors convert the stimulus into neural impulses, which are sent to the brain. For example, in hearing, tiny receptor cells in the inner ear convert mechanical vibrations from sound waves into electrochemical signals. These signals are carried by neurons to the brain, where specific sensory receptors detect and interpret the information. How does our brain differentiate between sensations, such as sounds and smells? Through a process known as **coding**, different physical stimuli are interpreted as distinct sensations because their neural impulses travel by different routes and arrive at different parts of the brain (**FIGURE 4.1**).

We also have structures that purposefully reduce the amount of sensory information we receive. Without this natural filtering of stimuli we would constantly hear blood rushing through our veins and feel our clothes brushing against our skin. Some level of filtering is needed so the brain is not overwhelmed with unnecessary information.

All species have evolved selective receptors that suppress or amplify information for survival (**FIGURE 4.2**). In the process of **sensory reduction**,

> **transduction** Process by which a physical stimulus is converted into neural impulses.

> **coding** Process that converts a particular sensory input into a specific sensation.

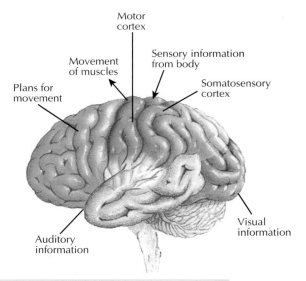

Sensory processing within the brain
FIGURE 4.1

Neural impulses travel from the sensory receptors to various parts of the brain.

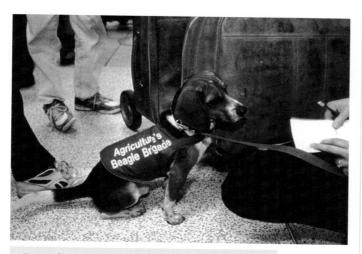

Amazing animal senses FIGURE 4.2

Many animals possess extraordinary sensory abilities. For example, dogs' sense of smell is far more sensitive than humans'. For this reason, specially trained dogs provide invaluable help in sniffing out dangerous plants, drugs, and explosives, tracking criminals, and assisting in search-and-rescue operations. Some researchers believe dogs can even detect chemical signs of certain illnesses, such as diabetes or cancer.

we filter incoming sensations and analyze the sensations that the body does register before a neural impulse is finally sent to the various parts of the brain. Humans, for example, cannot sense ultraviolet light, microwaves, the ultrasonic sound of a dog whistle, or infrared heat patterns from warm-blooded animals, as some other animals can. However, we can see a candle burning 30 miles away on a dark, clear night, hear a watch tick at 20 feet under quiet conditions, smell one drop of perfume in a six-room apartment, and taste 1 teaspoon of sugar dissolved in 2 gallons of water (FIGURE 4.3).

ADAPTATION: WEAKENING THE RESPONSE

Imagine that friends have invited you to come visit their beautiful new baby. As they greet you at the door, you are overwhelmed by the odor of a wet diaper. Why don't your friends do something about that smell? The answer lies in the previously mentioned sensory reduction and **sensory adaptation**. When a constant stimulus is presented for a length of time, sensation often fades or disappears. In sensory adaptation, receptors higher up in the sensory

system get "tired" and actually fire less frequently.

Sensory adaptation makes sense from an evolutionary perspective. We can't afford to waste attention and time on unchanging, normally unimportant stimuli. "Turning down the volume" on repetitive information helps the brain cope with an overwhelming amount of sensory stimuli and allows time to pay attention to change. Sometimes, however, adaptation can

sensory reduction Filtering and analyzing incoming sensations before sending a neural message to the cortex.

sensory adaptation Repeated or constant stimulation decreases the number of sensory messages sent to the brain, which causes decreased sensation.

Sensory thresholds
FIGURE 4.3

How do we know that humans can hear a watch ticking at 20 feet or smell one drop of perfume in a six-room apartment? The answer comes from research in **psychophysics**, the study of the relation between events in the physical world and our psychological experience of them. Researchers study how the strength or intensity of a stimulus affects an observer. Consider this example:

• To test for hearing loss, a hearing specialist uses a tone generator to produce sounds of differing pitches and intensities.

• You listen over earphones and indicate the earliest point at which you hear a tone. This is your *absolute threshold*, or the smallest amount of a stimulus needed to detect the stimulus.

• To test your *difference threshold*, or *just noticeable difference (JND)*, the examiner would gradually change the volume and ask you to respond when you notice a difference.

• Comparing your thresholds with those of people with normal hearing shows whether you have a hearing loss and, if so, the extent of the loss.

be dangerous, as when people stop paying attention to a gas leak in the kitchen.

Although some senses, like smell and touch, adapt quickly, we never completely adapt to visual stimuli or to extremely intense stimuli, such as the odor of ammonia or the pain of a bad burn. From an evolutionary perspective, these limitations on sensory adaptation aid survival, for example, by reminding us to avoid strong odors and heat or to take care of that burn.

If we don't adapt to pain, how do athletes keep playing despite painful injuries? In certain situations, the body releases natural painkillers called *endorphins* (see Chapter 2), which inhibit pain perception (**FIGURE 4.4**).

In addition to endorphin release, one of the most accepted explanations of pain perception is the **gate-control theory**, first proposed by Ronald Melzack and Patrick Wall (1965). According to this theory, the experience of pain depends partly on whether the neural message gets past a "gatekeeper" in the spinal cord. Normally, the gate is kept shut, either by impulses coming down from the brain or by messages coming from large-diameter nerve fibers that conduct most sensory signals, such as touch and pressure. However, when body tissue is damaged, impulses from smaller pain fibers open the gate.

gate-control theory Theory that pain sensations are processed and altered by mechanisms within the spinal cord.

According to the gate-control theory, massaging an injury or scratching an itch can temporarily relieve discomfort because pressure on large-diameter neurons interferes with pain signals. Messages from the brain can also control the pain gate, explaining how athletes and soldiers can carry on despite excruciating pain. When we are soothed by endorphins or distracted by competition or fear, our experience of pain can be greatly diminished. On the other hand, when we get anxious or dwell on our pain, we can intensify it (Sansone, Levengood, & Sellbom, 2004; Sullivan, Tripp, & Santor, 1998). Ironically, well-meaning friends who ask chronic pain sufferers about their pain may unintentionally reinforce and increase it (Jolliffe & Nicholas, 2004).

Research also suggests that the pain gate may be chemically controlled, that a neurotransmitter called *substance P* opens the pain gate, and that endorphins close it (Cesaro & Ollat, 1997; Liu, Mantyh, & Basbaum, 1997). Other research (Melzack, 1999; Vertosick, 2000) finds that when normal sensory input is disrupted, the brain can generate pain and other sensations on its own. Amputees sometimes continue to feel pain (and itching or tickling) long after a limb has been amputated. This *phantom limb pain* occurs because nerve cells send conflicting messages to the brain. The brain interprets this "static" as pain because it arises in the area of the spinal cord responsible for pain signaling. When amputees are fitted with prosthetic limbs and begin using them, phantom pain generally disappears (Gracely, Farrell, & Grant, 2002).

Sidelined? FIGURE 4.4

The body's ability to inhibit pain perception sometimes makes it possible for athletes to "play through" painful injuries.

CONCEPT CHECK STOP

How do we convert sensory information into signals the brain can understand?

What is the distinction between the absolute threshold and the difference threshold?

Why would we want to limit the amount of sensory information that we receive?

What factors govern pain perception?

How We See and Hear

WAVES OF LIGHT AND SOUND

Even the most complex visual and auditory experiences depend on our basic ability to detect light and sound. Both light and sound move in waves, similar to the movement of waves in the ocean (**FIGURE 4.5**).

Light waves are a form of electromagnetic energy, and different types of waves on the *electromagnetic spectrum* have different wavelengths.

In contrast to light waves, which are particles of electromagnetic energy, sound waves are produced by air molecules moving in a particular wave pattern. This occurs when an impact or vibrating objects, such as vocal cords or guitar strings, cause a sudden change in air pressure.

Differences between light and sound waves
FIGURE 4.5

Watching a fireworks show is only one of many ways you've probably learned that light travels faster than sound. But light also travels differently from sound, which must pass through a physical material to be heard. The speed of light is always 300 million meters (186,000 miles) per second no matter what it passes through. Sound travels through air at 344 meters (1100 ft) per second at 70°F. But if you were observing fireworks under water, you'd notice much less of a gap between the burst of light and the arrival of sound—in water sound will travel at a speed of about 1500 meters (5000 ft) per second, which is about five times its speed in air.

The human eye can see only visible light, a small part of the full spectrum. Visible light with a short wavelength is perceived as blue, visible light with a medium wavelength is green or yellow, and visible light with a long wavelength is red.

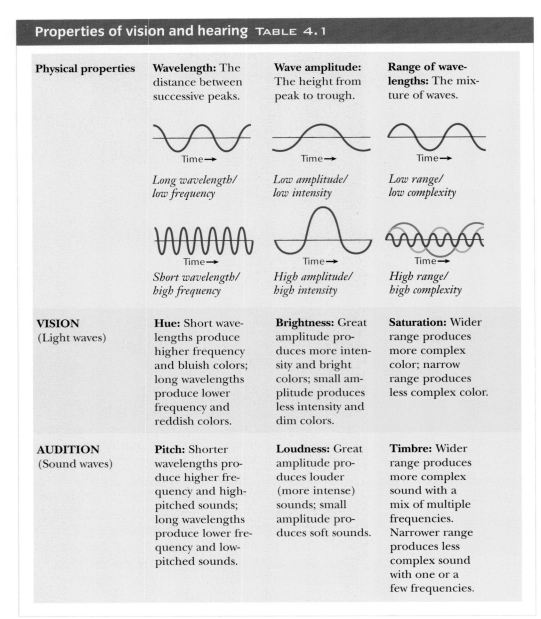

Properties of vision and hearing TABLE 4.1

Physical properties	Wavelength: The distance between successive peaks.	Wave amplitude: The height from peak to trough.	Range of wavelengths: The mixture of waves.
	Long wavelength/ low frequency	*Low amplitude/ low intensity*	*Low range/ low complexity*
	Short wavelength/ high frequency	*High amplitude/ high intensity*	*High range/ high complexity*
VISION (Light waves)	**Hue:** Short wavelengths produce higher frequency and bluish colors; long wavelengths produce lower frequency and reddish colors.	**Brightness:** Great amplitude produces more intensity and bright colors; small amplitude produces less intensity and dim colors.	**Saturation:** Wider range produces more complex color; narrow range produces less complex color.
AUDITION (Sound waves)	**Pitch:** Shorter wavelengths produce higher frequency and high-pitched sounds; long wavelengths produce lower frequency and low-pitched sounds.	**Loudness:** Great amplitude produces louder (more intense) sounds; small amplitude produces soft sounds.	**Timbre:** Wider range produces more complex sound with a mix of multiple frequencies. Narrower range produces less complex sound with one or a few frequencies.

Both light waves and sound waves vary in **wavelength**, **amplitude** (height), and **range**—each with a distinct effect on vision and hearing, or audition, as shown in TABLE 4.1.

VISION: THE EYES HAVE IT

Several structures in the eye are involved in capturing and focusing light and converting it into neural signals to be interpreted by the brain, as shown in FIGURE 4.6.

Thoroughly understanding the processes detailed in Figure 4.6 gives us clues for understanding some visual peculiarities. For example, small abnormalities in the eye sometimes cause images to be focused in front of the retina (**nearsightedness**, also called myopia) or behind it (**farsightedness**, or hyperopia). Corrective lenses or laser surgery can correct most such visual acuity problems. During middle age, most people's lenses lose elasticity and the ability to accommodate for near vision, a condition known as presbyopia that can normally be treated with corrective lenses.

If you walk into a dark movie theater on a sunny afternoon, you will at first be blinded. This is because in bright light, the pigment inside the rods is bleached, making them temporarily nonfunctional. It takes a second or two for the rods to become functional enough to see. This process of **dark adaptation** continues for 20 to 30 minutes. Light adaptation, the adjustment that takes place when you go from darkness to a bright setting, takes about 7 to 10 minutes and is the work of the cones.

HEARING: A SOUND SENSATION

The sense of hearing, or **audition**, has a number of important functions, from alerting us to dangers around us to helping us communicate with others. The ear

How the eye sees FIGURE 4.6

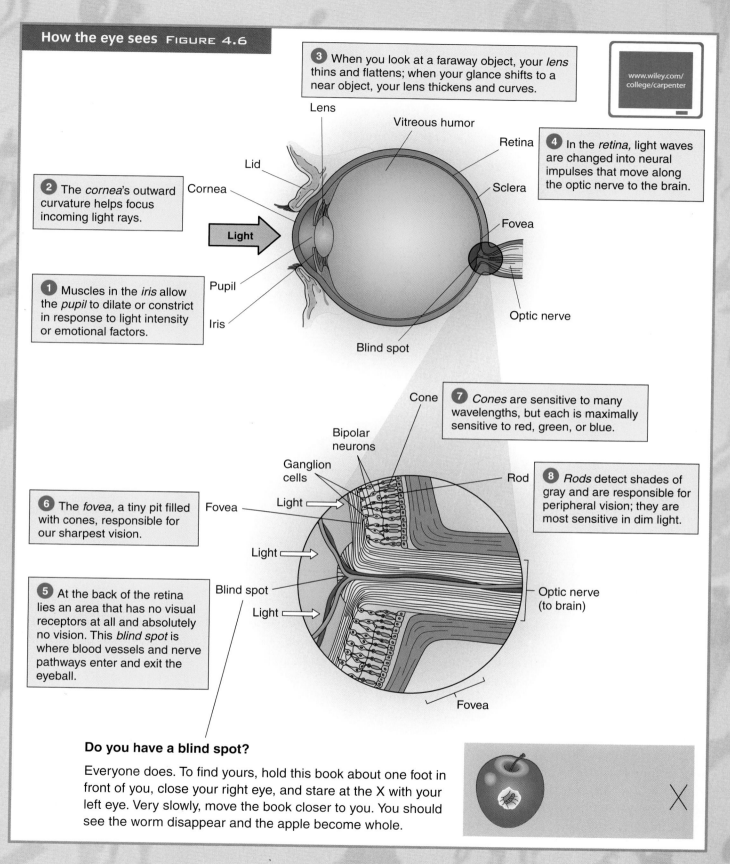

3 When you look at a faraway object, your *lens* thins and flattens; when your glance shifts to a near object, your lens thickens and curves.

www.wiley.com/college/carpenter

2 The *cornea*'s outward curvature helps focus incoming light rays.

4 In the *retina,* light waves are changed into neural impulses that move along the optic nerve to the brain.

1 Muscles in the *iris* allow the *pupil* to dilate or constrict in response to light intensity or emotional factors.

Lens
Vitreous humor
Retina
Lid
Cornea
Sclera
Fovea
Light
Pupil
Iris
Optic nerve
Blind spot

7 *Cones* are sensitive to many wavelengths, but each is maximally sensitive to red, green, or blue.

Cone
Bipolar neurons
Ganglion cells

6 The *fovea,* a tiny pit filled with cones, responsible for our sharpest vision.

Fovea
Light
Light
Blind spot
Light

8 *Rods* detect shades of gray and are responsible for peripheral vision; they are most sensitive in dim light.

Rod

5 At the back of the retina lies an area that has no visual receptors at all and absolutely no vision. This *blind spot* is where blood vessels and nerve pathways enter and exit the eyeball.

Optic nerve (to brain)

Fovea

Do you have a blind spot?

Everyone does. To find yours, hold this book about one foot in front of you, close your right eye, and stare at the X with your left eye. Very slowly, move the book closer to you. You should see the worm disappear and the apple become whole.

X

How the ear hears FIGURE 4.7

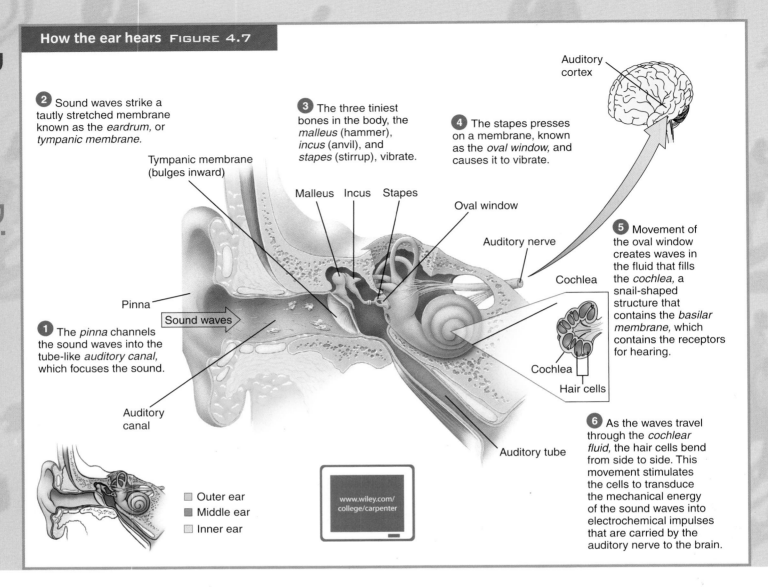

2 Sound waves strike a tautly stretched membrane known as the *eardrum,* or *tympanic membrane.*

3 The three tiniest bones in the body, the *malleus* (hammer), *incus* (anvil), and *stapes* (stirrup), vibrate.

4 The stapes presses on a membrane, known as the *oval window,* and causes it to vibrate.

Tympanic membrane (bulges inward)

Malleus   Incus   Stapes

Oval window

Auditory cortex

Auditory nerve

5 Movement of the oval window creates waves in the fluid that fills the *cochlea,* a snail-shaped structure that contains the *basilar membrane,* which contains the receptors for hearing.

Cochlea

Pinna

Sound waves

1 The *pinna* channels the sound waves into the tube-like *auditory canal,* which focuses the sound.

Auditory canal

Cochlea

Hair cells

Auditory tube

6 As the waves travel through the *cochlear fluid,* the hair cells bend from side to side. This movement stimulates the cells to transduce the mechanical energy of the sound waves into electrochemical impulses that are carried by the auditory nerve to the brain.

☐ Outer ear
☐ Middle ear
☐ Inner ear

www.wiley.com/college/carpenter

has three major sections that function as shown in FIGURE 4.7.

The mechanisms determining how we distinguish among sounds of different pitches (low to high) differ, depending on the sounds' **frequency**. According to **place theory**, different high-frequency sound waves (which produce high-pitched sounds) maximally stimulate the hair cells at different locations along the basilar membrane (FIGURE 4.8). Hearing for low-pitched sounds works differently. According to **frequency theory**, low-pitched sounds cause hair cells along the basilar membrane to bend and fire neural messages (action potentials) at the same rate as the frequency of that

sound. For example, a sound with a frequency of 90 hertz would produce 90 action potentials per second in the auditory nerve.

Whether we detect a sound as soft or loud depends on its intensity. Waves with high peaks and low valleys produce loud sounds; those that have relatively low peaks and shallow valleys produce soft sounds. The relative loudness or softness of sounds is measured on a scale of *decibels* (FIGURE 4.9).

Hearing loss can stem from two major causes: (1) **conduction deafness**, or middle-ear deafness, which results from problems with the mechanical system that conducts sound waves to the inner ear, and

(2) **nerve deafness**, or inner-ear deafness, which involves damage to the cochlea, hair cells, or auditory nerve (Figure 4.9).

Although most conduction deafness is temporary, damage to the auditory nerve or receptor cells is almost always irreversible. The only treatment for nerve deafness is a small electronic device called a **cochlear implant**. If the auditory nerve is intact, the implant bypasses hair cells to stimulate the nerve. Currently, cochlear implants produce only a crude approximation of hearing, but the technology is improving. It's best to protect your sense of hearing by avoiding exceptionally loud noises, wearing earplugs when such situations cannot be avoided, and paying attention to bodily warnings of possible hearing loss, including a change in your normal hearing threshold and *tinnitus*, a whistling or ringing sensation in your ears.

For whom the bell tolls FIGURE 4.8

Stealthy teenagers now have a biological advantage over their teachers: a cell phone ringtone that sounds at 17 kilohertz—too high for adult ears to detect. The ringtone is an ironic offshoot of another device using the same sound frequency. That invention, dubbed the Mosquito, was designed to help shopkeepers annoy and deter loitering teens.

How loud is too loud? FIGURE 4.9

The loudness of a sound is measured in *decibels* and the higher a sound's decibel reading, the more damaging it is to the ear. Chronic exposure to loud noise, such as loud music or heavy traffic—or brief exposure to really loud sounds, such as a stereo at full blast, a jackhammer, or a jet engine—can cause permanent nerve deafness. Disease and biological changes associated with aging can also cause nerve deafness.

CONCEPT CHECK STOP

What are the rods and cones responsible for?

What are the three major parts of the ear?

Why do people commonly need reading glasses beginning in middle age?

What causes deafness?

Our Other Senses

LEARNING OBJECTIVES

Explain the importance of smell and taste to survival.

Describe how the information contained in odor molecules reaches the brain.

Identify the locations of receptors for the body senses.

Explain the role of our vestibular and kinesthetic senses.

V ision and audition may be the most prominent of our senses, but the others—taste, smell, and the body senses—are also important for gathering information about our environment.

SMELL AND TASTE: SENSING CHEMICALS

Smell and taste are sometimes referred to as the *chemical senses* because they both involve chemoreceptors that are

NATIONAL GEOGRAPHIC

Process Diagram

How the nose smells FIGURE 4.10

Olfactory bulb

Sense stimulation

3 Each odorous chemical appears to excite a specific portion of the olfactory bulb and is coded according to the stimulated area (Dalton, 2002).

Frontal lobe

2 The molecules initiate a neural impulse that travels to the *olfactory bulb*, where most olfactory information is processed before being sent to other parts of the brain.

Bone

Olfactory bulb

Nasal cavity

Olfactory epithelium

1 Odor molecules stimulate the dendrites of receptors embedded in the *olfactory epithelium*, a mucus-coated membrane in the nasal cavity.

Olfactory receptor cell

sensitive to certain chemical molecules. Smell and taste receptors are located near each other and often interact so closely that we have difficulty separating the sensations.

Our sense of smell, **olfaction**, is remarkably useful and sensitive. We possess more than 1,000 types of olfactory receptors, allowing us to detect more than 10,000 distinct smells (**FIGURE 4.10**). The nose is more sensitive to smoke than any electronic detector, and—through practice—blind people can quickly recognize others by their unique odors.

Some research on pheromones—compounds found in natural body scents that may affect the behavior of others, including their sexual behavior—supports the idea that these chemical odors increase sexual behaviors in humans (McClintock, Zelano, & Ober, 2002; Pierce, Cohen, & Ulrich, 2004; Thornhill et al., 2003). However, other findings question the results (Hays, 2003), suggesting that human sexuality is far more complex than that of other animals—and more so than perfume advertisements would have you believe.

Today, the sense of taste, **gustation**, may be the least critical of our senses. In the past, however, it probably contributed to our survival. The major function of taste, aided by smell, is to help us avoid eating or drinking harmful substances. Because many plants that taste bitter contain toxic chemicals, an animal is more likely to survive if it avoids bitter-tasting plants (Cooper et al., 2002; Guinard et al., 1996). Humans and other animals have a preference for sweet foods, which are generally nonpoisonous and are good sources of energy. Children's taste buds are replaced more quickly than adults', so they often dislike foods with strong or unusual tastes. Many food and taste preferences are learned from childhood experiences and cultural influences, so that one person's delicacy can be a source of revulsion to others.

When we take away the sense of smell, there are five distinct tastes: sweet, sour, salty, bitter, and umami. Umami means "delicious" or "savory" and refers to sensitivity to an amino acid called glutamate. (Damak et al., 2003; Nelson et al., 2002). Glutamate is found in meats, meat broths, and monosodium glutamate (MSG).

Taste receptors respond differentially to food molecules of different shapes. The major taste receptors (*taste buds*) are clustered on our tongues within little bumps called *papillae* (**FIGURE 4.11**).

Taste sensation FIGURE 4.11

When we eat and drink, liquids and dissolved foods flow over bumps on our tongue called *papillae* and into the pores to the *taste buds*, which contain the receptors for taste.

Surface of the tongue

Cross section of a papilla

Taste bud

Taste pore

Taste receptors

Nerve

A taste bud

THE BODY SENSES: MORE THAN JUST TOUCH

The senses that tell the brain how the body is oriented, where and how the body is moving, and what it touches or is touched by are called the body senses. They include the **skin senses**, the **vestibular sense**, and **kinesthesia**.

Our skin is sensitive to touch (or pressure), temperature, and pain (**FIGURE 4.12A**). The concentration and depth of the receptors for each of these stimuli varies. For example, touch receptors are most concentrated on the face and fingers and least so in the back and legs. Some receptors respond to more than one type of stimulation—for example, itching, tickling, and vibrating sensations seem to be produced by light stimulation of both pressure and pain receptors. We have more

The body senses FIGURE 4.12

A The skin senses are vital. Not only do they protect our internal organs, but they provide our brains with basic attachment and survival information. Both humans and non-human animals are highly responsive to touch.

C Without her finely tuned kinesthetic sense to provide information about her bodily posture, orientation, and movement, U.S. gymnast Amy Chow would be on her way to the hospital rather than the Olympics.

B Part of the "thrill" of amusement park rides comes from our vestibular sense becoming confused. The vestibular sense is used by the eye muscles to maintain visual fixation and sometimes by the body to change body orientation. We can become dizzy or nauseated if the vestibular sense becomes "confused" by boat, airplane, or automobile motion. Children between ages 2 and 12 years have the greatest susceptibility to motion sickness.

skin receptors for cold than for warmth, and we don't seem to have any "hot" receptors at all. Instead, our cold receptors detect not only coolness but also extreme temperatures—both hot and cold (Craig & Bushnell, 1994).

The vestibular sense is responsible for balance—it informs the brain of how the body, and particularly the head, is oriented with respect to gravity and three-dimensional space. When the head moves, liquid in the *semicircular canals*, located in the inner ear, moves and bends hair cell receptors. At the end of the semicircular canals are the *vestibular sacs*, which contain hair cells sensitive to the specific angle of the head—straight up and down or tilted. Information from the semicircular canals and the *vestibular sacs* is converted to neural impulses that are then carried to the appropriate section of the brain (FIGURE 4.12B).

Kinesthesia is the sense that provides the brain with information about bodily posture, orientation, and

movement. Kinesthetic receptors are found throughout the muscles, joints, and tendons of the body. They tell the brain which muscles are being contracted or relaxed, how our body weight is distributed, where our arms and legs are in relation to the rest of our body, and so on (FIGURE 4.12C).

CONCEPT CHECK STOP

Why are our senses of smell and taste called the chemical senses?

How do the vestibular and kinesthetic senses differ?

What are the three basic skin sensations?

Understanding Perception

Document the relationship between selective attention, feature detectors, and habituation.

Summarize the factors involved in perceptual interpretation.

Describe the limitations of subliminal perception.

We are ready to move from sensation and the major senses to perception, the process of selecting, organizing, and interpreting incoming sensations into useful mental representations of the world.

SELECTION: EXTRACTING IMPORTANT MESSAGES

In almost every situation, we confront more sensory information than we can reasonably pay attention to. Three major factors help us focus on some stimuli and ignore others: **selective attention** (FIGURE 4.13), **feature detectors** (FIGURE 4.14), and **habituation** (FIGURE 4.15), all on the following page.

Certain basic mechanisms for perceptual selection are built into the brain. For example, through the process of selective attention (see Figure 4.13), the brain picks out the information that is important to us and discards the rest (Folk & Remington, 1998; Kramer et al., 2000).

In humans and other animals, the brain contains specialized cells, called feature detectors, that respond only to certain sensory information (see Figure 4.14). For example, humans have feature detectors in the temporal and occipital lobes that respond maximally to faces. Interestingly, people with a condition called prosopagnosia (*prospon* means "face" and *agnosia* means "failure to know") can recognize that they are looking at a face, but they cannot say whose face is reflected in a mirror, even if it is their own or that of a friend or relative.

Other examples of the brain's ability to filter

selective attention Filtering out and attending only to important sensory messages.

feature detectors Specialized brain cells that respond only to certain sensory information.

habituation Tendency of the brain to ignore environmental factors that remain constant.

With Your Own Eyes

Illusions

Normally, our perceptions agree with our sensations. When they do not, the result is an illusion—a false impression produced by errors in the perceptual process or by actual physical distortions, as in desert mirages. Illusions provide psychologists with a tool for studying the normal process of perception. For example, this drawing illustrates the *Müller–Lyer illusion.* The two vertical lines are the same length, but psychologists have learned that people who live in urban environments normally see the one on the right as longer. This is because they have learned to make size and distance judgments from perspective cues created by right angles and horizontal and vertical lines of buildings and streets.

Selective attention
FIGURE 4.13

When you are in a group of people, surrounded by various conversations, you can still select and attend to the voices of people you find interesting. Another example of selective attention occurs with the well-known "cocktail party phenomenon." Have you noticed how you can suddenly pick up on another group's conversation if someone in that group mentions your name?

Feature detectors FIGURE 4.14

Cats possess cells, known as feature detectors, that respond to specific lines and angles (Hubel & Wiesel, 1965, 1979). Researchers found that kittens reared in a vertical world fail to develop their innate ability to detect horizontal lines or objects. On the other hand, kittens restricted to only horizontal lines cannot detect vertical lines. A certain amount of interaction with the environment is apparently necessary for feature detector cells to develop normally (Crair, Gillespie, & Stryker, 1998).

Habituation FIGURE 4.15

These three girls' brains may "choose to ignore" their painful braces. (Sensory adaptation may have also occurred. Over time the girls' pressure sensors send fewer messages to the brain.)

experience are evidenced by habituation. Apparently, the brain is "prewired" to pay more attention to changes in the environment than to stimuli that remain constant. For example, when braces are first applied or when they are tightened, they can be very painful. After a while, however, awareness of the pain diminishes (see Figure 4.15).

As advertisers and political operatives well know, people tend to automatically select stimuli that are intense, novel, moving, contrasting, and repetitious. For sheer volume of sales (or votes), the question of whether you like the ad is irrelevant. If it gets your attention, that's all that matters.

ORGANIZATION: FORM, CONSTANCY, DEPTH, AND COLOR

Raw sensory data are like the parts of a watch—they must be assembled in a meaningful way before they are useful. We organize sensory data in terms of form, constancy, depth, and color.

Form perception   Gestalt psychologists were among the first to study how the brain organizes sensory impressions into a **gestalt**—a German word meaning "form" or "whole." They emphasized the importance of organization and patterning in enabling us to perceive the whole stimulus rather than perceiving its discrete parts as separate entities. The Gestaltists proposed several laws of organization that specify how people perceive form (**FIGURE 4.16**).

The most fundamental Gestalt principle of organization is our tendency to distinguish between

Figure-Ground: The ground is always seen as farther away than the figure.

Proximity: Objects that are physically close together are grouped together. (In this figure, we see 3 groups of 6 hearts, not 18 separate hearts.)

Continuity: Objects that continue a pattern are grouped together.

When we see this,

we normally see this

plus this.

Not this.

Closure: The tendency to see a finished unit (triangle, square, or circle) from an incomplete stimulus.

Similarity: Similar objects are grouped together (the green colored dots are grouped together and perceived as the number 5).

Gestalt principles of organization FIGURE 4.16

Figure-ground, proximity, continuity, closure, and similarity are shown here, but the Gestalt principle of contiguity cannot be shown because it involves nearness in time, not visual nearness. Although the examples of the Gestalt principles in this figure are visual, each principle applies to other modes of perception as well. You probably have experienced aural figure and ground effects, for example, at a movie or a concert, when there was a conversation going on close by during the performance. You probably experienced the difficulty of sorting out what sounds were the background and what you wanted to be your focus. It was probably especially difficult if the conversation was interesting or the performance was uninteresting or hard to hear.

A good example of visual closure happens every time you watch television. The picture on the TV screen appears to be a solid image, but it's really a very fast stream of small dots being illuminated one by one, "painting" tiny horizontal lines down the screen one line at a time. Your brain closes the momentary blank gaps on the screen.

www.wiley.com/college/carpenter

Reversible and impossible figures FIGURE 4.17

A This so-called *reversible figure* demonstrates alternating figure–ground relations. It can be seen as a woman looking in a mirror or as a skull, depending on what you see as figure or ground.

B When you first glance at the famous painting by the Dutch artist M. C. Escher, you detect specific features of the stimuli and judge them as sensible figures. But as you try to sort and organize the different elements into a stable, well-organized whole, you realize they don't add up—they're impossible. There is no one-to-one correspondence between your actual sensory input and your final perception.

figure (our main focus of attention) and ground (the background or surroundings).

Your sense of figure and ground is at work in what you are doing right now—reading. Your brain is receiving sensations of black lines and white paper, but your brain is organizing these sensations into black letters and words on a white background. You perceive the letters as the figure and the white as the ground. If you make a great effort, you might be able to force yourself to see the page reversed, as though a black background were showing through letter-shaped holes in a white foreground. There are times, however, when it is very hard to distinguish the figure from the ground as can be seen in **FIGURE 4.17A**. This is known as a "reversible figure." Your brain alternates between seeing the light areas as the figure and as the ground.

Like *reversible figures*, **impossible figures** help us understand perceptual principles—in this case, the principle of form organization (**FIGURE 4.17B**).

Depth perception   In our three-dimensional world, the ability to perceive the depth and distance of objects—as well as their height and width—is essential. We usually rely most heavily on vision to perceive distance and depth.

Depth perception is primarily learned through experience. However, research using an apparatus called the *visual cliff* (**FIGURE 4.18**) suggests that some depth perception is inborn.

One mechanism by which we perceive depth is the interaction of both eyes to produce **binocular cues**; the other involves **monocular cues**, which work with each eye separately.

One of the most important binocular cues for depth perception comes from **retinal disparity** (**FIGURE 4.19**). Because our eyes are about two and one-half inches apart, each retina receives a slightly different view of the world. (Watch what happens when you point at a distant object, closing one eye and then the other.) When both eyes are open, the brain fuses the different images into one, an effect known as *stereoscopic vision*.

As we move closer to an object, a second binocular cue, **convergence**, helps us judge depth. The closer the object, the more our eyes are turned inward. The

Visual cliff FIGURE 4.18

Crawling infants hesitate or refuse to move to the "deep end" of the visual cliff (Gibson & Walk, 1960), indicating that they perceive the difference in depth. (The same is true for baby animals that walk almost immediately after birth.) Even two-month-old infants show a change in heart rate when placed on the deep versus shallow side of the visual cliff (Banks & Salapatek, 1983).

Glass only

Shallow side

Floor as seen through the glass

resulting amount of eye-muscle strain helps the brain interpret distance.

The binocular (two eyes) cues of retinal disparity and convergence are inadequate in judging distances longer than a football field. Luckily, we have several monocular (one eye) cues available separately to each eye. See if you can identify each in the photo of the Taj Mahal in India, in FIGURE 4.20 on page 103.

Two additional monocular cues are **accommodation** of the lens (discussed earlier) and **motion parallax**. In accommodation, muscles that adjust the shape of the lens as it focuses on an object send neural impulses to the brain, which interprets the signal to perceive distance. Motion parallax refers to the fact that when we are moving, close objects appear to whiz by whereas farther objects seem to move more slowly or

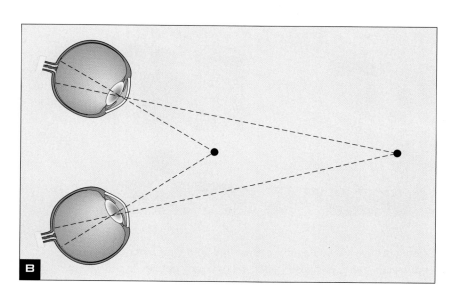

A

B

Retinal disparity FIGURE 4.19

A Stare at your two index fingers a few inches in front of your eyes with their tips an inch apart. Do you see the "floating finger"? Move it farther away and the "fin-ger" will shrink. Move it closer and it will enlarge. B Because of retinal disparity, objects at different distances (such as the "floating finger") project their images on different parts of the retina. Far objects project on the retinal area near the nose, whereas near objects project farther out, closer to the ears.

Perceptual Constancies

As noted earlier with sensory adaptation and habituation, we are particularly alert to change in our environment. However, we also are attuned to consistencies. Without this **perceptual constancy**, our world would be totally chaotic. We learn about perceptual constancies through prior experience and learning. Although the more distant zebras in this photograph appear much smaller than those nearby, we perceive them as similar in size because of *size constancy*. According to this principle, the perceived size of an object remains the same even though the size of its retinal image changes. Similarly, *shape constancy* allows us to perceive them as being the same kind of animal, even when they are turned in different orientations. Finally, although some animals are in shade and some in the sun, we know they are the same color because of *color* and *brightness constancy*.

Notice, however, how easily these constancies can be manipulated. This photograph illustrates the **Ames room illusion**. In the photo, the young boy on the right appears to be much larger than the woman on the left. How can this be?

Scientists have manipulated the room and your view of it, and your perceptual constancy has falsely filled in the wrong details. To the viewer, peering through the peephole, the room appears normal. But in this specially constructed room, the trapezoidal shape and sloping ceilings and floors provide misleading depth cues. Because our brains mistakenly assume the two people are the same distance away, we compensate for the apparent size difference by making the person on the left appear much smaller.

Monocular cues FIGURE 4.20

Linear perspective results when parallel lines converge in the distance.

Interposition occurs when a close object obscures part of a distant object.

Relative size is the result of close objects projecting a larger retinal image than distant objects.

Texture gradient results from the change in perceived texture with changes in distance.

Aerial perspective means that faraway objects look blurred compared with near objects because of intervening particles of dust or haze.

Light and shadow occurs when brighter objects are perceived as closer than darker objects.

remain stationary. This effect can easily be seen when traveling by car or train.

Color perception

Our color vision is as remarkable as our ability to perceive depth and distance. Humans may be able to discriminate among seven million different hues, and research conducted in many cultures suggests that we all seem to see essentially the same colored world (Davies, 1998). Furthermore, studies of infants old enough to focus and move their eyes show that they are able to see color nearly as well as adults (Knoblauch, Vital-Durand, & Barbur, 2000; Werner & Wooten, 1979).

Although we know color is produced by different wavelengths of light, the actual way in which we perceive color is a matter of scientific debate. Traditionally, there have been two theories of color vision, the trichromatic (three-color theory) and the opponent-process theory. The **trichromatic theory** (from the Greek word *tri*—meaning "three," and *chroma*—meaning "color") was first proposed by Thomas Young in the early nineteenth century and was later refined by Herman von Helmholtz and others. Apparently, we have three "color systems," as they called them—one system that is maximally sensitive to red, another maximally sensitive to green, and another maximally sensitive to blue (Young, 1802). The proponents of this theory demonstrated that mixing lights of these three colors could yield the full spectrum of colors we perceive. Unfortunately this theory has its flaws. One is that it doesn't explain *color after effects*, a phenomenon you can experience in With Your Own Eyes on page 106.

The **opponent-process theory**, proposed by Ewald Hering later in the nineteenth century, also suggested the three color systems, but he suggested that

trichromatic theory Theory that color perception results from mixing three distinct color systems—red, green, and blue.

opponent-process theory Theory that color perception is based on three systems of color receptors, each of which responds in an on-off fashion to opposite-color stimuli: blue-yellow, red-green, and black-white.

Color Aftereffects

Try staring at the dot in the middle of this color-distorted American flag for 60 seconds. Then stare at a plain sheet of white paper. You should get interesting color aftereffects—red in place of green, blue in place of yellow, and white in place of black: a "genuine" American flag.

What happened? This is a good example of the *opponent-process theory*. As you stared at the figure, the green stripes stimulated only the green channel of the red-green opponent color cells. Continuous stimulation fatigued the green channel, whereas the nonstimulated red channel was not fatigued. When you looked at the blank piece of paper, the white stimulated both the red and the green channels equally. Because the green channel was fatigued, the red channel fired at a higher rate and you therefore saw a red color aftereffect. Without first staring at the white dot, the red and green receptors would have canceled each other out and you would have seen white when you looked at the white paper.

each system is sensitive to two opposing colors—blue and yellow, red and green, black and white—in an "on-off" fashion. In other words, each color receptor responds either to blue or yellow or to red or green, with the black-or-white systems responding to differences in brightness levels. This theory makes a lot of sense because when different-colored lights are combined, people are unable to see reddish greens and bluish yellows. In fact, when red and green lights or blue and yellow lights are mixed in equal amounts, we see white.

In 1964, research by George Wald and colleague Paul Brown showed we do have three kinds of cones in the retina and confirmed the central proposition of the trichromatic theory. At nearly the same time, R. L. De-Valois (1965) was studying electrophysiological recording of cells in the optic nerve and optic pathways to the brain. He discovered that cells respond to color in an opponent fashion in the thalamus. The findings reconciled the once-competing trichromatic and opponent-process theories. Now, we know that color is processed in a trichromatic fashion at the level of the cones, in the retina, and in an opponent fashion at the level of the optic nerve and the thalamus, in the brain.

Color-deficient vision

Most people perceive three different colors—red, green, and blue—and are called *trichromats*. However, a small percentage of the population has a genetic deficiency in either the red-green system, the blue-yellow system, or both. Those who perceive only two colors are called *dichromats*. People who are sensitive to only the black-white system are called *monochromats*, and they are totally color blind. If you'd like to test yourself for red-green color blindness, see FIGURE 4.21.

INTERPRETATION: EXPLAINING OUR PERCEPTIONS

After selectively sorting through incoming sensory information and organizing it, the brain uses this information to explain and make judgments about the external world. This final stage of perception—interpretation—is influenced by several factors, including perceptual adaptation, perceptual set, frame of reference, and bottom-up or top-down processing.

Do you remember the upside-down photo and the discussion about the George Stratton (1897) experiment in the chapter opener? Stratton's experiment illustrates the critical role that **perceptual adaptation** plays in how we interpret the information that our brains gather. Without his ability to adapt his perceptions to a skewed environment, Stratton would not have been able to function. His brain's ability to "retrain" it-

Color-deficient vision FIGURE 4.21

Are you color blind? People who suffer red-green deficiency have trouble perceiving the number within this design. Although we commonly use the term "color blindness," most problems are color confusion rather than color blindness. Furthermore, most people who have some color blindness are not even aware of it.

self to his new surroundings allowed him to create coherence out of what would otherwise have been chaos.

Our previous experiences, assumptions, and expectations also affect how we interpret and perceive the world, by creating a **perceptual set**, or a readiness to perceive in a particular manner, based on expectations—in other words, we largely see what we expect to see (**FIGURE 4.22**). In one study involving participants who were members of a Jewish organization

Perceptual set? FIGURE 4.22

When you look at this drawing, do you see a young woman looking back over her shoulder, or an older woman with her chin buried in a fur collar? It may depend on your age. Younger students tend to first see a young woman and older students first see an older woman. Although basic sensory input stays the same, your brain's attempt to interpret ambiguous stimuli creates a type of perceptual dance, shifting from one interpretation to another (Gaetz et al., 1998).

Subliminal Perception

Is the public under siege by sneaky advertisers and politicians lobbing subliminal (literally, "below the threshold") messages that can undermine our intentions? Can you lose weight, stop smoking, or relieve stress by listening to "subliminal tapes" promising to solve your problems without your having to pay attention?

It is, in fact, possible to perceive something without conscious awareness (Luecken, Tartaro, & Appelhans, 2004; Núñez & de Vicente, 2004; Todorov & Bargh, 2002). For example, in one study the experimenter very briefly flashed one of two pictures subliminally (either a happy or an angry face) followed by a neutral face.

They found this subliminal presentation evoked matching unconscious facial expressions in the participants' own facial muscles (Dimberg, Thunberg, & Elmehed, 2000).

Despite this and much other evidence that *subliminal perception* occurs, that does not mean that such processes lead to *subliminal persuasion*. Subliminal stimuli are basically weak stimuli. At best, they have a modest (if any) effect on consumer behavior and absolutely no effect on citizens' voting behavior (Begg, Needham, & Bookbinder, 1993; Trappey, 1996). As for subliminal self-help tapes, you're better off with old-fashioned, conscious methods of self-improvement.

When first learning to read, you used bottom-up processing. You initially learned that certain arrangements of lines and "squiggles" represented specific letters. You later realized that these letters make up words.

Now, yuor aiblity to raed uisng top-dwon prcessoing mkaes it psosible to unedrstnad thsi sntenece desipte its mnay mssipllengis.

Bottom-up, top-down FIGURE 4.23

(Erdelyi & Applebaum, 1973), collections of symbols were briefly flashed on a screen. When a swastika was at the center, the Jewish participants were less likely to recognize and remember the other symbols. Can you see how the life experiences of the Jewish subjects led them to create a perceptual set for the swastika?

How we perceive people, objects, or situations is also affected by the **frame of reference**, or context. An elephant is perceived as much larger when it is next to a mouse than when it stands next to a giraffe.

Finally, recall that we began this chapter by discussing how we receive sensory information (sensation) and work our way upward to the top levels of perceptual processing (perception). Psychologists refer to this type of information processing as **bottom-up processing**. In contrast, **top-down processing** begins with "higher," "top"-level processing involving thoughts, previous experiences, expectations, language, and cultural background and works down to the sensory level (**FIGURE 4.23**).

Science and ESP

So far in this chapter, we have talked about sensations provided by our eyes, ears, nose, mouth, and skin. What about a so-called sixth sense? Can some people perceive things that cannot be perceived with the usual sensory channels, by using **extrasensory perception (ESP)**? People who claim to have ESP profess to be able to read other people's minds (telepathy), perceive objects or events that are inacces-

sible to their normal senses (clairvoyance), predict the future (precognition), or move or affect objects without touching them (psychokinesis).

Scientific investigations of ESP began in the early 1900s, and some work on the subject continues today. The most important criticism of both experimental and casual claims of ESP is their lack of stability and replicability—a core requirement for scientific acceptance (Hyman, 1996). A meta-analysis of 30 studies using strong scientific controls reported absolutely no evidence of ESP (Milton & Wiseman, 1999, 2001).

So why do so many people believe in ESP? One reason is that, as mentioned earlier in the chapter, our motivations and interests often influence our perceptions, driving us to selectively attend to things we want to see or hear. In addition, the subject of extrasensory perception (ESP) often generates strong emotional responses. When individuals feel strongly about an issue, they sometimes fail to recognize the faulty reasoning underlying their beliefs.

Belief in ESP is particularly associated with illogical or noncritical thinking. For example, people often fall victim to the *fallacy of positive instances,* noting and remembering events that confirm personal expectations and beliefs and ignoring nonsupportive evidence. Other times, people fail to recognize chance occurrences for what they are. Finally, human information processing often biases us to notice and remember the most vivid information—such as a detailed (and spooky) anecdote or a heartfelt personal testimonial.

CONCEPT CHECK STOP

Why do we experience perceptual illusions?

What are the processes that allow us to pay attention to some stimuli in our environments and ignore others?

What kinds of cues do we use to perceive depth and distance?

What factors entice some people to believe in ESP?

This man willingly endures what would normally be excruciating pain. What psychological and biological factors might make this possible for him?

Do you think this man would feel more pain, or less, if his friends and family members were frequently and solicitously asking how he was feeling?

VISUAL SUMMARY

1 Understanding Sensation

1. **Sensation** is the process by which we detect stimuli and convert them into neural signals (**transduction**). During **coding**, the neural impulses generated by different physical stimuli travel by separate routes and arrive at different parts of the brain. In **sensory reduction**, we filter and analyze incoming sensations.

2. In **sensory adaptation**, sensory receptors fire less frequently with repeated stimulation, so that over time, sensation decreases.

3. The absolute threshold is the smallest amount of a stimulus needed to detect a stimulus, and the difference threshold, or just noticeable difference (JND), is the smallest change in stimulus intensity that a person can detect.

4. According to the **gate-control theory**, our experience of pain depends partly on whether the neural message gets past a "gatekeeper" in the spinal cord, which researchers believe is chemically controlled.

❸ When you look at a faraway object, your *lens* thins and flattens; when your glance shifts to a near object, your lens thickens and curves.

Lens

Vitreous humor

Lid

Retina

❹ In the *retina*, light waves are changed into neural impulses that move along the optic nerve to the brain.

Cornea

Sclera

❷ The *cornea's* outward curvature helps focus incoming light rays.

Fovea

Light

❶ Muscles in the *iris* allow the *pupil* to dilate or constrict in response to light intensity or emotional factors.

Pupil

Iris

Optic nerve

Blind spot

Cone

❼ *Cones* are sensitive to many wavelengths, but each is maximally sensitive to red, green, or blue.

Bipolar neurons

Ganglion cells

Rod

❽ *Rods* detect shades of gray and are responsible for peripheral vision; they are most sensitive in dim light.

Fovea

❻ The *fovea*, a tiny pit filled with cones, responsible for our sharpest vision.

Light

Light

Blind spot

Optic nerve (to brain)

❺ At the back of the retina lies an area that has no visual receptors at all and absolutely no vision. This *blind spot* is where blood vessels and nerve pathways enter and exit the eyeball.

Light

Fovea

2 How We See and Hear

1. Light and sound move in waves. Light waves are a form of electromagnetic energy, and sound waves are produced when air molecules move in a particular wave pattern. Both light waves and sound waves vary in length, height, and range.

2. Light enters the eye at the front of the eyeball. The cornea protects the eye and helps focus light rays. The iris provides the eye's color, and muscles in the iris dilate or constrict the pupil. The lens further focuses light, adjusting to allow focusing on objects at different distances. At the back of the eye, incoming light waves reach the retina, which contains light-sensitive rods and cones. A network of neurons in the retina transmits neural information to the brain.

3. The outer ear gathers sound waves; the middle ear amplifies and concentrates the sounds; and the inner ear changes sounds' mechanical energy into neural impulses. Sounds' frequency and intensity determine how we distinguish among sounds of different pitches and loudness, respectively.

3 Our Other Senses

1. Smell and taste, sometimes called the chemical senses, involve chemoreceptors that are sensitive to certain chemical molecules. In olfaction, odor molecules stimulate receptors in the olfactory epithelium, in the nose. The resulting neural impulse travels to the olfactory bulb, where the information is processed before being sent elsewhere in the brain. Our sense of taste (gustation) involves five tastes: sweet, sour, salty, bitter, and umami (which means "savory" or "delicious"). The taste buds are clustered on our tongues within the papillae.

2. The body senses tell the brain how the body is oriented, where and how it is moving, and what it touches. They include the skin senses, the vestibular sense, and kinesthesia.

Surface of the tongue

Cross section of a papilla

Taste bud

Taste pore

Taste receptors

Nerve

A taste bud

4 Understanding Perception

1. Perception is the process of selecting, organizing, and interpreting incoming sensations into useful mental representations of the world. **Selective attention** allows us to filter out unimportant sensory messages. **Feature detectors** are specialized cells that respond only to certain sensory information. **Habituation** is the tendency to ignore stimuli that remain constant. People tend to automatically select stimuli that are intense, novel, moving, contrasting, and repetitious.

2. To be useful, sensory data must be assembled in a meaningful way. We organize sensory data in terms of form, constancy, depth, and color. Traditionally there have been two theories of color vision: the **trichromatic theory** and the **opponent-process theory**.

3. Perceptual adaptation, perceptual set, frame of reference, and bottom-up versus top-down processing affect our interpretation of what we sense and perceive. Subliminal stimuli, although perceivable, have a modest effect on behavior.

KEY TERMS

- sensation p. 86
- perception p. 86
- synesthesia p. 88
- transduction p. 88
- coding p. 88
- sensory reduction p. 88
- sensory adaptation p. 89
- psychophysics p. 89
- gate-control theory p. 90
- wavelength p. 92
- amplitude p. 92
- range p. 92
- nearsightedness p. 92

- farsightedness p. 92
- dark adaptation p. 92
- audition p. 92
- frequency p. 94
- place theory p. 94
- frequency theory p. 94
- conduction deafness p. 94
- nerve deafness p. 95
- cochlear implant p. 95
- olfaction p. 97
- gustation p. 97
- skin senses p. 97
- vestibular sense p. 97

- kinesthesia p. 97
- selective attention p. 99
- feature detectors p. 99
- habituation p. 99
- gestalt p. 101
- impossible figures p. 101
- depth perception p. 101
- binocular cues p. 101
- monocular cues p. 101
- retinal disparity p. 101
- convergence p. 101
- accommodation p. 103
- motion parallax p. 103

- perceptual constancy p. 104
- Ames room illusion p. 104
- trichromatic theory p. 105
- opponent-process theory p. 105
- perceptual adaptation p. 106
- perceptual set p. 107
- frame of reference p. 108
- bottom-up processing p. 108
- top-down processing p. 108
- extrasensory perception (ESP) p. 108

CRITICAL AND CREATIVE THINKING QUESTIONS

1. Sensation and perception are closely linked. What is the central distinction between the two?

2. If we sensed and attended equally to each stimulus in the world, the amount of information would be overwhelming. What sensory and perceptual processes help us lessen the din?

3. If we don't adapt to pain, why is it that people can sometimes "tune out" painful injuries?

4. What senses would likely be impaired if a person were somehow missing all of the apparatus of the ear (including the outer, middle, and inner ear)?

5. Can you explain how your own perceptual sets might create prejudice or discrimination?

SELF-TEST: SENSATION AND PERCEPTION

(Check your answers in Appendix B.)

1. Transduction is the process of converting _____ .

   a. neural impulses into mental representations of the world
   b. receptors into transmitters
   c. a physical stimulus into neural impulses
   d. receptors into neural impulses

2. Sensory reduction refers to the process of _____ .

   a. reducing your dependence on a single sensory system
   b. decreasing the number of sensory receptors that are stimulated
   c. filtering and analyzing incoming sensations before sending a neural message to the cortex
   d. reducing environmental sensations by physically preventing your sensory organs from seeing, hearing, etc.

3. HOW THE EYE SEES: Identify the parts of the eye, placing the appropriate label on the figure below:

cornea	rod
iris	cone
pupil	fovea
lens	blind spot
retina	

Light

4. A visual acuity problem that occurs when the cornea and lens focus an image in front of the retina is called _____ .

   a. farsightedness
   b. hyperopia
   c. myopia
   d. presbyopia

5. HOW THE EAR HEARS: Identify the parts of the ear, placing the appropriate label on the figure below:

pinna	stapes
tympanic membrane	oval window
malleus	cochlea
incus	

Sound waves

6. Chronic exposure to loud noise can cause permanent _____ .

   a. auditory illusions
   b. auditory hallucinations
   c. nerve deafness
   d. conduction deafness

7. Most information related to smell is processed in the _____ .

   a. nasal cavity
   b. temporal lobe
   c. olfactory bulb
   d. parietal lobe

8. Most of our taste receptors are found on the _____ .

   a. papillae
   b. gustatory cells
   c. taste buds
   d. frenulum

9. Identify which of these photos, 1 or 2, illustrates:

   a. vestibular sense: photo _____

   b. kinesthetic sense: photo _____

10. When the brain is sorting out and attending only to the most important messages from the senses, it is engaged in the process of _____ .

   a. sensory adaptation

   b. sensory habituation

   c. selective attention

   d. selective sorting

11. In a(n) _____ , the discrepancy between figure and ground is too vague and you may have difficulty perceiving which is figure and which is ground.

   a. illusion

   b. reversible figure

   c. optical illusion

   d. hallucination

12. The tendency for the environment to be perceived as remaining the same even with changes in sensory input is called

_____ .

   a. perceptual constancy

   b. the constancy of expectation

   c. an illusory correlation

   d. Gestalt's primary principle

13. The theory of color vision proposed by Thomas Young that says color perception results from mixing three distinct color systems is called the _____ .

   a. tricolor theory

   b. trichromatic theory

   c. tripigment theory

   d. opponent-process theory

14. The illustration below is an example of _____ .

   a. top-down processing

   b. frame of reference

   c. subliminal persuasion

   d. perceptual adaptation

15. A readiness to perceive in a particular manner is known as

_____ .

   a. sensory adaptation

   b. perceptual set

   c. habituation

   d. frame of reference

States of Consciousness

5

In December 1995, Jean-Dominique Bauby, the editor of French *Elle* magazine, suffered a massive stroke that left him almost completely paralyzed with a rare condition known as "locked-in syndrome." He was unable to speak or move his hands to write (or, for that matter, to even breathe unassisted or to swallow). He could communicate only by blinking his left eye. Yet Bauby remained alert and aware throughout his devastating ordeal—so much so that he was able to dictate a memoir on his experience, *The Diving Bell & the Butterfly*, by "winking" the text to a transcriptionist, one letter at a time.

What if Bauby had not been able to open or move his left eye? If he couldn't communicate, would he still be "conscious"? People commonly use the term consciousness, but what exactly does it mean? Is it simple awareness? What would it be like to be unaware? How can we study the contents of our consciousness when the only tool of discovery is the object itself?

In the late nineteenth century, when psychology first became a scientific discipline separate from philosophy, it defined itself as "the study of human consciousness." But the nebulousness of the area of study eventually led to great dissatisfaction within the field. One group, the behaviorists, led by John Watson, believed that behavior, not consciousness, was the proper focus of the new science.

More recently, psychology has renewed its original interest in consciousness, thanks to research in cognitive and cultural psychology. In addition, technological advances allow scientific study of brain activity during various alternate states of consciousness (ASCs). ASCs include sleep and dreaming, chemically induced changes from psychoactive drugs, daydreaming, fantasies, hypnosis, fasting, meditation, and even the so-called runner's high.

In this chapter, we begin with a general look at the definition and description of consciousness. Then we examine how consciousness changes because of circadian rhythms, sleep, and dreams. We also look at psychoactive drugs and their effects on consciousness. Finally, we explore alternative routes to altered consciousness, such as meditation and hypnosis.

Consciousness, Sleep, and Dreaming

LEARNING OBJECTIVES

Explain the difference between controlled and automatic processes.

Describe the effects of sleep deprivation and disruption of circadian rhythms.

Review the stages of sleep.

Compare and contrast the theories of why we sleep and dream.

Summarize the types of sleep disorders.

illiam James, the first American psychologist, likened **consciousness** to a stream that's constantly changing yet always the same. It meanders and flows, sometimes where the person wills and sometimes not. However, through the process of selective attention (Chapter 4), we can control our consciousness by deliberate concentration and full attention. For example, at the present moment you are, hopefully, awake and concentrating on the words on this page. At times, however, your control may weaken, and your stream of consciousness may drift to thoughts of a computer you want to buy, a job, or an attractive classmate.

In addition to meandering and flowing, your "stream of consciousness" also varies in depth. Consciousness is not an all-or-nothing phenomenon—conscious or unconscious. Instead, it exists along a continuum. As you can see in **FIGURE 5.1**, this continuum extends from high awareness and sharp, focused alertness at one extreme, to middle levels of awareness such as daydreaming, to unconsciousness and coma at the other extreme.

Two of the more common states of consciousness are sleep and dreaming. You may think of yourself as being unconscious while you sleep, but that's not true. Rather, you are in an **altered state of consciousness (ASC)**. In this chapter, you will learn about this state of consciousness and why we spend so much time in it.

consciousness
An organism's awareness of its own self and surroundings (Damasio, 1999).

To understand sleep and dreaming, we need to first explore the topic of **circadian rhythms**. Most animals have adapted to our planet's cycle of days and nights by developing a pattern of bodily functions that wax and wane over each 24-hour period. Our alertness, core body temperature, moods, learning efficiency, blood pressure, metabolism, and pulse rate all follow these circadian rhythms (Ariznavarreta et al., 2002; Ice et al., 2004; Kunz & Hermann, 2000; Lauc et al., 2004). Usually, these activities reach their peak during the day and their low point at night (**FIGURE 5.2** on page 118).

Disruptions in circadian rhythms cause increased fatigue, decreased concentration, sleep disorders, and other health problems (Bovbjerg, 2003; Garbarino et al., 2002; Valdez, Ramírez & García, 2003; Yesavage et al., 2004). Although many physicians, nurses, police, and others who have rotating work schedules (about 20 percent of employees in the United States) manage to function well, studies do find that shift work and sleep deprivation lead to decreased concentration and productivity—and increased accidents (Connor et al., 2002; Dement & Vaughan, 1999; Garbarino et al., 2002). Research suggests that shifting from days to evenings to nights makes it easier to adjust to rotating shift schedules—probably

altered state of consciousness (ASC) Mental states found during sleep, dreaming, psychoactive drug use, hypnosis, and so on.

circadian [ser-KAY-dee-an] rhythms Biological changes that occur on a 24-hour cycle (in Latin, *circa* means about, and *dies* means day).

Our varying levels of awareness occur through an integration of many processes in most of the brain. We can control our consciousness through deliberate concentration and full attention. **Controlled processes** demand focused attention and generally interfere with other ongoing activities. But once a task is well learned, your brain stores the information as an **automatic process**, which requires minimal attention and generally does not interfere with other ongoing activities (Geary, 2005). (Controlled processing explains why learning to drive is complicated, but later on, thanks to automatic processing, you can drive to school or work without consciously thinking about it.)

(perception, memory, attention)

Parietal lobe

Frontal lobe (working memory, temporal integration, abstract thinking, introspection, cognitive flexibility)

Temporal lobe

Occipital lobe

Limbic system (emotional memory, emotions, relational learning, memory consolidation)

Brain stem (arousal)

LEVELS OF CONSCIOUSNESS

High Awareness

High Level of Awareness

Controlled processes (high level of awareness, focused attention required)

Middle Awareness

Middle Level of Awareness

Automatic processes (awareness, but minimal attention required)

Daydreaming (low level of awareness and conscious effort, somewhere between active consciousness and dreaming while asleep)

Low Awareness

Minimal or No Awareness

Unconscious mind (a Freudian concept discussed in Chapter 12 consisting of unacceptable thoughts, feelings, and memories too painful or anxiety provoking to be admitted to consciousness)

Unconscious (biologically based lowest level of awareness due to head injuries, disease, anesthesia during surgery, or coma)

A Degree of alertness

Time of Day

B Core body temperature

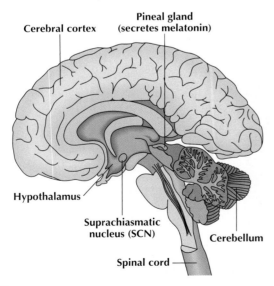

C The circadian "clock"

Like other animals, humans have adapted to our planet's rotation by developing a 24-hour day-night cycle. Variations in our degree of alertness (**A**) and core body temperature (**B**) are two prime examples of this daily cycle. **C** The "clock" that regulates our circadian rhythms is located in a part of the hypothalamus called the suprachiasmatic nucleus (SCN). The SCN also links to the pineal gland, which helps regulate sleep and arousal by secreting large quantities of the hormone melatonin at night, promoting relaxation and sleepiness.

because it's easier to go to bed later than the reverse. Productivity and safety also increase when shifts are rotated every three weeks versus every week.

Like shift work, flying across several time zones can also cause fatigue and irritability—symptoms of jet lag. Jet lag correlates with decreased alertness, mental agility, and efficiency, as well as exacerbation of psychiatric disorders (Dawson, 2004; Iyer, 2001; Katz et al., 2002). Jet lag tends to be worse when we fly eastward because our bodies adjust more easily to going to bed later.

What about long-term **sleep deprivation**? Exploring the scientific effects of severe sleep loss is limited by both ethical and practical concerns. For example, sleep deprivation increases stress, making it difficult to separate the effects of sleep deprivation from those of stress.

Nonetheless, researchers have learned that, like disrupted circadian cycles, sleep deprivation poses several hazards. These include reduced cognitive and motor performance, irritability and other mood alterations, decreased self-esteem, and increased *cortisol* levels (a sign of stress) (Bourgeois-Bougrine et al., 2003; Carskadon & Dement, 2002; Cho, 2001; Graw et al., 2004).

Are You Sleep Deprived?

Take the following test to determine whether you are sleep deprived.

Part 1 Set up a small mirror next to the text and see if you can trace the black star pictured here, using your nondominant hand while watching your hand in the mirror. The task is difficult, and sleep-deprived people typically make many errors. If you are not sleep deprived, it may be difficult to trace the star, but you'll probably trace it accurately

Part 2 Give yourself one point each time you answer yes to the following questions:

Do you often fall asleep …
watching TV?
during boring meetings or lectures or in warm rooms?
after heavy meals or after a small amount of alcohol?
while relaxing after dinner?
within five minutes of getting into bed?

In the morning, do you generally …
need an alarm clock to wake up at the right time?
struggle to get out of bed?
hit the snooze bar several times to get more sleep?

During the day, do you …
feel tired, irritable, and stressed out?
have trouble concentrating and remembering?
feel slow when it comes to critical thinking, problem solving, and being creative?
feel drowsy while driving?
need a nap to get through the day?
have dark circles around your eyes?

If you answered yes to three or more items, you probably are not getting enough sleep.

Source: Quiz adapted and reprinted from *Maas*, 1999, with permission.

The consequences of such impairments are wide-ranging, from threatening students' school performance to endangering physical health. Perhaps the most frightening danger is that lapses in attention among sleep-deprived pilots, physicians, truck drivers, and other workers cause serious accidents and cost thousands of lives each year (McCartt et al., 2000; Oeztuerk, Tufan, & Gueler, 2002; Paice et al., 2002).

STAGES OF SLEEP: HOW SCIENTISTS STUDY SLEEP

Surveys and interviews can provide some information about the nature of sleep, but researchers in sleep laboratories use a number of sophisticated instruments to study physiological changes during sleep.

Imagine that you are a participant in a sleep experiment. When you arrive at the sleep lab, you are assigned one of several bedrooms. The researcher hooks you up to various physiological recording devices (**FIGURE 5.3A** on the next page). You will probably need a night or two to adapt to the equipment before the researchers can begin to monitor your typical night's sleep. At first, you enter a relaxed, *presleep* state. As you continue relaxing, your brain's electrical activity slows even further. In the course of about an hour, you progress through four distinct stages of sleep (Stages 1 through 4), each progressively deeper (**FIGURE 5.3B** on the next page). Then the

A Sleep research participants wear electrodes on their heads and bodies to measure the brain's and body's responses during the sleep cycle.

B An **electroencephalogram (EEG)** records brain wave changes by means of small electrodes on the scalp. Other electrodes measure muscle activity and eye movements. The stages of sleep, defined by telltale changes in brain waves, are indicated by the green stepped lines. The compact brain waves of alertness gradually lengthen as we drift into Stages 1–4. By the end of Stage 4, a change in body position generally occurs and heart rate, blood pressure, and respiratory rates all decrease. The sleeper then reverses through Stages 3 and 2 before entering the first REM period of the night. Although the brain and body are giving many signs of active arousal, the musculature is deeply relaxed and unresponsive. Sleepers awakened from REM sleep often report vivid, bizarre dreams, indicated in the figure by red and yellow dots. Those awakened from Stages 1–4 sleep often have more peaceful thoughts (indicated by muted dots). The heavy green line on the graph showing all four stages of sleep indicates the approximate time spent in REM sleep. Note how the length of the REM period increases as the night progresses.

sequence begins to reverse itself. Although we don't necessarily go through all sleep stages in this sequence, during the course of a night, people usually complete four to five cycles of light to deep sleep and back, each lasting about 90 minutes.

REM and NREM sleep

Figure 5.3B also shows an interesting phenomenon that occurs at the end of the first sleep cycle (and subsequent cycles). You reverse back through Stages 3 and 2. Then your scalp recordings abruptly display a pattern of small-amplitude, fast-wave activity, similar to an awake, vigilant person's brain waves. Your breathing and pulse rates become fast and irregular, and your genitals likely show signs of arousal. Yet your musculature is deeply relaxed and unresponsive. Because of these contradictory qualities, this stage is sometimes referred to as *paradoxical sleep*.

During this stage of paradoxical sleep, rapid eye movements occur under your closed eyelids. Researchers refer to this stage as **rapid-eye-movement (REM) sleep**. When awakened from REM sleep, people almost always report dreaming. Because REM sleep is so different from the other periods of sleep, Stages 1 through 4 are often collectively referred to as **non-rapid-eye-movement (NREM) sleep**. Dreaming also occurs during NREM sleep, but less frequently, and the dreams usually contain a simple experience, such as "I dreamed of a house" (Hobson, 2002; Squier & Domhoff, 1998).

Scientists believe that REM sleep is important for learning and consolidating new memories (Kavanau, 2000; Maquet et al., 2003; Squier & Domhoff, 1998). Further evidence of the importance of REM sleep for complex brain functions comes from the fact that the amount of REM sleep increases after periods of stress or intense learning and that fetuses, infants, and young children spend a large percentage of their sleep time in this stage. In addition, REM sleep occurs only in mammals of higher intelligence (Rechtschaffen & Siegel, 2000).

REM sleep meets an important biological need. When deprived of REM sleep, most people "catch up" later by spending more time than usual in REM sleep (Dement & Vaughan, 1999).

NREM sleep may be even more important to our biological functioning than REM sleep. When people are deprived of *total* sleep, they spend more time in NREM sleep during their first uninterrupted night of sleep (Borbely, 1982). Also, it is only after our need for NREM sleep has been satisfied each night that we begin to devote more time to REM sleep. Further, studies show that adults who sleep five or fewer hours each night spend less time in REM sleep than do those who sleep nine or more hours. Similarly, infants get much more sleep and have a higher percentage of REM sleep (close to 50 percent of total daily sleep during the first six months of life) than do adults (22 percent of total daily sleep at age 19, declining to about 14 percent in old age) (**FIGURE 5.4**).

The effect of aging on the sleep cycle FIGURE 5.4

Our biological need for sleep changes throughout our lifetimes. The pie charts in this figure show the relative amounts of REM sleep (dark blue), non-REM sleep (medium blue), and awake time (light blue) that the average person experiences as an infant, as an adult, and as an elderly person. An infant sleeps 14 hours and spends 40 percent of that time in REM. An adult sleeps about 7.5 hours, with 20 percent of that in REM. The average 70-year-old sleeps only about six hours, with 14 percent of that in REM.

Why do we sleep? There are two prominent theories about why we sleep. The **evolutionary/circadian theory** emphasizes the relationship of sleep to basic circadian rhythms. Sleep keeps animals still when predators are active (Hirshkowitz, Moore, & Minhoto, 1997) (FIGURE 5.5).

In contrast, the **repair/restoration theory** suggests that sleep helps us recuperate—physically, emotionally, and intellectually—from depleting daily activities (Maas, 1999).

It may be that both theories of sleep are correct—that sleep initially served to conserve energy and keep us out of trouble and that, over time, it has evolved to allow for repair and restoration.

Why do we dream? The question of why we dream—and whether dreams carry special meaning or information—has fascinated and perplexed psychologists at least as much as the question of why we sleep. Let's look at three theories of why we dream.

One of the oldest and most scientifically controversial explanations for why we dream is Freud's **psychoanalytic view**. Freud proposed that unacceptable desires, which are normally repressed, rise to the sur-

■ evolutionary/ circadian theory As a part of circadian rhythms, sleep evolved to conserve energy and to serve as protection from predators.

■ repair/restoration theory Sleep serves a recuperative function, allowing organisms to repair or replenish key factors.

face of consciousness during dreaming. We avoid anxiety, Freud believed, by disguising our forbidden unconscious needs (what Freud called the dream's latent content) as symbols (manifest content). For example, a journey is supposed to symbolize death; horseback riding and dancing would symbolize sexual intercourse; and a gun might represent a penis.

Most modern scientific research does not support Freud's view (Domhoff, 2003; Fisher & Greenberg, 1996). Critics also say that Freud's theory is highly subjective and that the symbols can be interpreted according to the particular analyst's view or training.

In contrast to Freud, a biological view called the **activation–synthesis hypothesis** suggests that dreams are a by-product of random stimulation of brain cells during REM sleep (Hobson, 1988, 2005). Alan Hobson and Robert McCarley (1977) proposed that specific neurons in the brainstem fire spontaneously during REM sleep and that the cortex struggles to "synthesize" or make sense out of this random stimulation by manufacturing dreams. This is *not* to say that dreams are totally meaningless. Hobson (1988, 2005) suggests that even if dreams begin with essentially random brain activity, your individual personality, motivations, memories, and life experiences guide how your brain constructs the dream.

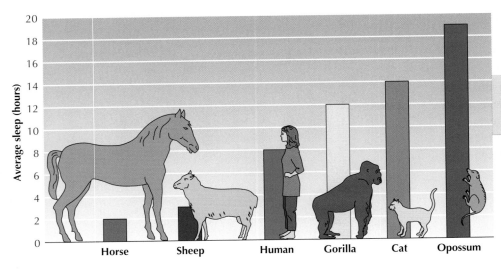

Average daily hours of sleep for different mammals

FIGURE 5.5

According to the evolutionary/circadian theory, differences in diet and number of predators affect different species' sleep habits.

Finally, some researchers support the **cognitive view** that dreams are simply another type of information processing. That is, our dreams help us to periodically sift and sort our everyday experiences and thoughts. For example, some research reports strong similarities between dream content and waking thoughts, fears, and concerns (Domhoff, 1999, 2005; Erlacher & Schredl, 2004).

SLEEP DISORDERS: WHEN SLEEP BECOMES A PROBLEM

An estimated two-thirds of American adults suffer from sleep problems, and about 25 percent of children under age 5 have a sleep disturbance (Dement & Vaughan, 1999; Doghramji, 2000; Hobson, 2005; National Sleep Foundation, 2006). Each year, Americans spend more than $98 million on nonprescription sleep medications, and about half that amount on caffeine tablets for daytime use. One in five adults is so sleepy during the day that it interferes with daily activities, and 20 percent of all drivers have fallen asleep for a few seconds (a phenomenon called microsleep) at the wheel.

Mental health professionals divide sleep disorders into two major diagnostic categories: dyssomnias, which describe problems sleeping, and parasomnias, which describe abnormal sleep disturbances.

The most common **dyssomnia** is **insomnia**. Although it's normal to have trouble sleeping before an exciting event, as many as one in ten people has persistent difficulty falling asleep or staying asleep, or wakes up too early. Nearly everybody has insomnia at some time in their life (Doghramji, 2000; Riemann & Voderholzer, 2003). A telltale sign of insomnia is that the person feels poorly rested the next day. Most people with serious insomnia have other medical or psychological disorders as well (Riemann & Voderholzer, 2003; Taylor, Lichstein, & Durrence, 2003).

dyssomnias

Problems in the amount, timing, and quality of sleep, including insomnia, sleep apnea, and narcolepsy.

Unfortunately, nonprescription insomnia pills generally don't work. Prescription tranquilizers and barbiturates do help people sleep, but they decrease Stage 4 and REM sleep, seriously affecting sleep quality. In the short term, drugs such as Ambien, Dalmane, Xanax, Halcion, and Lunesta may be helpful in treating sleep problems related to anxiety and acute, stressful situations. However, chronic users run the risk of psychological and physical drug dependence (Leonard, 2003; McKim, 2002).

Narcolepsy is another serious dyssomnia, characterized by sudden and irresistible onsets of sleep during normal waking hours. Narcolepsy afflicts about 1 person in 2,000 and generally runs in families (Kryger, Wald, & Manfreda, 2002; Siegel, 2000). During an attack, REM-like sleep suddenly intrudes into the waking state of consciousness. Victims may experience sudden, incapacitating attacks of muscle weakness or paralysis (known as cataplexy). Such people may fall asleep while walking, talking, or driving a car. Although long naps each day and stimulant or antidepressant drugs may help reduce the frequency of attacks, both the causes and cure of narcolepsy are still unknown (FIGURE 5.6).

Narcolepsy FIGURE 5.6

William Dement and his colleagues at Stanford University's Sleep Disorders Center have bred a group of narcoleptic dogs, which has increased our understanding of the genetics of this disorder. Research on these specially bred dogs has found degenerated neurons in certain areas of the brain (Siegel, 2000). Whether human narcolepsy results from similar degeneration is a question for future research.

Perhaps the most serious dyssomnia is **sleep apnea**. People with sleep apnea may fail to breathe for a minute or longer and then wake up gasping for breath. When they do breathe during their sleep, they often snore. Although people with sleep apnea are often unaware of it, repeated awakenings result in insomnia and leave the person feeling tired and sleepy during the day. Sleep apnea seems to result from blocked upper airway passages or from the brain's ceasing to send signals to the diaphragm, thus causing breathing to stop. This disorder can lead to high blood pressure, stroke, and heart attack (National Sleep Foundation, 2004; Miller, 2004; Young, Skatrud, & Peppard, 2006).

Treatment for sleep apnea depends partly on its severity. If the problem occurs only when you're sleeping on your back, sewing tennis balls to the back of your pajama top may help remind you to sleep on your side. Because obstruction of the breathing passages is related to obesity and heavy alcohol use (Christensen, 2000), dieting and alcohol restriction are often recommended. For others, surgery, dental appliances that reposition the tongue, or machines that provide a stream of air to keep the airway open may be the answer (FIGURE 5.7).

Recent findings suggest that even "simple" snoring (without the breathing stoppage characteristic of sleep apnea) can lead to heart disease and possible death (Peppard et al., 2000). Although occasional mild snoring remains somewhat normal, chronic snoring is a possible "warning sign that should prompt people to seek help" (Christensen, 2000, p. 172).

The second major category of sleep disorders, **parasomnias**, includes abnormal sleep disturbances such as **nightmares** and **night terrors** (FIGURE 5.8).

Sleep apnea FIGURE 5.7

A During sleep apnea, airways are blocked, causing breathing to be severely restricted. To treat this disorder, researchers and doctors have created equipment to promote respiration.

B For some patients, there is help from dental devices that reposition the tongue and open the airway.

A

C Another treatment for sleep apnea is a machine that provides a steady supply of air to keep the airway open.

Nightmare or night terrors? FIGURE 5.8

Nightmares, or bad dreams, occur toward the end of the sleep cycle, during REM sleep. Less common but more frightening are night terrors, which occur early in the cycle, during Stage 3 or Stage 4 of NREM sleep. Like the child in this photo, the sleeper may sit bolt upright, screaming and sweating, walk around, and talk incoherently, and the person may be almost impossible to awaken.

parasomnias
Abnormal disturbances occurring during sleep, including nightmares, night terrors, sleepwalking, and sleep talking.

Sleepwalking, which tends to accompany night terrors, usually occurs during NREM sleep. (Recall that large muscles are paralyzed during REM sleep, which explains why sleepwalking is possible). **Sleep talking** can occur during any stage of sleep, but appears to arise most commonly during NREM sleep. It can consist of single indistinct words or long, articulate sentences. It is even possible to engage some sleep talkers in a limited conversation.

Nightmares, night terrors, sleepwalking, and sleep talking are all more common among young children, but they can also occur in adults, usually during times of stress or major life events (Hobson & Silvestri, 1999; Muris et al., 2000). Patience and soothing reassurance at the time of the sleep disruption are usually the only treatment recommended for both children and adults.

A recent large-scale study reported that behavior therapy had good success in treating some sleep problems (Smith et al., 2003). You can use similar techniques in your own life. For example:

- When you're having a hard time going to sleep, don't keep checking the clock and worrying about your loss of sleep. Instead, remove all TVs, stereos, and books, and limit the use of the bedroom to sleep.

- Work off tension through exercise (but not too close to bedtime).

- Avoid stimulants such as caffeine and nicotine.

- Avoid late meals and heavy drinking.

- Follow the same presleep routine every evening. It might include listening to music, writing in a diary, or meditating.

- Use *progressive muscle relaxation*. Alternately tense and relax various muscle groups.

- Practice yoga or deep breathing, or take a warm bath to help you relax.

If you need additional help, try the relaxation techniques from the Better Sleep Council, a nonprofit education organization in Burtonsville, Maryland. Also, check out these websites: www.sleepfoundation.org and www.stanford.edu/~dement.

CONCEPT CHECK STOP

What are the different levels of consciousness?

What are the elements of the circadian "clock"?

What are some of the theories about why we sleep and dream?

How do dyssomnias and parasomnias differ?

Psychoactive Drugs

LEARNING OBJECTIVES

Explain the difference between psychological and physical drug dependence.

Summarize the differences among the four major types of psychoactive drugs.

Compare how different psychoactive drugs affect the nervous system.

Have you noticed how difficult it is to have a logical, nonemotional discussion about drugs? In our society, where the most popular **psychoactive drugs** are caffeine, tobacco, and ethyl alcohol, people often become defensive when these drugs are grouped with illicit drugs such as marijuana and cocaine. Similarly, marijuana users are disturbed that their drug of choice is grouped with "hard" drugs like heroin. Most scientists believe that there are good and bad uses of all drugs. How drug use differs from drug abuse and how chemical alterations in consciousness affect a person, psychologically and physically, are important topics in psychology.

Psychoactive drugs influence the nervous system in a variety of ways. Alcohol, for example, has a diffuse effect on neural membranes throughout the nervous system. Most psychoactive drugs, however, act in a more specific way: by either enhancing a particular neurotransmitter's effect (an **agonistic** drug action) or inhibiting it (an **antagonistic** drug action) (**FIGURES 5.9** and **5.10**).

Is drug abuse the same as drug addiction? The term **drug abuse** generally refers to drug taking that causes emotional or physical harm to oneself or others. The drug consumption is also typically compulsive, frequent, and intense. **Addiction** is a broad term referring

> **psychoactive drugs**
> Chemicals that change conscious awareness, mood, or perception.

Cocaine: an agonist drug in action FIGURE 5.9

After releasing neurotransmitter into the synapse, the sending neuron normally reabsorbs (or reuptakes) excess neurotransmitter back into the terminal buttons. However, if cocaine is present in the synapse, it will block the reuptake of dopamine, serotonin, and norepinephrine. This intensifies the normal mood-altering effect of these three neurotransmitters.

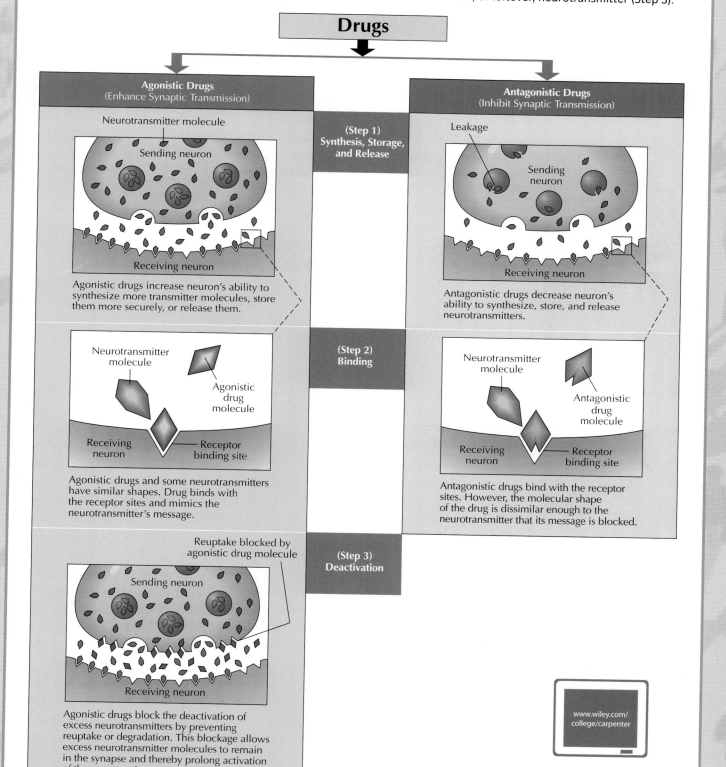

How agonistic and antagonistic drugs produce their psychoactive effect FIGURE 5.10

Most psychoactive drugs produce their mood-, energy-, and perception-altering effects by changing the body's supply of neurotransmitters. They can alter synthesis, storage, and release of neurotransmiters (Step 1). Psychoactive drugs can also alter the effect of neurotransmittters on the receiving site of the receptor neuron (Step 2). After neurotransmitters carry their messages across the synapse, the sending neuron normally deactivates the excess, or leftover, neurotransmitter (Step 3).

Drugs

Agonistic Drugs
(Enhance Synaptic Transmission)

Antagonistic Drugs
(Inhibit Synaptic Transmission)

**(Step 1)
Synthesis, Storage,
and Release**

Neurotransmitter molecule

Sending neuron

Receiving neuron

Agonistic drugs increase neuron's ability to synthesize more transmitter molecules, store them more securely, or release them.

Leakage

Sending neuron

Receiving neuron

Antagonistic drugs decrease neuron's ability to synthesize, store, and release neurotransmitters.

**(Step 2)
Binding**

Neurotransmitter molecule

Agonistic drug molecule

Receiving neuron

Receptor binding site

Agonistic drugs and some neurotransmitters have similar shapes. Drug binds with the receptor sites and mimics the neurotransmitter's message.

Neurotransmitter molecule

Antagonistic drug molecule

Receiving neuron

Receptor binding site

Antagonistic drugs bind with the receptor sites. However, the molecular shape of the drug is dissimilar enough to the neurotransmitter that its message is blocked.

**(Step 3)
Deactivation**

Reuptake blocked by agonistic drug molecule

Sending neuron

Receiving neuron

Agonistic drugs block the deactivation of excess neurotransmitters by preventing reuptake or degradation. This blockage allows excess neurotransmitter molecules to remain in the synapse and thereby prolong activation of the receptor site.

Process Diagram

Effects of the major psychoactive drugs FIGURE 5.11

	Category	Desired effects	Undesirable effects
	Depressants (Sedatives) Alcohol, barbiturates, anti-anxiety drugs (Valium), Rohypnol (roofies), Ketamine (special K), CoHB	Tension reduction, euphoria, disinhibition, drowsiness, muscle relaxation	Anxiety, nausea, disorientation, impaired reflexes and motor functioning, amnesia, loss of consciousness, shallow respiration, convulsions, coma, death
	Stimulants Cocaine, amphetamine, methamphetamine (crystal meth), MDMA (Ecstasy)	Exhilaration, euphoria, high physical and mental energy, reduced appetite, perceptions of power, sociability	Irritability, anxiety, sleeplessness, paranoia, hallucinations, psychosis, elevated blood pressure and body temperature, convulsions, death
	Caffeine	Increased alertness	Insomnia, restlessness, increased pulse rate, mild delirium, ringing in the ears, rapid heartbeat
	Nicotine	Relaxation, increased alertness, sociability	Irritability, raised blood pressure, stomach pains, vomiting, dizziness, cancer, heart disease, emphysema
	Opiates (Narcotics) Morphine, heroin, codeine	Euphoria, "rush" of pleasure, pain relief, prevention of withdrawal discomfort	Nausea, vomiting, constipation, painful withdrawal, shallow respiration, convulsions, coma, death
	Hallucinogens (Psychedelics) LSD (lysergic acid diethylamide)	Heightened aesthetic responses, euphoria, mild delusions, hallucinations, distorted perceptions and sensations	Panic, nausea, longer and more extreme delusions, hallucinations, perceptual distortions ("bad trips"), psychosis
	Marijuana	Relaxation, mild euphoria, increased appetite	Perceptual and sensory distortions, hallucinations, fatigue, lack of motivation, paranoia, possible psychosis

to a condition in which a person feels compelled to use a specific drug. People now use the term to describe almost any type of compulsive activity, from working to surfing the Internet (Coombs, 2004).

For the sake of clarity, many researchers use the term **psychological dependence** to refer to the mental desire or craving to achieve a drug's effects. They use the term **physical dependence** to refer to changes in bodily processes that make a drug necessary for minimum daily functioning. Physical dependence appears most clearly when the drug is withheld and the user undergoes painful **withdrawal** reactions, including physical pain and intense cravings. After repeated use of a drug, many of the body's physiological processes adjust to higher and higher levels of the drug, producing a decreased sensitivity called **tolerance**.

Tolerance leads many users to escalate their drug use and to experiment with other drugs in an attempt to re-create the original pleasurable altered state. Sometimes, using one drug increases tolerance for another. This is known as **cross-tolerance**. Developing tolerance or cross-tolerance does not prevent drugs from seriously damaging the brain, heart, liver, and other organs.

Psychological dependence is no less damaging than physical dependence. The craving in psychological dependence can be strong enough to keep the user in a constant drug-induced state—and to lure an "addict" back to a drug habit long after he or she has overcome physical dependence.

PSYCHOACTIVE DRUGS: FOUR CATEGORIES

For convenience, psychologists divide psychoactive drugs into four broad categories: depressants, stimulants, opiates, and hallucinogens. FIGURE 5.11 provides examples of each and their effects.

Depressants (sometimes called "downers") act on the central nervous system to suppress or slow bodily processes and to reduce overall responsiveness. Because tolerance and dependence (both physical and psychological) are rapidly acquired with these drugs, there is strong potential for abuse.

Although alcohol is primarily a depressant, at low doses it has stimulating effects, thus explaining its reputation as a "party drug." As consumption increases, symptoms of drunkenness appear (FIGURE 5.12).

Alcohol's effects are determined primarily by the amount that reaches the brain. Because the liver breaks down alcohol at the rate of about one ounce per hour, the number of drinks and the speed of consumption are both very important. People can die after drinking large amounts of alcohol in a short period of time (Chapter 3). In addition, men's bodies are more efficient at breaking down alcohol. Even after accounting for differences in size and muscle-to-fat ratio, women have a higher blood alcohol level than men following equal doses of alcohol.

Alcohol should not be combined with any other drug, but combining alcohol and barbiturates—both depressants—is particularly dangerous. Together, they

Alcohol's effect on the body and behavior FIGURE 5.12

Number of drinks in two hours	Blood alcohol content (%)	Effect
(2)	0.05	Relaxed state; increased sociability
(3)	0.08	Everyday stress lessened
(4)	0.10[a]	Movements and speech become clumsy
(7)	0.20	Very drunk; loud and difficult to understand; emotions unstable
(12)	0.40	Difficult to wake up; incapable of voluntary action
(15)	0.50	Coma and/or death

[a]A drink refers to one 12-ounce beer, a 4-ounce glass of wine, or a 1.25-ounce shot of hard liquor.
[b]In America, the legal blood alcohol level for "drunk driving" varies from 0.05 to 0.12.

can relax the diaphragm muscles to such a degree that the person literally suffocates.

Depressants suppress central nervous system activity, whereas **stimulants** (uppers) increase the overall activity and responsiveness of the central nervous system. Even legal stimulants can lead to serious problems. For example, the U.S. Public Health Service considers cigarette smoking the single most preventable cause of death and disease in the United States. Researchers have found that nicotine activates the same brain areas (nucleus accumbens) as cocaine—a dangerous stimulant well known for its addictive potential (Pich et al., 1997; Grilly, 2006). Nicotine's effects (relaxation, increased alertness, diminished pain and appetite) are so powerfully reinforcing that some people continue to smoke even after having a cancerous lung removed.

Opiates (or narcotics), which are derived from the opium poppy, are used medically to relieve pain (Kuhn, Swartzwelder, & Wilson, 2003). They mimic the brain's natural endorphins (Chapter 2), which numb pain and elevate mood. This creates a dangerous pathway to drug abuse. After repeated flooding with artificial opiates, the brain eventually reduces or stops the production of its own opiates. If the user later attempts to stop, the brain lacks both the artificial and normal level of painkilling chemicals, and withdrawal becomes excruciatingly painful (**FIGURE 5.13**). Interestingly, when opiates are used medically to relieve intense pain, they are very seldom habit-forming. However, when taken recreationally, they are strongly addictive (Butler, Joel, & Jeffries, 2000; Coombs, 2004).

So far, we have discussed three of the four types of psychoactive drugs: depressants, stimulants, and opiates. One of the most intriguing alterations of consciousness comes from **hallucinogens**, drugs that produce sensory or perceptual distortions, including visual, auditory, and kinesthetic hallucinations (**FIGURE 5.14**). Some cultures have used hallucinogens for religious purposes, as a way to experience "other realities" or to communicate with the supernatural. In Western societies, most people use hallucinogens for their reported "mind-expanding" potential.

Hallucinogens are commonly referred to as psychedelics (from the Greek for "mind manifesting"). They include mescaline (derived from the peyote cactus), psilocybin (derived from mushrooms), phencyclidine (chemically derived), and LSD (lysergic acid diethylamide, derived from ergot, a rye mold).

LSD, or acid, is a synthetic substance that produces dramatic alterations in sensation and perception. LSD use by high school and college students has been increasing (Connolly, 2000; Hedges & Burchfield, 2006; Yacoubian, Green, & Peters, 2003). LSD can be an extremely dangerous drug. Bad LSD trips can be terrifying and may lead to accidents, deaths, or suicide. Dangerous flashbacks may unpredictably recur long after the initial ingestion. They can be brought on by stress, fatigue, marijuana use, illness, or occasionally by the individual's intentional effort (Grilly, 2006).

As you may know from television or newspapers, psychoactive drugs like Rohypnol (the date rape drug) and MDMA (3-4 methylenedioxymethampheta-

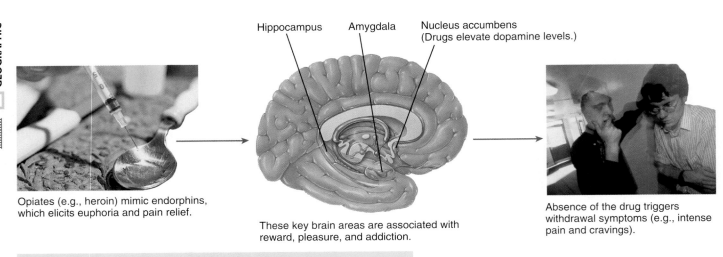

Hippocampus   Amygdala   Nucleus accumbens
(Drugs elevate dopamine levels.)

Opiates (e.g., heroin) mimic endorphins, which elicits euphoria and pain relief.

These key brain areas are associated with reward, pleasure, and addiction.

Absence of the drug triggers withdrawal symptoms (e.g., intense pain and cravings).

How opiates create physical dependence FIGURE 5.13

Ecstasy stimulates serotonin

LSD mimics serotonin

GHB & Rohypnol depress CNS, increase dopamine

"Date rape" drugs
• colorless
• odorless
• tasteless

Euphoria, empathy, clouded thinking

Sensory distortions, hallucinations

Sedation, muscle relaxation, memory blackouts

mine, commonly known as Ecstasy) are fast becoming some of our nation's most popular drugs of abuse, especially at all-night dance parties. Other "club drugs," like GHB (gamma-hydroxybutyrate), ketamine (Special K), methamphetamine (crystal meth), and LSD, are also gaining in popularity (Agar & Reisinger, 2004; Martins, Mazzotti, & Chilcoat, 2005; Yacoubian, Green, & Peters, 2003).

Although these drugs can produce desirable effects (e.g., Ecstasy's feeling of great empathy and connectedness with others), almost all psychoactive drugs may cause serious health problems—in some cases, even death (National Institute on Drug Abuse, 2005). Club drugs (like most psychoactive drugs) affect motor coordination, perceptual skills, and reaction time necessary for safe driving. As with all illicit drugs, there are no truth-in-packaging laws to protect club drug buyers from unscrupulous practices. Sellers often substitute unknown, cheaper, and possibly even more dangerous substances for the ones they claim to be selling.

Club drug use may lead to risky sexual behaviors and an increased risk of contracting AIDS (acquired immunodeficiency syndrome) and other sexually transmitted diseases. Some drugs, such as Rohypnol, are odorless, colorless, and tasteless, and they can easily be added to beverages by individuals who want to intoxicate or sedate others, so the dangers of club drug use go far beyond the drug itself (Fernández et al., 2005; National Institute on Drug Abuse, 2005). For more information on the specific dangers and effects of club drugs, check out http://www.drugabuse.gov/ClubAlert/Clubdrugalert.html.

Marijuana is also classified as a hallucinogen even though it has some properties of a depressant (it induces drowsiness and lethargy) and some of a narcotic (it acts as a weak painkiller). In low doses, marijuana produces mild euphoria; moderate doses lead to an intensification of sensory experiences and the illusion that time is passing slowly. High doses may produce hallucinations, delusions, and distortions of body image (Hedges & Burchfield, 2006). The active ingredient in marijuana (cannabis) is THC, or tetrahydrocannabinol, which attaches to receptors that are abundant throughout the brain.

Some research has found marijuana to be therapeutic in the treatment of glaucoma (an eye disease), in alleviating the nausea and vomiting associated with chemotherapy, and with other health problems (Darmini & Crim, 2005; Iversen, 2003; Zagnoni & Albano, 2002).

Chronic marijuana use can lead to throat and respiratory disorders, impaired lung functioning and immune response, declines in testosterone levels, reduced sperm count, and disruption of the menstrual cycle and ovulation (Hedges & Burchfield, 2006; Nahas et al., 2002; Iversen, 2003; Roth et al., 2004). Some studies suggest that marijuana's effects on the brain are disturbingly similar to those produced by such drugs as heroin, cocaine, alcohol, and nicotine (Blum et al., 2000; Tanda & Goldberg, 2004). However, other studies have found little or no lasting negative health effects (e.g., Eisen et al., 2002). Marijuana also can be habit-forming, but few users experience intense cravings and withdrawal symptoms because the drug leaves the body very slowly.

CONCEPT CHECK STOP

What are the four types of psychoactive drugs?

What are the differences between psychological and physical drug dependence?

How do psychoactive drugs affect nervous system functioning?

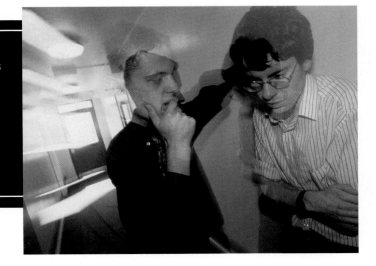

Altering Consciousness Through Meditation and Hypnosis

MEDITATION: A HEALTHY "HIGH"

"Suddenly, with a roar like that of a waterfall, I felt a stream of liquid light entering my brain through the spinal cord . . . I experienced a rocking sensation and then felt myself slipping out of my body, entirely enveloped in a halo of light. I felt the point of consciousness that was myself growing wider, surrounded by waves of light" (Krishna, 1999, pp. 4–5).

This is how spiritual leader Gopi Krishna described his experience with **meditation**. Although most people in the beginning stages of meditation report a simpler, mellow type of relaxation followed by a mild euphoria, some advanced meditators report experiences of profound rapture and joy or strong hallucinations.

The highest functions of consciousness occur in the frontal lobe, particularly in the cerebral cortex. Scientists are seeing increasing evidence that the altered state of consciousness one experiences during meditation occurs when one purposely changes how the prefrontal cortex (the area immediately behind your eyes) functions. Typically, your prefrontal cortex is balancing your working memory, temporal integration, and higher-order thinking, among other tasks. Scientists theorize, based on brain imaging, that when you focus on a single object, emotion, or word, you diminish the amount of brain cells that must be devoted to these multiple tasks, and they instead become involved in the singular focus of your meditation. This narrow focus allows other areas of the brain to be affected, and since the neurons devoted to time have changed focus, you experience a sense of timelessness and mild euphoria (Dietrich, 2003) (**FIGURE 5.15**).

meditation A group of techniques designed to refocus attention, block out all distractions, and produce an alternate state of consciousness.

1. Focusing on a single object or word...

2. ...engages neurons in the prefrontal cortex.

Cerebral cortex

Cerebral white matter

3. Attention and focus are strengthened. The cortex thickens over time.

4. Neural emphasis on time and worries decreases, resulting in a sensation of peace and timelessness.

How meditation alters consciousness FIGURE 5.15

Meditation and the Brain

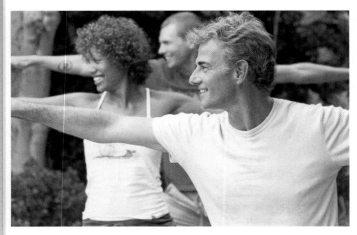

Some meditation techniques, such as t'ai chi and hatha yoga, involve body movements and postures, while in other techniques the meditator remains motionless, chanting or focusing on a single point. Some practitioners of meditation believe that it offers a more enlightened form of consciousness, superior to all other levels, and that it allows the meditator to have remarkable control over his or her bodily processes (Kim et al., 2005; Surawy, Roberts, & Silver, 2005).

Through the use of sophisticated equipment, researchers can gather images of the activity in the brain during meditation. According to this research, a wider area of the brain responds to sensory stimuli during meditation, suggesting that meditation enhances the coordination between the brain hemispheres (see graphic, right; and Lyubimov, 1992). As well, researchers have found not only that those who meditated used a larger portion of their brain, but also that faster and more powerful gamma waves existed in those individuals who meditated regularly than in those who did not (Lutz et al., 2004). Past studies have shown that the type of increased coordination associated with more powerful gamma waves correlates with improvements in focus, memory, learning, and consciousness.

Top view of head

−2.6

0

2.6

Before meditation   During meditation νV

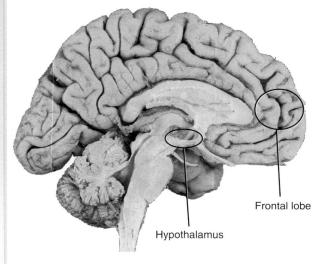

Hypothalamus

Frontal lobe

Research has verified that meditation can produce dramatic changes in basic physiological processes, including heart rate, oxygen consumption, sweat gland activity and brain activity. Meditation has also been somewhat successful in reducing stress and lowering blood pressure (Kim et al., 2005; Harrison, 2005; Labiano & Brusasca, 2002). Studies have even implied that meditation can change the body's parasympathetic response (Young & Taylor, 1998) and increase structural support for the sensory, decision-making, and attention processing centers of the brain (Lazar et al., 2005). It seems that during meditation, the part of the brain that is responsible for both sympathetic and parasympathetic responses, the hypothalamus, diminishes the sympathetic response and increases the parasympathetic response. Shutting down the so-called "fight or flight" response in this way allows for deep rest, slower respiration, and increased and more coordinated use of the brain's two hemispheres. At the same time, meditation engages the part of the brain that is responsible for decision making and reasoning, the frontal lobe.

HYPNOSIS: USES AND MYTHS

Relax . . . your eyelids are so very heavy . . . your muscles are becoming more and more relaxed . . . your breathing is becoming deeper and deeper . . . relax . . . your eyes are closing . . . let go . . . relax.

Hypnotists use suggestions like these to begin **hypnosis**. Once hypnotized, some people can be convinced that they are standing at the edge of the ocean lis-

hypnosis A trance-like state of heightened suggestibility, deep relaxation, and intense focus.

tening to the sound of the waves and feeling the ocean mist on their faces. Invited to eat a delicious apple that is actually an onion, the hypnotized person may relish the flavor. Told they are watching a very funny or sad movie, hypnotized people may begin to laugh or cry at their self-created visions.

From the 1700s to modern times, entertainers and quacks have used (and abused) hypnosis (**FIGURE 5.16**); but physicians, dentists, and therapists have also long employed it as a respected clinical tool.

Hypnosis myths and facts FIGURE 5.16

Myth	Fact
People can be involuntarily hypnotized or hypnotically "brainwashed."	Hypnosis requires a willing, conscious choice to relinquish control of one's consciousness to someone else. The best potential subjects are those who are able to focus attention, are open to new experiences, and are capable of imaginative involvement or fantasy (Barber, 2000; Liggett, 2000).
Hypnosis can make people behave immorally or take dangerous risks against their will.	Hypnotized people retain awareness and control of their behavior, and they can refuse to comply with the hypnotist's suggestions (Kirsch & Braffman, 2001; Kirsch & Lynn, 1995).
Hypnosis participants are "faking it," playing along with the hypnotist.	Although most participants are not consciously faking hypnosis, some researchers believe the effects result from a blend of conformity, relaxation, obedience, suggestion, and role playing (Baker, 1996, 1998; Lynn et al., 2002; Stafford & Lynn, 2002). Other theorists believe that hypnotic effects result from a special altered state of consciousness (Bowers & Woody, 1996; Hilgard, 1978, 1992). A group of "unified" theorists suggests that hypnosis is a combination of both relaxation/role playing and a unique alternate state of consciousness.
Hypnotized people can perform acts of superhuman strength.	When nonhypnotized people are simply asked to try their hardest on tests of physical strength, they generally can do anything that a hypnotized person can (Druckman & Bjork, 1994).
Hypnotized people can recall things they otherwise could not.	Although the heightened relaxation and focus that hypnosis engenders improves recall for some information, people also are more willing to guess (Perry, Orne, London, & Orne, 1996; Stafford & Lynn, 2002). Because memory is normally filled with fabrication and distortion (Chapter 7), hypnosis generally increases the potential for error.

www.wiley.com/college/carpenter

Modern scientific research has removed much of the mystery surrounding hypnosis. A number of features characterize the hypnotic state (Barber, 2000; David & Brown, 2002; Gay, Philoppot, & Luminet, 2002; Hilgard, 1986, 1992; Spiegel, 1999):

- Narrowed, highly focused attention (ability to "tune out" competing sensory stimuli)

- Increased use of imagination and hallucinations

- A passive and receptive attitude

- Decreased responsiveness to pain

- Heightened suggestibility, or a willingness to respond to proposed changes in perception ("This onion is an apple")

Today, even with available anesthetics, hypnosis is occasionally used in surgery and for the treatment of chronic pain and severe burns (Harandi, Esfandani, & Shakibaei, 2004; Montgomery et al., 2002). Hypnosis has found its best use in medical areas, such as dentistry and childbirth, in which patients have a high degree of anxiety, fear, and misinformation. Because tension and anxiety strongly affect pain, any technique that helps the patient relax is medically useful.

In psychotherapy, hypnosis can help patients relax, remember painful memories, and reduce anxiety. Despite the many myths about hypnosis, it has been used with modest success in the treatment of phobias and in helping people to lose weight, stop smoking, and improve study habits (Ahijevych, Yerardi, & Nedilsky, 2000; Dobbin et al., 2004; Gemignani et al., 2000).

CONCEPT CHECK **STOP**

What are some major features of meditation?

Who might benefit from hypnosis?

Why can't people be hypnotized against their will?

What is happening in this picture ?

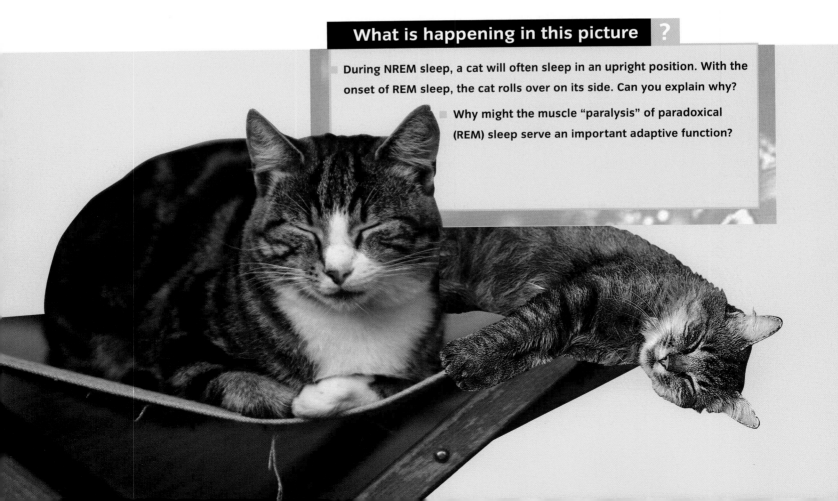

During NREM sleep, a cat will often sleep in an upright position. With the onset of REM sleep, the cat rolls over on its side. Can you explain why?

Why might the muscle "paralysis" of paradoxical (REM) sleep serve an important adaptive function?

VISUAL SUMMARY

1 Consciousness, Sleep, and Dreaming

1. **Consciousness**, an organism's awareness of its own self and surroundings, exists along a continuum, from high awareness to unconsciousness and coma. Sleep is a particular **altered state of consciousness (ASC)**.

2. Controlled processes demand focused attention and generally interfere with other ongoing activities. Automatic processes require minimal attention and generally do not interfere with other ongoing activities.

(perception, memory, attention)

Frontal lobe (working memory, temporal integration, abstract thinking, introspection, cognitive flexibility)

Parietal lobe

Temporal lobe

Occipital lobe

Limbic system (emotional memory, emotions, relational learning, memory consolidation)

Brain stem (arousal)

AWAKE NREM REM OLD AGE

AWAKE NREM REM ADULT

AWAKE NREM REM INFANCY

3. Many physiological funtions follow 24-hour **circadian rhythms**. Disruptions in circadian rhythms, as well as long-term sleep deprivation, cause increased fatigue, cognitive and mood disruptions, and other health problems.

4. The electroencephalogram (EEG) detects and records electrical changes in the nerve cells of the cerebral cortex. People progress through four distinct stages of non-REM (NREM) sleep, with periods of rapid-eye-movement (REM) sleep occurring at the end of each sleep cycle. Both REM and NREM sleep are important for our biological functioning.

5. The **evolutionary/circadian theory** proposes that sleep evolved to conserve energy and as protection from predators. The **repair/restoration theory** suggests that sleep helps us recuperate from the day's events. Three major theories for why we dream are Freud's psychoanalytic view, the activation–synthesis hypothesis, and the information-processing view.

6. **Dyssomnias** are problems in the amount, timing, and quality of sleep; they include insomnia, sleep apnea, and narcolepsy. **Parasomnias** are abnormal disturbances occurring during sleep; they include nightmares, night terrors, sleepwalking, and sleep talking. Although drugs are the most common method of treating sleep disorders, a recent large-scale study reported that behavior therapy had good success in treating some sleep problems.

2 Psychoactive Drugs

1. Psychoactive drugs influence the nervous system in a variety of ways. Alcohol affects neural membranes throughout the entire nervous system. Most psychoactive drugs act in a more specific way, by either enhancing a particular neurotransmitter's effect (an *agonistic* drug action) or inhibiting it (an *antagonistic* drug action). Drugs can interfere with neurotransmission at any of four stages: production or synthesis; storage and release; reception; or removal.

Sending neuron

Synapse

Cocaine

Neurotransmitter

Hippocampus   Amygdala   Nucleus accumbens
(Drugs elevate dopamine levels.)

Opiates (e.g., heroin) mimic endorphins, which elicits euphoria and pain relief.

These key brain areas are associated with reward, pleasure, and addiction.

Absence of the drug triggers withdrawal symptoms (e.g., intense pain and cravings).

2. The term drug abuse refers to drug-taking behavior that causes emotional or physical harm to oneself or others. Addiction refers to a condition in which a person feels compelled to use a specific drug. Psychological dependence refers to the mental desire or craving to achieve a drug's effects. Physical dependence refers to biological changes that make a drug necessary for minimum daily functioning. Repeated use of a drug can produce decreased sensitivity, or tolerance. Sometimes, using one drug increases tolerance for another (cross-tolerance).

3. Psychologists divide psychoactive drugs into four categories: depressants (such as alcohol, barbituates, Rohypnol, and Ketamine), stimulants (such as caffeine, nicotine, cocaine, amphetamine, methamphetamine, and Ecstasy), opiates (such as morphine, heroin, and codeine), and hallucinogens (such as marijuana and LSD). Almost all psychoactive drugs may cause serious health problems and, in some cases, even death.

3 Altering Consciousness Through Meditation and Hypnosis

1. The term **meditation** refers to techniques designed to refocus attention, block out distractions, and produce an alternate state of consciousness. Some followers believe that meditation offers a more enlightened form of consciousness, and researchers have verified that it can produce dramatic changes in basic physiological processes.

2. Modern research has removed the mystery surrounding **hypnosis**, a trancelike state of heightened suggestibility, deep relaxation, and intense focus. It is used in surgery and medicine, and it is especially useful in medical areas in which patients have a high degree of anxiety, fear, and misinformation. In psychotherapy, hypnosis can help patients relax, remember painful memories, and reduce anxiety. It is also used to help with a number of behavioral issues, such as efforts to quit smoking, lose weight, or overcome phobias.

1. Focusing on a single object or word…

Cerebral cortex

Cerebral white matter

3. Attention and focus are strengthened. The cortex thickens over time.

2. …engages neurons in the prefrontal cortex.

4. Neural emphasis on time and worries decreases, resulting in a sensation of peace and timelessness.

KEY TERMS

- consciousness p. 116
- altered state of consciousness (ASC) p. 116
- circadian rhythms p. 116
- controlled processes p. 117
- automatic processes p. 117
- sleep deprivation p. 118
- electroencephalogram (EEG) p. 120
- rapid-eye-movement (REM) sleep p. 121
- non-rapid-eye-movement (NREM) sleep p. 121
- evolutionary/circadian theory p. 122
- repair/restoration theory p. 122
- psychodynamic view p. 122

- activation-synthesis hypothesis p. 122
- cognitive view p. 123
- dyssomnias p. 123
- insomnia p. 123
- narcolepsy p. 123
- sleep apnea p. 123
- parasomnias p. 124
- nightmares p. 124
- night terrors p. 124
- sleepwalking p. 125
- sleep talking p. 125
- psychoactive drugs p. 126
- agonistic p. 126
- antagonistic p. 126

- drug abuse p. 126
- addiction p. 126
- psychological dependence p. 129
- physical dependence p. 129
- withdrawal p. 129
- tolerance p. 129
- cross-tolerance p. 129
- depressants p. 129
- stimulants p. 130
- opiates p. 130
- hallucinogens p. 130
- meditation p. 133
- hypnosis p. 135

CRITICAL AND CREATIVE THINKING QUESTIONS

1. Do you believe that people have an unconscious mind? If so, how does it affect thoughts, feelings, and behavior?

2. Which of the three main theories of dreams do you most agree with, and why?

3. Do you think marijuana use should be legal? Why or why not?

4. Why might hypnosis help treat people who suffer from chronic pain?

(Check your answers in Appendix B.)

1. *Consciousness* is defined in this text as _____ .
   a. ordinary and extraordinary wakefulness
   b. an organism's awareness of its own self and surroundings
   c. mental representations of the world in the here and now
   d. any mental state that requires thinking and processing of sensory stimuli

2. Mental activities that require focused attention are called _____ .
   a. thinking processes
   b. controlled processes
   c. alert states of consciousness
   d. conscious awareness

3. *Automatic processes* require _____ attention.
   a. internal
   b. unconscious
   c. minimal
   d. delta wave

4. *Circadian rhythms* are _____ .
   a. patterns that repeat themselves on a twice-daily schedule
   b. physical and mental changes associated with the cycle of the moon
   c. rhythmical processes in your brain
   d. biological changes that occur on a 24-hour cycle

5. Identify the main areas of the brain involved in the operation of circadian rhythms.
   a. hypothalamus
   b. pineal gland
   c. suprachiasmatic nucleus

6. This figure displays brain wave changes tracked by means of small electrodes on the scalp. This research tool is called a(n) _____ .
   a. EKG
   b. PET scan
   c. EEG recording
   d. EMG

7. The sleep stage marked by irregular breathing, eye movements, high-frequency brain waves, and dreaming is called _____ sleep.
   a. beta
   b. hypnologic
   c. REM
   d. transitional

8. Which of the following people is clearly experiencing insomnia?
   a. Joan frequently cannot fall asleep the night before a final exam.
   b. Cliff regularly sleeps less than eight hours per night.
   c. Consuela persistently has difficulty falling or staying asleep.
   d. All of these persons are clearly experiencing insomnia.

9. Dieting, surgery, dental appliances, and tennis balls are all recommended treatments for _____ , a dyssomnia.
   a. insomnia
   b. parasomnia
   c. nightmares
   d. sleep apnea

10. *Psychoactive drugs* _____ .

   a. change conscious awareness, mood, or perception
   b. are addictive, mind altering, and dangerous to your health
   c. are illegal unless prescribed by a medical doctor
   d. all of these options

11. Match the 4 major categories of psychoactive drugs with the correct photo:

   a. depressants, photo _____
   b. stimulants, photo _____
   c. opiates, photo _____
   d. hallucinogens, photo _____

12. *Depressants* include all of the following **EXCEPT**
_____ .

   a. Downers such as sedatives, barbiturates, antianxiety drugs
   b. Alcohol
   c. Tobacco
   d. Valium, Seconal

13. Altered states of consciousness can be achieved in which of the following ways?

   a. during sleep and dreaming
   b. via chemical channels
   c. through hypnosis and meditation
   d. all of these options

14. _____ is a group of techniques designed to refocus attention and produce an alternate state of consciousness.

   a. Hypnosis
   b. Scientology
   c. Parapsychology
   d. Meditation

15. _____ is an alternate state of heightened suggestibility characterized by deep relaxation and intense focus.

   a. Meditation
   b. Amphetamine psychosis
   c. Hypnosis
   d. Daydreaming

Learning

O n June 7, 1998, James Byrd, a disabled 49-year-old African American, was walking home along Martin Luther King Boulevard in Jasper, Texas. Three young men offered him a ride, but they had no intention of taking him home. Instead, they chained Mr. Byrd by his ankles to the back of their pickup and dragged him along an old country road until his head and right arm were ripped from his body.

What brought about this grisly murder? Byrd's murderers—two of whom were sentenced to death and the third given life imprisonment without the possibility of parole—were "white supremacists." Although white supremacists form only a tiny fraction of the American population, hate crimes are a serious and growing problem around the world. People are ridiculed, attacked, and even murdered simply because of their ethnicity, sexual orientation, gender, or religious preference. Where does such hatred come from? Is prejudice learned?

We usually think of learning as classroom activities, such as math and reading, or as motor skills, like riding a bike or playing the piano. Psychologists define learning more broadly, as *a relatively permanent change in behavior or mental processes because of practice or experience.* This relative permanence applies not only to useful behaviors (using a spoon or writing great novels) but also to bad habits, racism, and hatred. The good news is that what is learned can be unlearned—through retraining, counseling, and self-reflection.

In this chapter, we discuss several types of conditioning, the most basic form of learning. Then we look at social-cognitive learning and the biological factors involved in learning. Finally, we explore how learning theories and concepts touch on everyday life.

Classical Conditioning

THE BEGINNINGS OF CLASSICAL CONDITIONING

Why does your mouth water when you see a large slice of chocolate cake or a juicy steak? The answer to this question was accidentally discovered in the laboratory of Russian physiologist Ivan Pavlov (1849–1936). Pavlov's work focused on the role of saliva in digestion, and one of his experiments involved measuring salivary responses in dogs, using a tube attached to the dogs' salivary glands.

One of Pavlov's students noticed that many dogs began salivating at the sight of the food or the food dish, the smell of the food, or even the sight of the person who delivered the food long before receiving the actual food. This "unscheduled" salivation was intriguing. Pavlov recognized that an involuntary reflex (salivation) that occurred before the appropriate stimulus (food) was presented could not be inborn and biological. It had to have been acquired through experience—through **learning**.

Excited by their accidental discovery, Pavlov and his students conducted several experiments. Their most basic method involved sounding a tone on a tuning fork just before food was placed in the dogs' mouths. After several pairings of tone and food, the dogs would salivate on hearing the tone alone. Pavlov and others went on to show that many things can be conditioned stimuli for salivation if they are paired with food—the ticking of a metronome, a buzzer, a light, and even the sight of a circle or triangle drawn on a card.

> **learning** A relatively permanent change in behavior or mental processes because of practice or experience.

The type of learning that Pavlov described came to be known as **classical conditioning** (FIGURE 6.1). To understand classical conditioning, you first need to realize that **conditioning** is just another word for learning. You also need to know that some responses are inborn and don't require conditioning. For example, the inborn salivary reflex consists of an **unconditioned stimulus (UCS)** and an **unconditioned response (UCR)**. That is, the UCS (food) elicits the UCR (salivation) without previous conditioning.

Before conditioning occurs, a **neutral stimulus (NS)** does not naturally elicit a relevant or consistent response. For example, as shown in Figure 6.1, Pavlov's dogs did not naturally salivate when a tone sounded. Similarly, as the figure shows, the sight of a cardboard box (neutral stimulus) doesn't naturally make a person hungry for a slice of pizza.

Pavlov's discovery was that learning occurs when a neutral stimulus such as a tone (or the cardboard box), is regularly paired with an unconditioned stimulus (food, or the smell of pizza in the cardboard box). The neutral stimulus (tone, or the cardboard box) then becomes a **conditioned stimulus (CS)**, which elicits a **conditioned response (CR)**—salivation.

> **classical conditioning** Learning that occurs when a neutral stimulus (NS) becomes paired (associated) with an unconditioned stimulus (UCS) to elicit a conditioned response (CR).

> **conditioning** The process of learning associations between environmental stimuli and behavioral responses.

Pavlov's classical conditioning FIGURE 6.1

Pavlov Example

Modern-day Example

Step 1
Before conditioning
The neutral stimulus produces no relevant response. The unconditioned (unlearned) *stimulus* elicits the unconditioned (unlearned) *response*.

Step 2
During conditioning
The neutral stimulus is repeatedly paired with the unconditioned (unlearned) *stimulus* to produce the unconditioned (unlearned) *response*.

Step 3
After conditioning
The neutral stimulus has become a conditioned (learned) *stimulus*, and it now produces a conditioned (learned) *response* that is usually similar to the unconditioned (unlearned) response.

Summary
An originally neutral stimulus comes to elicit a response that it did not previously elicit.

www.wiley.com/
college/carpenter

What does a salivating dog have to do with your life? Classical conditioning is the most fundamental way that all animals, including humans, learn most new responses, emotions, and attitudes. Your love for your parents (or significant other), the hatred and racism that led to the murder of James Byrd, and your drooling at the sight of chocolate cake or pizza are largely the result of classical conditioning.

In a famous experiment, John Watson and Rosalie Rayner (1920, 2000) demonstrated how the emotion of fear could be classically conditioned (FIGURE 6.2).

Watson and Rayner's experiment could not be performed today because it violates several ethical guidelines for scientific research (Chapter 1). Moreover, Watson and Rayner ended their experiment without extinguishing (removing) Albert's fear, although they knew that it could endure for a long period. Watson and Rayner also have been criticized because they did not measure Albert's fear objectively. Their subjective evaluation raises doubt about the degree of fear conditioned (Paul & Blumenthal, 1989).

Despite such criticisms, John Watson made important and lasting contributions to psychology. Unlike other psychologists of his time, Watson emphasized strictly observable behaviors, and he founded the school of *behaviorism*, which explains behavior as a result of observable stimuli and observable responses.

Watson's study of Little Albert also has had legendary significance for many psychologists. Watson showed us that many of our likes, dislikes, prejudices, and fears are **conditioned emotional responses**. Watson's research in producing Little Albert's fears also led to powerful clinical tools for eliminating extreme, irrational fears known as *phobias* (Chapter 13).

FINE-TUNING CLASSICAL CONDITIONING

Next, we outline six important principles of classical conditioning: stimulus generalization, stimulus discrimination, extinction, spontaneous recovery, reconditioning, and higher-order conditioning (FIGURE 6.3). These processes complicate classical conditioning, but they also underscore its fundamental nature.

Stimulus generalization occurs when an event similar to the originally conditioned stimulus triggers the same conditioned response. The more the stimulus resembles the conditioned stimulus, the stronger the conditioned response (Hovland, 1937). For example, after first conditioning dogs to salivate at the sound of low-pitched tones, Pavlov later demonstrated that the dogs would also salivate in response to higher-pitched tones. Similarly, after conditioning, the infant in Wat-

Conditioning and the case of Little Albert FIGURE 6.2

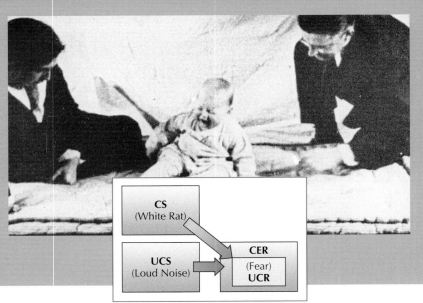

In the famous "Little Albert" study, a healthy 11-month-old child was first allowed to play with a white laboratory rat. Like most infants, Albert was curious and reached for the rat, showing no fear. Using the fact that infants are naturally frightened (UCR) by loud noises (UCS), Watson stood behind Albert and again put the rat (NS) near him. When the infant reached for the rat, Watson banged a steel bar with a hammer. The loud noise frightened Albert and made him cry. The white rat (NS) was paired with the loud noise (UCS) only seven times before the white rat alone produced a *conditioned emotional response* (CER) in Albert, fear of the rat.

CS
(White Rat)

UCS
(Loud Noise)

CER
(Fear)
UCR

"I don't care if she is a tape dispenser. I love her."

Which of the six basic principles of classical conditioning explain(s) this cartoon?

Answer: Stimulus generalization

Six principles of classical conditioning FIGURE 6.3

son and Rayner's experiment ("Little Albert") feared not only rats but also a rabbit, a dog, and a bearded Santa Claus mask.

Eventually, through the process of **stimulus discrimination** (a term that refers to a learned response to a specific stimulus, but not to other similar stimuli), Albert presumably learned to recognize differences between rats and other stimuli. As a result, he probably overcame his fear of Santa Claus, even if he remained afraid of white rats. Similarly, Pavlov's dogs learned to distinguish between the tone that signaled food and those that did not.

Most behaviors that are learned through classical conditioning can be weakened or suppressed through **extinction**. Extinction occurs when the unconditioned stimulus (UCS) is repeatedly withheld whenever the conditioned stimulus (CS) is presented, which gradually weakens the previous association. When Pavlov repeatedly sounded the tone without presenting food, the dogs' salivation gradually declined. Similarly, if you have a classically conditioned fear of cats and later start to work as a veterinary assistant, your fear would gradually diminish.

However, extinction is not unlearning—it does not "erase" the learned connection between the stimulus and the response (Bouton, 1994). In fact, on occasion an extinguished response may "spontaneously" reappear (**FIGURE 6.4**). This **spontaneous recovery**

The more often the UCS is withheld whenever the CS is presented, the lower an individual's response rate to the UCS, until extinction occurs.

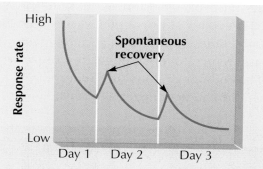

But an extinguished response may spontaneously reappear, which can trigger a rush of feelings and emotions as though the UCS were present.

Extinction and spontaneous recovery FIGURE 6.4

Children are not born salivating upon seeing the McDonald's golden arches. So why do they beg their parents to stop at "Mickey D's" after simply seeing a billboard for the restaurant? It is because of **higher-order conditioning**, which occurs when a neutral stimulus (NS) becomes a conditioned stimulus (CS) through re-

peated pairings with a previously conditioned stimulus (CS).

If you wanted to demonstrate higher-order conditioning in Pavlov's dogs, you would first condition the dogs to salivate in response to the sound of the tone (a). Then you would pair a flash of light with the tone (b). Eventually, the dogs would salivate in

response to the flash of light alone (c). Similarly, children first learn to pair McDonald's restaurants with food and later learn that two golden arches are a symbol for McDonald's. Their salivation and begging to eat at the restaurant upon seeing the arches are a classic case of higher-order conditioning (and successful advertising).

helps explain why you might suddenly feel excited at the sight of a former girlfriend or boyfriend, even though years have passed (and extinction has occurred). It also explains why couples who've recently broken up sometimes misinterpret a sudden flare-up of feelings and return to unhappy relationships. Furthermore, if a conditioned stimulus is reintroduced after extinction, the conditioning occurs much faster the

second time around—a phenomenon known as **reconditioning**. Both spontaneous recovery and reconditioning help underscore why it can be so difficult for us to break bad habits (such as eating too much chocolate cake) or internalize new beliefs (such as egalitarian racial beliefs). The phenomenon of **higher-order conditioning** (FIGURE 6.5) further expands and complicates our learned habits and associations.

CONCEPT CHECK STOP

Why did Pavlov's dogs salivate before being presented with any food?

When does stimulus discrimination occur?

Under what conditions might you expect to see spontaneous recovery?

Operant Conditioning

LEARNING OBJECTIVES

Explain how reinforcement and punishment influence behavior.

Describe Thorndike's and Skinner's contributions to operant conditioning research.

Identify examples of primary and secondary reinforcers.

Explain how different schedules of reinforcement affect behavior.

Describe the negative side effects of punishment.

---

operant conditioning
Learning in which voluntary responses are controlled by their consequences (also known as instrumental or Skinnerian conditioning).

reinforcement A consequence that strengthens a response and makes it more likely to recur.

punishment A consequence that weakens a response and makes it less likely to recur.

Consequences are the heart of **operant conditioning**. In classical conditioning, consequences are irrelevant—Pavlov's dogs still got to eat whether they salivated or not. But in operant conditioning, the organism performs a behavior (an **operant**) that produces either reinforcement or punishment. **Reinforcement** strengthens the response, making it more likely to recur. **Punishment** weakens the response, making it less likely to recur.

Classical and operant conditioning also differ in another important way. In classical conditioning, the organism's response is generally passive and involuntary. In operant conditioning, the organism's response is generally active and voluntary. The learner "operates" on the environment and produces consequences that influence whether the behavior will be repeated. (If your friends smile and laugh when you tell a joke, you are likely to joke more. If they frown, groan, or ridicule you, you are likely to joke less.)

THE BEGINNINGS OF OPERANT CONDITIONING

Edward Thorndike (1874–1949), a pioneer of operant conditioning, determined that the frequency of a behavior is modified by its consequences. He developed the **law of effect** (Thorndike, 1911), a first step in understanding how consequences can modify active, voluntary behaviors (**FIGURE 6.6**).

B. F. Skinner (1904–1990) extended Thorndike's law of effect to more complex behaviors. He emphasized that reinforcement and punishment always occur after the behavior of interest has occurred. In addition, Skinner cautioned that the only way to know how we have influenced someone's behavior is to check whether it increases or decreases.

law of effect
Thorndike's rule that the probability of an action being repeated is strengthened when followed by a pleasant or satisfying consequence.

Thorndike box FIGURE 6.6

In one famous experiment, Thorndike put a cat inside a specially built puzzle box. When the cat stepped on a pedal inside the box (at first, through trial-and-error), the door opened and the cat could get out to eat. With each additional success, the cat's actions became more purposeful, and it soon learned to open the door immediately (from Thorndike, 1898).

Sometimes, he noted, we think we're reinforcing or punishing, when we're actually doing the opposite. For example, a professor may think she is encouraging shy students to talk by repeatedly praising them each time they speak up in class. But what if shy students are embarrassed by this attention? If so, they may decrease the number of times they talk in class.

REINFORCEMENT: STRENGTHENING A RESPONSE

Reinforcers, which strengthen a response, can be grouped into two types, primary and secondary. Primary reinforcers satisfy an intrinsic, unlearned biological need (food, water, sex). Secondary reinforcers are not intrinsic; the value of this reinforcer is learned (money, praise, attention). Each type of reinforcer can produce **positive reinforcement** or **negative reinforcement** (TABLE 6.1), depending on whether certain stimuli are added or taken away.

It's easy to confuse negative reinforcement with punishment, but the two concepts are actually completely opposite. Reinforcement (either positive or negative) strengthens a behavior, whereas punishment weakens a behavior. If the terminology seems confusing, it may help to think of positive and negative reinforcement in the mathematical sense (that is, in terms of something being added [+] or taken away [−]) rather than in terms of good and bad.

When you make yourself do some necessary but tedious task—say, paying bills—before letting yourself go to the movies, you are not only using positive reinforcement. You are also using the **Premack principle**, named after psychologist David Premack. Recognizing that you love to go to movies, you intuitively tie your less-desirable, low-frequency activities (paying bills) to your high-frequency behavior (going to the movies).

What are the best circumstances for using reinforcement? It depends on the desired outcome. To make this decision, you need to understand various **schedules of reinforcement** (Terry, 2003). This term refers to the rate or interval at which responses are reinforced. Although there are numerous schedules of reinforcement, the most important distinction is whether they are continuous or partial. When Skinner was training his animals, he found that learning was most rapid if the response was reinforced every time it occurred—a procedure called **continuous reinforcement**.

How reinforcement strengthens and increases behaviors TABLE 6.1		
	Positive reinforcement *Adds to (+) and strengthens behavior*	**Negative reinforcement** *Takes away (−) and strengthens behavior*
Primary reinforcers Satisfy an unlearned biological need	You hug your baby and he smiles at you. The "addition" of his smile strengthens the likelihood that you will hug him again.	Your baby is crying, so you hug him and he stops crying. The "removal" of crying strengthens the likelihood that you will hug him again.
	You do a favor for a friend and she buys you lunch in return.	You take an aspirin for your headache, which takes away the pain.
Secondary reinforcers Value is learned, not intrinsic	You increase profits and receive $200 as a bonus.	After high sales, your boss says you won't have to work on weekends.
	You study hard and receive a good grade on your psychology exam.	Professor says you won't have to take the final exam because you did so well on your unit exam.

As you have probably noticed, real life seldom provides continuous reinforcement. Yet your behavior persists because your efforts are occasionally rewarded. Most everyday behavior is rewarded on a **partial (or intermittent) schedule of reinforcement**, which involves reinforcing only some responses, not all (Sangha et al., 2002).

Once a task is well learned, it is important to move to a partial schedule of reinforcement. Why? Because under partial schedules, behavior is more resistant to extinction. There are four partial schedules of reinforcement: **fixed ratio** (FR), **variable ratio** (VR), **fixed interval** (FI), and **variable interval** (VI). The type of partial schedule selected depends on the type of behavior being studied and on the speed of learning desired (Neuringer, Deiss, & Olson, 2000). A fixed ratio leads to the highest overall response rate, but each of the four types of partial schedules has different advantages and disadvantages (**TABLE 6.2**).

Each of the four schedules of partial reinforcement is important for *maintaining* behavior. But how would you teach someone to play the piano or to speak a foreign language? For new and complex behaviors such as these, which aren't likely to occur naturally, **shaping** is an especially valuable tool. Skinner believed that shaping explains a variety of abilities that each of us possess, from eating with a fork, to playing a musical instrument, to driving a stick-shift car. Parents, athletic coaches,

■ **shaping** Reinforcement by a series of successively improved steps leading to desired response.

Four schedules of reinforcement TABLE 6.2

		Definitions	Response rates	Examples
Ratio schedules (response based)	**Fixed ratio (FR)**	Reinforcement occurs after a predetermined set of responses; the ratio (number or amount) is fixed	Produces a high rate of response, but a brief dropoff just after reinforcement	Car wash employee receives $10 for every 3 cars washed. In a laboratory, a rat receives a food pellet every time it presses the bar 7 times.
	Variable ratio (VR)	Reinforcement occurs unpredictably; the ratio (number or amount) varies	High response rates, no pause after reinforcement, and very resistant to extinction	Slot machines are designed to pay out after an average number of responses (maybe every 10 times), but any one machine may pay out on the first response, then seventh, then the twentieth.
Interval schedules (time based)	**Fixed interval (FI)**	Reinforcement occurs after a predetermined time has elapsed; the interval (time) is fixed	Responses tend to increase as the time for the next reinforcer is near, but drop off after reinforcement and during interval	You can get a monthly paycheck. Rat's behavior is reinforced with a food pellet when (or if) it presses a bar after 20 seconds have elapsed.
	Variable interval (VI)	Reinforcement occurs unpredictably; the interval (time) varies	Relatively low response rates, but they are steady because the non-human animal or person cannot predict when reward will come	In a class with pop quizzes, you study at a slow but steady rate because you can't anticipate the next quiz. Rat's behavior is reinforced with a food pellet after a response and a variable, unpredictable interval of time.

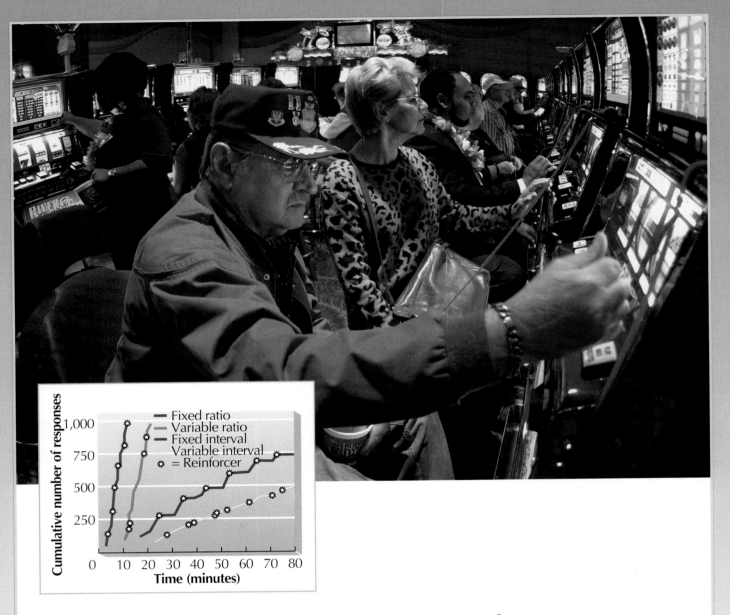

Partial Reinforcement Keeps 'em Coming Back

Have you noticed that people spend long hours pushing buttons and pulling levers on slot machines in hopes of winning the jackpot? This compulsion to keep gambling in spite of significant losses is evidence of the strong resistance to extinction with partial (or intermittent) schedules of reinforcement. Machines in Nevada casinos, by law, have payout rates of at least 75%—that is, for every dollar spent, a player "wins" $0.75. Different machines are programmed different ways. Some meet the percentage by giving very large, but infre-

quent payouts to a lucky winner. Others give frequent smaller payouts to all players. In either case, people are reinforced just often enough to keep them coming back.

This type of partial reinforcement also helps parents maintain children's positive behaviors, such as toothbrushing and bed making. After the child has initially learned these behaviors with continuous reinforcement, occasional, partial reinforcement is the most efficient way to maintain the behavior. (Chart adapted from Skinner, 1961.)

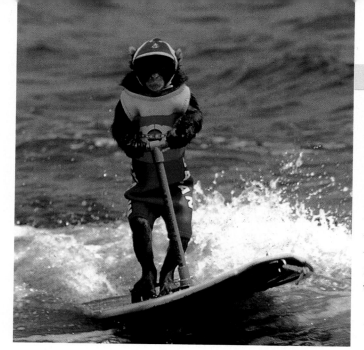

Momoko, a female monkey, is famous in Japan for her water-skiing, deep-sea diving, and other amazing abilities. Animal trainers use several principles of reinforcement to shape her behavior in many successive steps. First, they reinforced Momoko (with a small food treat) for standing or sitting on the water ski. Then they reinforced her each time she put her hands on the pole. Next, they slowly dragged the water ski on dry land and reinforced her for staying upright and holding the pole. Then they took Momoko to a shallow and calm part of the ocean and reinforced her for staying upright and holding the pole as the ski moved in the water. Finally, they took her into the deep part of the ocean.

teachers, and animal trainers all use shaping techniques (FIGURE 6.7).

PUNISHMENT: WEAKENING A RESPONSE

Unlike reinforcement, punishment *decreases* the strength of a response. As with reinforcement, there are two kinds of punishment—positive and negative (Gordon, 1989; Skinner, 1953)

Positive punishment is the addition (+) of a stimulus that decreases (or weakens) the likelihood of the response occurring again. **Negative punishment** is the taking away (−) of a reinforcing stimulus, which decreases (or weakens) the likelihood of the response occurring again (TABLE 6.3). (To check your understanding of the principles of reinforcement and punishment, see FIGURE 6.8 on the next page.)

Punishment is a tricky business, and it isn't always intentional. Any process that adds or takes away something and causes a behavior to decrease is punishment. Thus, if parents ignore all the A's on their child's report card ("taking away" encouraging comments) and ask repeated questions about the B's and C's, they may unintentionally be punishing the child's excellent grade achievement and weakening the likelihood of future A's. Similarly, dog owners who yell at or spank their

How punishment weakens and decreases behaviors TABLE 6.3

Positive punishment *Adds stimulus (+) and weakens the behavior*	Negative punishment *Takes stimulus away (−) and weakens the behavior*
You must run 4 extra laps in your gym class because you were late.	You're excluded from gym class because you were late.
A parent adds chores following a child's poor report card.	A parent takes away a teen's cell phone following a poor report card.
Your boss complains about your performance.	Your boss reduces your expense account after a poor performance.

The Skinner box application FIGURE 6.8

To test his behavioral theories, Skinner used an animal, usually a pigeon or a rat, and an apparatus that has come to be called a **Skinner Box**. In Skinner's basic experimental design, an animal such as a rat received a food pellet each time it pushed a lever, and the number of responses was recorded. Note in this drawing that an electric grid on the cage floor could be used to deliver small electric shocks.

Test yourself

Is this positive reinforcement, negative reinforcement, positive punishment, or negative punishment? Fill in the name of the appropriate learning principle in the spaces provided in each box.

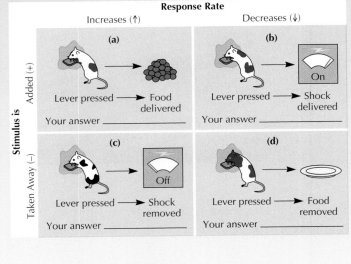

Answers: (a) positive reinforcement, (b) positive punishment, (c) negative reinforcement, (d) negative punishment.

---

dogs for finally coming to them ("adding" verbal or physical aggression) after being called several times are actually punishing the desired behavior—coming when called.

Punishment plays an unavoidable role in our social world. Dangerous criminals must be stopped and possibly removed from society. Parents must stop their teenagers from drinking and driving. Teachers must stop disruptive students in the classroom and bullies on the playground. Yet punishment can be problematic (Javo et al., 2004; Reis et al., 2004; Saadeh, Rizzo, & Roberts, 2002).

To be effective, punishment should be immediate and consistent. However, in the real world, this is extremely hard to do. Police officers cannot stop every driver every time they speed. To make matters worse, when punishment is not immediate, during the delay the behavior is likely to be reinforced on a partial schedule, which makes it highly resistant to extinction. Think about gambling. It should be a punishing situation—gamblers usually lose far more than they win. However, the fact that they occasionally win keeps gamblers "hanging in there."

Even if punishment immediately follows the misbehavior, the recipient may learn what not to do but not necessarily what he or she should do. It's much more efficient to teach someone by giving him or her clear examples of correct behavior, rather than simply punishing the incorrect behavior. Finally, punishment can have serious side effects (FIGURE 6.9).

Side effects of punishment FIGURE 6.9

1. **Increased aggression.** Because punishment often produces a decrease in undesired behavior, at least for the moment, the punisher is in effect rewarded for applying punishment. Thus, a vicious circle may be established in which both the punisher and recipient are reinforced for inappropriate behavior—the punisher for punishing and the recipient for being fearful and submissive. This side effect partially explains the escalation of violence in family abuse and bullying (Javo et al., 2004; Larzelere & Johnson, 1999). In addition to fear and submissiveness, the recipient also might become depressed and/or respond with his or her own form of aggression.

2. **Passive aggressiveness.** For the recipient, punishment often leads to frustration, anger, and eventually aggression. But most of us have learned from experience that retaliatory aggression toward a punisher (especially one who is bigger and more powerful) is usually followed by more punishment. We therefore tend to control our impulse toward open aggression and instead resort to more subtle techniques, such as showing up late or forgetting to mail a letter for someone. This is known as *passive aggressiveness* (Gilbert, 2003; Stormshak et al., 2000).

3. **Avoidance behavior.** No one likes to be punished, so we naturally try to avoid the punisher. If every time you come home a parent or spouse starts yelling at you, you will delay coming home or find another place to go.

4. **Modeling.** Have you ever seen a parent spank or hit a child for hitting another child? The punishing parent may unintentionally serve as a "model" for the same behavior he or she is attempting to stop.

5. **Temporary suppression.** Do you notice that car drivers quickly slow down when they see a police car but quickly resume their previous speed once the police officer is out of sight? Punishment generally suppresses the behavior only temporarily during the presence of the punishing person or circumstances.

6. **Learned helplessness.** Why do some people stay in abusive homes or marital situations? Research shows that if you repeatedly fail in your attempts to control your environment, you acquire a general sense of powerlessness or *learned helplessness* and you may make no further attempts to escape (Alloy & Clements, 1998; Seligman, 1975; Shors, 2004; Zhukov & Vinogradova, 2002).

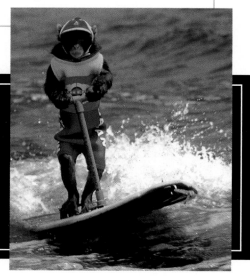

CONCEPT CHECK STOP

What is the difference between negative reinforcement and punishment?

What are the law of effect and the Premack principle?

Why does a fixed ratio lead to the highest rate of response, compared with other schedules of reinforcement?

When is shaping most useful in learning?

Cognitive-Social Learning

LEARNING OBJECTIVES

Explain how insight and latent learning differ from the principles of classical and operant conditioning.

Summarize the evidence that animals (including humans) form internal, cognitive maps.

Identify Bandura's necessary conditions for observational learning.

So far, we have examined learning processes that involve associations between a stimulus and an observable behavior. Although some behaviorists believe that almost all learning can be explained in such stimulus–response terms, other psychologists feel that there is more to learning than can be explained solely by operant and classical conditioning. **Cognitive-social theory** (also called cognitive-social learning or cognitive-behavioral theory) incorporates the general concepts of conditioning, but rather than relying on a simple S–R (stimulus and response) model, this theory emphasizes the interpretation or thinking that occurs within the organism: S–O–R (stimulus–organism–response). According to this view, animals (including humans) have attitudes, beliefs, expectations, motivations, and emotions that affect learning. Furthermore, both human and nonhuman animals are social creatures that are capable of learning new behaviors through observation and imitation of others. We begin with a look at the *cognitive* part of cognitive-social theory, followed by an examination of the *social* aspects of learning.

cognitive-social theory A perspective that emphasizes the roles of thinking and social learning in behavior.

INSIGHT AND LATENT LEARNING: WHERE ARE THE REINFORCERS?

Early behaviorists likened the mind to a "black box" whose workings could not be observed directly, but German psychologist Wolfgang Köhler wanted to look inside the box. He believed that there was more to learning—especially learning to solve a complex problem—than responding to stimuli in a trial-and-error fashion. In one of a series of experiments, he placed a banana just outside the reach of a caged chimpanzee (Figure 6.10). To reach the banana, the chimp had to

Is this Insight? FIGURE 6.10

Grande, one of Köhler's chimps, has just solved the problem of how to get the banana. (Also, the chimp in the foreground is engaged in observational learning, our next topic.)

use a stick to extend its reach. Köhler noticed that the chimp did not solve this problem in a random trial-and-error fashion but, instead, seemed to sit and think about the situation for a while. Then, in a flash of insight, the chimp picked up the stick and maneuvered the banana to within its grasp (Köhler, 1925).

Another of Köhler's chimps, an intelligent fellow named Sultan, was put in a similar situation. This time there were two sticks available to him, and the banana was placed even farther away, too far to reach with a single stick. Sultan seemingly lost interest in the banana, but he continued to play with the sticks. When he later discovered that the two sticks could be interlocked, he instantly used the now longer stick to pull the banana to within his grasp. Köhler designated this type of learning **insight learning** because some internal mental event that he could only describe as "insight" went on between the presentation of the banana and the use of the stick to retrieve it.

Like Köhler, Edward C. Tolman (1898–1956) believed that previous researchers underestimated animals' cognitive processes and cognitive learning. He noted that, when allowed to roam aimlessly in an experimental maze with no food reward at the end, rats seemed to develop a **cognitive map**, or mental representation of the maze.

insight Sudden understanding of a problem that implies the solution.

latent learning
Hidden learning that exists without behavioral signs.

To test the idea of cognitive learning, Tolman allowed one group of rats to aimlessly explore a maze with no reinforcement. A second group was reinforced with food whenever they reached the end of the maze. The third group was not rewarded during the first 10 days of the trial, but starting on day 11 they found food at the end of the maze. As expected from simple operant conditioning, the first and third groups were slow to learn the maze, whereas the second group, which had reinforcement, showed fast, steady improvement. However, when the third group started receiving reinforcement (on the 11th day), their learning quickly caught up to the group that had been reinforced every time (Tolman & Honzik, 1930). This showed that the nonreinforced rats had been thinking and building cognitive maps of the area during their aimless wandering and that their **latent learning** only showed up when there was a reason to display it (the food reward).

Cognitive learning is not limited to rats. For example, a chipmunk will pay little attention to a new log in its territory (after initially checking it for food). When a predator comes along, however, the chipmunk heads directly for and hides beneath the log. Recent experiments (Burgdorf, Knutson, & Panksepp, 2000) provide additional clear evidence of latent learning and the existence of internal, cognitive maps.

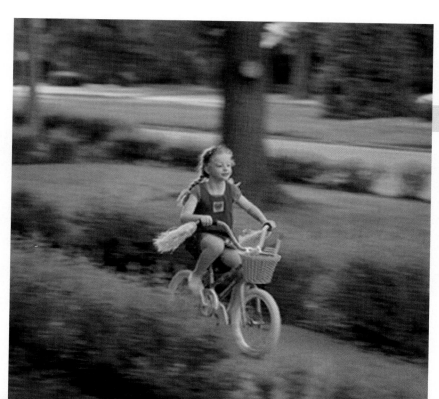

Is this learning? FIGURE 6.11

This child often rides through her neighborhood for fun, without a specific destination. Could she tell her mother where the nearest mailbox is located? What type of learning is this?

OBSERVATIONAL LEARNING: WHAT WE SEE IS WHAT WE DO

In addition to classical and operant conditioning and cognitive processes (such as insight and latent learning), we also learn many things through **observational learning**. From birth to death, observational learning is very important to our biological, psychological, and social survival (the *biopsychosocial model*). Watching others helps us avoid dangerous stimuli in our environment, teaches us how to think and feel, and shows us how to act and interact socially.

observational learning Learning new behavior or information by watching others (also known as social learning or modeling).

Some of the most compelling examples of observational learning come from the work of Albert Bandura and his colleagues (Bandura, 2003; Bandura, Ross, & Ross, 1961; Bandura & Walters, 1963). In several experiments, children watched an adult kick, punch, and shout at an inflated Bobo doll. Later, children who had seen the aggressive adult were much more aggressive with the Bobo doll than children who had not seen the aggression. In other words, "monkey see, monkey do" (FIGURE 6.12).

According to Bandura, observational learning requires at least four separate processes: attention, retention, motor reproduction, and reinforcement (FIGURE 6.13 on facing page).

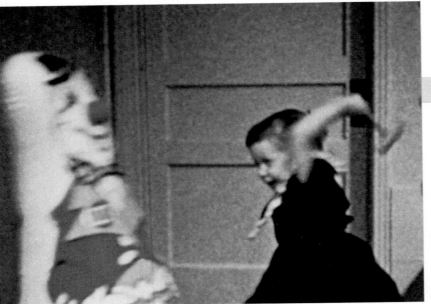

Bandura's classic Bobo doll studies FIGURE 6.12

Bandura's "Bobo doll" study is considered a classic in psychology. It proved that children will imitate models they observe. Why is this new or important? Are there circumstances where observational learning (for example, gleaned from television) could have positive effects?

CONCEPT CHECK STOP

Why were Köhler's studies of insight in chimpanzees important?

Why was Bandura's work on modeling of aggressiveness important?

What is a cognitive map?

Attention. Observational learning requires attention. This is why teachers insist on having students watch their demonstrations.

Retention. To learn a complex new dance step, we need to carefully note and remember the instructor's directions and demonstrations.

Motor reproduction. Observational learning cannot occur if we lack the motor skills necessary to imitate the model.

Reinforcement. We are more likely to repeat a modeled behavior if the model was reinforced for the behavior.

The Biology of Learning

LEARNING OBJECTIVES

Explain how an animal's environment might affect learning and behavior.

Identify an example of biological preparedness.

Describe how instinctive drift constrains learning.

So far in this chapter, we have considered how external forces—from reinforcing events to our observations of others—affect learning. But we also know that for changes in behavior to persist over time, lasting biological changes must occur within the organism. In this section, we will examine neurological and evolutionary influences on learning.

NEUROSCIENCE AND LEARNING: THE ADAPTIVE BRAIN

Each time we learn something, either consciously or unconsciously, that experience creates new synaptic connections and alterations in a wide network of brain structures, including the cortex, cerebellum, hypothal-amus, thalamus, and amygdala (Debaere et al., 2004; Fanselow & Poulos, 2005; Pelletier & Paré, 2004; Thompson, 2005).

Evidence that experience changes brain structure first emerged in the 1960s from studies of animals raised in enriched versus deprived environments. Compared with rats raised in a stimulus-poor environment, those raised in a colorful, stimulating "rat Disneyland" had a thicker cortex, increased nerve growth factor (NGF), more fully developed synapses, more dendritic branching, and improved performance on many tests of learning (Guilarte et al., 2003; Pham et al., 2002; Rosenzweig & Bennett, 1996).

Admittedly, it is a big leap from rats to humans, but research suggests that the human brain also responds to environmental conditions (**FIGURE 6.14**). For example, older adults who are exposed to stimulat-

Daily enrichment FIGURE 6.14

For humans and non-human animals alike, environmental conditions play an important role in enabling learning. How might a classroom rich with stimulating toys, games, and books foster intellectual development in young children?

ing environments generally perform better on intellectual and perceptual tasks than those who are in restricted environments (Schaie, 1994).

EVOLUTION AND LEARNING: BIOLOGICAL PREPAREDNESS AND INSTINCTIVE DRIFT

Humans and other animals are born with various innate reflexes and instincts. Although these biological tendencies help ensure evolutionary survival, they are inherently inflexible. Only through learning are we able to react to important environmental cues—such as spoken words and written symbols—that our innate reflexes and instincts do not address. Thus, from an evolutionary perspective, learning is an adaptation that enables organisms to survive and prosper in a constantly changing world.

Because animals can be operantly conditioned to perform a variety of novel behaviors (like waterskiing), learning theorists initially believed that the fundamental laws of conditioning would apply to almost all species and all behaviors. However, researchers have identified several biological constraints that limit the generality of conditioning principles. These include biological preparedness and instinctive drift.

Years ago, a young woman named Rebecca unsuspectingly bit into a Butterfinger candy bar filled with small, wiggling maggots. Horrified, she ran gagging and screaming to the bathroom. Many years later, Rebecca still feels nauseated when she sees a Butterfinger candy bar (but, fortunately, she doesn't feel similarly nauseated by the sight of her boyfriend, who bought her the candy).

Rebecca's graphic (and true!) story illustrates an important evolutionary process. When a food or drink is associated with nausea or vomiting, that particular food or drink can become a conditioned stimulus (CS) that triggers a conditioned **taste aversion**. Like other classically conditioned responses, taste aversions develop involuntarily (**FIGURE 6.15**).

Can you see why this automatic response would be adaptive? If one of our cave-dwelling ancestors became ill after eating a new plant, it would increase his or her chances for survival if he or she immediately

Taste aversion in the wild FIGURE 6.15

In applied research, Garcia and his colleagues used classical conditioning to teach coyotes not to eat sheep (Gustavson & Garcia, 1974). The researchers began by lacing freshly killed sheep with a chemical that caused extreme nausea and vomiting in the coyotes that ate the tainted meat. The conditioning worked so well that the coyotes would run away from the mere sight and smell of sheep. This taste aversion developed involuntarily. This research has since been applied many times in the wild and in the laboratory with coyotes and other animals (Aubert & Dantzer, 2005; Cooper et al., 2002; Domjan, 2005; Nakajima & Masaki, 2004).

biological preparedness Built-in (innate) readiness to form associations between certain stimuli and responses.

developed an aversion to that plant—but not to other family members who might have been present at the time. Similarly, people tend to develop phobias of snakes, darkness, spiders, and heights more easily than of guns, knives, and electrical outlets. We apparently inherit a built-in (innate) readiness to form associations between certain stimuli and responses. This is known as **biological preparedness**.

Laboratory experiments have provided general support for both taste aversion and biological preparedness. For example, Garcia and his colleagues (1966) produced taste aversion in lab rats by pairing flavored water (NS) and a drug (UCS) that produced gastrointestinal distress (UCR). After being conditioned and then recovering from the illness, the rats refused to drink the flavored water (CS) because of the conditioned taste aversion. Remarkably, however, Garcia discovered that only certain neutral stimuli could produce the nausea. Pairings of a noise (NS) or a shock (NS) with the nausea-producing drug (UCS) produced no

taste aversion. Garcia suggested that when we are sick to our stomachs, we have a natural, evolutionary tendency to attribute it to food or drink. Being biologically prepared to quickly associate nausea with food or drink is adaptive because it helps us avoid that or similar food or drink in the future (Cooper et al., 2002; Domjan, 2005; Garcia, 2003).

Just as Garcia couldn't produce noise-nausea associations, other researchers have found that an animal's natural behavior pattern can interfere with operant conditioning. For example, the Brelands (1961) tried to teach a chicken to play baseball. Through shaping and reinforcement, the chicken first learned to pull a loop that activated a swinging bat and then learned to actually hit the ball. But instead of running to first base, it would chase the ball as if it were food. Regardless of the lack of reinforcement for chasing the

instinctive drift The tendency of some conditioned responses to shift (or drift) back toward innate response pattern.

ball, the chicken's natural behavior took precedence. This biological constraint is known as **instinctive drift**.

CONCEPT CHECK **STOP**

What changes occur in rats' brains when they are raised in an enriched (versus deprived) environment?

Why is taste aversion evolutionarily adaptive?

Why don't the fundamental laws of conditioning apply to all species and all behaviors?

Conditioning and Learning in Everyday Life

 n this section, we discuss several everyday applications for classical conditioning, operant conditioning, and cognitive-social learning.

CLASSICAL CONDITIONING: FROM PREJUDICE TO PHOBIAS

One common—and very negative—instance of classical conditioning is prejudice. In a classic study in the 1930s, Kenneth Clark and Mamie P. Clark (1939) found that given a choice, both black and white children preferred white dolls to black dolls. When asked which doll was good and which was bad, both groups of children responded that the white doll was good and nice and that the black doll was bad, dirty, and ugly.

The Clarks reasoned that the children had learned to associate inferior qualities with darker skin and more positive qualities with lighter skin. You may think that this 1930s study no longer applies, but follow-up research in the late 1980s found that 65 percent of the African American children and 74 percent of the white children still preferred the white doll (Powell-Hopson & Hopson, 1988) (**FIGURE 6.16**).

Classical conditioning is also a primary tool for marketing and advertising professionals, filmmakers and musicians, medical practitioners, psychotherapists, and politicians who want to influence our purchases, motivations, emotions, health behavior, and votes.

Instances of classical conditioning are also found in the medical field. For example, for alcohol-addicted patients, some hospitals pair the smell and taste of alcohol with a nausea-producing drug. Afterward, just the smell or taste of alcohol makes the person sick. Some patients, but not all, have found this treatment successful (Chapter 14).

Finally, researchers have found that classically conditioned emotional responses explain most everyday fears and even most *phobias*, which are exaggerated and irrational fears of a specific object or situation (Rauhut, Thomas, & Ayres, 2001; Ressler & Davis, 2003). The good news is that extreme fears—for example, of heights, spiders, or public places—can be effectively treated with *behavior therapy* (Chapter 14).

Conditioned race prejudice FIGURE 6.16

The Clarks' findings played a pivotal role in the famous *Brown v. Board of Education of Topeka* decision in 1954, which ruled that segregation of public facilities was unconstitutional.

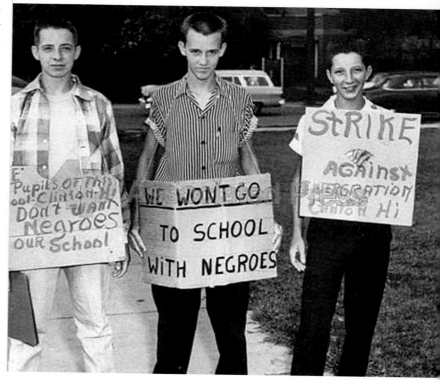

Classical Conditioning as a Marketing Tool

Marketers have mastered numerous classical conditioning principles. For example, TV commercials, magazine ads, and business promotions often pair a company's products or logo (the neutral stimulus/NS) with pleasant images, such as attractive models and celebrities (the conditioned stimulus/CS), which, through higher-order conditioning, automatically trigger favorable responses (the conditioned response/CR). Advertisers hope that after repeated viewings, the previously neutral stimulus (the company's products or logo) will become a conditioned stimulus that elicits favorable responses (CR)—we buy the advertised products. Researchers caution that these ads also help produce visual stimuli that trigger conditioned responses, such as urges to smoke, overeat, and drink alcohol (Dols et al., 2000; Martin et al., 2002; Tirodkar & Jain, 2003; Wakefield et al., 2003).

To appreciate the influences of classical conditioning on your own life, try this:

- Look through a popular magazine and find several advertisements. What images are used as the unconditioned stimulus (UCS) or conditioned stimulus (CS)? Note how you react to these images.

- While watching a movie or a favorite TV show, note what sounds and images are used as conditioned stimuli (CS). (Hint: certain types of music are used to set the stage for happy stories, sad events, and fearful situations.) What are your conditioned emotional responses (CERs)?

OPERANT CONDITIONING: PREJUDICE, BIOFEEDBACK, AND SUPERSTITION

Just as people can learn prejudice through classical conditioning, we also can learn prejudice through operant conditioning. Demeaning others gains attention and sometimes approval from others, as well as increases one's self-esteem (at the expense of the victim), positively reinforcing prejudice and discrimination (Fein & Spencer, 1997; Rudman & Fairchild, 2004) (**FIGURE 6.17A**). People also may have a single negative (punishing) experience with a specific member of a group, which they then generalize and apply to all members of the group (Vidmar, 1997)—an example of stimulus generalization.

But what explains behavior that is as extreme as that of the men who murdered James Byrd? Why would people do anything that they know could bring the death penalty? Punishment does weaken and suppress behavior, but as mentioned before, to be effective it must be consistent and immediate. Unfortunately, this seldom happens. Instead, when a criminal gets away with one or more crimes, that criminal behavior is put on a partial (intermittent) schedule of reinforcement, making it more likely to be repeated and to become more resistant to *extinction*.

Biofeedback is another example of operant conditioning in action (**FIGURE 6.17B**). Something is added (feedback) that increases the likelihood that the behavior will be repeated—positive reinforcement. The biofeedback itself is a secondary reinforcer because of the learned value of the relief from pain or other aversive stimuli (primary reinforcer). Finally, biofeedback involves shaping. A person using biofeedback watches a monitor screen (or other instrument) that provides graphs or numbers indicating some bodily state. Like a mirror, the biofeedback reflects back the results of the various strategies that the participant uses to gain control. Through trial and error, the participant gets progressively better at making the desired changes.

Prejudice and Discrimination

What reinforcement might these boys receive for teasing this girl? Can you see how operant conditioning, beginning at an early age, can promote unkind or prejudiced behavior?

Biofeedback

In biofeedback, internal bodily processes (like blood pressure, muscle tension, or brain wave patterns) are electrically recorded and reported back to the patient. This information helps the person gain control over processes that are normally involuntary. Researchers have successfully used biofeedback (usually in conjunction with other techniques, such as behavior modification) to treat hypertension, anxiety, epilepsy, urinary incontinence, cognitive functioning, chronic pain, and headache (Hammond, 2004; Moss, 2004; Penzien, Rains, & Andrasik, 2002; Stetter & Kupper, 2002; Tatrow, Blanchard, & Silverman, 2003).

Superstition

Like Skinner's pigeons, we humans also believe in many superstitions that may have developed from accidental reinforcement—from wearing "something old" during one's wedding, to knocking on wood for good fortune, to performing some bizarre ritual before every athletic competition.

Even accidental reinforcement can exert a powerful effect, sometimes causing superstitious behavior (FIGURE 6.17C). In a fascinating experiment, B. F. Skinner (1948, 1992) set the feeding mechanism in the cages of eight pigeons to release food once every 15 seconds. No matter what the birds did, they were reinforced at 15-second intervals. Six of the pigeons acquired behaviors that they repeated over and over, even though the behaviors were not necessary to receive the food. For example, one pigeon kept turning in counterclockwise circles, and another kept making jerking movements with its head. Why did this happen?

Although Skinner was not using the food to reinforce any particular behavior, the pigeons associated the food with whatever behavior they were engaged in when the food was originally dropped into the cage.

COGNITIVE-SOCIAL LEARNING: WE SEE, WE DO?

We use cognitive-social learning in many ways in our everyday lives, yet two of the most powerful areas of learning are frequently overlooked—prejudice and media influences. One of James Byrd's murderers, Bill King, had numerous tattoos that proudly proclaimed his various prejudices (FIGURE 6.18). King's family and friends insist that he was a pleasant and quiet man until he began serving an eight-year prison sentence for burglary (Galloway, 1999). What did he learn about prejudice during his prison sentence? Did he model his killing of Byrd after his uncle's well-known killing of a gay traveling salesman a number of years earlier? Or did he learn his prejudices during his numerous years of active membership with the Ku Klux Klan?

The media also propagates some forms of prejudice. When children watch television, go to the movies, and read books and magazines that portray minorities and women in demeaning and stereotypical roles, they learn to expect these behaviors and to accept them as "natural." Exposure of this kind initiates and reinforces the learning of prejudice (Blaine & McElroy, 2002; Neto & Furnham, 2005).

Both children and adults may be learning other destructive behaviors through observational learning and the media, as well. Correlational evidence from

John William "Bill" King FIGURE 6.18

King was sentenced to death for the murder of James Byrd. How might King and his two accomplices have learned some of their hatred and prejudice through observation and modeling?

Video games and aggression FIGURE 6.19

Researchers hypothesize that video games are more likely to model aggressive behavior because, unlike TV and other media, they are interactive, engrossing, and require the player to identify with the aggressor. What forms of learning might be involved in this phenomenon?

more than 50 studies indicates that observing violent behavior is related to later desensitization and the performance of violent behavior (Anderson, 2004; Coyne, Archer & Elsea, 2004; Kronenberger et al., 2005). In addition, more than 100 experimental studies have also shown a causal link between observing violence and later performing it (Primavera & Herron, 1996).

Researchers are just beginning to study how video games (and virtual reality games) affect behavior (FIGURE 6.19). For example, studies have found that students who play more violent video games in junior high and high school also engage in more aggressive behaviors (Anderson & Bushman, 2001; Bartholow & Anderson, 2002; Carnagey & Anderson, 2004). Furthermore, psychologists Craig Anderson and Karen Dill (2000) first assigned 210 students to play either a violent or a nonviolent video game and later allowed them to punish their opponents with a loud sound blast. The researchers found that those who played the violent game punished their opponents not only longer but also with greater intensity.

CONCEPT CHECK STOP

HOW do classical conditioning, operant conditioning, and cognitive-social learning all contribute to prejudice?

HOW can "accidental reinforcement" affect behavior?

Why might video games (and virtual reality games) foster aggressiveness?

What is happening in this picture ?

For political candidates, kissing babies seems to be as critical as pulling together a winning platform and airing persuasive campaign ads. What principle of learning explains why politicians kiss babies?

What other stimuli or symbols might elicit similar effects?

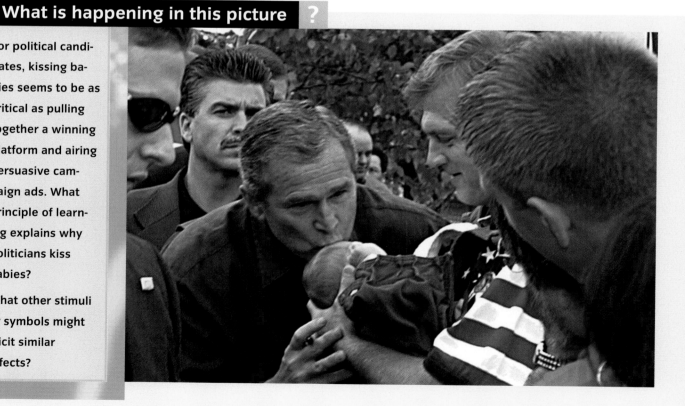

Conditioning and Learning in Everyday Life   167

VISUAL SUMMARY

1 Classical Conditioning

1. Learning is a relatively permanent change in behavior or mental processes as a result of practice or experience. Pavlov discovered a fundamental form of **conditioning** (learning) called **classical conditioning**, in which a neutral stimulus becomes associated with an unconditioned stimulus to elicit a conditioned response. In the "Little Albert" experiment, Watson and Rayner demonstrated how many of our likes, dislikes, prejudices, and fears are conditioned emotional responses.

2. Stimulus generalization occurs when an event similar to the originally conditioned stimulus triggers the same conditioned response. With experience, animals learn to distinguish between an original conditioned stimulus and similar stimuli—stimulus discrimination. Most learned behaviors can be weakened through extinction. Extinction is a gradual weakening or suppression of a previously conditioned response. However, if a conditioned stimulus is reintroduced after extinction, an extinguished response may spontaneously recover.

2 Operant Conditioning

1. In **operant conditioning**, an organism performs a behavior that produces either reinforcement or punishment. **Reinforcement** strengthens the response, while **punishment** weakens the response.

2. Thorndike developed the **law of effect**, a first step in understanding how consequences can modify voluntary behaviors. Skinner extended Thorndike's law of effect to more complex behaviors.

3. Reinforcers can be either primary or secondary, and reinforcement can be either positive or negative.

4. "Schedules of reinforcement" refers to the rate or interval at which responses are reinforced. Most behavior is rewarded on one of four partial schedules of reinforcement: fixed ratio, variable ratio, fixed interval, or variable interval.

5. Organisms learn complex behaviors through **shaping**, in which reinforcement is delivered for successive approximations of the desired response.

6. Punishment weakens the response. To be effective, punishment must be immediate and consistent. Even then, the recipient may learn only what not to do. Punishment can have serious side effects: increased aggression, avoidance behavior, passive aggressiveness, and learned helplessness.

3 Cognitive-Social Learning

1. Cognitive-social theory emphasizes cognitive and social aspects of learning. Köhler discovered that animals sometimes learn through sudden **insight**, rather than through trial and error. Tolman provided evidence of **latent learning** and internal cognitive maps.

2. Bandura's research found that children who watched an adult behave aggressively toward an inflated Bobo doll were later more aggressive than those who had not seen the aggression. According to Bandura, **observational learning** requires attention, retention, motor reproduction, and reinforcement.

4 The Biology of Learning

1. Learning creates structural changes in the brain. Early evidence for such changes came from research on animals raised in enriched versus deprived environments.

2. Learning is an evolutionary adaptation that enables organisms to survive and prosper in a constantly changing world. Researchers have identified biological constraints that limit the generality of conditioning principles: **biological preparedness** and **instinctive drift**.

5 Conditioning and Learning in Everyday Life

1. Prejudice is a negative instance of classical conditioning. Influential research by Clark and Clark indicated that children learn to associate inferior qualities with darker skin and more positive qualities with lighter skin. Classical conditioning is a primary tool for advertisers and others who want to manipulate beliefs, emotions, or behavior. Classical conditioning is also used in medicine and in the development and treatment of phobias.

2. Operant conditioning plays a role in the development of prejudice. Operant conditioning also underlies biofeedback. Even accidental reinforcement can cause superstitious behavior.

3. Cognitive-social learning plays a role in many areas of everyday life, including the development of prejudice and media influences on our behavior.

KEY TERMS

CRITICAL AND CREATIVE THINKING QUESTIONS

1. How might Watson and Rayner, who conducted the famous "Little Albert" study, have designed a more ethical study of conditioned emotional responses?

2. What are some examples of ways in which observational learning has benefited you in your life? Are there instances in which observational learning has worked to your disadvantage?

3. Do you have any taste aversions? How would you use information in this chapter to remove them?

4. Most classical conditioning is involuntary. Considering this, is it ethical for politicians and advertisers to use classical conditioning to influence our thoughts and behavior? Why or why not?

SELF-TEST: LEARNING

(Check your answers in Appendix B.)

1. _____ *conditioning* occurs when a neutral stimulus becomes associated with an unconditioned stimulus to elicit a conditioned response.

    a. Reflex
    b. Instinctive
    c. Classical
    d. Basic

2. The process of learning associations between environmental stimuli and behavioral responses is known as

    _____.

    a. maturation
    b. contiguity learning
    c. conditioning
    d. latent learning

3. In John Watson's demonstration of classical conditioning with Little Albert, the unconditioned **STIMULUS** was

    _____.

    a. symptoms of fear
    b. a rat
    c. a bath towel
    d. a loud noise

4. Which of the following is an example of the use of classical conditioning in everyday life?

    a. treating alcoholism with a drug that causes nausea when alcohol is consumed
    b. the use of seductive women to sell cars to men
    c. politicians associating themselves with home, family, babies, and the American flag
    d. all of these options

5. *Extinction* _____.

    a. is a gradual weakening or suppression of a previously conditioned behavior
    b. occurs when a CS is repeatedly presented without the UCS
    c. is a weakening of the association between the CS and the UCS
    d. all of these options

6. When a neutral stimulus is paired with a previously conditioned stimulus to become a conditioned stimulus as well, this is called _____ conditioning.

    a. operant
    b. classical
    c. higher-order
    d. secondary

7. Learning in which voluntary responses are controlled by their consequences is called _____.

    a. self-efficacy
    b. classical conditioning
    c. operant conditioning
    d. all of these options

8. Match the illustration with the correct label:

    a Thorndike box: law of effect, photo _____
    b. Skinner box: reinforcement, photo _____

1.

2.

9. The addition of a(n) _____ stimulus results in
positive reinforcement; whereas the subtraction of
a(n) _____ stimulus results in *negative
reinforcement*.
   a. desirable; painful or annoying
   b. primary; secondary
   c. operant; classical
   d. higher order; lower order

10. The chimpanzee in Köhler's *insight* experiment
_____.
   a. used trial-and-error to reach a banana placed just out of
      reach
   b. turned its back on the banana out of frustration
   c. sat for a while, then used a stick to bring the banana within
      reach
   d. didn't like bananas

11. Learning new behavior or information by watching others is
known as _____.
   a. social learning
   b. observational learning
   c. modeling
   d. all of the above

12. Garcia and his colleagues taught coyotes to avoid sheep by
pairing a nausea-inducing drug with freshly killed sheep eaten
by the coyotes. This is an example of _____.
   a. classical conditioning
   b. operant conditioning
   c. positive punishment
   d. negative punishment

13. Mamie has an EMG attached to her forehead. When tension
rises, a computer says, "That's too high." When it drops to normal or lower levels, it says, "That's very good." Mamie is using
_____ to decrease her tension headaches.
   a. primary reinforcement
   b. computerized reinforcement
   c. biofeedback
   d. electromyography monitoring

14. According to this text, research has shown that exposure to
media portrayals of demeaning and stereotypical roles for minorities and women _____.
   a. increases critical thinking about minorities and women
   b. initiates and reinforces the learning of prejudice
   c. increases empathy for minorities and women
   d. decreases a child's own stereotypical gender-role behavior

15. This is a factor that increases the modeling effects of violent
video games.
   a. Videos are interactive.
   b. Videos are engrossing to the player.
   c. Video players identify with the aggressor in the game.
   d. all of these options

Memory

W hen Elizabeth was 14 years old, her mother drowned in their backyard pool. As she grew older, the details surrounding her mother's death became increasingly vague for Elizabeth. Decades later, a relative told Elizabeth that she had been the one to find her mother's body. Despite her initial shock, the memories soon slowly started coming back.

> *I could see myself, a thin, dark-haired girl, looking into the flickering blue-and-white pool. My mother, dressed in her nightgown, is floating face down. I start screaming. I remember the police cars, their lights flashing, and the stretcher with the clean, white blanket tucked in around the edges of the body. The memory had been there all along, but I just couldn't reach it.* (Loftus & Ketcham, 1994, p. 45).

This is the true story of Elizabeth Loftus, who today is a well-known psychologist specializing in the study of memory. Compare her story with this one:

When H.M. was 27 years old, portions of his temporal lobes and limbic system were removed as treatment for severe epileptic seizures. The surgery was successful, but something was clearly wrong with H.M.'s long-term memory. Two years after the surgery, he still believed he was 27. He had to be repeatedly reminded of events that had occurred after his surgery. Today, more than 50 years later, H.M. cannot recognize the people who care for him daily, the scientists who have studied him for decades, or even the room where he lives. He reads the same books and magazines repeatedly and laughs at the same old jokes—each time as though it were the first. He thinks he is still 27, and no longer recognizes a photograph of his own face (Corkin, 2002).

What should we make of these two stories? How could a child forget finding her mother's body? What would it be like to be unable to form new memories? How can we remember our second grade teacher's name but forget the name of someone we just met? In this chapter, you'll discover answers to these and other fascinating questions about memory.

The Nature of Memory

LEARNING OBJECTIVES

Review the principles of the four major memory models.

Describe the different purposes of sensory memory, short-term memory, and long-term memory.

Identify ways of extending the duration and capacity of short-term memory.

Describe the various types of long-term memory.

Explain how organization, elaborative rehearsal, and retrieval cues improve long-term memory.

emory allows us to learn from our experiences and to adapt to ever-changing environments—without it, we would have no past or future. Yet our memories are also highly fallible. Although some people think of **memory** as a gigantic library or an automatic tape recorder, our memories are not exact recordings of events. Instead, memory is a *constructive process* through which we actively organize and shape information. As you might expect, this construction often leads to serious errors and biases, which we'll discuss throughout this chapter.

memory An internal record or representation of some prior event or experience.

FOUR MODELS OF MEMORY: AN OVERVIEW

Over the years, psychologists have developed numerous models of how memory operates. According to the **information processing model**, the barrage of information that we encounter every day goes through three basic operations—*encoding, storage,* and *retrieval.* These three processes are closely analogous to the operation of a computer (FIGURE 7.1).

The information processing model is limited because it describes memories as literal, "hard" data stored on a computer disk or hard drive. But human memories are often fuzzy and fragile. Furthermore, computers process one piece of data at a time (sequentially). In contrast, human memory can process a lot of information at the same time (simultaneously).

Memory as a computer FIGURE 7.1

In the information processing model of memory, the brain *encodes* sensory information such as sounds or visual images into a neural code that it can understand; *stores* the information for later use; and *retrieves* information by searching for the appropriate stored "files" and bringing them into short-term memory, where they can be used. These operations are analogous to incoming information being typed on a keyboard and translated into computer language, stored on a disk, and retrieved and brought to the computer screen to be viewed.

Due to the limitations of the information processing model of memory, some cognitive scientists prefer a different way of describing memory: the **parallel distributed processing (PDP)**, or *connectionist,* **model** of memory (McClelland, 1995; Rogers et al., 2004). This model holds that the brain performs multiple, *parallel* operations all at once. In addition, memory is spread (distributed) throughout a weblike network of processing units (FIGURE 7.2).

The PDP model FIGURE 7.2

The PDP model seems consistent with what scientists know about brain activity, perception, language, and decision making. And survival in our environment does require instantaneous information processing, which the PDP model allows. However, the information processing model remains better at explaining the processing of new information and memory for single events.

Fergus Craik and Robert Lockhart's **levels of processing model** (1972) takes a different approach to understanding memory. It suggests that memory relies on how *deeply* we process information. During shallow processing, we're only aware of basic incoming sensory information, and little or no memory is formed. However, when we process the information more deeply by adding meaning, developing organizations and associations, or relating it to things we already know, it can be stored for a lifetime.

Since the late 1960s, one of the most widely used models in memory research has been the **traditional three-stage memory model** (Atkinson & Shiffrin, 1968; Healy & McNamara, 1996). According to this model, three different storage "boxes," or memory stages (*sensory, short-term,* and *long-term*), hold and process information. Each stage has a different purpose, duration, and capacity (FIGURE 7.3). The three-stage memory model remains the leading paradigm in memory research because it offers a convenient way to organize the major findings. Let's discuss the model in more detail.

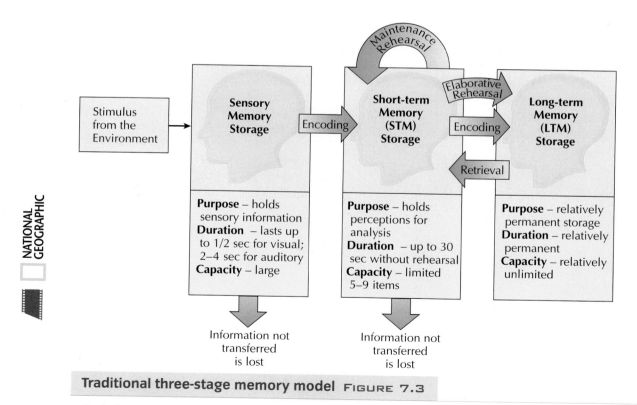

Traditional three-stage memory model FIGURE 7.3

Each "box" represents a separate memory system that differs in purpose, duration, and capacity.

SENSORY MEMORY

sensory memory This first memory stage holds sensory information. It has a relatively large capacity, but duration is only a few seconds.

Everything we see, hear, touch, taste, and smell first enters our **sensory memory**. Information remains in sensory memory just long enough to locate relevant bits of information and transfer them to the next stage of memory. For visual information, known as *iconic memory*, the visual image (icon) lasts about one-half of a second (**FIGURE 7.4**). Auditory information (what we hear) is held in sensory memory about the same length of time as visual information, one-quarter to one-half of a second, but a weaker "echo," or *echoic memory*, of this auditory information can last up to four seconds (Lu, Williamson, & Kaufman, 1992; Neisser, 1967) (**FIGURE 7.5**).

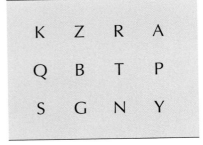

How do researchers test sensory memory?

FIGURE 7.4

In an early study of sensory memory, George Sperling (1960) flashed an arrangement of letters like these for 1/20 of a second. Most people, he found, could recall only 4 or 5 letters. But when instructed to report just the top, middle, or bottom row, depending on whether they heard a high, medium, or low tone, they reported almost all the letters correctly. Apparently, all 12 letters are held in sensory memory right after they are viewed, but only those that are immediately attended to are noted and processed.

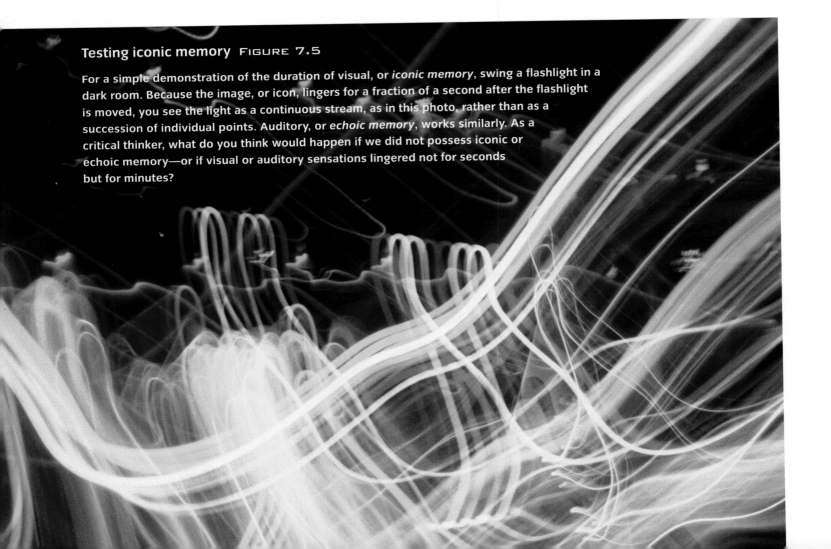

Testing iconic memory FIGURE 7.5

For a simple demonstration of the duration of visual, or *iconic memory*, swing a flashlight in a dark room. Because the image, or icon, lingers for a fraction of a second after the flashlight is moved, you see the light as a continuous stream, as in this photo, rather than as a succession of individual points. Auditory, or *echoic memory*, works similarly. As a critical thinker, what do you think would happen if we did not possess iconic or echoic memory—or if visual or auditory sensations lingered not for seconds but for minutes?

Early researchers believed that sensory memory had an unlimited capacity. However, later research suggests that sensory memory does have limits and that stored images are fuzzier than once thought (Best, 1999; Grondin, Ouellet, & Roussel, 2004).

SHORT-TERM MEMORY

The second stage of memory processing, **short-term memory (STM)**, temporarily stores and processes the sensory stimuli. If the information is important, STM organizes and sends this information along to relatively permanent storage, called **long-term memory (LTM)**. Otherwise, it decays and is lost.

The *capacity* and *duration* of STM are limited (Best, 1999; Kareev, 2000). To extend the *capacity* of STM, you can use a technique called **chunking** (Boucher & Dienes, 2003; Miller, 1956). Have you noticed that credit card, Social Security, and telephone numbers are all grouped into three or four units separated by hyphens? This is because it's easier to remember numbers in "chunks" rather than as a string of single digits.

Chunking in Chess

To the inexpert eye, a chess game in progress looks like little more than a random assembly of black and white game pieces. Accordingly, novice chess players can remember the positions of only a few pieces when a chess game is underway. But expert players generally remember all the positions. To the experts, the scattered pieces form meaningful patterns—classic arrangements that recur often. Just as you group the letters of this sentence into meaningful words and remember them long enough to understand the meaning of the sentence, expert chess players group the chess pieces into easily recalled patterns (or chunks) (Amidzic et al., 2001; Huffman, Matthews, & Gagne, 2001).

Source: Chess moves from 1986 Kasparov–Karpov game.

What a Scientist Sees

You can extend the *duration* of your STM almost indefinitely by consciously "juggling" the information, a process called **maintenance rehearsal**. You are using maintenance rehearsal when you look up a phone number and repeat it over and over until you dial the number. People who are good at remembering names also know to take advantage of maintenance rehearsal. They repeat the name of each person they meet, aloud or silently, to keep it active in STM. They also make sure that other thoughts (such as their plans for what to say next) don't intrude.

> ■ **maintenance rehearsal** Repeating information to maintain it in short-term memory (STM).

Short-term memory as a "working memory"

Short-term memory is more than just a passive, temporary "holding area." Most current researchers (Baddeley, 1992, 2000; Bleckley et al., 2003; DeStefano & LeFevre, 2004) realize that active processing of information also occurs in STM. Because the short-term memory is active, or working, we think of STM as a three-part **working memory**. Let's look at each of the three working parts of short-term memory (**FIGURE 7.6**).

The *central executive* supervises and coordinates two subsystems, the phonological rehearsal loop and the visuospatial sketchpad, while also sending and retrieving information from LTM.

The working memory's *phonological rehearsal loop* holds and manipulates verbal (phonological) information (Best, 1999; Jiang, Olson, & Chun, 2000; Noël et al., 2001). Picture yourself as a food server in a busy restaurant, where a couple has just given you a complicated food order. Your phonological loop allows you to subvocally repeat all their specific requests while you write a brief description on your order pad.

In contrast to the phonological loop, the *visuospatial sketchpad* holds and manipulates visual images and spatial information (Best, 1999; Kemps et al, 2004). Again, imagine yourself as the food server who's delivering the food to the same customers. Using your mind's visuospatial sketchpad, you can mentally visualize where to fit all the entrees, side dishes, and dinnerware on their small table.

Working memory as a central executive FIGURE 7.6

According to this model, working memory serves as a central executive that coordinates the visuospatial sketchpad and the phonological loop.

Central executive
(coordinates material phonologically and visuospatially to help memory)

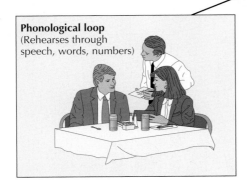

Phonological loop
(Rehearses through speech, words, numbers)

Visuospatial sketchpad
(Mentally imagines visual and spatial material)

LONG-TERM MEMORY

Think back to the opening story of H.M. Although his surgery was successful in stopping the severe epileptic seizures, it also apparently destroyed the mechanism that transfers information from short-term to long-term memory. (That is why he cannot remember the people he sees every day and doesn't recognize his own, now aged face.)

Once information is transferred from STM, it is organized and integrated with other information in *long-term memory (LTM)*. LTM serves as a storehouse for information that must be kept for long periods. When we need the information, it is sent back to STM for our use. Compared with sensory memory and short-term memory, long-term memory has relatively unlimited *capacity* and *duration* (Klatzky, 1984). But, just as with any other possession, the better we label and arrange our memories, the more readily we'll be able to retrieve them.

How do we store the vast amount of information that we collect over a lifetime? Several types of LTM exist (**FIGURE 7.7**). The two major systems are *explicit/declarative memory* and *implicit/nondeclarative memory*.

Explicit/declarative memory refers to intentional learning or conscious knowledge. If asked to remember your phone number or your mother's name, you can state (*declare*) the answers directly (*explicitly*). Explicit/declarative memory can be further subdivided into two parts. *Semantic* memory is memory for general knowledge, rules, events, facts, and specific information. It is our mental encyclopedia. In contrast, *episodic* memory is like a mental diary. It records the major events (*episodes*) in our lives. Some of our episodic memories are short-lived, whereas others can last a lifetime.

Have you ever wondered why most adults can recall almost nothing of the years before age 3? Research suggests that a concept of "self," sufficient language development, and growth of the frontal lobes of the cortex (along with other structures) may be necessary before these early events can be encoded and retrieved many years later (Simcock & Hayne, 2002; Sluzenski, Newcombe, & Ottinger, 2004; Suzuki & Amaral, 2004; Uehara, 2000).

> **explicit/ declarative memory** The subsystem within long-term memory that consciously stores facts, information, and personal life experiences.

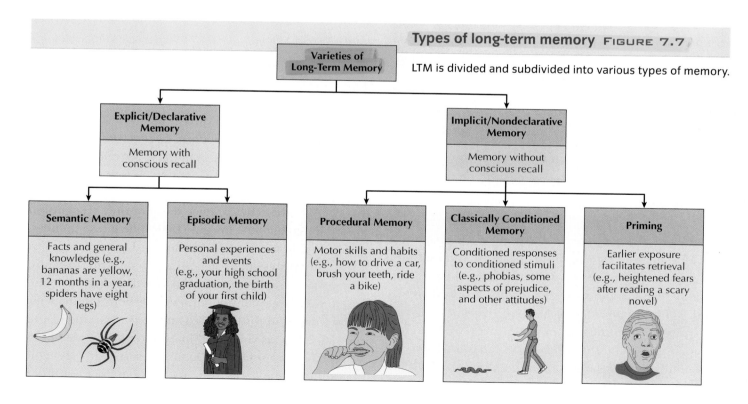

Types of long-term memory FIGURE 7.7

LTM is divided and subdivided into various types of memory.

Varieties of Long-Term Memory

Explicit/Declarative Memory — Memory with conscious recall

Implicit/Nondeclarative Memory — Memory without conscious recall

Semantic Memory — Facts and general knowledge (e.g., bananas are yellow, 12 months in a year, spiders have eight legs)

Episodic Memory — Personal experiences and events (e.g., your high school graduation, the birth of your first child)

Procedural Memory — Motor skills and habits (e.g., how to drive a car, brush your teeth, ride a bike)

Classically Conditioned Memory — Conditioned responses to conditioned stimuli (e.g., phobias, some aspects of prejudice, and other attitudes)

Priming — Earlier exposure facilitates retrieval (e.g., heightened fears after reading a scary novel)

implicit/ nondeclarative memory The subsystem within long-term memory that consists of unconscious procedural skills, simple classically conditioned responses (Chapter 6), and priming.

Implicit/nondeclarative memory refers to unintentional learning or unconscious knowledge. Try telling someone else how you tie your shoelaces without demonstrating the actual behavior. Because your memory of this skill is unconscious and hard to describe (*declare*) in words, this type of memory is sometimes referred to as *nondeclarative*.

Implicit/nondeclarative memory consists of *procedural* motor skills, like tying your shoes or riding a bike, as well as *classically conditioned memory* responses, such as fears or taste aversions.

Implicit/nondeclarative memory also includes *priming*, where prior exposure to a stimulus (*prime*) facilitates or inhibits the processing of new information (Burton et al., 2004; McCarley et al., 2004; Tulving, 2000). For example, you might feel nervous being home alone after reading a Stephen King novel, and watching a romantic movie might kindle your own romantic feelings. Priming can occur even when we do not consciously remember being exposed to the prime.

IMPROVING LONG-TERM MEMORY: ORGANIZATION, REHEARSAL, RETRIEVAL

Organization To successfully encode information for LTM, we need to *organize* material into hierarchies. This involves arranging a number of related items into broad categories that are further divided and subdivided. (This organization strategy for LTM is similar to the strategy of chunking material in STM.) For instance, by grouping small subsets of ideas together (as subheadings under larger, main headings, and within diagrams, tables, and so on), we hope to make the material in this book more understandable and *memorable*.

Admittedly, organization takes time and work. But you'll be happy to know that some memory organization and filing is done automatically while you sleep (Maquet et al., 2003; Smith & Smith, 2003; Wagner et al., 2003). (Unfortunately, despite claims to the contrary, research shows that we can't recruit our sleeping hours to memorize new material, such as a foreign language.)

Rehearsal Like organization, *rehearsal* also improves encoding for both STM and LTM. If you need to hold information in STM for longer than 30 seconds, you can simply keep repeating it (maintenance rehearsal). But storage in LTM requires *deeper levels of processing* called **elaborative rehearsal** (FIGURE 7.8A).

The immediate goal of elaborative rehearsal is to *understand*—not to memorize. Fortunately, this attempt to understand is one of the best ways to encode new information into long-term memory.

Retrieval Finally, effective *retrieval* is critical to improving long-term memory. There are two types of **retrieval cues**. *Specific* cues require you only to *recognize* the correct response. *General* cues require you to *recall* previously learned material by searching through all possible matches in LTM—a much more difficult task (FIGURE 7.8B and C).

Whether cues require recall or only recognition is not all that matters. Imagine that while house hunting, you walk into a stranger's kitchen and are greeted with the unmistakable smell of freshly baked bread. Instantly, the aroma transports you to your grandmother's kitchen, where you spent many childhood afternoons doing your homework. You find yourself suddenly thinking of the mental shortcuts your grandmother taught you to help you learn your multiplication tables. You hadn't thought about these little tricks for years, but somehow a whiff of baking bread brought them back to you. Why?

elaborative rehearsal The process of linking new information to previously stored material.

retrieval cue A clue or prompt that helps stimulate recall and retrieval of a stored piece of information from long-term memory.

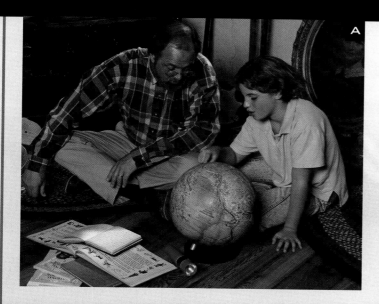

A Elaborative rehearsal

To improve elaborative rehearsal, you can expand (or elaborate on) the information to try to understand it better, actively explore and question new information, and search for meaningfulness. For example, a student might compare what she reads in an anthropology or history textbook with what she knows about world geography. Can you see how this might deepen her understanding and memory for the subject?

B Recall versus recognition

It's more difficult to recall, in order, the names of the planets in our solar system (general cue) than to recognize them when cued with the first three letters of each planet's name (specific retrieval cue): Mer-, Ven-, Ear-, Mar-, Jup-, Sat-, Ura-, Nep-, Plu-.

C Recognition memory

Both name recognition and picture recognition for high school classmates remain high even many years after graduation, whereas recall memory would be expected to drop significantly over time.

In this imagined episode, you have stumbled upon the **encoding specificity principle** (Tulving & Thompson, 1973). One important contextual cue for retrieval is *location*. In a clever study, Godden and Baddeley (1975) had underwater divers learn a list of 40 words either on land or underwater. The divers had better recall for lists that they had encoded underwater if they were also underwater at the time of retrieval; similarly, lists that were encoded above water were better recalled above water.

People also remember information better if their moods during learning and retrieval match (Kenealy, 1997). This phenomenon, called *mood congruence*, occurs because a given mood tends to evoke memories that are consistent with that mood. When you're sad (or happy or angry), you're more likely to remember events and circumstances from other times when you were sad (or happy or angry).

> **encoding specificity principle** Retrieval of information is improved when the conditions of recovery are similar to the conditions that existed when the information was encoded.

Finally, as generations of coffee-guzzling university students have discovered, if you learn something while under the influence of a drug, such as caffeine, you will remember it more easily when you take that drug again than at other times (Baddeley, 1998). This is called *state-dependent retrieval*.

CONCEPT CHECK **STOP**

What happens to information after it leaves sensory memory?

What is working memory?

What is the difference between recognition and recall?

How do contextual cues such as location, mood, and physiological state affect memory retrieval?

Biological Bases of Memory

LEARNING OBJECTIVES

Describe two kinds of biological changes that occur when we learn something new.

Identify some brain areas involved in memory.

Explain how injury and disease can affect memory.

NEURONAL AND SYNAPTIC CHANGES IN MEMORY

We know that learning modifies the brain's neural networks (Chapters 2 and 6). As you learn to play tennis, for example, repeated practice builds specific neural "pathways" that make it easier and easier for you to get the ball over the net. This

> **long-term potentiation (LTP)** Long-lasting increase in neural excitability believed to be a biological mechanism for learning and memory.

long-term potentiation (LTP) happens in at least two ways.

First, as early research with rats raised in "enriched" environments showed (Rosenzweig, Bennett, & Diamond, 1972; Chapter 6), repeated stimulation of a synapse can strengthen the synapse by causing the dendrites to grow more spines (Barinaga, 1999). This results in more synapses, more receptor sites, and more sensitivity.

Second, learning affects a particular neuron's ability to release its neurotransmitters. This is demonstrated in research with *Aplysia* (sea slugs), which can be classically conditioned to reflexively withdraw their gills when squirted with water (FIGURE 7.9).

Further evidence comes from research with genetically engineered "smart mice," which have extra receptors for a neurotransmitter named NMDA (N-methyl-d-aspartate). These mice performed significantly better on memory tasks than did normal mice (Tang et al., 2001; Tsien, 2000).

Although it is difficult to generalize from sea slugs and mice, research on long-term potentiation (LTP) in humans has been widely supportive (Kikusui, Aoyagi, & Kaneko, 2000; Wixted, 2004).

How does a sea slug learn and remember?
FIGURE 7.9

After repeated squirting with water, followed by a mild shock, the sea slug, *Aplysia*, releases more neurotransmitters at certain synapses. These synapses then become more efficient at transmitting signals that allow the slug to withdraw its gills when squirted. As a critical thinker, can you explain why this ability might provide an evolutionary advantage?

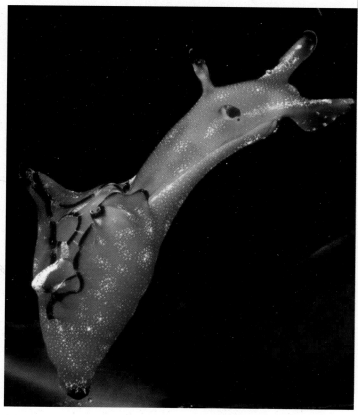

HORMONAL CHANGES AND MEMORY

When stressed or excited, we naturally produce hormones that arouse the body, such as *epinephrine* and *cortisol* (Chapter 3). These hormones in turn affect the amygdala (a brain structure involved in emotion), which then stimulates the hippocampus and cerebral cortex (parts of the brain that are important for memory storage). Research has shown that direct injections of epinephrine or cortisol or electrical stimulation of the amygdala will increase the encoding and storage of new information (Akirav & Richter-Levin, 1999; Buchanan & Lovallo, 2001; McGaugh & Roozendaal, 2002). However, prolonged or excessive stress (and increased levels of cortisol) can interfere with memory (Al'absi, Hugdahl, & Lovallo, 2002; Heffelfinger & Newcomer, 2001; McAllister-Williams & Rugg, 2002).

The powerful effect of hormones on memory also can be seen in what are known as *flashbulb memories*—vivid images of circumstances associated with surprising or strongly emotional events (Brown & Kulik, 1977; Edery-Halpern & Nachson, 2004). In such situations, we secrete fight-or-flight hormones when we initially hear of the event and then replay the event in our minds again and again, which makes for stronger memories. Despite their intensity, flashbulb memories are not as accurate as you might think (Squire, Schmolck, & Buffalo, 2001). For example, when asked how he heard the news of the September 11 attacks, President George W. Bush's answers contained numerous inconsistencies (Greenberg, 2004). Thus, not even flashbulb memories are immune to alteration.

WHERE ARE MEMORIES LOCATED?

Early memory researchers believed that memory was *localized*, or stored in a particular brain area. Later research suggests that, in fact, memory tends to be localized not in a single area but in many separate areas throughout the brain.

Today, research techniques are so advanced that we can experimentally induce and measure

Brain and memory formation FIGURE 7.10

Damage to any one of these areas can affect encoding, storage, and retrieval of memories.

Amygdala	Emotional memory (Blair et al., 2005; McGaugh & Roozendaal, 2002)
Basal Ganglia and Cerebellum	Creation and storage of the basic memory trace and implicit (nondeclarative) memories (such as skills, habits, and simple classical conditions responses) (Christian & Thompson, 2005; Frank, 2005; Thompson, 2005)
Hippocampal Formation (hippocampus and surrounding area)	Memory recognition; implicit, explicit, spatial, episodic memory; declarative long-term memory; sequences of events (Bachevalier & Varga-Khadem, 2005; Thompson, 2005)
Thalamus	Formation of new memories and spatial and working memory (Mendrek et al., 2005; Ridley et al., 2005; Sarnthein et al., 2005)
Cortex	Encoding of explicit (declarative) memories; storage of episodic and semantic memories; skill learning; printing; working memory (Pastemak & Greenlee, 2005; Reed et al., 2005; Thompson, 2005)

memory-related brain changes as they occur—on-the-spot reporting! For example, James Brewer and his colleagues (1998) used functional magnetic resonance imaging (fMRI) to locate areas of the brain responsible for encoding memories of pictures. They showed 96 pictures of indoor and outdoor scenes to participants while scanning their brains, and then later tested participants' ability to recall the pictures. Brewer and his colleagues identified the *right prefrontal cortex* and the *parahippocampal cortex* as being the most active during the encoding of the pictures. These are only two of several brain regions involved in memory storage (FIGURE 7.10).

BIOLOGICAL CAUSES OF MEMORY LOSS: INJURY AND DISEASE

The leading cause of neurological disorders—including memory loss—among Americans between the ages of 15 and 25 is *traumatic brain injury*. These injuries most commonly result from car accidents, falls, blows, and gunshot wounds.

Traumatic brain injury happens when the skull suddenly collides with another object. The compression, twisting, and distortion of the brain inside the skull all cause serious and sometimes permanent damage to the brain. The frontal and temporal lobes often take the heaviest hit because they directly impact against the bony ridges inside the skull.

Loss of memory as a result of brain injury is called *amnesia*. Two major types of amnesia are **retrograde amnesia** and **anterograde amnesia** (FIGURE 7.11). Usually, retrograde amnesia is temporary. Unfortunately, anterograde amnesia is usually permanent, but patients show surprising abilities to learn and remember implicit/nondeclarative tasks (such as procedural motor skills).

Like traumatic brain injuries, disease can alter the physiology of the brain and nervous system, affecting memory processes. For example, **Alzheimer's disease (AD)** is a progressive mental deterioration that occurs most commonly in old age

Alzheimer's disease (AD) Progressive mental deterioration characterized by severe memory loss.

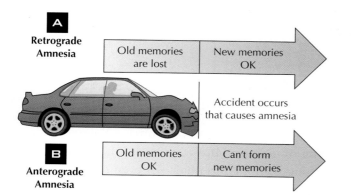

Retrograde Amnesia	Old memories are lost	New memories OK

Accident occurs that causes amnesia

Anterograde Amnesia	Old memories OK	Can't form new memories

Two types of amnesia FIGURE 7.11

A In *retrograde amnesia*, the person loses memories of events that occurred *before* the accident yet has no trouble remembering things that happened afterward (old, "retro" memories are lost). **B** In *anterograde amnesia*, the person cannot form new memories for events that occur *after* the accident. Anterograde amnesia also may result from a surgical injury or from diseases such as chronic alcoholism.

(FIGURE 7.12). The most noticeable early symptoms are disturbances in memory, which become progressively worse until, in the final stages, the person fails to recognize loved ones, needs total nursing care, and ultimately dies.

Alzheimer's does not attack all types of memory equally. A hallmark of the disease is an extreme decrease in *explicit/declarative memory* (Ballesteros & Manuel Reales, 2004; Mitchell et al., 2002). Alzheimer's patients fail to recall facts, information, and personal life experiences, yet they still retain some *implicit/nondeclarative* memories, such as simple classically condi-

tioned responses and procedural tasks, like brushing their teeth.

Brain autopsies of people with Alzheimer's show unusual *tangles* (structures formed from degenerating cell bodies) and *plaques* (structures formed from degenerating axons and dendrites). Hereditary Alzheimer's generally strikes its victims between the ages of 45 and 55. Some experts believe that the cause of Alzheimer's is primarily genetic, but others think that genetic makeup may make some people more susceptible to environmental triggers (Bernhardt, Seidler, & Froelich, 2002; Khachaturian et al., 2004; Vickers et al., 2000).

The effect of Alzheimer's disease on the brain FIGURE 7.12

Note the high amount of red and yellow color (signs of brain activity) in the positron emission tomography scans of the normal brain **A** and the reduced activity in the brain of the Alzheimer's disease patient **B**. The loss is most significant in the temporal and parietal lobes, which indicates that these areas are particularly important for storing memories.

CONCEPT CHECK STOP

Why would animals raised in enriched environments develop different neuronal connections than those raised in deprived environments?

How are hormones involved in memory?

What is the difference between retrograde and anterograde amnesia?

Forgetting

THEORIES OF FORGETTING

If you couldn't forget, your mind would be filled with meaningless data, such as what you ate for breakfast every morning of your life. Similarly, think of the incredible pain you would continuously endure if you couldn't distance yourself from tragedy through forgetting. The ability to forget is essential to the proper functioning of memory. But what about those times when forgetting is an inconvenience or even dangerous?

There are five major theories that explain why forgetting occurs (**FIGURE 7.13**): decay, interference, motivated forgetting, encoding failure, and retrieval failure. Each theory focuses on a different stage of the memory process or a particular type of problem in processing information.

Process Diagram

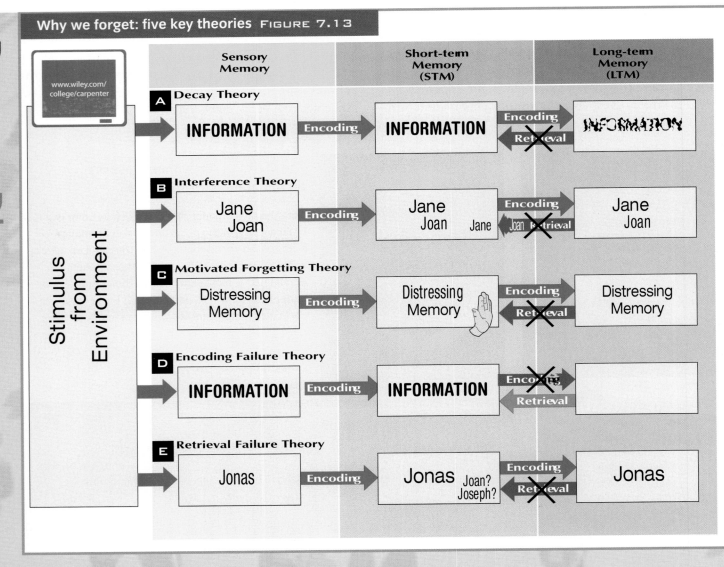

Why we forget: five key theories FIGURE 7.13

How Quickly We Forget

Hermann Ebbinghaus first introduced the experimental study of learning and forgetting in 1885. Using himself as a subject, he calculated how long it took to learn a list of three-letter *nonsense syllables* such as *SIB* and *RAL*. He found that one hour after he knew a list perfectly, he remembered only 44 percent of the syllables. A day later, he recalled 35 percent, and a week later, only 21 percent. This figure shows his famous "forgetting curve."

Depressing as these findings may seem, keep in mind that meaningful material is much more memorable than nonsense syllables. Even so, we all forget some of what we have learned.

On a more cheerful note, after some time passed and he had forgotten the list, Ebbinghaus found that *relearning* a list took less time than the initial learning did. This research suggests that we often retain some memory for things that we have learned, even when we seem to have forgotten them completely.

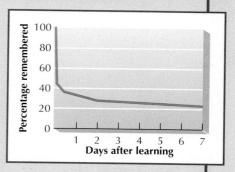

A *Decay Theory.* Because memory is processed and stored in a physical form—for example, in a network of neurons—connections between neurons probably deteriorate over time. We know that skills and memory degrade if they go unused—in other words, "use it or lose it" (Rosenzweig, Barnes, & McNaughton, 2002; Villarreal et al., 2002). However, conclusive experimental support for decay theory is difficult to obtain—how can you prove that a previously stored memory does *not* exist?

B *Interference Theory.* Forgetting is caused by two competing memories (Anderson, Bjork, & Bjork, 1994; Conway & Pleydell-Pearce, 2000; Wixted, 2004). Interference is particularly strong among memories with similar qualities. **Retroactive interference** occurs when new information (such as one's new phone number) interferes with old information (such as one's former number). In contrast, **proactive interference** occurs when old information interferes with new information. Calling a new partner by an old partner's name is proactive interference.

C *Motivated Forgetting Theory.* We forget some information—such as a dental appointment or an embarrassing remark we once made—for a reason. According to Freudian theory, people forget unpleasant or anxiety-producing information either consciously (*suppression*) or unconsciously (*repression*).

D *Encoding Failure Theory.* Our sensory memory receives the information and passes it to STM. But during STM, we probably decide there was no need to remember the precise details, so we do not encode it and pass it on for storage in LTM. For example, few people can correctly recognize the details on the U.S. penny (the location of the mint date, the direction Lincoln's head is facing, and so on), despite having seen it many times (Nickerson & Adams, 1979).

E *Retrieval Failure Theory.* Memories stored in LTM aren't forgotten. They're just momentarily inaccessible—perhaps because of interference, faulty cues, or emotional states. For example, the *tip-of-the-tongue phenomenon*—the feeling that at any second, a word or event you are trying to remember and perhaps can *almost* remember—will pop out from the "tip of your tongue." Although it is difficult to distinguish retrieval failure from encoding failure, most memory failures probably stem from poor encoding, not retrieval failure (Howe & O'Sullivan, 1997).

Stepping Out of The Shadows of Emphysema

That was daily life for Brenda Wilson. Sidelined with severe emphysema, the Bristol, Va., woman struggled with everyday tasks like dressing, bathing and walking even short distances. With conventional therapies no longer providing relief, Wilson's pulmonologist referred her to the University of Virginia Health System, a leading treatment center for Chronic Obstructive Pulmonary Disease (COPD), a group of progressive lung disorders that includes emphysema and chronic bronchitis. While her prognosis seemed poor, UVa's experts determined that Wilson could benefit from Lung Volume Reduction Surgery (LVRS), an innovative procedure that removes up to 30 percent of the lung tissue most damaged by emphysema.

The results were dramatic. Within a week of her surgery, Wilson no longer needed supplemental oxygen. Once too breathless to walk a mile, "I can't describe how well Wilson...

Capitalizing on the sleeper effect? FIGURE 7.14

The sleeper effect can be a significant problem when reliable and unreliable information are intermixed—for example, when advertising or PR are disguised to look like more objective reports. Can you see how, after a time, we may forget the source and no longer discount the information in the ad?

FACTORS INVOLVED IN FORGETTING

Since Ebbinghaus's original research, scientists have discovered numerous factors that contribute to forgetting. Four of the most important are the *serial position effect, source amnesia, the sleeper effect,* and *spacing of practice.*

When study participants are given lists of words to learn and are allowed to recall them in any order they choose, they remember the words at the beginning (*primacy effect*) and the end of the list (*recency effect*) better than those in the middle, which are quite often forgotten (Burns et al., 2004; Golob & Starr, 2004; Suhr, 2002). This effect, including both the primacy effect and recency effect, is known as the **serial position effect**.

The reasons for this effect are complex, but they do have interesting real-life implications. For example, a potential employer's memory for you might be enhanced if you are either the first or last candidate interviewed.

Each day we read, hear, and process an enormous amount of information, and it's easy to confuse "who said what to whom" and in what context. Forgetting the true source of a memory is known as **source amnesia** (Drummey & Newcombe, 2002; Leichtman & Ceci, 1995; Oakes & Hyman, 2001).

When we first hear something from an unreliable source, we tend to disregard that information in favor of a more reliable source. However, as the source of the information is forgotten (source amnesia), the unreliable information is no longer discounted. This is called the **sleeper effect** (Kumkale & Albarracín, 2004; Underwood & Pezdek, 1998) (FIGURE 7.14).

Finally, if we try to memorize too much information at once (as when students "cram" before an exam), we're not likely to learn and remember as much as we would with more distributed study (Donovan & Radosevich, 1999). **Distributed practice** refers to spacing your learning periods, with rest periods between sessions. Cramming is called **massed practice** because the time spent learning is massed into long, unbroken intervals.

CONCEPT CHECK STOP

How does previous learning affect relearning?

What is the difference between proactive and retroactive interference?

Why do we sometimes fall prey to those who offer erroneous information, even when we know it is suspect?

Which is more effective, distributed practice or massed practice?

In many societies, tribal leaders pass down vital information through orally related stories. As a result, children living in these cultures have better memories for information that is related through stories than do other children.

Memory Distortions

One of my first memories would date, if it were true, from my second year. I can still see, most clearly, the following scene, in which I believed until I was about fifteen. I was sitting in my pram, which my nurse was pushing in the Champs-Élysées, when a man tried to kidnap me. I was held in by the strap fastened round me while my nurse bravely tried to stand between the thief and me. She received various scratches, and I can still see vaguely those on her face. Then a crowd gathered, a policeman with a short cloak and a white baton came up, and the man took to his heels. I can still see the whole scene, and can even place it near the tube station. When I was about fifteen, my parents received a letter from my former nurse saying that she had been converted to the Salvation Army. She wanted to confess her past faults, and in particular to return the watch she had been given as a reward on this occasion. She had made up the whole story, faking the scratches. I, therefore, must have heard, as a child, the account of this story, which my parents believed, and projected it into the past in the form of a visual memory, which was a memory of a memory, but false (Piaget, 1962, pp. 187–188).

A Memory Test

Carefully read through all the words in the following list.

Bed	Drowse
Awake	Nurse
Tired	Sick
Dream	Lawyer
Wake	Medicine
Snooze	Health
Snore	Hospital
Rest	Dentist
Blanket	Physician
Doze	Patient
Slumber	Stethoscope
Nap	Curse
Peace	Clinic
Yawn	Surgeon

Now cover the list and write down all the words you remember.

Number of correctly recalled words:

21 to 28 words = excellent memory

16 to 20 words = better than most

12 to 15 words = average

8 to 11 words = below average

7 or fewer words = you might need a nap

How did you do? Do you have a good or excellent memory? Did you recall seeing the words *sleep* and *doctor?* Look back over the list. These words are not there. However, over 65 percent of students commonly report seeing these words. Why? As mentioned in the introduction to this chapter, memory is not a faithful duplicate of an event; it is a *constructive process.* We actively shape and build on information as it is encoded and retrieved.

This is a self-reported childhood memory of Jean Piaget, a brilliant and world-famous cognitive and developmental psychologist. Why did Piaget create such a strange and elaborate memory for something that never happened?

There are several reasons why we shape, rearrange, and distort our memories. One of the most common is our need for *logic* and *consistency.* When we're initially forming new memories or sorting through the old ones, we fill in missing pieces, make "corrections," and rearrange information to make it logical and consistent with our previous experiences. If Piaget's beloved nurse said someone attempted to kidnap him, it was only logical for the boy to "remember" the event.

We also shape and construct our memories for the sake of *efficiency.* We summarize, augment, and tie new information in with related memories in LTM. Similarly, when we need to retrieve the stored information, we leave out seemingly unimportant elements or misremember the details.

Despite all their problems and biases, our memories are normally quite accurate and serve us well in most situations. Our memories have evolved to encode, store, and retrieve vital information. Even while sleeping we process and store important memories. However, when faced with tasks like remembering precise details in a scholarly text, the faces and names of potential clients, or where we left our house keys, our brains are simply not as well equipped.

MEMORY AND EYEWITNESS TESTIMONY

When our memory errors involve the criminal justice system, they may lead to wrongful judgments of guilt or innocence and even life-or-death decisions.

In the past, one of the best forms of trial evidence a lawyer could have was an *eyewitness*—"I was there; I saw it with my own eyes." Unfortunately for lawyers, research has identified several problems with eyewitness testimony (Lindsay et al., 2004; Loftus, 2000, 2001; Yarmey, 2004). In fact, researchers have demonstrated that it is relatively easy to create false memories (Dodd & McLeod, 2004; Greenberg, 2004; Loftus, 1993, 1997, 2001).

In one classic study, participants watched a film of a car driving through the countryside. Later, those who were asked to estimate how fast the car was going when it passed the barn (actually nonexistent) were six times as likely to later report that they had seen a barn in the film than participants who hadn't been asked about a barn (Loftus, 1982).

Problems with eyewitness recollections are so well established and important that judges now allow expert testimony on the unreliability of eyewitness testimony and routinely instruct jurors on its limits (Durham & Dane, 1999; Ramirez, Zemba, & Geiselman, 1996). If you serve as a member of a jury or listen to accounts of crimes in the news, remind yourself of these problems. Also, keep in mind that research participants in eyewitness studies generally report their inaccurate memories with great self-assurance and strong conviction (Migueles & Garcia-Bajos, 1999). Eyewitnesses to an actual crime may similarly identify—with equally high confidence—an innocent person as the perpetrator (**FIGURE 7.15**).

Like eyewitness testimony, false memories can have serious legal and social implications. Do you recall the opening story of Elizabeth, who suddenly remembered finding her mother's drowned body decades after it had happened? Elizabeth's recovery of these gruesome childhood memories, although painful, initially brought great relief. It also seemed to explain why she had always been fascinated by the topic of memory.

How often are eyewitnesses mistaken? FIGURE 7.15

In one experiment, participants watched people committing a staged crime. Only an hour later, 20 percent of the eyewitnesses identified innocent people from mug shots, and a week later, 8 percent identified innocent people in a lineup (Brown, Deffenbacher, & Sturgill, 1977). What memory processes might have contributed to the eyewitnesses' errors?

Elizabeth Loftus FIGURE 7.16

Then, her brother called to say there had been a mistake! The relative who told Elizabeth that she had been the one to discover her mother's body later remembered—and other relatives confirmed—that it had actually been Aunt Pearl, not Elizabeth Loftus. Loftus, an expert on memory distortions, had unknowingly created her own *false memory*. (FIGURE 7.16)

REPRESSED MEMORIES

Creating false memories may be somewhat common, but can we recover true memories that are buried in childhood? **Repression** is the supposed unconscious coping mechanism by which we prevent anxiety-provoking thoughts from reaching consciousness. According to some research, repressed memories are *actively* and *consciously* "forgotten" in an effort to avoid the pain of their retrieval (Anderson et al., 2004). Others suggest that some memories are so painful that they exist only in an *unconscious* corner of the brain, making them inaccessible to the individual (Karon & Widener, 1998). In these cases, therapy would be necessary to unlock the hidden memories (Davies, 1996).

This is a complex and controversial topic in psychology. No one doubts that some memories are forgotten and later recovered. What is questioned is the concept of *repressed memories* of painful experiences (especially childhood sexual abuse) and their storage in the unconscious mind (Goodman et al., 2003; Kihlstrom, 2004; Loftus & Polage, 1999).

Critics suggest that most people who have witnessed or experienced a violent crime, or are adult survivors of childhood sexual abuse, have intense, persistent memories. They have trouble *forgetting*, not remembering. Some critics also wonder whether therapists sometimes inadvertently create false memories in their clients during therapy. Some worry that if a clinician even suggests the possibility of abuse, the client's own *constructive processes* may lead him or her to create a false memory. The client might start to incorporate portrayals of abuse from movies and books into his or her own memory, forgetting their original sources and eventually coming to see them as reliable.

This is not to say that all psychotherapy clients who recover memories of sexual abuse (or other painful incidents) have invented those memories. Indeed, the repressed memory debate has grown increasingly bitter, and research on both sides is hotly contested. The stakes are high because some lawsuits and criminal prosecutions of sexual abuse are sometimes based on recovered memories of childhood sexual abuse. As researchers continue exploring the mechanisms underlying delayed remembering, we must be careful not to ridicule or condemn people who recover true memories of abuse. In the same spirit, we must protect the innocent from wrongful accusations that come from false memories. We look forward to a time when we can justly balance the interests of the victim with those of the accused.

CONCEPT CHECK STOP

HOW can our desire for logic, consistency, and efficiency thwart accurate memory?

Why might Elizabeth Loftus have formed a false memory about her mother's death?

Tips for Memory Improvement

As you review the key points from this chapter, think about how you might exploit basic principles of memory, using them to your own advantage. One additional "trick" for giving your memory a boost is to use **mnemonic devices** to encode items in a special way. (But be warned—you may get more "bang for your buck" using the well-researched principles discussed throughout this chapter.)

Three popular mnemonic techniques are listed here.

A **Method of loci** Greek and Roman orators developed the *method of loci* to keep track of the many parts of their long speeches. Orators would imagine the parts of their speeches attached to places in a courtyard. For example, if an opening point in a speech was the concept of *justice*, they might visualize a courtroom placed in the first corner of their garden. As they mentally walked around their garden during their speech, they would encounter, in order, each of the points to be made.

B **Peg words** To use the *peg-word* mnemonic, you first need to memorize a set of 10 images that you can use as "pegs" on which to hang ideas. For example, if you learn 10 items that rhyme with the numbers they stand for, you can then use the images as pegs to hold the items of any list. Try it with items you might want to buy on your next trip to the grocery store: milk, eggs, and bread.

C **Method of word association** The *method of word associations* creates verbal associations for items to be learned. To recall the names of the Great Lakes, think of *HOMES on a great lake* (*H*uron, *O*ntario, *M*ichigan, *E*rie, *S*uperior).

A Method of loci

1. Milk
2. Eggs
3. Bread

One is a bun

Two is a shoe

B Peg words

Three is a tree

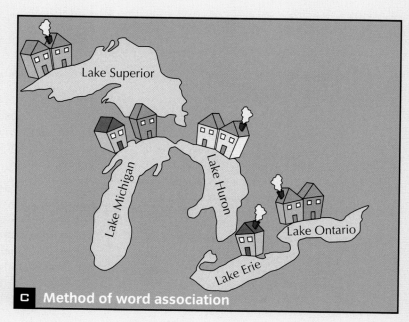

C Method of word association

- What phenomenon accounts for why many people feel that they can remember the September 11, 2001, terrorist attacks (or the explosion of the space shuttle Challenger, or the assassination of President John F. Kennedy) with perfect clarity?

- What biological process accounts for this kind of intense memory?

- Are memories like this impervious to distortion?

VISUAL SUMMARY

1 The Nature of Memory

1. **Memory** is an internal representation of some prior event or experience. Major perspectives on memory include the information processing model, the parallel distributed processing (PDP) model, the levels of processing model, and the three-stage memory model.

2. The three-stage memory model proposes that information is stored and processed in **sensory memory**, **short-term memory (STM)**, and **long-term memory (LTM)**; these differ in purpose, duration, and capacity.

3. **Chunking** and **maintenance rehearsal** improve STM's duration and capacity. Researchers think of STM as a three-part working memory.

4. LTM is an almost unlimited storehouse for information that must be kept for long periods. The two major types of LTM are **explicit/declarative memory** and **implicit/nondeclarative memory**. Organization and **elaborative rehearsal** improve encoding for LTM. **Retrieval cues** help stimulate retrieval of information from LTM. According to the **encoding specificity principle**, retrieval is improved when conditions of recovery are similar to encoding conditions.

2 Biological Bases of Memory

1. Learning modifies the brain's neural networks through **long-term potentiation (LTP)**, strengthening particular synapses and affecting neurons' ability to release their neurotransmitters.

2. Stress hormones affect the amygdala, which stimulates brain areas that are important for memory storage. Heightened arousal increases the encoding and storage of new information. Secretion of fight-or-flight hormones can contribute to "flashbulb" memories.

3. Research using advanced techniques has indicated that several brain regions are involved in memory storage.

A   B

4. Traumatic brain injuries and disease, such as **Alzheimer's disease (AD)**, can cause memory loss. Two major types of amnesia are retrograde and anterograde amnesia.

3 Forgetting

1. Researchers have proposed that we forget information through decay, retroactive and proactive interference, motivated forgetting, encoding failure, and retrieval failure.

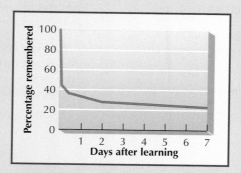

2. Early research by Ebbinghaus showed that we tend to forget newly learned information quickly but that we relearn the information more readily the second time.

3. Four factors that contribute to forgetting are the serial position effect (primacy and recency effects), source amnesia, the sleeper effect, and spacing of practice (distributed versus massed practice).

4 Memory Distortions

1. People shape, rearrange, and distort memories in order to create logic, consistency, and efficiency. Despite all their problems and biases, our memories are normally accurate and usually serve us well.

2. When memory errors involve the criminal justice system, they can have serious legal and social consequences. Problems with eyewitness recollections are well established. Judges often allow expert testimony on the unreliability of eyewitnesses.

3. Memory **repression** (especially of childhood sexual abuse) is a complex and controversial topic. Critics note that most people who have witnessed or experienced a violent or traumatic event have intense, persistent memories. Critics worry that if a clinician suggests the possibility of abuse, the client's *constructive processes* may lead him or her to create a false memory of being abused. Researchers continue to explore delayed remembering.

4. **Mnemonic devices** help us remember lists and facts. Three popular mnemonic techniques are method of loci, the pegword mnemonic, and the method of word associations.

KEY TERMS

- memory p. 174
- information processing model p. 174
- parallel distributed processing (PDP) model p. 174
- levels of processing model p. 175
- traditional three-stage memory model p. 175
- sensory memory p. 176
- short-term memory (STM) p. 177
- long-term memory (LTM) p. 177
- chunking p. 177

- maintenance rehearsal p. 178
- working memory p. 178
- explicit/declarative memory p. 179
- implicit/nondeclarative memory p. 180
- elaborative rehearsal p. 180
- retrieval cue p. 180
- encoding specificity principle p. 182
- long-term potentiation (LTP) p. 182
- retrograde amnesia p. 184
- anterograde amnesia p. 184

- Alzheimer's disease p. 184
- retroactive interference p. 187
- proactive interference p. 187
- serial position effect p. 188
- source amnesia p. 188
- sleeper effect p. 188
- distributed practice p. 188
- massed practice p. 188
- repression p. 192
- mnemonic devices p. 193

CRITICAL AND CREATIVE THINKING QUESTIONS

1. If you were forced to lose one type of memory—sensory, short-term, or long-term—which would you select? Why?

2. Why might students do better on a test if they take it in the same seat and classroom where they originally studied the material?

3. What might be the evolutionary benefit of heightened (but not excessive) arousal enhancing memory?

4. Why might advertisers of shoddy services or products benefit from "channel surfing" if the television viewer is skipping from news programs to cable talk shows to infomercials?

5. As an eyewitness to a crime, how could you use information in this chapter to improve your memory for specific details? If you were a juror, what would you say to the other jurors about the reliability of eyewitness testimony?

SELF-TEST

(Check your answers in Appendix B.)

1. Label the following *terms* on the figure below: In a computer model of memory, (a)_____ would happen at the keyboard, (b)_____ on the monitor, and (c)_____ on the hard drive.

a. _____
b. _____
c. _____

2. Label the three-stage memory model in the correct sequence on the figure below:

a. _____
b. _____
c. _____

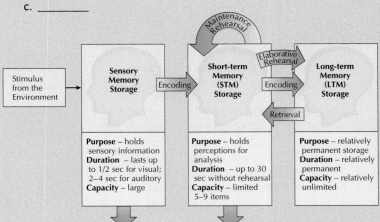

3. The following descriptions are characteristic of _____: information lasts only a few seconds or less, and a relatively large (but not unlimited) storage capacity.

   a. perceptual processes
   b. short-term storage
   c. working memory
   d. sensory memory

4. _____ is the process of grouping separate pieces of information into a single unit.

   a. Chunking   b. Cheating
   c. Collecting   d. Dual-coding

5. The two major systems of long-term memory are _____.

   a. _____
   b. _____

Varieties of Long-Term Memory

Memory with conscious recall

Memory without conscious recall

6. In answering this question, the correct multiple choice option may serve as a _____ for recalling accurate information from your long-term memory.

   a. specificity code   b. priming pump
   c. retrieval cue   d. flashbulb stimulus

7. The encoding specificity principle says that information retrieval is improved when _____.

   a. both maintenance and elaborative rehearsal are used
   b. reverberating circuits consolidate information
   c. conditions of recovery are similar to encoding conditions
   d. long-term potentiation is accessed

8. The long-lasting increase in neural excitability believed to be a biological mechanism for learning and memory is called _____.

   a. maintenance rehearsal
   b. adrenaline activation
   c. long-term potentiation
   d. the reverberating response

9. A progressive mental deterioration characterized by severe memory loss that occurs most commonly in the elderly is called _____.

   a. retrieval loss
   b. prefrontal cortex deterioration
   c. Alzheimer's disease
   d. age-related amnesia

10. Label the two types of amnesia on the figure below:

   a. _____
   b. _____

A
Retrograde Amnesia
Old memories are lost | New memories OK
Accident occurs that causes amnesia
B
Anterograde Amnesia
Old memories OK | Can't form new memories

11. List the 5 major theories of forgetting:

   a. _____
   b. _____
   c. _____
   d. _____
   e. _____

12. Distributed practice is a learning technique in which _____.

   a. subjects are distributed across equal study sessions
   b. learning periods alternate with nonlearning rest periods
   c. learning decays faster than it can be distributed
   d. several students study together, distributing various subjects according to their individual strengths

13. Researchers have demonstrated that it is _____ to create false memories.

   a. relatively easy
   b. moderately difficult
   c. rarely possible
   d. never possible

14. _____ memories are related to anxiety-provoking thoughts or events that are prevented from reaching consciousness.

   a. Suppressed
   b. Flashback
   c. Motivated
   d. Repressed

15. _____ devices improve memory by encoding items in a special way.

   a. Eidetic imagery
   b. Mnemonic
   c. Reverberating circuit
   d. ECS

Thinking, Language, and Intelligence

8

On July 9, 2005, professional skateboarder Danny Way rocketed down a 120-foot ramp at almost 50 miles per hour and leapt a 61-foot gap across the Great Wall of China. He did it not once but five times. Do you think that this feat reflects merely athletic skill and daring, or does it also speak to Way's intelligence?

When we think of intelligence, many of us think of Nobel Prize winners, great inventors, or chess champions. But success as a professional skateboarder also requires intelligence—perhaps of a different kind than people generally associate with being "smart." Skateboarding enthusiasts admire the creativity and technical innovation that Way brings to the sport. He has broken many skateboarding records and has devised stunts—including a jump from a helicopter— that few others would try.

The three topics of this chapter—thinking, language, and intelligence—are often studied together under the broader umbrella of cognition, the mental activities involved in acquiring, storing, retrieving, and using knowledge. In a real sense, we discuss cognition throughout this text because psychology is "the scientific study of behavior and *mental processes*" (Chapter 1). For example, the chapters on sensation and perception (Chapter 4) consciousness (Chapter 5), learning (Chapter 6), and memory (Chapter 7) all focus on cognition. In this chapter, we emphasize thinking, language, and intelligence. As you will discover, each is a complex phenomenon that is greatly affected by numerous factors.

Thinking

LEARNING OBJECTIVES

Explain the role of the prefrontal cortex in thinking.

Describe how mental images, concepts, and prototypes enable thinking.

Identify the three steps involved in problem solving.

Summarize how biases and heuristics can hamper problem solving.

Explain Sternberg and Lubart's investment theory of creativity.

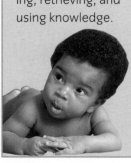

cognition Mental activities involved in acquiring, storing, retrieving, and using knowledge.

THE THINKING BRAIN: MAKING CONNECTIONS

Thinking, language, and intelligence are closely related facets of **cognition**. Every time you take information and mentally act on it, you are thinking. Our thought processes are distributed throughout our brains in networks of neurons. However, they are also localized. For example, during problem solving or decision making, our brains are active in the **prefrontal cortex**. This region associates complex ideas; makes plans; forms, initiates, and allocates attention; and supports multitasking. The prefrontal cortex also links to other areas of the brain, such as the limbic system (Chapter 2), to synthesize information from several senses (Heyder, Suchan, & Daum, 2004; Saab & Willis, 2003).

COGNITIVE BUILDING BLOCKS

Imagine yourself lying relaxed in the warm, gritty sand on an ocean beach. Do you see palms swaying in the wind? Can you smell the salty sea and taste the dried salt on your lips? Can you hear children playing in the surf? What you've just created is a **mental image** (FIGURE 8.1),

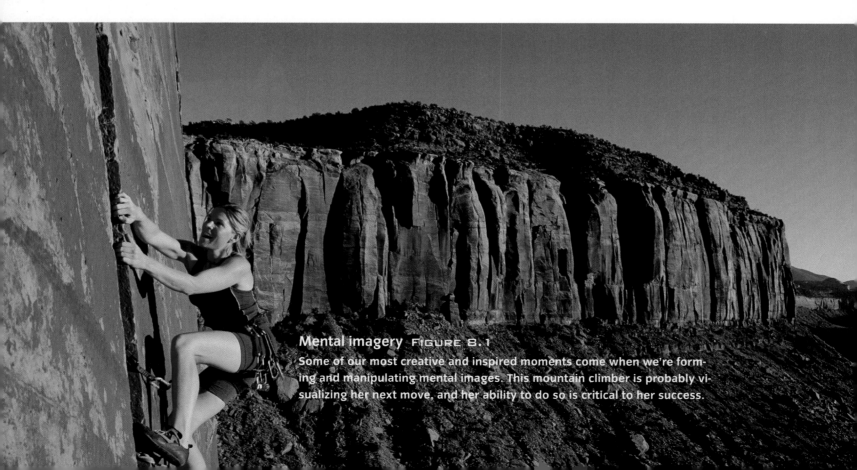

Mental imagery FIGURE 8.1
Some of our most creative and inspired moments come when we're forming and manipulating mental images. This mountain climber is probably visualizing her next move, and her ability to do so is critical to her success.

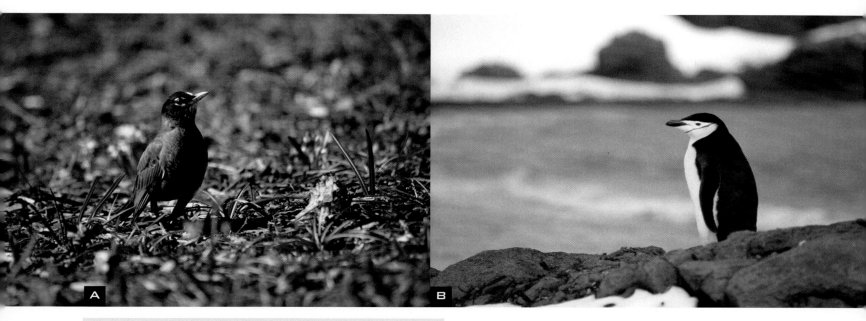

Some birds are "birdier" than others FIGURE 8.2

A Most people have a *prototype* bird that captures the essence of "birdness" and allows us to quickly classify flying animals correctly.

B When we encounter an example that doesn't quite fit our prototype, we need time to review our artificial concept. Because the penguin doesn't fly, it's harder to classify than a robin.

a mental representation of a previously stored sensory experience, which includes visual, auditory, olfactory, tactile, motor, and gustatory imagery (McKellar, 1972). We all have a mental space where we visualize and manipulate our sensory images (Brewer & Pani, 1996; Hamm, Johnson, & Corballis, 2004).

In addition to mental images, our thinking also involves forming **concepts**, or mental representations of a group or category (Smith, 1995). Concepts can be concrete (*car, concert*) or abstract (*intelligence, pornography*). They are essential to thinking and communication because they simplify and organize information. Normally, when you see a new object or encounter a new situation, you relate it to your existing conceptual structure and categorize it according to where it fits. For example, if you see a metal box with four wheels driving on the highway, you know it is a car, even if you've never seen that particular model before.

We create some of our concepts from logical rules or definitions. Some concepts, such as *triangle*, are called "artificial" (or "formal") because the rules for inclusion are sharply defined.

Artificial concepts are often found in science and academe, but in real life, we seldom use precise, artificial definitions. When we see birds in the sky, we don't think *warm-blooded animals that fly, have wings, and lay eggs*—an artificial concept. Instead, we use natural concepts, or **prototypes**, which are based on a typical representative of that concept (Rosch, 1973) (**FIGURE 8.2**).

Some of our concepts also develop when we create **hierarchies**, or group specific concepts as subcategories within broader concepts. This process makes mastering new material faster and easier. For example, when you learn that all animals have mitochondria in their cells, you don't need to relearn that fact every time you learn a new animal species.

When we first learn something, we rely primarily on basic-level concepts (Rosch, 1978). Thus, children tend to learn *bird* before they learn the higher-order concept *animal* or the lower-order concept *parakeet*. Even as adults, when shown a picture of a parakeet, we classify it as a *bird* first.

SOLVING PROBLEMS

Several years ago in Los Angeles, a 12-foot-high tractor-trailer got stuck under a bridge that was 6 inches too low. After hours of towing, tugging, and pushing, the police and transportation workers were stumped. Then a young boy happened by and asked, "Why don't you let some air out of the tires?" It was a simple, creative suggestion—and it worked.

Our lives are filled with problems, some simple, some difficult. In all cases, **problem solving** requires moving from a given state (the problem) to a goal state (the solution), a process that usually involves three steps (Bourne, Dominowski, & Loftus, 1979) (FIGURE 8.5 on the following page).

BARRIERS TO PROBLEM SOLVING

We all encounter barriers to solving problems. We stick to problem-solving strategies (**mental sets**) that worked in the past, rather than trying new, possibly more effective ones (FIGURE 8.3). Or we fail to let our inventive instincts run free, thinking of objects as functioning only in their prescribed, customary way—a phenomenon called **functional fixedness** (FIGURE 8.4).

"Repurposing" as art FIGURE 8.4

Some people look at a pile of garbage and see only trash. Others see possibility. This photo shows how an artist recycled old hubcaps. Have you ever "repurposed" a discarded object such as an old wheelbarrow for use as a garden planter? If so, you have overcome *functional fixedness*.

Other barriers to effective problem solving stem from our tendency to ignore important information. Have you ever caught yourself agreeing with friends who support your political opinions, and discounting conflicting opinions? This inclination to seek confirmation for our preexisting beliefs and to overlook contradictory evidence is known as the **confirmation bias** (Nickerson, 1998; Reich, 2004).

British researcher Peter Wason (1968) first demonstrated the confirmation bias. He asked participants to generate a list of numbers that conformed to the same rule that applied to this set of numbers:

2   4   6

Hypothesizing that the rule was "numbers increasing by two," most participants generated sets such as (4, 6, 8) or (1, 3, 5). Each time, Wason assured them that their sets of numbers conformed to the rule but that the rule "numbers increasing by two" was incorrect. The problem was that the participants were searching only for information that confirmed their hypothesis. Proposing a series such as (1, 3, 4) would have led them to reject their initial hypothesis and discover the correct rule: "numbers in increasing order of magnitude."

Cognitive psychologists Amos Tversky and Daniel Kahneman found that heuristics, as handy as

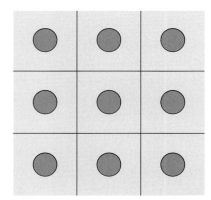

The nine-dot problem FIGURE 8.3

Draw no more than four lines that run through all nine dots on this page without lifting your pencil from the paper (see page 220 for the solution).

Three steps to the goal FIGURE 8.5

Step 1: Preparation

Imagine that you are determined to find a new home close to work.
There are at least three separate components to successful preparation.

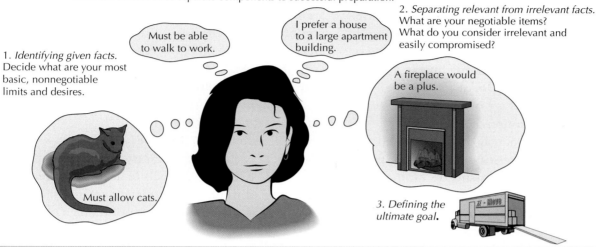

1. Identifying given facts.
Decide what are your most
basic, nonnegotiable
limits and desires.

Must be able
to walk to work.

I prefer a house
to a large apartment
building.

2. Separating relevant from irrelevant facts.
What are your negotiable items?
What do you consider irrelevant and
easily compromised?

A fireplace would
be a plus.

Must allow cats.

*3. Defining the
ultimate goal.*

Step 2: Production

During the *production step*, the problem solver produces possible solutions, called *hypotheses*.
Two major approaches to generating hypotheses are *algorithms* and *heuristics*.

Algorithm:
Answers
every
ad.

An **algorithm** is a logical, step-by-step
procedure (well suited for computers) that
will always produce the solution. (For
example, an algorithm for solving the
problem 2×4 is 2 + 2 + 2 + 2.) For complex
problems, algorithms may take a long
time.

A **heuristic** is a simple rule or
shortcut that does not guarantee a
solution. Heuristics include working
backward from the solution (a known
condition) and *creating subgoals*,
stepping stones to the original goal.

CITY MAP

Heuristic:
Works backward—start
by drawing a 1-mile
radius around work to
narrow search.

■ **algorithm** A set
of steps that, if
followed correctly,
will eventually
solve the
problem.

■ **heuristic** A sim-
ple rule used in
problem solving
and decision mak-
ing that does not
guarantee a solu-
tion but offers a
likely shortcut to it.

Step 3: Evaluation

1. Did your hypotheses (possible solutions) solve the problem?
If not, then you must return to the production stage
and produce more possible solutions.

2. Take action. Once you've identified the right
apartment, sign the lease and start packing.

Process Diagram

they can be, can lead us to ignore relevant information (Kahneman, 2003; Tversky & Kahneman, 1974, 1993). When we use the **availability heuristic**, we judge the likelihood of an event based on how easily recalled (available) other instances of the event are (McKelvie & Drumheller, 2001; Oppenheimer, 2004). Shortly after the September 11, 2001, terrorist attacks, one study found that the average American believed that he or she had a 20.5 percent chance of being hurt in a terrorist attack within a year (Lerner et al., 2003). Can you see how intense media coverage of the attacks created this erroneously high perception of risk?

Tversky and Kahneman also demonstrated that the **representativeness heuristic** sometimes hinders problem solving. Using this heuristic, we estimate the probability of something based on how well the circumstances match (represent) our prototype. For example, if John is six feet, five inches tall, we may guess that he is an NBA basketball player rather than, say, a bank president. But in this case, the representative heuristic ignores *base rate information*—the probability of a characteristic occurring in the general population. In the United States, bank presidents outnumber NBA players by about 50 to 1. So despite his height, John is much more likely to be a bank president.

CREATIVITY: FINDING UNIQUE SOLUTIONS

What makes a person creative? Conceptions of creativity are subject to cultural relevance and to whether a solution or performance is considered useful at the time. In general, three characteristics are associated with **creativity**: originality, fluency, and flexibility. Thomas Edison's invention of the light bulb offers a prime example of each of these characteristics (TABLE 8.1).

> **creativity** The ability to produce valued outcomes in a novel way.

Most tests of creativity focus on **divergent thinking**, a type of thinking where many possibilities are developed from a single starting point (Baer, 1994). For example, in the Unusual Uses Test, people are asked to think of as many uses as possible for an object, such as a brick. In the Anagrams Test, people are asked to reorder the letters in a word to make as many new words as possible.

One prominent theory of creativity is Robert J. Sternberg and Todd Lubart's **investment theory** (1992, 1996). According to this theory, creative people tend to

With Your Own Eyes

Are You Creative?

Everyone exhibits a certain amount of creativity in some aspects of life. Even when doing ordinary tasks, like planning an afternoon of errands, you are being somewhat creative. Similarly, if you've ever tightened a screw with a penny or used a telephone book on a chair as a booster seat for a child, you've found creative solutions to problems.

Would you like to test your own creativity?

- Find 10 coins and arrange them in the configuration shown here. By moving only two coins, form two rows that each contain 6 coins (see page 220 for the solution).

- In five minutes, see how many words you can make using the letters in the word *hippopotamus*.

- In five minutes, list all the things you can do with a paper clip.

Three elements of creative thinking TABLE 8.1

	Explanations	Thomas Edison Examples
Originality	Seeing unique or different solutions to a problem	After noting that electricity passing through a conductor produces a glowing red or white heat, Edison imagined using this light for practical uses.
Fluency	Generating a large number of possible solutions	Edison tried literally hundreds of different materials to find one that would heat to the point of glowing white heat without burning up.
Flexibility	Shifting with ease from one type of problem-solving strategy to another	When he couldn't find a long-lasting material, Edison tried heating it in a vacuum—thereby creating the first light bulb.

"buy low" in the realm of ideas, championing ideas that others dismiss (much like a bold entrepreneur might invest in low-priced, unpopular stocks, believing that their value will rise). Once their creative ideas are highly valued, they "sell high" and move on to another unpopular but promising idea. Investment theory also suggests that creativity requires the coming together of six interrelated resources (Kaufman, 2002; Sternberg & Lubart, 1996). These resources are summarized in TABLE 8.2. One way to improve your personal creativity is to study this list and then strengthen yourself in those areas you think need improvement.

Resources of creative people TABLE 8.2

Intellectual Ability	Enough intelligence to see problems in a new light
Knowledge	Sufficient basic knowledge of the problem to effectively evaluate possible solutions
Thinking Style	Novel ideas and ability to distinguish between the worthy and worthless
Personality	Willingness to grow and change, take risks, and work to overcome obstacles
Motivation	Sufficient motivation to accomplish the task and more internal than external motivation
Environment	An environment that supports creativity

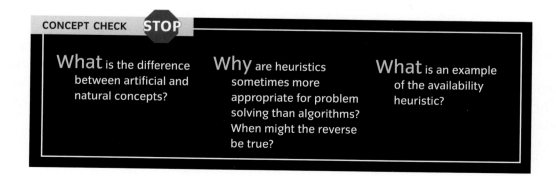

CONCEPT CHECK STOP

What is the difference between artificial and natural concepts?

Why are heuristics sometimes more appropriate for problem solving than algorithms? When might the reverse be true?

What is an example of the availability heuristic?

Language

language A form of communication using sounds and symbols combined according to specified rules.

phoneme [FO-neem] The smallest basic unit of speech or sound. The English language has about 40 phonemes.

morpheme [MOR-feem] The smallest meaningful unit of language, formed from a combination of phonemes.

WHAT IS LANGUAGE?

Language enables us to mentally manipulate symbols, thereby expanding our thinking, and to communicate our thoughts, ideas, and feelings. To produce language, we first build words using **phonemes** and **morphemes**. Then we string words into sentences using rules of **grammar** (syntax and semantics) (FIGURE 8.6).

LANGUAGE AND THOUGHT: A COMPLEX INTERACTION

Does the fact that you speak English instead of Spanish —or Chinese instead of Swahili—determine how you reason, think, and perceive the world? Linguist Benjamin Whorf (1956) believed so. As evidence for his **linguistic relativity hypothesis**, Whorf offered a now classic example: Because Eskimos (Inuits) have many words for snow (*apikak* for "first snow falling," *pukak* for "snow for drinking water," and so on), they can perceive and

think about snow differently from English speakers, who have only one word—*snow.*

Although intriguing, Whorf's hypothesis has not fared well. He apparently exaggerated the number of Inuit words for snow (Pullum, 1991) and ignored the fact that English speakers have a number of terms to describe various forms of snow, such as *slush, sleet, hard pack,* and *powder.* Other research has directly contradicted Whorf's theory. For example, Eleanor Rosch (1973) found that although people of the Dani tribe in New Guinea possess only two color names—one indicating cool, dark colors, and the other describing warm, bright colors—they discriminate among multiple hues as well as English speakers do.

Whorf apparently was mistaken in his belief that language determines thought. But there is no doubt that language *influences* thought (Lillard, 1998). People who speak both Chinese and English report that the language they're currently using affects their sense of self (Matsumoto, 2000). When using Chinese, they tend to conform to Chinese cultural norms; when speaking English, they tend to adopt Western norms.

Our words also influence the thinking of those who hear them. That's why companies avoid *firing* employees; instead, employees are *outplaced* or *nonrenewed.* Similarly, the military uses terms like *preemptive strike* to cover the fact that they attacked first and *tactical redeployment* to refer to a retreat.

grammar Rules that specify how phonemes, morphemes, words, and phrases should be combined to express thoughts. These rules include syntax and semantics.

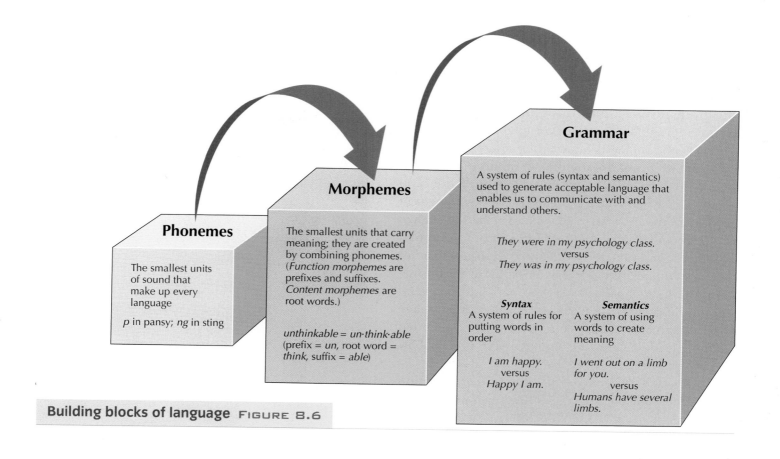

Phonemes

The smallest units of sound that make up every language

p in pansy; *ng* in sting

Morphemes

The smallest units that carry meaning; they are created by combining phonemes. (*Function morphemes* are prefixes and suffixes. *Content morphemes* are root words.)

unthinkable = un·think·able (prefix = *un*, root word = *think*, suffix = *able*)

Grammar

A system of rules (syntax and semantics) used to generate acceptable language that enables us to communicate with and understand others.

They were in my psychology class.
versus
They was in my psychology class.

Syntax
A system of rules for putting words in order

I am happy.
versus
Happy I am.

Semantics
A system of using words to create meaning

I went out on a limb for you.
versus
Humans have several limbs.

Building blocks of language FIGURE 8.6

LANGUAGE DEVELOPMENT: FROM CRYING TO TALKING

From birth, a child communicates through facial expressions, eye contact, and body gestures. Babies only hours old begin to "teach" their caregivers when and how they want to be held, fed, and played with.

Eventually, children also communicate verbally, progressing through several distinct stages of language acquisition. These stages are summarized in

Baby Signs

Many babies and toddlers—beginning as young as nine months of age—learn to communicate by using a modified form of sign language, sometimes called "Baby Sign." Symbolic gestures for basic ideas such as "more," "milk," or "love" enhance parents' and caregivers' interactions with children who cannot yet talk. Many parents find that signing with their infant or toddler gives them a fascinating "window" into the baby's mind—and eliminates a lot of frustration for both them and baby! Some researchers believe that teaching babies to sign helps foster better language comprehension and can speed up the process of learning to talk.

TABLE 8.3. By age 5, most children have mastered basic grammar and typically use about 2,000 words (a level of mastery considered adequate for getting by in any given culture). Past this point, vocabulary and grammar gradually improve throughout life.

Some theorists believe that language capability is innate, primarily a matter of maturation. Noam Chomsky (1968, 1980) suggests that children are "pre-wired" with a neurological ability known as a **language acquisition device (LAD)** that enables a child to analyze language and to extract the basic rules of grammar. This mechanism needs only minimal exposure to adult speech to unlock its potential. As evidence for this nativist position, Chomsky observes that children everywhere progress through the same stages of language development at about the same ages. He also notes that babbling is the same in all languages and that deaf babies babble just like hearing babies.

"Nurturists" argue that the nativist position doesn't fully explain individual differences in language development. They hold that children learn language through a complex system of rewards, punishments, and imitation. For example, parents smile and encourage any vocalizations from a very young infant. Later, they respond even more enthusiastically when the infant babbles "mama" or "dada." In this way, parents unknowingly use *shaping* (Chapter 6) to help babies learn language (FIGURE 8.7).

Early nonverbal communication Even before they begin to communicate verbally, infants use nonverbal signals to "speak" to others. What emotions are each of these infants expressing? Infants as young as 2½ months express basic emotions such as interest, joy, anger, and sadness. Children who are born blind and deaf exhibit the same emotional expressions as sighted and hearing children, supporting the contention that these expressions are universal and innate.

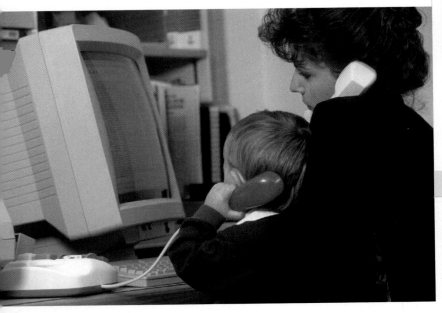

Nature or nurture? FIGURE 8.7

Both sides of the nature-versus-nurture debate have staunch supporters. However, most psychologists believe that language acquisition is a combination of both biology (nature) and environment (nurture) (Casti, 2000; Van Hulle, Goldsmith, & Lemery, 2004). Can you see how both might contribute to this baby's pretend phone conversation?

Developmental Stage	Age	Language Features	Example
Prelinguistic stage	Birth to ~12 months	Crying (reflexive in newborns; soon, crying becomes more purposeful)	hunger cry anger cry pain cry
	2 to 3 months	Cooing (vowel-like sounds)	"ooooh" "aaaah"
	4 to 6 months	Babbling (consonants added)	"bahbahbah" "dahdahdah"
Linguistic stage	~12 months	Babbling begins to resemble language of the child's home. Child seems to understand that sounds relate to meaning.	
		At first, speech is limited to one-word utterances.	"Mama" "juice" "Daddy" "up"
		Expressive ability more than doubles once child begins to join words into short phrases.	"Daddy, milk" "no night-night!"
	~ 2 years	Child sometimes *overextends* (using words to include objects that do not fit the word's meaning).	all men = "Daddy" all furry animals = doggy
	~2 years to ~5 years	Child links several words to create short but intelligible sentences. This speech is called *telegraphic* because (like telegrams) it omits nonessential connecting words.	"Me want cookie" "Grandma go bye-bye?"
		Vocabulary increases at a phenomenal rate.	
		Child acquires a wide variety of rules for grammar.	Adding *–ed* for past tense adding *s* to form plurals
		Child sometimes *overgeneralizes* (applying the basic rules of grammar even to cases that are exceptions to the rule).	"I goed to the zoo" "Two mans"

CAN HUMANS TALK TO NONHUMAN ANIMALS?

Without question, nonhuman animals communicate, sending warnings, signaling sexual interest, sharing locations of food sources, and so on. But can nonhuman animals master the complexity of human language? Since the 1930s, many language studies have attempted to answer this question by probing the language abili-

ties of chimpanzees and gorillas (Itakura, 1992; Savage-Rumbaugh, 1990).

One of the most successful early studies was conducted by Beatrice and Allen Gardner (1969), who recognized chimpanzees' manual dexterity and ability to imitate gestures. The Gardners used American Sign Language (ASL) with a chimp named Washoe. By the time Washoe was 4 years old, she had learned 132 signs and was able to combine them into simple sentences

Signing FIGURE 8.8

According to her teacher, Penny Patterson, Koko has used ASL to converse with others, talk to herself, joke, express preferences, and even lie (Linden, 1993; Patterson, 2002).

such as "Hurry, gimme toothbrush" and "Please tickle more." The famous gorilla Koko also uses ASL to communicate; she reportedly uses more than 1,000 words (FIGURE 8.8).

In another well-known study, a chimp named Lana learned to use symbols on a computer to get things she wanted, such as food, a drink, a tickle from her trainers, and having her curtains opened (Rumbaugh et al., 1974) (FIGURE 8.9).

Dolphins are also the subject of interesting language research. Communication with dolphins is done by means of hand signals or audible commands transmitted through an underwater speaker system. In one typical study, trainers gave dolphins commands made up of

Computer-aided communication FIGURE 8.9

Apes lack the necessary anatomical structures to vocalize the way humans do. For this reason, language research with chimps and gorillas has focused on teaching the animals to use sign language or to "speak" by pointing to symbols on a keyboard.

two- to five-word sentences, such as "Big ball—square—return," which meant that they should go get the big ball, put it in the floating square, and return to the trainer (Herman, Richards, & Woltz, 1984). By varying the syntax (for example, the order of the words) and specific content of the commands, the researchers showed that dolphins are sensitive to these aspects of language.

Psychologists disagree about how to interpret these findings on apes and dolphins. Most psychologists believe that nonhuman animals communicate but that their ideas are severely limited. Critics claim that apes and dolphins are unable to convey subtle meanings, use language creatively, or communicate at an abstract level (Jackendoff, 2003).

Other critics claim that these animals do not truly understand language but are simply operantly conditioned (Chapter 6) to imitate symbols to receive rewards (Savage-Rumbaugh, 1990; Terrace, 1979). Finally, other critics suggest that data regarding animal language has not always been well documented (Lieberman, 1998; Willingham, 2001).

Proponents of animal language respond that apes can use language creatively and have even coined some words of their own. For example, Koko signed "finger bracelet" to describe a ring and "eye hat" to describe a mask (Patterson & Linden, 1981). Proponents also argue that, as demonstrated by the dolphin studies, animals can be taught to understand basic rules of sentence structure.

Still, the gap between human and nonhuman animals' language is considerable. Current evidence suggests that, at best, nonhuman animal language is less complex, less creative, and has fewer rules than any language used by humans.

CONCEPT CHECK STOP

What are phonemes, morphemes, and grammar?

Why do some psychologists believe that language is an innate ability, whereas others believe that it is learned through imitation and reinforcement?

What are some differences between human and nonhuman animal language?

Intelligence

Many people equate intelligence with "book smarts." For others, what is intelligent depends on the characteristics and skills that are valued in a particular social group or culture (Sternberg, 2005; Sternberg & Hedlund, 2002; Suzuki & Aronson, 2005). For example, the Mandarin word that corresponds most closely to the word "intelligence" is a character meaning "good brain and talented" (Matsumoto, 2000). The word is associated with traits like imitation, effort, and social responsibility (Keats, 1982).

Even among Western psychologists there is debate over the definition of intelligence. In this discussion, we rely on a formal definition developed by psychologist David Wechsler (pronounced "WEX-ler") (1944, 1977). Wechsler defined **intelligence** as the global capacity to think rationally, act purposefully, and deal effectively with the environment.

intelligence The global capacity to think rationally, act purposefully, and deal effectively with the environment.

DO WE HAVE ONE OR MANY INTELLIGENCES?

One of the central debates in research on intelligence concerns whether intelligence is a single ability or a collection of many specific abilities.

In the 1920s, British psychologist Charles Spearman first observed that high scores on separate tests of mental abilities tend to correlate with each other. Spearman (1923) thus proposed that intelligence is a single factor, which he termed **general intelligence (g)**. He believed that g underlies all intellectual behavior, including reasoning, solving problems, and performing well in all areas of cognition. Spearman's work laid the foundations for today's standardized intelligence tests (Deary et al., 2004; Johnson, Bouchard, et al., 2004; Lubinski & Dawis, 1992).

About a decade later, L. L. Thurstone (1938) proposed seven primary mental abilities: verbal comprehension, word fluency, numerical fluency, spatial visualization, associative memory, perceptual speed, and reasoning. J. P. Guilford (1967) later expanded this number, proposing that as many as 120 factors were involved in the structure of intelligence.

Around the same time, Raymond Cattell (1963, 1971) reanalyzed Thurstone's data and argued against the idea of multiple intelligences. He believed that two subtypes of g exist:

- **Fluid intelligence (gf)** refers to innate, inherited reasoning abilities, memory, and speed of information processing. Fluid intelligence is relatively independent of education and experience, and like other biological capacities it declines with age (Burns, Nettelbeck, & Cooper, 2000; Li et al., 2004; Rozencwajg et al., 2005).

- **Crystallized intelligence (gc)** refers to the store of knowledge and skills gained through experience and education (Facon & Facon-Bollengier, 1999). Crystallized intelligence tends to increase over the life span.

Today there is considerable support for the concept of g as a measure of academic smarts. However, many contemporary cognitive theorists believe that intelligence is not a single general factor but a collection of many separate specific abilities.

One of these cognitive theorists, Howard Gardner, believes that people have many kinds of intelligences. The fact that brain-damaged patients often lose some intellectual abilities while retaining others suggests to Gardner that different intelligences are located in discrete areas throughout the brain. According to Gardner's (1983, 1998, 1999) **theory of multiple intelligences**, people have different profiles of intelligence because they are stronger in some areas than others (TABLE 8.4). They also use their intelligences differently to learn new material, perform tasks, and solve problems.

Robert Sternberg's **triarchic theory of successful intelligence** also involves multiple abilities. As discussed in TABLE 8.5, Sternberg theorized that three separate, learned aspects of intelligence exist: (1) analytic intelligence, (2) creative intelligence, and (3) practical intelligence (Grigorenko, Jarvin, & Sternberg, 2002; Sternberg, 1999).

Sternberg (1985, 1999) emphasizes the process underlying thinking rather than just the product. He also stresses the importance of applying mental abilities to real-world situations rather than testing mental abilities in isolation (e.g., Sternberg 2005; Sternberg & Hedlund, 2002). Sternberg (1998) introduced the term *successful intelligence* to describe the ability to adapt to, shape, and select environments in order to accomplish personal and societal goals.

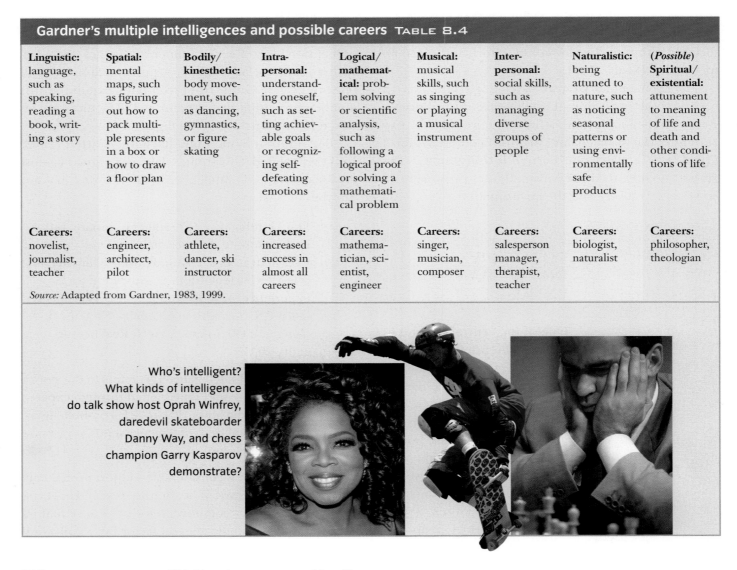

Gardner's multiple intelligences and possible careers TABLE 8.4

Linguistic: language, such as speaking, reading a book, writing a story	**Spatial:** mental maps, such as figuring out how to pack multiple presents in a box or how to draw a floor plan	**Bodily/ kinesthetic:** body movement, such as dancing, gymnastics, or figure skating	**Intra- personal:** understanding oneself, such as setting achievable goals or recognizing self-defeating emotions	**Logical/ mathematical:** problem solving or scientific analysis, such as following a logical proof or solving a mathematical problem	**Musical:** musical skills, such as singing or playing a musical instrument	**Inter- personal:** social skills, such as managing diverse groups of people	**Naturalistic:** being attuned to nature, such as noticing seasonal patterns or using environmentally safe products	(*Possible*) **Spiritual/ existential:** attunement to meaning of life and death and other conditions of life
Careers: novelist, journalist, teacher	**Careers:** engineer, architect, pilot	**Careers:** athlete, dancer, ski instructor	**Careers:** increased success in almost all careers	**Careers:** mathematician, scientist, engineer	**Careers:** singer, musician, composer	**Careers:** salesperson manager, therapist, teacher	**Careers:** biologist, naturalist	**Careers:** philosopher, theologian

Source: Adapted from Gardner, 1983, 1999.

Who's intelligent? What kinds of intelligence do talk show host Oprah Winfrey, daredevil skateboarder Danny Way, and chess champion Garry Kasparov demonstrate?

Sternberg's triarchic theory of successful intelligence TABLE 8.5			
	Analytical Intelligence	**Creative Intelligence**	**Practical Intelligence**
Sample skills	Good at analysis, evaluation, judgment, and comparison skills	Good at invention, coping with novelty, and imagination skills	Good at application, implementation, execution, and utilization skills
Methods of assessment	These skills are assessed by intelligence or scholastic aptitude tests. Questions ask about meanings of words based on context and how to solve number-series problems.	These skills are assessed in many ways, including open-ended tasks, writing a short story, drawing a piece of art, or solving a scientific problem requiring insight.	Although these skills are more difficult to assess, they can be measured by asking for solutions to practical and personal problems.

MEASURING INTELLIGENCE

Many college admissions offices, scholarship committees, and employers use scores from intelligence tests as a significant part of their selection criteria. How well do these tests predict student and employee success? Different IQ tests approach the measurement of intelligence from different perspectives. However, most are designed to predict grades in school. Let's look at the most commonly used IQ tests.

The **Stanford-Binet Intelligence Scale** is loosely based on the first IQ tests developed in France around the turn of the last century by Alfred Binet. In the United States, Lewis Terman (1916) developed the Stanford-Binet (at Stanford University) to test the intellectual ability of U.S.-born children ages 3 to 16. The test is revised periodically—most recently in 1985. The test is administered individually and consists of such tasks as copying geometric designs, identifying similarities, and repeating number sequences.

In the original version of the Stanford-Binet, results were expressed in terms of a mental age. For example, if a 7-year-old's score equaled that of an average 8-year-old, the child was considered to have a mental age of eight. To determine the child's **intelligence quotient (IQ)**, mental age was divided by the child's chronological age (actual age in years) and multiplied by 100.

Today, most intelligence test scores are expressed as a comparison of a single person's score to a national sample of similar-aged people (FIGURE 8.10).

> **intelligence quotient (IQ)** A subject's mental age divided by his or her chronological age and multiplied by 100.

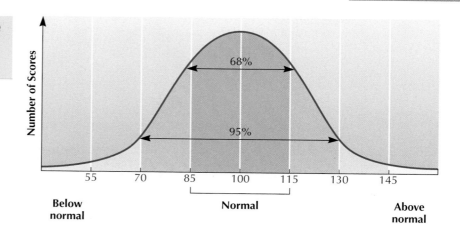

The distribution of scores on the Stanford-Binet Intelligence Test
FIGURE 8.10

Sixty-eight percent of children score within one standard deviation (16 points) above or below the national average, which is 100 points. About 16 percent score above 116, and about 16 percent score below 84.

A modern intelligence test FIGURE 8.11

The most widely used intelligence test, the Wechsler Adult Intelligence Scale (WAIS), was developed by David Wechsler in the early 1900s. He later created a similar test for school-age children, the Wechsler Intelligence Scale for Children, now in its fourth edition (WISC-III), and one for preschool children, the Wechsler Preschool and Primary Scale of Intelligence (WPPSI).

Like the Stanford-Binet, Wechsler's tests yield an overall intelligence score, but they have separate scores for verbal intelligence (such as vocabulary, comprehension, and knowledge of general information) and performance (such as arranging pictures to tell a story or arranging blocks to match a given pattern).

Verbal Subtests

Information	How many senators are elected from each state?
Similarities	How are computers and books alike?
Arithmetic	If one baseball card costs three cents, how much will five baseball cards cost?
Vocabulary	Define *lamp*.
Comprehension	What should you do if you accidentally break a friend's toy?

Picture Completion
What is missing from this ambulance?

Coding
Write the appropriate number above each symbol.

Picture Arrangement
Arrange these pictures in chronological order.

Block Design
Copy this design with blocks.

Object Assembly
Assemble this small jigsaw puzzle.

Wechsler's approach has three advantages:

1. The WAIS was specifically designed for adults, not children.

2. Different abilities can be evaluated either separately or together.

3. Non-English speakers can still be tested, with the verbal portion omitted.

Even though the actual IQ is no longer calculated using the original formula comparing mental and chronological ages, the term *IQ* remains as a shorthand expression for intelligence test scores.

What makes a good test? How are the tests developed by Binet and Wechsler any better than those published in popular magazines and presented on television programs? To be scientifically acceptable, all psychological tests must fulfill three basic requirements:

- **Standardization.** Intelligence tests (as well as personality, aptitude, and most other tests)

standardization Establishment of the norms and uniform procedures for giving and scoring a test.

must be standardized in two senses (Hogan, 2003). First, every test must have norms, or average scores, developed by giving the test to a representative sample of people (a diverse group of people who resemble those for whom the test is intended). Second, testing procedures must be standardized. All test takers must be given the same instructions, questions, and time limits, and all test administrators must follow the same objective score standards.

- **Reliability.** To be trustworthy, a test must be consistent, or reliable. Reliability is usually determined by retesting subjects

reliability A measure of the consistency and stability of test scores when a test is readministered.

to see whether their test scores change significantly (Hogan, 2003, p. 142). Retesting can be done via the **test-retest method**, in which participants' scores on two separate administrations of the same test are compared, or via the **split-half method**, which involves splitting a test into two equivalent parts (e.g., odd and even questions) and determining the degree of similarity between the two halves.

- **Validity.** Validity is the ability of a test to measure what it is designed to measure. The most important type of validity is **criterion-related validity**, or the accuracy with which test scores can be used to predict another variable of interest (known as the criterion). Criterion-related validity is expressed as the *correlation* (Chapter 1) between the test score and the criterion. If two variables are highly correlated, then one variable can be used to predict the other. Thus, if a test is valid, its scores will be useful in predicting people's behavior in some other specified situation. One example of this is using intelligence test scores to predict grades in college.

validity The ability of a test to measure what it was designed to measure.

Can you see why a test that is standardized and reliable but not valid is worthless? For example, a test for skin sensitivity may be easy to standardize (the instructions specify exactly how to apply the test agent), and it may be reliable (similar results are obtained on each retest). But it certainly would not be valid for predicting college grades.

CONCEPT CHECK **STOP**

What is the difference between Gardner's and Sternberg's theories of intelligence?

Why is criterion validity important?

The Intelligence Controversy

LEARNING OBJECTIVES

Explain why extremes in intelligence provide support for the validity of IQ testing.

Review research on how brain functioning is related to intelligence.

Describe how genetics and environment interact to shape intelligence.

Summarize the debate over ethnicity and intelligence.

EXTREMES IN INTELLIGENCE: MENTAL RETARDATION AND GIFTEDNESS

One of the best methods for judging the validity of a test is to compare people who score at the extremes. Despite the uncertainties discussed in the previous section, intelligence tests provide one of the major criteria for assessing mental ability at the extremes—specifically, for diagnosing **mental retardation** and **giftedness**.

The clinical label *mentally retarded* is applied when someone is significantly below average in intellectual functioning and has significant deficits in adaptive functioning (such as communicating, living independently, social or occupational functioning, or maintaining safety and health) (American Psychiatric Association, 2000).

savant syndrome A condition in which a person who has mental retardation exhibits exceptional skill or brilliance in some limited field.

Fewer than 3 percent of people are classified as having mental retardation. Of this group, 85 percent have only mild retardation and many become self-supporting, integrated members of society. Furthermore, people can score low on some measures of intelligence and still be average or even gifted in others. The most dramatic examples are people with **savant syndrome** (FIGURE 8.12).

Some forms of retardation stem from genetic abnormalities, such as Down syndrome, fragile-X syndrome, and phenylketonuria (PKU). Other causes are environmental, including prenatal exposure to alcohol and other drugs, extreme deprivation or neglect in early life, and brain damage from accidents. However, in many cases, there is no known cause of retardation.

An unusual form of intelligence FIGURE 8.12

Although people with *savant syndrome* score very low on IQ tests (usually between 40 and 70), they demonstrate exceptional skills or brilliance in specific areas, such as rapid calculation, art, memory, or musical ability (Bonnel et al., 2003; McMahon, 2002; Pring & Hermelin, 2002). Seventeen-year-old Brittany Maier, who has autism and severe visual impairment, is a gifted composer and pianist who performs publicly and has recorded a CD. Her musical repertoire includes more than 15,000 songs.

At the other end of the intelligence spectrum are people with especially high IQs (typically defined as being in the top 1 or 2 percent).

In 1921, Lewis Terman identified 1,500 gifted children—affectionately nicknamed the "Termites"—with IQs of 140 or higher and tracked their progress through adulthood. The number who became highly successful professionals was many times the number a random group would have provided (Leslie, 2000; Terman, 1954). Those who were most successful tended to have extraordinary motivation, and they also had someone at home or school who was especially encouraging (Goleman, 1980). There were some notable failures, and the "Termites" became alcoholics, got divorced, and committed suicide at close to the national rate (Leslie, 2000). A high IQ is no guarantee of success in every endeavor; it only offers more intellectual opportunities.

THE BRAIN'S INFLUENCE ON INTELLIGENCE

A basic tenet of neuroscience is that all mental activity (including intelligence) results from neural activity in the brain. Most recent research on the biology of intelligence has focused on brain functioning. For example, neuroscientists have found that people who score highest on intelligence tests also respond more quickly on tasks involving perceptual judgments (Bowling & Mackenzie, 1996; Deary & Stough, 1996, 1997; Posthuma et al., 2001).

Other research using positron emission tomography (PET) to measure brain activity (Chapter 1) suggests that intelligent brains work smarter, or more efficiently, than less-intelligent brains (Haier et al., 1995; Neubauer et al., 2004; Posthuma, Neale, Boomsma, & de Geus, 2001) (FIGURE 8.13).

Does size matter? It makes logical sense that bigger brains would be smarter—after all, humans have larger brains than less-intelligent species, such as dogs. (Some animals, such as whales and dolphins, have larger brains than humans, but our brains are larger relative to our body size.) In fact, brain-imaging studies have found a significant correlation between brain size (adjusted for body size) and intelligence

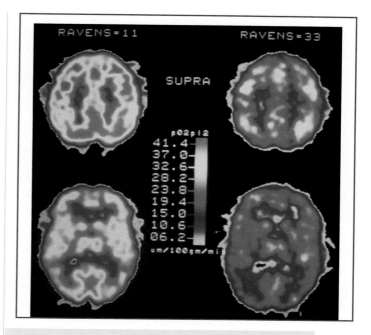

Do intelligent brains work more efficiently?
FIGURE 8.13

When given problem-solving tasks, people of low intelligence (PET scans on the left) show more activity (red and yellow indicate more brain activity) in relevant brain areas than people of higher intelligence (PET scans on right).

"I don't know how to tell you this, but it looks like you have a brain the size of a walnut."

(Ivanovic et al. 2004; Posthuma et al., 2002; Stelmack, Knott, & Beauchamp, 2003). On the other hand, Albert Einstein's brain was no larger than normal (Witelson, Kigar, & Harvey, 1999). In fact, some of Einstein's brain areas were actually smaller than average, but the area responsible for processing mathematical and spatial information was 15 percent larger than average.

GENETIC AND ENVIRONMENTAL INFLUENCES ON INTELLIGENCE

Similarities in intelligence between family members are due to a combination of hereditary (shared genetic material) and environmental factors (similar living arrangements and experiences). Researchers who are interested in the role of heredity in intelligence often focus on identical (monozygotic) twins because they share 100 percent of their genetic material.

The long-running Minnesota Study of Twins, an investigation of identical twins raised in different homes and reunited only as adults (Bouchard et al., 1994, 1998, 1999; Markon et al., 2002), found that genetic factors appear to play a surprisingly large role in the IQ scores of identical twins.

However, such results are not conclusive. Adoption agencies tend to look for similar criteria in their choice of adoptive parents. Therefore, the homes of these "reared apart" twins were actually quite similar. In addition, these twins also shared the same nine-month prenatal environment, which also might have influenced their brain development and intelligence (White, Andreasen, & Nopoulos, 2002).

What a Scientist Sees

	Correlations in IQ Scores
Siblings reared apart	.21
Unrelated individuals reared together	.32
Siblings reared together	.45
Fraternal twins reared together	.57
Identical twins reared apart	.75
Identical twins reared together	.86

Scale: .00 .10 .20 .30 .40 .50 .60 .70 .80 .90 1.00

Family Studies of Intelligence

Why are some people intellectually gifted while others struggle? As twin studies demonstrate, genetics plays an important role. In the figure here, note the higher correlations between identical twins' IQ scores compared to the correlations between all other pairs (Bouchard & McGue, 1981; Bouchard et al., 1998; McGue et al., 1993). Although heredity equips each of us with innate intellectual capabilities, the environment significantly influences whether a person will reach his or her full intellectual potential (Dickens & Flynn, 2001; Sangwan, 2001). For example, early malnutrition can retard a child's brain development, which in turn affects the child's curiosity, responsiveness to the environment, and motivation for learning—all of which can lower the child's IQ. The reverse is also true: An enriching early environment can set the stage for intellectual growth.

ETHNICITY AND INTELLIGENCE: ARE IQ TESTS CULTURALLY BIASED?

How would you answer the following questions?

1. A symphony is to a composer as a book is to a(n) _____. (a) musician; (b) editor; (c) novel; (d) author

2. If you throw a pair of dice and they land with a 7 on top, what is on the bottom? (a) snake eyes; (b) box cars; (c) little Joes; (d) 11

Can you see how the content and answers to these questions might reflect cultural bias? Which of these two questions do you think is most likely to appear on standard IQ tests?

One of the most controversial issues in psychology involves group differences in intelligence test scores and what they really mean. In 1969, Arthur Jensen began a heated debate when he argued that genetic factors are "strongly implicated" as the cause of ethnic differences in intelligence. Richard J. Herrnstein and Charles Murray's book *The Bell Curve: Intelligence and Class Structure in American Life* reignited this debate in 1994; in it, the authors claimed that African Americans score below average in IQ because of their "genetic heritage."

Herrnstein and Murray's book provoked considerable discussion and research. Although there is no clear answer in the debate, psychologists have made several important points:

- IQ tests may be culturally biased, making them an inaccurate measure of true capability (Manly et al., 2004; Naglieri & Ronning, 2000).

- **Stereotype threat** can significantly reduce the test scores of people in stereotyped groups (Gonzales, Blanton, & Williams, 2002; Josephs et al., 2003; Steele, 2003).

> ■ **stereotype threat** Negative stereotypes about minority groups cause some members to doubt their abilities.

Stereotype Threat

In the first study of stereotype threat, Claude Steele and Joshua Aronson (1995) recruited African American and white Stanford University students (with similar ability levels) to complete a "performance exam" that supposedly measured intellectual abilities. The exam's questions were similar to those on the Graduate Record Exam (GRE). Results showed that African American students performed far below white students. In contrast, when the researchers told students it was a "laboratory task," there was no difference between African American and white scores.

Subsequent work showed that stereotype threat occurs because members of stereotyped groups are anxious that they will fulfill their group's negative stereotype. This anxiety in turn hinders their performance on tests. Some people cope with stereotype threat by **disidentifying**, telling themselves they don't care about the test scores (Major et al., 1998). Unfortunately, this attitude lessens motivation, decreasing performance.

Stereotype threat affects many social groups, including African Americans, women, Native Americans, Latinos, low-income people, elderly people, and white male athletes (e.g., Ford et al., 2004; Gonzales, Blanton, & Williams, 2002; Major & O'Brien, 2005; Steele, 2003; Steele, James, & Barnett, 2002). This research helps explain some group differences in intelligence and achievement tests. As such, it underscores why relying solely on such tests to make critical decisions affecting individual lives—for example, in hiring, college admissions, or clinical application—is unwarranted and possibly even unethical.

- African Americans and most other minorities are especially likely to live in poverty, hampering intellectual development (Solan & Mozlin, 2001).

- Race and ethnicity, like intelligence itself, are almost impossible to define. For example, depending on the definition that you use, there are between 3 and 300 races, and no race is pure in a biological sense (Beutler et al., 1996; Yee et al., 1993).

- Intelligence is not a fixed characteristic. Around the world, IQ scores have increased from one generation to the next (Flynn, 1987, 2000, 2003; Neisser, 1998; Resing & Nijland, 2002). Interestingly, there has been a steady increase in the verbal ability scores of African Americans over time and a slight decline for white Americans (Huang & Hauser, 1998). James Flynn (2000, 2003) believes that such increases reflect the fact that intelligence tests might not actually measure intelligence but rather some weak link to intelligence. Three possible contributors to the **Flynn effect** are improved nutrition (leading to increasing intelligence), better public education, and more proficient test-taking skills. No matter what the cause, the finding suggests that the environment is a significant contributor to group differences in intelligence.

- Differences in IQ scores reflect motivational and language factors. In some ethnic groups, a child who excels in school is ridiculed for trying to be different from his or her classmates. Moreover, if children's own language and dialect do not match their education system or the IQ tests they take, they are obviously at a disadvantage (Tanner-Halverson, Burden, & Sabers, 1993).

- Different groups' distribution of IQ scores overlap considerably, and IQ scores and intelligence have their greatest relevance in terms of individuals, not groups (Garcia & Stafford, 2000; Myerson, Rank, Raines, & Schnitzler, 1998; Reifman, 2000). Many individual African Americans receive higher IQ scores than many individual white Americans.

- Traditional IQ tests do not measure many of our multiple intelligences (Gardner, 1999; Sternberg & Hedlund, 2002).

The ongoing debate over the nature of intelligence and its measurement highlights the complexities of studying *cognition*. In this chapter, we've explored three cognitive processes: thinking, language, and cognition. As you've seen, all three processes are greatly affected by numerous interacting factors.

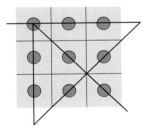

Solution to the nine-dot problem

People find this puzzle difficult because they see the arrangement of dots as a square—a *mental set* that limits possible solutions.

CONCEPT CHECK STOP

Why might more-intelligent people show less activity in cognitive-processing areas than less-intelligent people?

How have twin studies improved researchers' understanding of intelligence?

What is stereotype threat, and why does it occur?

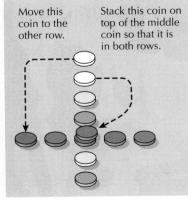

Move this coin to the other row.

Stack this coin on top of the middle coin so that it is in both rows.

Coin problem solution

Stack one coin on top of the middle coin so that it shares both the row and the column.

■ Jerry Levy and Mark Newman, twins separated at birth, first met as adults at a firefighter's convention. What factors might explain why they both became firefighters?

■ Does the brothers' choosing the same uncommon profession seem like a case of "telepathy"? How might the *confirmation bias* contribute to this perception?

VISUAL SUMMARY

1 Thinking

1. Thinking is a central aspect of **cognition**. Thought processes are distributed throughout the brain in neural networks. Mental images, concepts (both artificial and natural), and hierarchies aid our thought processes.

2. Problem solving usually involves three steps: preparation, production, and evaluation. **Algorithms** and **heuristics** help us produce solutions.

3. Barriers to problem solving include mental sets, functional fixedness, confirmation bias, availability heuristic, and representativeness heuristic.

4. **Creativity** is the ability to produce valued outcomes in a novel way. Tests of creativity usually focus on divergent thinking. One prominent theory of creativity is investment theory.

2 Language

1. **Language** supports thinking and enables us to communicate. To produce language, we use **phonemes, morphemes**, and **grammar** (syntax and semantics).

2. According to Whorf's linguistic relativity hypothesis, language determines thought. Generally, Whorf's hypothesis is not supported. However, language does strongly influence thought.

3. Children communicate nonverbally from birth. Their verbal communication proceeds in stages: prelinguistic (crying, cooing, babbling) and linguistic (single utterances, telegraphic speech, and acquisition of rules of grammar).

4. According to Chomsky, humans are "prewired" with a language acquisition device that enables language development. "Nurturists" hold that children learn language through rewards, punishments, and imitation. Most psychologists hold an intermediate view.

5. Research with apes and dolphins suggests that these animals can learn and use basic rules of language. However, nonhuman animal language is less complex, less creative, and not as rule-laden as human language.

3 Intelligence

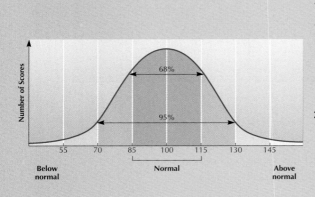

1. There is considerable debate over the meaning of **intelligence**. Here it is defined as the global capacity to think rationally, act purposefully, and deal effectively with the environment.

2. Spearman proposed that intelligence is a single factor, which he termed general intelligence (g). Thurstone and Guilford argued that intelligence included distinct abilities. Cattell proposed two subtypes of g: fluid intelligence and crystallized intelligence. Many contemporary cognitive theorists, including Gardner and Sternberg, believe that intelligence is a collection of many separate specific abilities.

3. Early intelligence tests involved computing a person's mental age to arrive at an **intelligence quotient (IQ)**. Today, two of the most widely used intelligence tests are the Stanford-Binet Intelligence Scale and the Wechsler Adult Intelligence Scale (WAIS).

4. To be scientifically acceptable, all psychological tests must fulfill three basic requirements: **standardization**, **reliability**, and **validity**.

4 The Intelligence Controversy

1. Intelligence tests provide one of the major criteria for assessing mental retardation and giftedness. Mental retardation exists on a continuum. In **savant syndrome**, a person who has mental retardation is exceptional in some limited field. Studies of people who are intellectually gifted found that they had more intellectual opportunities and tended to excel professionally. However, a high IQ is no guarantee of success in every endeavor.

2. Most recent research on the biology of intelligence has focused on brain functioning, not size. Research indicates that intelligent people's brains respond especially quickly and efficiently.

3. Both hereditary and environmental factors contribute to intelligence. Researchers interested in the role of heredity on intelligence often focus on identical twins. Twin studies have found that genetics plays an important role in intelligence. However, the environment significantly influences whether a person will reach his or her full intellectual potential.

4. Claims that genetic factors underlie ethnic differences in intelligence have caused heated debate and have received intense scrutiny. Among other arguments, some psychologists argue that IQ tests may be culturally biased; that **stereotype threat** can significantly reduce test scores of people in stereotyped groups; that socioeconomic factors heavily influence intellectual development; and that traditional IQ tests do not measure many of our multiple intelligences.

KEY TERMS

- **cognition** p. 200
- **prefrontal cortex** p. 200
- **mental image** p. 200
- **concepts** p. 201
- **prototypes** p. 201
- **hierarchies** p. 201
- **problem solving** p. 202
- **mental sets** p. 202
- **functional fixedness** p. 202
- **confirmation bias** p. 202
- **algorithm** p. 203
- **heuristic** p. 203
- **availability heuristic** p. 204
- **representative heuristic** p. 204
- **creativity** p. 204

- **divergent thinking** p. 204
- **investment theory** p. 204
- **language** p. 206
- **phoneme** p. 206
- **morpheme** p. 206
- **grammar** p. 206
- **linguistic relativity hypothesis** p. 206
- **language acquisition device (LAD)** p. 208
- **intelligence** p. 211
- **general intelligence (g)** p. 211
- **fluid intelligence (gf)** p. 211
- **crystallized intelligence (gc)** p. 211
- **theory of multiple intelligences** p. 212
- **triarchic theory of successful intelligence** p. 212

- **Stanford-Binet scale of intelligence** p. 213
- **Intelligence quotient (IQ)** p. 213
- **standardization** p. 215
- **reliability** p. 215
- **test-retest method** p. 215
- **split-half method** p. 215
- **validity** p. 215
- **criterion-related validity** p. 215
- **mental retardation** p. 216
- **giftedness** p. 216
- **savant syndrome** p. 216
- **stereotype threat** p. 219
- **disidentifying** p. 219
- **Flynn effect** p. 220

CRITICAL AND CREATIVE THINKING QUESTIONS

1. During problem solving, do you use primarily algorithms or heuristics? What are the advantages of each?

2. Would you like to be more creative? Can you do anything to acquire more of the resources of creative people?

3. Do you believe that we are born with an innate "language acquisition device," or is language development a function of our environments?

4. Do you think apes and dolphins have true language? Why or why not?

5. Physicians, teachers, musicians, politicians, and people in many other occupations may become more successful with age and can continue working well into old age. Which kind of general intelligence might explain this phenomenon?

6. If Gardner's and Sternberg's theories of multiple intelligences are correct, what are the implications for intelligence testing and for education?

(Check your answers in Appendix B.)

1. The mental activities involved in acquiring, storing, retrieving, and using knowledge are collectively known as _____ .

   a. perception
   b. cognition
   c. consciousness
   d. awareness

2. Mental representations of previously stored sensory experiences are called _____ .

   a. illusions
   b. psychoses
   c. mental images
   d. mental propositions

3. Label the three stages of problem-solving on the figure below:

4. This is a logical step-by-step procedure that, if followed, will always produce the solution.

   a. algorithm   b. heuristic
   c. problem-solving set   d. brainstorming

5. _____ is the ability to produce valuable outcomes in a novel way.

   a. Problem-solving   b. Incubation
   c. Functional flexibility   d. Creativity

6. _____ is the set of rules that specify how phonemes, morphemes, words, and phrases should be combined to express meaningful thoughts.

   a. Syntax   b. Pragmatics
   c. Semantics   d. Grammar

7. Label the three building blocks of language on the figure below:

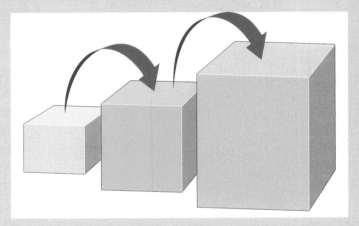

8. According to Chomsky, the innate mechanism that enables a child to analyze language is known as _____ .

   a. telegraphic understanding device (TUD)
   b. language acquisition device (LAD)
   c. language and grammar translator (LGT)
   d. overgeneralized neural net (ONN)

9. The definition of *intelligence* stated in your textbook stresses the global capacity to _____ .

   a. perform in school and on the job
   b. read, write, and make computations
   c. perform verbally and physically
   d. think rationally, act purposefully, and deal effectively with the environment

10. The IQ test sample in the figure below is from the _____ , the most widely used intelligence test.

   a. Wechsler Intelligence Scale for Children
   b. Wechsler Adult Intelligence Scale
   c. Stanford-Binet Intelligence Scale
   d. Binet-Terman Intelligence Scale

Picture Completion
What is missing from this ambulance?

Coding
Write the appropriate number above each symbol.

Picture Arrangement
Arrange these pictures in chronological order.

Block Design
Copy this design with blocks.

Object Assembly
Assemble this small jigsaw puzzle.

11. The development of standard procedures for administering and scoring a test is called _____ .

   a. norming
   b. standardization
   c. procedural protocol
   d. normalization

12. People with mental retardation who demonstrate exceptional ability in specific areas are called _____ .

   a. savants
   b. idiot geniuses
   c. mildly retarded
   d. connoisseurs

13. Speed of response is _____ correlated with IQ scores.

   a. negatively
   b. positively
   c. highly
   d. not

14. Ethnic group differences in IQ scores can be due to _____ .

   a. genetic influences
   b. socioeconomic differences
   c. cultural biases in IQ tests
   d. all of these options

15. Stereotype threat affects the IQ scores of which of the following groups?

   a. women
   b. white male athletes
   c. the elderly
   d. all of these options

Life Span Development I

9

If you have ever spent time at a daycare center, the word *pandemonium* may have sprung to mind. You were probably struck by the sight of many exuberant children laughing, shrieking, and jostling for attention from those entrusted with their care. But amid this barely controlled chaos, you may also have noticed something remarkable. Within the walls of a typical childcare center, babies learn to crawl, toddlers learn not to bite one another, and preschoolers learn their ABC's. Day by day, every child grows a little stronger, a little more independent—in short, a little readier to navigate the world.

Over a lifetime, every person develops a vast store of skills, knowledge, habits, and beliefs—not to mention the many physical changes that we undergo. These changes are most striking in early childhood because they happen so rapidly and are so visible. But everyone is in a state of constant change and development throughout his or her entire life. The typical human will be many different people in his or her lifetime—infant, child, teenager, adult, and senior citizen.

Would you like to know more about yourself at each of these ages? In the next two chapters, we will explore research in developmental psychology. We will begin this chapter by studying how developmental psychologists conduct their research. Then we will look at changes in our physical and cognitive development from conception to death. In Chapter 10, we will examine important aspects of our social, moral, and personality development across the life span. To emphasize that development is an ongoing, lifelong process, throughout the next two chapters we will trace physical, cognitive, social, moral, and personality development—one at a time—from conception to death. This topical approach will allow us to see how development affects an individual over the entire life span.

Studying Development

THEORETICAL ISSUES: ONGOING DEBATES

The three most important debates or questions in **developmental psychology** are about nature versus nurture, continuity versus stages, and stability versus change.

developmental psychology The study of age-related changes in behavior and mental processes from conception to death.

Nature or nurture

The issue of "nature versus nurture" has been with us since the beginning of psychology (Chapter 1). Even the ancient Greeks had the same debate—Plato argued that humans are born with innate knowledge and abilities, while Aristotle held that learning occurs through the five senses. Early philosophers proposed that at birth our minds are a *tabula rasa* (blank slate) and that the environment determines what messages are written on that slate.

According to the nature position, human behavior and development are governed by automatic, genetically predetermined signals in a process known as **maturation**. Just as a flower unfolds in accord with its genetic blueprint, we humans crawl before we walk and walk before we run. Furthermore, there is an optimal period shortly after birth, one of several **critical periods** during our lifetime, when an organism is especially sensitive to certain experiences that shape the capacity for future development. On the other side of the debate, those who hold an extreme nurturist position argue that development occurs by learning through personal experience and observation of others.

maturation Development governed by automatic, genetically predetermined signals.

critical period A period of special sensitivity to specific types of learning that shapes the capacity for future development.

Continuity or stages

Continuity proponents believe that development is continuous, with new abilities, skills, and knowledge being gradually added at a relatively uniform pace. Therefore, the continuity model suggests that adult thinking and intelligence differ quantitatively from a child's. That is, adults simply have more math and verbal skills than children do. Stage theorists, on the other hand, believe that development occurs at different rates, alternating between periods of little change and periods of abrupt, rapid change. In this chapter, we discuss one stage theory, Piaget's theory of cognitive development. In Chapter 10, we discuss Erikson's psychosocial theory of personality development and Kohlberg's theory of moral development.

Stability or change

Have you generally maintained your personal characteristics as you matured from infant to adult (stability)? Or does your current personality bear little resemblance to the personality you displayed during infancy (change)? Psychologists who emphasize stability in development hold that measurements of personality taken during childhood are important predictors of adult personality. Of course, psychologists who emphasize change disagree.

Which of these positions is more correct? Most psychologists do not take a hard line either way. Rather, they prefer an **interactionist perspective**. For example, in the age-old nature versus nurture debate, psychologists generally agree that development emerges both from each individual's unique genetic predisposition and from individual experiences in the environment (Allik & McCrae, 2004; Gottesman & Hanson, 2005; McCrae, 2004; Olson, 2004; Sullivan, Kendler, & Neale, 2003). More recently, the interactionist position has evolved into the *biopsychosocial model* mentioned throughout this text. In this model, biological factors (genetics, brain functions, biochemistry, and evolution), psychological influences (learning, thinking, emotion, personality, and motivation), and social forces (family, school, culture, ethnicity, social class, and politics) all affect and are affected by one another.

Like the nature versus nurture debate, the debates about continuity versus stages and stability versus change are not a matter of "either-or." Physical development and motor skills, for example, are believed to be primarily continuous in nature, whereas cognitive skills usually develop in discrete stages. Similarly, some traits are stable, whereas others vary greatly across the life span.

RESEARCH METHODS: TWO BASIC APPROACHES

cross-sectional method Research design that measures individuals of various ages at one point in time and gives infomation about age differences.

longitudinal method Research design that measures a single individual or a group of same-aged individuals over an extended period and gives information about age changes.

To study development, psychologists use either a cross-sectional or longitudinal method. The **cross-sectional method** examines individuals of various ages (e.g., 20, 40, 60, and 80) at one point in time and gives information about age differences. The **longitudinal method** follows a single individual or a group of same-aged individuals over an extended period and gives information about age changes (**FIGURE 9.1**).

Imagine that you are a developmental psychologist interested in studying intelligence in adults. Which method would you choose—cross-sectional or longitudinal? Before you decide, note the different research results shown in **FIGURE 9.2**.

Why do the two methods show such different results? Researchers suggest that the different results may reflect a central problem with cross-sectional studies. These studies often confuse genuine age differences with **cohort effects**, differences that result from specific histories of the age group studied (Elder, 1998). As Figure 9.2 shows, the 81-year-olds measured by the cross-sectional method have dramatically lower scores than the 25-year-olds. But is this due to aging or to broad environmental differences, such as less formal education or poorer nutrition? Because the different age groups, called "cohorts," grew up in different historical periods, their results may not apply to people growing up at other times. With the cross-sectional method, age effects and cohort effects are hopelessly tangled.

Longitudinal studies also have their limits. They are expensive in terms of time and money, and

Cross-sectional versus longitudinal research FIGURE 9.1

Note that cross-sectional research uses different participants and is interested in age-related differences, whereas longitudinal research studies the same participants over time to find age-related changes.

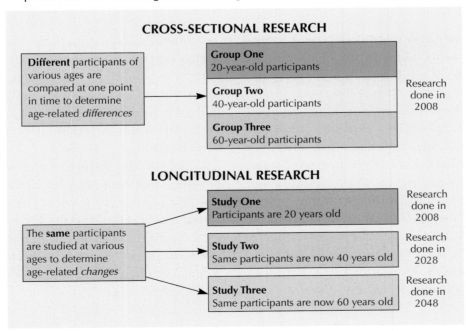

CROSS-SECTIONAL RESEARCH

Different participants of various ages are compared at one point in time to determine age-related *differences*

Group One 20-year-old participants

Group Two 40-year-old participants

Group Three 60-year-old participants

Research done in 2008

LONGITUDINAL RESEARCH

The **same** participants are studied at various ages to determine age-related *changes*

Study One Participants are 20 years old — Research done in 2008

Study Two Same participants are now 40 years old — Research done in 2028

Study Three Same participants are now 60 years old — Research done in 2048

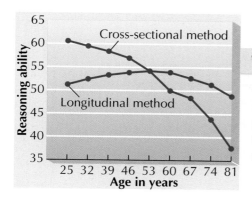

Which results are "true"? FIGURE 9.2

Cross-sectional studies have shown that reasoning and intelligence reach their peak in early adulthood and then gradually decline. In contrast, longitudinal studies have found that a marked decline does not begin until about age 60. (Adapted from Schaie, 1994, with permission.)

their results are restricted in generalizability. Because participants often drop out or move away during the extended test period, the experimenter may end up with a self-selected sample that differs from the general population in important ways. Each method of research has its own strengths and weaknesses (TABLE 9.1). Keep these differences in mind when you read the findings of developmental research.

Human development, like most areas of psychology, cannot be studied outside its sociocultural context.

In fact, researchers have found that culture may be the most important determinant of development. If a child grows up in an individualistic/independent culture (such as North America or most of Western Europe), we can predict that this child will probably be competitive and question authority as an adult. If this same child were reared in a collectivist/interdependent culture (common in Africa, Asia, and Latin America), she or he would most likely grow up to be cooperative and respectful of others (Delgado-Gaitan, 1994; Berry et al., 2002).

Advantages and disadvantages of cross-sectional and longitudinal research designs TABLE 9.1

	Cross-sectional	*Longitudinal*
Advantages	Gives information about age differences Quick Less expensive Typically larger sample	Gives information about age changes Increased reliability More in-depth information per participant
Disadvantages	Cohort effects are difficult to separate Restricted generalizability (measures behaviors at only one point in time)	More expensive Time consuming Restricted generalizability (typically smaller sample and dropouts over time)

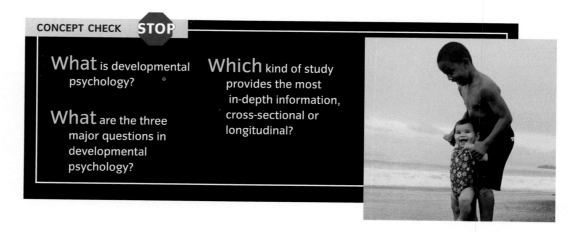

CONCEPT CHECK **STOP**

What is developmental psychology?

What are the three major questions in developmental psychology?

Which kind of study provides the most in-depth information, cross-sectional or longitudinal?

Physical Development

LEARNING OBJECTIVES

Describe the three phases of prenatal physical development.

Identify some important teratogens and their effects on prenatal development.

Summarize physical development during early childhood.

Describe the physical changes that occur during adolescence and adulthood.

 n this section, we will explore the fascinating processes of physical development from conception through adulthood.

PRENATAL AND EARLY CHILDHOOD: A TIME OF RAPID CHANGE

The early years of development are characterized by rapid and unparalleled change. In fact, if you continued to develop at the same rapid rate that marked your first two years of life, you would weigh several tons and be over 12 feet tall as an adult! Thankfully, physical development slows, yet it is important to note that change continues until the very moment of death. Let's look at some of the major physical changes that occur throughout the life span.

Your prenatal development began at conception, when your mother's egg, or ovum, united with your father's sperm cell (FIGURE 9.3). At that time, you were a single cell barely 1/175 of an inch in diameter—smaller than the period at the end of this sentence. This new cell, called a **zygote**, then began a process of rapid cell division that resulted in a multi-million-celled infant (you) some nine months later.

The vast changes that occur during the nine months of a full-term pregnancy are usually divided into three stages: the **germinal period**, the **embryonic period**, and the **fetal period** (FIGURE 9.4). Prenatal growth, as well as growth during the first few years after birth, is **proximodistal** (near to far), with the head and upper body developing before the lower body.

During pregnancy, the **placenta** (the vascular organ that unites the fetus to the mother's uterus) serves as the link for food and excretion of wastes. It also screens out some, but not all, harmful substances. Environmental hazards such as X-rays or toxic waste, drugs, and diseases such as rubella (German measles) can cross the placental barrier. These influences generally have their most devastating effects during the first three months of pregnancy, making this a critical period in development.

Conception FIGURE 9.3

A Note the large number of sperm surrounding the ovum (egg).

B Although a joint effort is required to break through the outer coating, only one sperm will actually fertilize the egg.

Prenatal development FIGURE 9.4

A **Germinal period: From ovulation to implantation.** After discharge from either the left or right ovary (1), the ovum travels to the opening of the fallopian tube.

If fertilization occurs (2), it normally takes place in the first third of the fallopian tube. The fertilized ovum is referred to as a zygote.

When the zygote reaches the uterus, it implants itself in the wall of the uterus (3) and begins to grow tendril-like structures that intertwine with the rich supply of blood vessels located there. After implantation, the organism is known as an embryo.

B **Embryonic period.** This stage lasts from implantation to eight weeks. At eight weeks, the major organ systems have become well differentiated. Note that at this stage, the head grows at a faster rate than other parts of the body.

C **Fetal period.** This is the period from the end of the second month to birth. At four months, all the actual body parts and organs are established. The fetal stage is primarily a time for increased growth and "fine detailing."

germinal period The first stage of prenatal development, which begins with conception and ends with implantation in the uterus (the first two weeks).

embryonic period The second stage of prenatal development, which begins after uterine implantation and lasts through the eighth week.

fetal period The third, and final, stage of prenatal development (eight weeks to birth), which is characterized by rapid weight gain in the fetus and the fine detailing of bodily organs and systems.

www.wiley.com/college/carpenter

Sample environmental conditions that endanger the child TABLE 9.2

Maternal factors	Possible effects on embryo, fetus, newborn, or young child
Malnutrition	Low birth weight, malformations, less developed brain, greater vulnerability to disease
Stress exposure	Low birth weight, hyperactivity, irritability, feeding difficulties
Exposure to X-rays	Malformations, cancer
Legal and illegal drugs	Inhibition of bone growth, hearing loss, low birth weight, fetal alcohol syndrome, mental retardation, attention deficits in childhood, death
Diseases: German measles (rubella), herpes, AIDS, and toxoplasmosis	Blindness, deafness, mental retardation, heart and other malformations, brain infection, spontaneous abortion, premature birth, low birth weight, death

Sources: Abadinsky, 2004; Hyde & DeLamater, 2006; Lee, Mattson, & Riley, 2004; Rybacki & Long, 1999; Watson, Mednick, Huttunen, & Wang, 1999.

The pregnant mother plays a primary role in prenatal development because her health directly influences the child she is carrying (TABLE 9.2). Almost everything she ingests can cross the placental barrier (a better term might be *placental sieve*). However, the father also plays a role (other than just fertilization). Environmentally, the father's smoking may pollute the air the mother breathes, and genetically, he may transmit heritable diseases. In addition, research suggests that alcohol, opiates, cocaine, various gases, lead, pesticides, and industrial chemicals can damage sperm (Bandstra et al., 2002; Grilly, 2006; Richardson et al., 2002).

Perhaps the most important—and generally avoidable—danger to the fetus comes from drugs, both legal and illegal. Nicotine and alcohol are major **teratogens**, environmental agents that cause damage during prenatal development. Mothers who smoke tobacco during pregnancy have significantly higher rates of premature births, low-birth-weight infants, and fetal deaths (American Cancer Society, 2004; Bull, 2003; Oliver, 2002). Their children also show increased behavioral abnormalities and cognitive problems (Roy, Seidler, & Slotkin, 2002; Thapar et al., 2003). Alcohol also readily crosses the placenta, affects fetal development, and can result in a neurotoxic syndrome called **fetal alcohol syndrome (FAS)** (FIGURE 9.5). About one in a hundred babies in the United States is born with FAS or other birth defects resulting from the mother's alcohol use during pregnancy (National Organization on Fetal Alcohol Syndrome, 2006).

Fetal alcohol syndrome FIGURE 9.5

Compare the healthy newborn infant brain (left) to the brain of a newborn (right) whose mother drank while pregnant. Prenatal exposure to alcohol can cause facial abnormalities and stunted growth. But the most disabling features of FAS are neurobehavioral problems, ranging from hyperactivity and learning disabilities to mental retardation, depression, and psychoses (Korkman, Kettunen, & Autti-Rämö, 2003; Lee, Mattson, & Riley, 2004; Sokol, Delaney-Black, & Nordstrom, 2003).

Body proportions FIGURE 9.6

Notice how body proportions change as we grow older. At birth, an infant's head is one-fourth its total body's size, whereas in adulthood, the head is one-eighth.

Like the prenatal period, early childhood is also a time of rapid physical development. Although Shakespeare described newborns as capable of only "mewling and puking in the nurse's arms," they are actually capable of much more. Let's explore three key areas of change in early childhood: brain, motor, and sensory/perceptual development.

Brain development

The brain and other parts of the nervous system grow faster than any other part of the body during both prenatal development and the first two years of life. At birth, a healthy newborn's brain is one-fourth its full adult size, and it will grow to about 75 percent of its adult weight and size by the age of 2. At age 6, the child's brain is nine-tenths its full adult weight (FIGURE 9.6).

Rapid brain growth during infancy and early childhood slows down in later childhood. Further brain development and learning occur primarily because neurons grow in size and because the number of axons and dendrites, as well as the extent of their connections, increases (DiPietro, 2000) (FIGURE 9.7).

| At birth | 1 month | 3 months | 15 months | 24 months |

Brain growth in the first two years FIGURE 9.7

As children learn and develop, synaptic connections between active neurons strengthen, and dendritic connections become more elaborate. *Synaptic pruning* (reduction of unused synapses) helps support this process. *Myelination,* the accumulation of fatty tissue coating the axons of nerve cells, continues until early adulthood. Myelin increases the speed of neural impulses, and the speed of information processing shows a corresponding increase (Chapter 2). In addition, synaptic connections in the frontal lobes and other parts of the brain continue growing and changing throughout the entire life span (Chapters 2 and 6).

Motor development   Compared to the hidden, internal changes in brain development, the orderly emergence of active movement skills, known as motor development, is easily observed and measured. The newborn's first motor abilities are limited to *reflexes,* or involuntary responses to stimulation. For example, the rooting reflex occurs when something touches a baby's cheek: the infant will automatically turn its head, open its mouth, and root for a nipple.

In addition to simple reflexes, the infant soon begins to show voluntary control over the movement of various body parts (see box below). Thus, a helpless newborn who cannot even lift her head is soon transformed into an active toddler capable of crawling, walking, and climbing. Keep in mind that motor devel-opment is largely due to natural maturation, but it can also be affected by environmental influences like dis-ease and neglect.

Sensory and perceptual development   At birth, a newborn can smell most odors and distinguish between sweet, salty, and bitter tastes. Breast-fed new-borns also recognize and show preference for the odor and taste of their mother's milk over another mother's (DiPietro, 2000). In addition, the newborn's sense of touch and pain is highly developed, as evidenced by re-actions to heel pricks for blood testing and to circum-cision (Williamson, 1997).

The newborn's sense of vision, however, is poorly developed. At birth, an infant is estimated to

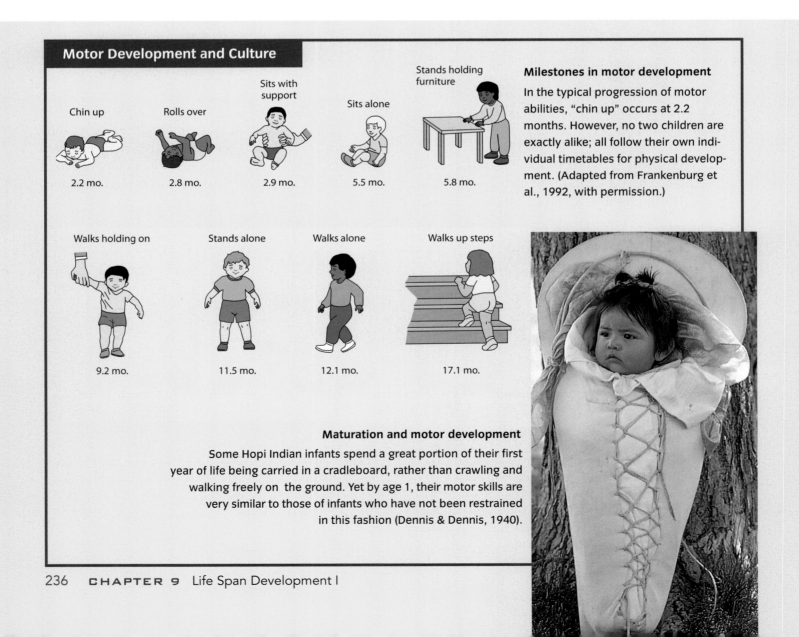

Motor Development and Culture

Chin up — 2.2 mo.

Rolls over — 2.8 mo.

Sits with support — 2.9 mo.

Sits alone — 5.5 mo.

Stands holding furniture — 5.8 mo.

Milestones in motor development
In the typical progression of motor abilities, "chin up" occurs at 2.2 months. However, no two children are exactly alike; all follow their own indi-vidual timetables for physical develop-ment. (Adapted from Frankenburg et al., 1992, with permission.)

Walks holding on — 9.2 mo.

Stands alone — 11.5 mo.

Walks alone — 12.1 mo.

Walks up steps — 17.1 mo.

Maturation and motor development
Some Hopi Indian infants spend a great portion of their first year of life being carried in a cradleboard, rather than crawling and walking freely on the ground. Yet by age 1, their motor skills are very similar to those of infants who have not been restrained in this fashion (Dennis & Dennis, 1940).

have vision between 20/200 and 20/600 (Haith & Benson, 1998). Imagine what the infant's visual life is like: the level of detail you see at 200 or 600 feet (if you have 20/20 vision) is what they see at 20 feet. Within the first few months, vision quickly improves, and by six months it is 20/100 or better. At two years, visual acuity reaches a near-adult level of 20/20 (Courage & Adams, 1990).

One of the most interesting findings in infant sensory and perceptual research concerns hearing. Not only can the newborn hear quite well at birth (Matlin & Foley, 1997) but also, during the last few months in the womb, the fetus can apparently hear sounds outside the mother's body (Vaughan, 1996). This raises the interesting possibility of fetal learning, and some have advo-

cated special stimulation for the fetus as a way of increasing intelligence, creativity, and general alertness (e.g., Van de Carr & Lehrer, 1997).

Studies on possible fetal learning have found that newborn infants easily recognize their own mother's voice over that of a stranger (Kisilevsky et al., 2003). They also show a preference for children's stories (such as *The Cat in the Hat* or *The King, the Mice, and the Cheese*) that were read to them while they were still in the womb (DeCasper & Fifer, 1980; Karmiloff & Karmiloff-Smith, 2002). On the other hand, some experts caution that too much or the wrong kind of stimulation before birth can be stressful for both the mother and fetus. They suggest that the fetus gets what it needs without any special stimulation.

How an Infant Perceives the World

Because infants cannot talk or follow directions, researchers have had to create ingenious experiments to measure their perceptual abilities and preferences. One of the earliest experimenters, Robert Fantz (1956, 1963), designed a "looking chamber" to measure how long infants stared at stimuli. Research using this apparatus indicates that infants prefer complex rather than simple patterns and pictures of faces rather than nonfaces.

Researchers also use newborns' heart rates and innate abilities, such as the sucking reflex, to study learning and perceptual development. To study the sense of smell, researchers measure changes in the newborns' heart rates when odors are presented. Presumably, if they can smell one odor but not another, their heart rates will change in the presence of the first but not the second. From research such as this, we now know that the senses develop very early in life.

ADOLESCENCE AND ADULTHOOD: A TIME OF BOTH DRAMATIC AND GRADUAL CHANGE

Think back for a moment to your teen years. Were you concerned about the physical changes you were going through? Did you worry about how you differed from your classmates? Changes in height and weight, breast development and menstruation for girls, and a deepening voice and beard growth for boys are important milestones for adolescents. **Puberty**, the period of adolescence when a person becomes capable of reproduction, is a major physical milestone for everyone. It is a clear biological signal of the end of childhood.

Although commonly associated with puberty, **adolescence** is the loosely defined psychological period of development between childhood and adulthood. In the United States, it roughly corresponds to the teenage years. The concept of adolescence and its meaning varies greatly across cultures (FIGURE 9.8).

The clearest and most dramatic physical sign of puberty is the **growth spurt**, which is characterized by rapid increases in height, weight, and skeletal growth (FIGURE 9.9), and by significant changes in reproductive structures and sexual characteristics. Maturation and hormone secretion cause rapid development of the ovaries, uterus, and vagina and the onset of menstruation (**menarche**) in the adolescent fe-

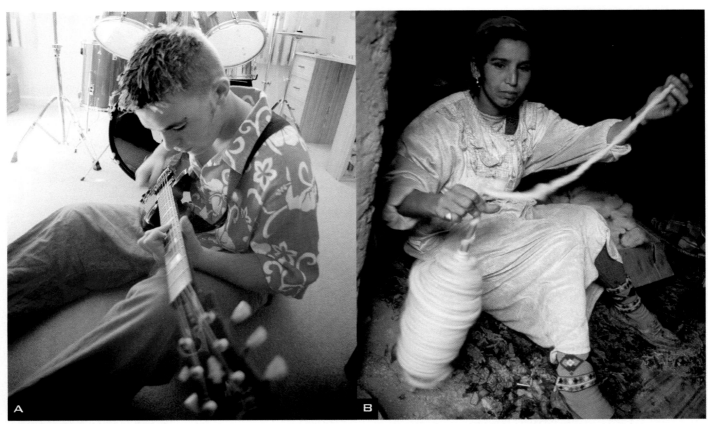

A teenage American boy plays his guitar.

A Berber girl in the Atlas Mountains of Morocco spins yarn on a spindle. Can you see how this cultural difference is reflected in the two photos above?

Ready for responsibility? FIGURE 9.8

Adolescence is not a universal concept. Unlike in the United States and other Western nations, some nonindustrialized countries have no need for a slow transition from childhood to adulthood; children simply assume adult responsibilities as soon as possible.

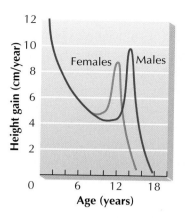

Adolescent growth spurt FIGURE 9.9

Note the gender differences in height gain during puberty. Most girls are about two years ahead of boys in their growth spurt and are therefore taller than most boys between the ages of 10 and 14.

male. In the adolescent male, the testes, scrotum, and penis develop, and he experiences his first ejaculation (**spermarche**). The ovaries and testes in turn produce hormones that lead to the development of secondary sex characteristics, such as the growth of pubic hair, deepening of the voice and growth of facial hair in men, and the growth of breasts in women (**FIGURE 9.10**).

Age-related physical changes that occur after the obvious pubertal changes are less dramatic. Other than some modest increases in height and muscular development during the late teens and early twenties,

most individuals experience only minor physical changes until middle age.

For women, **menopause**, the cessation of the menstrual cycle, which occurs somewhere between the ages of 45 and 55, is the second most important life milestone in physical development. The decreased production of estrogen (the dominant female hormone) produces certain physical changes. However, the popular belief that menopause (or "the change of life") causes serious psychological mood swings, loss of sexual interest, and depression is not supported by current research (Hvas, 2001; Matlin, 2003; Morrison & Tweedy,

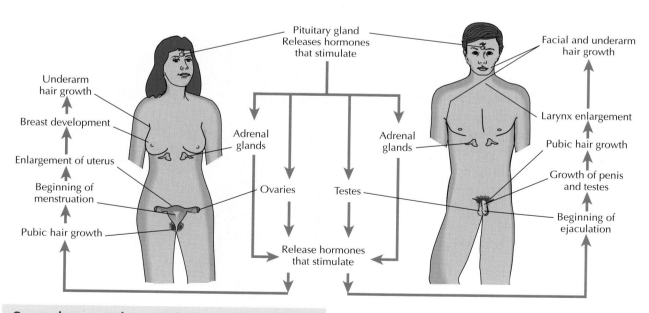

Secondary sex characteristics FIGURE 9.10

Complex physical changes in puberty primarily result from hormones secreted from the ovaries and testes, the pituitary gland in the brain, and adrenal glands near the kidneys.

2000). In fact, one large-scale study of postmenopausal American women found that almost two-thirds felt relief that their periods had stopped, and more than half did not experience hot flashes (Brim, 1999). When psychological problems do arise, they may reflect the social devaluation of aging women, not the physiological process of menopause. Given that in Western society women are highly valued for their youth and beauty, a biological process such as aging can be difficult for some women. Women in cultures that derogate aging tend to experience more anxiety and depression during menopause (Mingo, Herman, & Jasperse, 2000; Sampselle et al., 2002; Winterich, 2003).

For men, youthfulness is less important and the physical changes of middle age are less obvious. Beginning in middle adulthood, men experience a gradual decline in the production of sperm and testosterone (the dominant male hormone), although they may remain capable of reproduction into their eighties or nineties. Physical changes such as unexpected weight gain, decline in sexual responsiveness, loss of muscle strength, and graying or loss of hair may lead some men (and women as well) to feel depressed and to question their life progress. They often see these alterations as a biological signal of aging and mortality. Such physical and psychological changes in some men are reportedly known as the **male climacteric**.

After middle age, most physical changes in development are gradual and occur in the heart and arteries and in the sensory receptors. For example, cardiac output (the volume of blood pumped by the heart each minute) decreases, whereas blood pressure increases due to the thickening and stiffening of arterial walls. Visual acuity and depth perception decline, hearing acuity lessens (especially for high-frequency sounds), and smell sensitivity decreases (Atchley & Kramer, 2000; Kiessling et al., 2003; Wahl et al., 2004).

Television, magazines, movies, and advertisements generally portray aging as a time of balding and graying hair, sagging parts, poor vision, hearing loss, and, of course, no sex life. Such negative portrayals contribute to our society's widespread **ageism**—prejudice or discrimination based on physical age. However, as advertising companies pursue the revenue of the huge aging baby boomer population, there has been a recent shift toward a more accurate portrayal of aging as a time of vigor, interest, and productivity (FIGURES 9.11 and 9.12).

What about memory problems and inherited genetic tendencies toward Alzheimer's disease and other serious diseases of old age? The public and most researchers have long thought that aging is accompanied by widespread death of neurons in the brain. Although this decline does happen with degenerative disorders like Alzheimer's disease, this is no longer believed to be a part of normal aging (Chapter 2). It is also important to remember that age-related memory problems are not on a continuum with Alzheimer's disease (Wilson et al., 2000). That is, normal forgetfulness does not mean that serious dementia is around the corner.

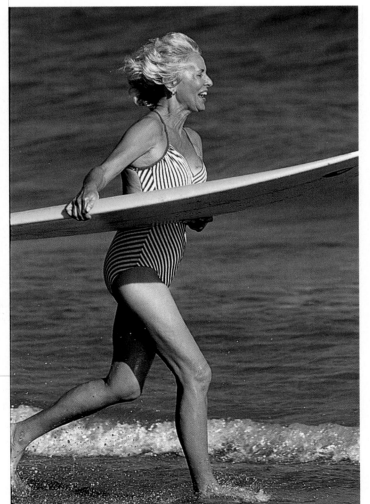

Use it or lose it? FIGURE 9.11

Recent research shows that cognitive functioning in older adults can be greatly enhanced with simple aerobic training (Benloucif et al., 2004; Lytle et al., 2004).

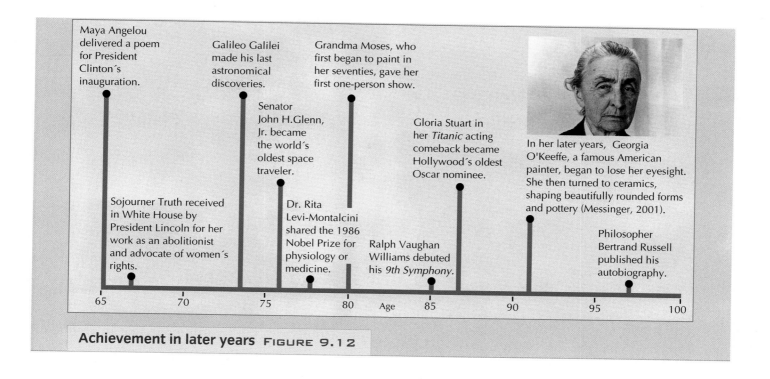

Maya Angelou delivered a poem for President Clinton's inauguration.

Galileo Galilei made his last astronomical discoveries.

Grandma Moses, who first began to paint in her seventies, gave her first one-person show.

Senator John H. Glenn, Jr. became the world's oldest space traveler.

Gloria Stuart in her *Titanic* acting comeback became Hollywood's oldest Oscar nominee.

In her later years, Georgia O'Keeffe, a famous American painter, began to lose her eyesight. She then turned to ceramics, shaping beautifully rounded forms and pottery (Messinger, 2001).

Sojourner Truth received in White House by President Lincoln for her work as an abolitionist and advocate of women's rights.

Dr. Rita Levi-Montalcini shared the 1986 Nobel Prize for physiology or medicine.

Ralph Vaughan Williams debuted his *9th Symphony*.

Philosopher Bertrand Russell published his autobiography.

65   70   75   80   Age   85   90   95   100

Achievement in later years FIGURE 9.12

Aging does seem to take its toll on the speed of information processing (Chapter 7). Decreased speed of processing may reflect problems with encoding (putting information into long-term storage) and retrieval (getting information out of storage). If memory is like a filing system, older people may have more filing cabinets, and it may take them longer to initially file and later retrieve information. Although mental speed declines with age, general information processing and memory ability is largely unaffected by the aging process (Lachman, 2004; Whitbourne, 2005).

What causes us to age and die? If we set aside contributions from **secondary aging** (changes resulting from disease, disuse, or neglect), we are left to consider **primary aging** (gradual, inevitable age-related changes in physical and mental processes). There are two main theories explaining primary aging and death: programmed theory and damage theory (Cristofalo, 1996; Medina, 1996; Wallace, 1997).

According to the **programmed theory**, aging is genetically controlled. Once the ovum is fertilized, the program for aging and death is set and begins to run. Researcher Leonard Hayflick (1977, 1996) found that human cells seem to have a built-in life span. He observed that after doubling about 50 times, laboratory-cultured cells ceased to divide—they reached the Hayflick limit. The other explanation of primary aging is **damage theory**, which proposes that an accumulation of damage to cells and organs over the years ultimately causes death.

Whether aging is genetically controlled or caused by accumulated damage over the years, scientists generally agree that humans appear to have a maximum life span of about 110 to 120 years. Although we can try to control secondary aging in an attempt to reach that maximum, so far we have no means to postpone primary aging.

CONCEPT CHECK STOP

What role does the placenta play in prenatal development?

What are some important milestones in motor development in early childhood?

How does the developmental concept of adolescence differ between cultures?

What physical changes occur during middle age and late adulthood?

Cognitive Development

LEARNING OBJECTIVES

Explain the role of schemas, assimilation, and accommodation in cognitive development.

Describe the major characteristics of Piaget's four stages of cognitive development.

Identify two critiques of Piaget's theory.

The following fan letter was written to Shari Lewis (1963), a children's television performer, about her puppet Lamb Chop:

Dear Shari:
All my friends say Lamb Chop isn't really a little girl that talks. She is just a puppet you made out of a sock. I don't care even if it's true. I like the way Lamb Chop talks. If I send you one of my socks will you teach it how to talk and send it back?

Randi

Randi's understanding of fantasy and reality is certainly different from an adult's. Just as a child's body and physical abilities change, his or her way of knowing and perceiving the world also grows and changes. This seems intuitively obvious, but early psychologists—with one exception—focused on physical, emotional, language, and personality development. The one major exception was Jean Piaget (pronounced Pee–ah– ZHAY).

Piaget demonstrated that a child's intellect is fundamentally different from an adult's. He showed that an infant begins at a cognitively "primitive" level and that intellectual growth progresses in distinct stages, motivated by an innate need to know. Piaget's theory, developed in the 1920s and 1930s, has proven so comprehensive and insightful that it remains the major force in the cognitive area of developmental psychology today.

To appreciate Piaget's contributions, we need to consider three major concepts: schemas, assimilation, and accommodation. **Schemas** are the most basic units of intellect. They act as patterns that organize our interactions with the environment, like an architect's drawings or a builder's blueprints.

In the first few weeks of life, for example, the infant apparently has several schemas based on the innate reflexes of sucking, grasping, and so on. These schemas are primarily motor and may be little more than stimulus-and-response mechanisms—the nipple is presented, and the baby sucks. Soon, other schemas emerge. The infant develops a more detailed schema for eating solid food, a different schema for the concepts of "mother" and "father," and so on. Schemas, our tools for learning about the world, expand and change throughout our lives. For example, music lovers who were previously accustomed to LP records and cassette tapes have had to develop schemas for playing CDs and MP3s.

Assimilation and **accommodation** are the two major processes by which schemas grow and change over time. Assimilation is the process of absorbing new information into existing schemas. For instance, infants use their sucking schema not only in sucking nipples but also in sucking blankets and fingers.

> **schemas** Cognitive structures or patterns consisting of a number of organized ideas that grow and differentiate with experience.
>
> **assimilation** In Piaget's theory, the process of absorbing new information into existing schemas.
>
> **accommodation** In Piaget's theory, the process of adjusting old schemas or developing new ones to better fit with new information.

Accommodation occurs when new information or stimuli cannot be assimilated and new schemas are developed or old schemas are changed to better fit with the new information. An infant's first attempt to eat solid food with a spoon is a good example of accommodation. When the spoon first enters her mouth, the child attempts to assimilate it by using the previously successful sucking schema—shaping the lips and tongue around the spoon as if they were around a nipple. After repeated trials, she accommodates by adjusting her lips and tongue in a way that moves the food off the spoon and into her mouth.

STAGES OF COGNITIVE DEVELOPMENT: BIRTH TO ADOLESCENCE

According to Piaget, all children go through approximately the same four stages of cognitive development, regardless of the culture in which they live (**TABLE 9.3**). Stages cannot be skipped because skills acquired at earlier stages are essential to mastery at later stages. Let's take a closer look at these four stages: sensorimotor, preoperational, concrete operational, and formal operational.

Piaget's four stages of cognitive development TABLE 9.3

Birth to 2

Sensorimotor

Abilities: Uses senses and motor skills to explore and develop cognitively.

Limits: Beginning of stage lacks *object permanence* (understanding things continue to exist even when not seen, heard, or felt).

Age 2–7

Preoperational

Abilities: Has significant language and thinks symbolically.

Limits:
- Cannot perform "operations."
- *Egocentric* thinking (inability to consider another's point of view).
- *Animistic* thinking (believing all things are living).

Age 7–11

Concrete Operational

Abilities:
- Can perform "operations" on concrete objects.
- Understands *conservation* (realizing changes in shape or appearance can be reversed).

Limits: Cannot think abstractly and hypothetically.

Age 11 and up

Formal Operational

Abilities: Can think abstractly and hypothetically.

Limits: *A*dolescent egocentrism at the beginning of this stage, with related problems of the *personal fable* and *imaginary audience*.

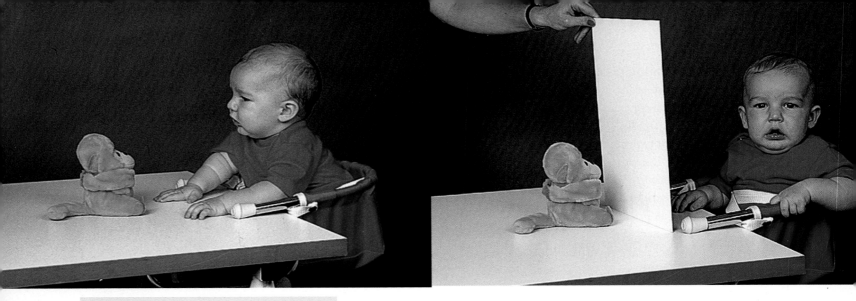

Object permanence FIGURE 9.13

At birth and for the next three or four months, children lack object permanence. They seem to have no schemas for objects they cannot see, hear, or touch—out of sight is truly out of mind.

sensorimotor stage Piaget's first stage (birth to approximately age 2), in which schemas are developed through sensory and motor activities.

preoperational stage Piaget's second stage (roughly ages 2 to 7 years), which is characterized by the ability to employ significant language and to think symbolically, though the child lacks operations (reversible mental processes), and thinking is egocentric and animistic.

During the **sensorimotor stage**, which lasts from birth until "significant" language acquisition (about age 2), children explore the world and develop their schemas primarily through their senses and motor activities—hence the term *sensorimotor*. One important concept that infants acquire during the sensorimotor stage is **object permanence** (FIGURE 9.13).

During the **preoperational stage** (roughly ages 2 to 7), language advances significantly, and the child begins to think symbolically —using symbols, such as words, to represent concepts. Three other qualities characterize this stage:

Concepts are not yet operational Piaget labeled this period "preoperational" because the child lacks **opera-tions**, or reversible mental processes. For instance, if a preoperational boy who has a brother is asked, "Do you have a brother?" he will easily respond, "Yes." However, when asked, "Does your brother have a brother?" he will answer, "No!" To understand that his brother has a brother, he must be able to reverse the concept of "having a brother."

Thinking is egocentric Children at this stage have difficulty understanding that there are points of view other than their own. **Egocentrism** refers to the preoperational child's limited ability to distinguish between his or her own perspective and someone else's. (It does not mean "selfishness" in the ordinary sense of the word.) The preschooler who moves in front of you to get a better view of the TV or repeatedly asks questions while you are talking on the telephone is demonstrating egocentrism. They assume that others see, hear, feel, and think exactly as they do. Consider the following telephone conversation between a 3-year-old, who is at home, and her mother, who is at work:

> Mother: Emma is that you?
> Emma: (Nods silently.)
> Mother: Emma, is Daddy there? May I speak to him?
> Emma: (Twice nods silently.)

Egocentric preoperational children fail to understand that the phone caller cannot see their nodding head. Charming as this is, the preoperational child's egocentrism also sometimes leads them to believe that their "bad thoughts" caused their sibling or parent to get sick or that their misbehavior caused their parents' marital problems. Because they think the world centers on them, they often cannot separate reality from what goes on inside their own heads.

Thinking is animistic

Children in the preoperational stage believe that objects such as the sun, trees, clouds, and bars of soap have motives, feelings, and intentions (for example, "dark clouds are angry" and "soap sinks because it is tired"). **Animism** refers to the belief that all things are living (or animated). Our earlier example of Randi's letter asking puppeteer Shari Lewis to teach her sock to talk like Lamb Chop is also an example of animistic thinking.

Can preoperational children be taught how to use operations and to avoid egocentric and animistic thinking? Although some researchers have reported success in accelerating the preoperational stage, Piaget did not believe in pushing children ahead of their own developmental schedules. He believed that children should grow at their own pace, with minimal adult interference (Elkind, 1981, 2000). In fact, Piaget thought that Americans were particularly guilty of pushing their children, calling American childhood the "Great American Kid Race."

Following the preoperational stage, at approximately age 7, children enter the **concrete operational stage**. During this stage, many important thinking skills emerge. Unlike the preoperational stage, concrete operational children perform operations on concrete objects. Because they understand the concept of reversibility, they recognize that certain physical attributes (such as volume) remain unchanged when the outward appearance of an object is altered, a process known as **conservation** (FIGURE 9.14).

■ **concrete operational stage** Piaget's third stage (roughly ages 7 to 11), in which the child can perform mental operations on concrete objects and understand reversibility and conservation, though abstract thinking is not yet present.

Test for conservation FIGURE 9.14

A In the classic conservation of liquids test, the child is first shown two identical glasses with liquid at the same level.

B The liquid is poured from one of the short, wide glasses into the tall, thin one.

C When asked whether the two glasses have the same amount or if one has more, the preoperational child replies that the tall, thin glass has more. This demonstrates a failure to conserve volume.

Putting Piaget to the Test

www.wiley.com/college/carpenter

This is a sample of experiments used to test Piaget's different types of conservation. If you know children in the preoperational or concrete operational stages, you may enjoy testing their grasp of conservation by trying some of these experiments. The equipment is easily obtained, and you will find their responses fascinating. Keep in mind that this should be done as a game. Do not allow the child to feel that he or she is failing a test or making a mistake.

Type of Conservation Task (Average age at which concept is grasped)	Your task as experimenter...	Child is asked...
Length (ages 6–7)	**Step 1** Center two sticks of equal length. Child agrees that they are of equal length. **Step 2** Move one stick.	**Step 3** *"Which stick is longer?"* Preoperational child will say that one of the sticks is longer. Child in concrete stage will say that they are both the same length.
Substance amount (ages 6–7)	**Step 1** Center two identical clay balls. Child acknowledges that the two have equal amounts of clay. **Step 2** Flatten one of the balls.	**Step 3** *"Do the two pieces have the same amount of clay?"* Preoperational child will say that the flat piece has more clay. Child in concrete stage will say that the two pieces have the same amount of clay.
Area (ages 8–10)	**Step 1** Center two identical sheets of cardboard with wooden blocks placed on them in identical positions. Child acknowledges that the same amount of space is left open on each piece of cardboard. **Step 2** Scatter the blocks on one piece of the cardboard.	**Step 3** *"Do the two pieces of cardboard have the same amount of open space?"* Preoperational child will say that the cardboard with scattered blocks has less open space. Child in concrete stage will say that both pieces have the same amount of open space.

formal operational stage Piaget's fourth stage (around age 11 and beyond), which is characterized by abstract and hypothetical thinking.

The final stage in Piaget's theory is the **formal operational stage**, which typically begins around age 11. In this stage, children begin to apply their operations to abstract concepts in addition to concrete objects. They also become capable of hypothetical thinking ("What if ?"), which allows systematic formulation and testing of concepts.

For example, before filling out applications for part-time jobs, adolescents may think about possible conflicts with school and friends, the number of hours they want to work, and the kind of work for which they are qualified. Formal operational thinking also allows the adolescent to construct a well-reasoned argument based on hypothetical concepts and logical processes. Consider the following argument:

1. If you hit a glass with a feather, the glass will break.

2. You hit the glass with a feather.

What is the logical conclusion? The correct answer, "The glass will break," is contrary to fact and direct experience. Therefore, the child in the concrete operational stage would have difficulty with this task, whereas the formal operational thinker understands that this problem is about abstractions that need not correspond to the real world.

Along with the benefits of this cognitive style come several problems. Adolescents in the early stages of the formal operational period demonstrate a type of egocentrism different from that of the preoperational child. Although adolescents recognize that others have unique thoughts and perspectives, they often fail to differentiate between what they are thinking and what others are thinking (**FIGURE 9.15**). This adolescent egocentrism has two characteristics that may affect social interactions as well as problem solving:

- *Personal fable.* Because of their unique form of egocentrism, adolescents may conclude that they alone are having insights or difficulties and that no one else understands or sympathizes with them. David Elkind (1967, 2000)

described this as the formation of a **personal fable**, an intense investment in an adolescent's own thoughts and feelings and a belief that these thoughts are unique. For example, one young woman remembered being very upset in middle school when her mother tried to comfort her over the loss of an important relationship. "I felt like she couldn't possibly know how it felt—no one could. I couldn't believe that anyone had ever suffered like this or that things would ever get better."

Several forms of risk taking, such as engaging in sexual intercourse without contraception, driving dangerously, and experimenting with drugs, seem to arise from the personal fable (Coley & Chase-Lansdale, 1998; Flavell, Miller, & Miller, 2002; Greene et al., 2000). Adolescents have a sense of uniqueness, invulnerability, and immortality. They recognize the dangers of these activities but think the rules don't apply to them.

- *Imaginary audience.* In early adolescence, people tend to believe that they are the center of others' thoughts and attentions, instead of considering that everyone is equally wrapped up in his or her own concerns and plans. In other words, adolescents feel that all eyes are focused on their behaviors. Elkind referred to this as the **imaginary audience**.

Self-conscious FIGURE 9.15

What developmental phenomenon might explain why adolescents sometimes display what seem like extreme forms of self-consciousness and concern for physical appearance?

If the imaginary audience results from an inability to differentiate the self from others, the personal fable is a product of differentiating too much. Thankfully, these two forms of adolescent egocentrism tend to decrease during later stages of the formal operational period.

ASSESSING PIAGET'S THEORY: CRITICISMS AND CONTRIBUTIONS

As influential as Piaget's account of cognitive development has been, it has received significant criticisms. Let's look briefly at two major areas of concern: underestimated abilities and underestimated genetic and cultural influences.

Research shows that Piaget may have underestimated young children's cognitive development (FIGURES 9.16 and 9.17). For example, researchers report that very young infants have a basic concept of how objects move, are aware that objects continue to exist even when screened from view, and can recognize speech sounds (Baillargeon, 2000; Streri et al., 2004).

Non-egocentric responses also appear in the earliest days of life. For example, newborn babies tend to cry in response to the cry of another baby (Dondi, Simion, & Caltran, 1999). Also, preschoolers will adapt their speech by using shorter, simpler expressions when talking to 2-year-olds than to adults.

Piaget's model, like other stage theories, has also been criticized for not sufficiently taking into account genetic and cultural differences (Flavell, Miller, & Miller, 2002; Matusov & Hayes, 2000; Maynard & Greenfield, 2003). During Piaget's time, the genetic influences on cognitive abilities were poorly understood, but there has been a rapid explosion of information in this field in recent years. In addition, formal education and specific cultural experiences can also significantly affect cognitive development. Consider the following example from a researcher attempting to test the formal operational skills of a farmer in Liberia (Scribner, 1977):

Visualizing

Infant imitation FIGURE 9.16

In a series of well-known studies, Andrew Meltzoff and M. Keith Moore (1977, 1985, 1994) found that newborns could imitate such facial movements as tongue protrusion, mouth opening, and lip pursing. At 9 months, infants will imitate facial actions a full day after seeing them (Heimann & Meltzoff, 1996). Can you see how this early infant facial expression raises questions about Piaget's estimates of early infant abilities?

When an adult models a facial expression, even very young infants will respond with a similar expression. Is this true imitation or a simple stimulus-response reflex?

Are preoperational children always egocentric? FIGURE 9.17

Some toddlers and preschoolers clearly demonstrate empathy for other people. How does this ability to take others' perspective contradict Piaget's beliefs about egocentrism in very young children?

Researcher: All Kpelle men are rice farmers. Mr. Smith is not a rice farmer. Is he a Kpelle man?

Kpelle farmer: I don't know the man. I have not laid eyes on the man myself.

Instead of reasoning in the "logical" way of Piaget's formal operational stage, the Kpelle farmer reasoned according to his specific cultural and educational training, which apparently emphasized personal knowledge. Not knowing Mr. Smith, the Kpelle farmer did not feel qualified to comment on him. Thus, Piaget's theory may have underestimated the effect of culture on a person's cognitive functioning.

Despite criticisms, Piaget's contributions to psychology are enormous. As one scholar put it, "assessing the impact of Piaget on developmental psychology is like assessing the impact of Shakespeare on English literature or Aristotle on philosophy—impossible." (Beilin, 1992, p. 191)

CONCEPT CHECK STOP

When does accommodation occur?

How do egocentrism and animism limit children's thinking during the preoperational stage?

Why might the personal fable explain risky behavior among adolescents?

What criticisms has Piaget's theory received?

What is happening in this picture ?

- What stage of cognitive development does this child's behavior typify?

- What schemas might the child build by "exploring" her food in this way?

VISUAL SUMMARY

1 Studying Development

1. **Developmental psychology** is the study of age-related changes in behavior and mental processes from conception to death. Development is an ongoing, life-long process.

2. The three most important debates or questions in human development are about nature versus nurture (including studies of **maturation** and **critical periods**), continuity versus stages, and stability versus change. For each question, most psychologists prefer an interactionist perspective.

3. Developmental psychologists use two research methods: the **cross-sectional method** and the **longitudinal method**. Although both have valuable attributes, each also has disadvantages. Cross-sectional studies can confuse true developmental effects with cohort effects. On the other hand, longitudinal studies are expensive and time-consuming, and their results are restricted in generalizability.

2 Physical Development

1. The early years of development are characterized by rapid change. Prenatal development begins at conception and is divided into three stages: the **germinal period**, the **embryonic period**, and the **fetal period**. During pregnancy, the placenta serves as the link for food and the excretion of wastes, and it screens out some harmful substances. However, some environmental hazards (teratogens), such as alcohol and nicotine, can cross the placental barrier and endanger prenatal development.

2. Early childhood is also a time of rapid physical development, including brain, motor, and sensory/perceptual development.

3. During adolescence, both boys and girls undergo dramatic changes in appearance and physical capacity. Puberty is the period of adolescence when a person becomes capable of reproduction. The clearest and most dramatic physical sign of puberty is the growth spurt, characterized by rapid increases in height, weight, and skeletal growth and by significant changes in reproductive structures and sexual characteristics.

4. During adulthood, most individuals experience only minor physical changes until middle age. Around age 45–55, women experience menopause, the cessation of the menstrual cycle. At the same time, men experience a gradual decline in the production of sperm and testosterone, as well as other physical changes, known as the male climacteric.

5. After middle age, most physical changes in development are gradual and occur in the heart and arteries and in the sensory receptors. Most researchers, as well as the public in general, have long thought that aging is accompanied by widespread death of neurons in the brain. Although this decline does happen with degenerative disorders like Alzheimer's disease, this is no longer believed to be a part of normal aging. Although mental speed declines with age, general information processing and much of memory ability is largely unaffected by the aging process. There are two main theories explaining primary aging and death— programmed theory and damage theory.

3 Cognitive Development

1. In the 1920s and 1930s, Piaget conducted groundbreaking work on children's cognitive development. Piaget's theory remains the major force in the cognitive area of developmental psychology today.

2. Three major concepts are central to Piaget's theory: **schemas**, **assimilation**, and **accommodation**. According to Piaget, all children progress through four stages of cognitive development: the **sensorimotor stage**, the **preoperational stage**, the **concrete operational stage**, and the **formal operational stage**. As they progress through these stages, children acquire progressively more sophisticated ways of thinking.

3. Piaget's account of cognitive development has been enormously influential, but it has also received significant criticisms. In particular, research shows that Piaget may have underestimated infants' and young children's cognitive abilities, and he may have underestimated genetic and cultural influences on cognitive development.

CRITICAL AND CREATIVE THINKING QUESTIONS

1. Based on what you have learned about the advantages and disadvantages of cross-sectional and longitudinal methods, can you think of a circumstance when each might be preferable over the other?

2. If a mother knowingly ingests a quantity of alcohol that causes her child to develop fetal alcohol syndrome (FAS), is she guilty of child abuse? Why or why not?

3. From an evolutionary perspective, why might babies prefer looking at more complex patterns and at faces, rather than simple patterns?

4. Based on what you have learned about development during late adulthood, do you think that this period is necessarily a time of physical and mental decline?

5. Piaget's theory states that all children progress through all of the discrete stages of cognitive development in order and without skipping any. Do you agree with this theory? Do you know any children who seem to contradict this theory?

(Check your answers in Appendix B.)

1. The study of age-related changes in behavior and mental processes from conception to death is called _____.

   a. thanatology
   b. neo-gerontology
   c. developmental psychology
   d. longitudinal psychology

2. This is development governed by automatic, genetically predetermined signals.

   a. growth
   b. natural progression
   c. maturation
   d. tabula rasa

3. Label the two basic types of research studies in the figure below:

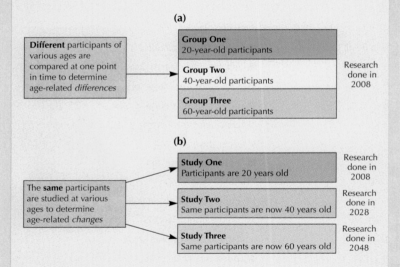

(a)

Different participants of various ages are compared at one point in time to determine age-related *differences*	Group One 20-year-old participants	Research done in 2008
	Group Two 40-year-old participants	
	Group Three 60-year-old participants	

(b)

The **same** participants are studied at various ages to determine age-related *changes*	Study One Participants are 20 years old	Research done in 2008
	Study Two Same participants are now 40 years old	Research done in 2028
	Study Three Same participants are now 60 years old	Research done in 2048

4. This is the first stage of prenatal development which begins with conception and ends with implantation in the uterus (the first two weeks).

   a. embryosis
   b. zygote stage
   c. critical period
   d. germinal period

5. At birth, an infant's head is _____ its body's size, whereas in adulthood, the head is _____ its body's size.

   a. 1/3; 1/4
   b. 1/3; 1/10
   c. 1/4; 1/10
   d. 1/4; 1/8

| 2 months (fetal) | 5 months (fetal) | Newborn | 2 years | 6 years | 12 years | 25 years |

6. Which of the following is **NOT** true regarding infant sensory and perceptual development?

   a. Vision is almost 20/20 at birth.
   b. A newborn's sense of pain is highly developed at birth.
   c. An infant can recognize, and prefers, its own mother's breast milk by smell.
   d. An infant can recognize, and prefers, its own mother's breast milk by taste.

7. The clearest and most physical sign of puberty is the _____, characterized by rapid increases in height, weight, and skeletal growth.

   a. menses
   b. spermarche
   c. growth spurt
   d. age of fertility

8. Some employers are reluctant to hire 50- to 60-year-old workers because of a generalized belief that they are sickly and will take too much time off. This is an example of _____.

   a. discrimination
   b. prejudice
   c. ageism
   d. all of these options

9. _____ was one of the first scientists to demonstrate that a child's intellect is fundamentally different from an adult's.

   a. Baumrind
   b. Beck
   c. Piaget
   d. Elkind

10. _____ occurs when existing schemas are used to interpret new information, whereas _____ involves changes and adaptations of the schemas.

   a. Adaptation; accommodation
   b. Adaptation; reversibility
   c. Egocentrism; postschematization
   d. Assimilation; accommodation

11. Label the four stages of Piaget's cognitive development model on the table below.

Piaget's four stages of cognitive development

12. The photos below are an example of _____ .

   a. sensory permanence
   b. perceptual constancy
   c. perceptual permanence
   d. object permanence

13. The ability to think abstractly or hypothetically occurs in Piaget's _____ stage.

   a. egocentric
   b. post-operational
   c. formal operational
   d. concrete operational

14. Extreme forms of self-consciousness and concerns for physical appearance are common in adolescents, who tend to exhibit the early formal operational characteristic called _____ .

   a. the personal fable
   b. adolescent egocentrism
   c. the imaginary audience
   d. the egocentric fable

15. Critics have suggested that Piaget underestimated _____ .

   a. the cognitive abilities of infants and young children
   b. genetic influences on cognition
   c. the effect of cultural experiences on cognition
   d. all of these options

Life Span Development II

Imagine that you are standing on a bridge over a railroad track when you see that a runaway train is about to kill five people. Coincidentally, you are standing next to a switching mechanism, and you realize that by simply throwing a switch, you can divert the train onto a spur, allowing the five to survive. But here is the catch: diverting the train will condemn *one* person, who is standing on the spur, to death. What would you do? Would you allow one person to die in order to save five others? Would your answer be different if that person were your mother, father, or some other much-loved person? What might lead different people to make different decisions?

It's unlikely that you will ever have to make such a gruesome choice. Yet everyone encounters moral dilemmas from time to time. (Should you remind the cable company that they forgot to disconnect the cable after you discontinued the service? Should you give spare change to a friendly panhandler in your neighborhood?) Likewise, everyone has personal relationships and particular character traits that color our decisions. How we approach moral dilemmas—as well as many other events and circumstances throughout our lives—reflects several key facets of our personal growth: our social, moral, and personality development. In this chapter, we'll look at these aspects of development. We'll also examine how sex, gender, and culture influence our development. Finally, we'll explore several key developmental challenges during adulthood.

Social, Moral, and Personality Development

In addition to physical and cognitive development (Chapter 9), developmental psychologists study social, moral, and personality development by looking at how social forces and individual differences affect development over the life span. The poet John Donne wrote, "No man is an island, entire of itself." In this section, we focus on three major facets of development that shed light on how we affect each other: attachment, Kohlberg's stages of moral development, and Erikson's psychosocial stages.

SOCIAL DEVELOPMENT: THE IMPORTANCE OF ATTACHMENT

An infant arrives in the world with a multitude of behaviors that encourage a strong bond of **attachment** with primary caregivers.

In studying attachment behavior, researchers are often divided along the lines of the nature-versus-nurture debate. Those who advocate the nativist, or innate, position cite John Bowlby's work (1969, 1989, 2000). He proposed that newborn infants are biologically equipped with verbal and nonverbal behaviors (such as crying, clinging, and smiling) and with "following" behaviors (such as crawling and walking after the caregiver) that elicit instinctive nurturing responses from the caregiver. Konrad Lorenz's (1937) early studies of **imprinting** further support the biological argument for attachment (FIGURE 10.1).

attachment
A strong affectional bond with special others that endures over time.

Imprinting FIGURE 10.1

Lorenz's studies on imprinting demonstrated that baby geese attach to, and then follow, the first large moving object they see during a certain critical period in their development.

Attachment: The Power of Touch

According to Freud, infants become attached to the caregiver who provides oral pleasure (Chapter 12). But is there scientific evidence to support Freud's claim? In a classic experiment involving infant rhesus monkeys, Harry Harlow and Robert Zimmerman (1959) investigated the variables that might affect attachment. They created two types of wire-framed surrogate (substitute) "mother" monkeys: one covered by soft terry cloth and one left uncovered. The infant monkeys were fed by either the cloth or the wire mother, but they otherwise had access to both mothers. The researchers found that monkeys "reared" by a cloth mother clung frequently to the soft material of their surrogate mother and developed greater emotional security and curiosity than did monkeys assigned to the wire mother.

In later research (Harlow & Harlow, 1966), monkey babies were exposed to rejection. Some of the "mothers" contained metal spikes that would suddenly protrude from the cloth covering and push the babies away; others had air jets that would sometimes blow the babies away. Nevertheless, the infant monkeys waited until the rejection was over and then clung to the cloth mothers as tightly as before. From these and related findings, Harlow concluded that **contact comfort**, the pleasurable tactile sensations provided by a soft and cuddly "parent," is a powerful contributor to attachment. The satisfaction of other physical needs, such as food, is not enough

A Even when the wire "mother" provided all their food, monkeys showed strong attachment behaviors toward the cloth mother.

Several studies suggest that contact comfort between human infants and mothers is similarly important. For example, touching and massaging premature infants produce significant physical and emotional benefits (Diener et al., 2003; Field & Hernandez-Reif, 2001; Gitau et al., 2002). Mothers around the world tend to kiss, nuzzle, nurse, comfort, clean, and respond to their children with lots of physical contact. (**B**) For example, Japanese mothers touch their infants to communicate with them, breast-feed, carry them around on their backs, take baths with them, and sleep in the same bed with them.

Maternal contact comfort

Fathers matter too.

C Although almost all research on attachment and contact comfort has focused on mothers and infants, recent research shows that the same results also apply to fathers and other caregivers (Diener et al., 2002; Grossmann et al., 2002; van IJzendoorn & De Wolff, 1997).

What a Scientist Sees

What if a child does not form an attachment? Research shows that infants raised in impersonal surroundings (such as in institutions that do not provide the stimulation and love of a regular caregiver) or under abusive conditions suffer from a number of problems. They seldom cry, coo, or babble; they become rigid when picked up; and they have few language skills. They also tend to form shallow or anxious relationships. Some appear forlorn, withdrawn, and uninterested in their caretakers, whereas others seem insatiable in their need for affection (Zeanah, 2000). Finally, they also tend to show intellectual, physical, and perceptual retardation; increased susceptibility to infection; and neurotic "rocking" and isolation behaviors. In some cases, they die from lack of attachment (Belsky & Cassidy, 1994; Bowlby, 1973, 1982, 2000; Spitz & Wolf, 1946).

Although most children are never exposed to such extreme institutional conditions, developmental psychologist Mary Ainsworth and her colleagues (1967, 1978) have found significant differences in the typical levels of attachment between infants and their mothers. In addition, level of attachment affects long-term behaviors. Using a method called the **strange situation procedure**, in which a researcher observes infants in the presence or absence of their mother and a stranger, Ainsworth found that children could be divided into three groups: securely attached, avoidant, and anxious/ambivalent (**FIGURE 10.2**).

Ainsworth found that infants with a secure attachment style had caregivers who were sensitive and responsive to their signals of distress, happiness, and fatigue (Ainsworth et al., 1967, 1978; van IJzendoorn & De Wolff, 1997). On the other hand, avoidant infants had caregivers who were aloof and distant, and anxious/ambivalent infants had inconsistent caregivers who alternated between strong affection and indifference. Follow-up studies found that, over time, securely attached children were the most sociable, emotionally aware, enthusiastic, cooperative, persistent, curious, and competent (De Rosnay & Harris, 2002; Goldberg, 2000; Jacobsen & Hofmann, 1997).

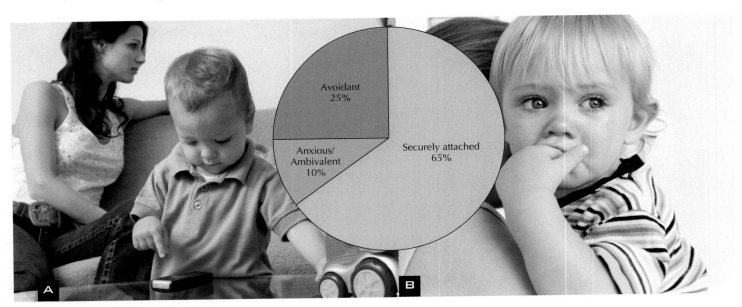

Degrees of attachment FIGURE 10.2

Ainsworth's research identified three different degrees of attachment.

- *Securely attached (65 percent).* When exposed to the stranger, the infant seeks closeness and contact with the mother, uses the mother as a safe base from which to explore, (**A**) shows moderate distress on separation from the mother, (**B**) and is happy when the mother returns.

- *Avoidant (25 percent).* The infant does not seek closeness or contact with the mother, treats the mother much like a stranger, and rarely cries when the mother leaves the room.
- *Anxious/Ambivalent (10 percent).* The infant becomes very upset when the mother leaves the room. When she returns, the infant seeks close contact and then squirms angrily to get away.

Attachment Across the Life Span

If you've been around young children, you've probably noticed how often they share toys and discoveries with a parent and how they seem much happier when a parent is nearby. You may also have thought how cute and sweet it is when infants and parents coo and share baby talk with each other. But have you noticed that these very same behaviors often occur between adults in romantic relationships?

Intrigued by these parallels, several researchers have studied the relationship between an infant's attachment to a parent figure and an adult's love for a romantic partner (Bachman & Zakahi, 2000; Diamond, 2004; Myers & Vetere, 2002). In one study, Cindy Hazan and Phillip Shaver (1987, 1994) discovered that adults who had an avoidant pattern in infancy find it hard to trust others and to self-disclose, and they rarely report finding "true love" (Cooper, Shaver, & Collins, 1998; Fraley & Shaver, 1997). In short, they block intimacy by being emotionally aloof and distant.

Anxious/ambivalent infants tend to be obsessed with their romantic partners as adults, fear-

ing that their intense love will not be reciprocated. As a result, they tend to smother intimacy by being possessive and emotionally demanding.

In contrast, individuals who are securely attached as infants easily become close to others, expect intimate relationships to endure, and perceive others as generally trustworthy. As you may expect, the securely attached lover has intimacy patterns that foster long-term relationships and is the most desired partner by the majority of adults, regardless of their own attachment styles (Klohnen & Bera, 1998; Pietromonaco & Carnelley, 1994).

As you consider these correlations between infant attachment and adult romantic love styles, remember that it is always risky to infer causation from correlation. Accordingly, the relationship between romantic love style and early infant attachment is subject to several alternative explanations. Also, be aware that early attachment experiences may predict the future, but they do not determine it. Throughout life, we can learn new social skills and different attitudes toward relationships.

Early attachment experiences may predict romantic love style.

MORAL DEVELOPMENT: KOHLBERG'S STAGES

Developing a sense of right and wrong, or morality, is a part of psychological development. Consider the following situation in terms of what you would do.

> In Europe, a woman was near death from a special kind of cancer. There was one drug that doctors thought might save her. It was a form of radium that a druggist in the same town had recently discovered. The drug was expensive to make, but the druggist was charging 10 times what the drug cost him to make. He paid $200 for the radium and charged $2,000 for a small dose of the drug. The sick woman's husband, Heinz, went to everyone he knew to borrow the money, but he could gather together only about $1,000, half of what it cost. He told the druggist that his wife was dying and asked him to sell it cheaper or let him pay later. But the druggist said, "No, I discovered the drug, and I'm going to make money from it." So Heinz got desperate and broke into the man's store to steal the drug for his wife (Kohlberg, 1964, pp. 18–19).

Was Heinz right to steal the drug? What do you consider moral behavior? Is morality "in the eye of the beholder," or are there universal truths and principles? Whatever your answer, your ability to think, reason, and respond to Heinz's dilemma demonstrates another type of development that is very important to psychology—morality.

One of the most influential researchers in moral development was Lawrence Kohlberg (1927–1987). He presented what he called "moral stories" like the Heinz dilemma to people of all ages, and on the basis of his findings, he developed a model of moral development (1964, 1984).

What is the right answer to Heinz's dilemma? Kohlberg was interested not in whether participants judged Heinz to be right or wrong but in the reasons they gave for their decisions. On the basis of participants' responses, Kohlberg proposed three broad levels in the evolution of moral reasoning, each composed of two distinct stages (TABLE 10.1, page 262). Individuals at each stage and level may or may not support Heinz's stealing of the drug, but their reasoning changes from level to level.

Kohlberg believed that, like Piaget's stages of cognitive development (Chapter 9), his stages of moral development are universal and invariant. That is, they supposedly exist in all cultures, and everyone goes through each of the stages in a predictable fashion. The age trends that are noticed tend to be rather broad.

Preconventional level

(Stages 1 and 2—birth to adolescence). At the **preconventional level**, moral judgment is self-centered. What is right is what one can get away with or what is personally satisfying. Moral understanding is based on rewards, punishments, and the exchange of favors. This level is called "preconventional" because children have not yet accepted society's (conventional) rule-making processes.

> **preconventional level**
> Kohlberg's first level of moral development, in which morality is based on rewards, punishment, and the exchange of favors.

- *Stage 1 (punishment and obedience orientation).* Children at this stage focus on self-interest—obedience to authority and avoidance of punishment. Because they also have difficulty considering another's point of view, they ignore people's intentions in their moral judgments. Thus, a 5-year-old will often say that accidentally breaking 15 cups is "badder" and should receive more punishment than intentionally breaking 1 cup.

- *Stage 2 (instrumental-exchange orientation).* During this stage, children become aware of others' perspectives, but their morality is based on reciprocity—an equal exchange of favors. "I'll share my lunch with you because if I ever forget mine you'll share yours with me." The guiding philosophy is "You scratch my back and I'll scratch yours."

Conventional level (Stages 3 and 4—adolescence and young adulthood). At the **conventional level**, moral reasoning advances from being self-centered to other-centered. The individual personally accepts conventional societal rules because they help ensure social order and judges morality in terms of compliance with these rules and values.

- *Stage 3 ("good child" orientation).* At Stage 3, the primary moral concern is with being nice and gaining approval. People are also judged by their intentions and motives ("His heart was in the right place").

- *Stage 4 (law-and-order orientation).* During this stage, the individual takes into account a larger perspective—societal laws. Stage 4 individuals understand that if everyone violated laws, even with good intentions, there would be chaos. Thus, doing one's duty and respecting law and order are highly valued. According to Kohlberg, Stage 4 is the highest level attained by most adolescents and adults.

Postconventional level (Stages 5 and 6—adulthood). At the **postconventional level**, individuals develop personal standards for right and wrong. They also define morality in terms of abstract principles and values that apply to all situations and societies. A 20-year-old who judges the "discovery" and settlement of North America by Europeans as immoral because it involved the theft of land from native peoples is thinking in postconventional terms.

- *Stage 5 (social-contract orientation).* Individuals at Stage 5 appreciate the underlying purposes served by laws. When laws are consistent with interests of the majority, they are obeyed because of the "social contract." However, laws can be morally disobeyed if they fail to express the will of the majority or fail to maximize social welfare (**FIGURE 10.3**).

Postconventional moral reasoning
FIGURE 10.3

Would you travel hundreds of miles to participate in a political demonstration? Would you be willing to be arrested for violating the law to express your moral convictions?

Kohlberg's stages of moral development TABLE 10.1

Name of stage	Moral reasoning	Heinz's dilemma responses	
		Pro	Con
Preconventional level			
Stage 1: punishment—obedience orientation	Morality is what you can get away with	"If you let your wife die, you will get in trouble. You'll be blamed for not spending the money to save her and there'll be an investigation of you and the druggist for your wife's death."	"You shouldn't steal the drug, because you'll be caught and sent to jail. If you do get away, your conscience would bother you thinking how the police will catch up with you at any minute."
Stage 2: instrumental-exchange orientation	Obey rules to obtain rewards or favors	"If you do happen to get caught, you could give the drug back and you wouldn't get much of a sentence. It wouldn't bother you much to serve a little jail term if you had your wife when you got out."	"You may not get much of a jail term if you steal the drug, but your wife will probably die before you get out, so it won't do you much good. If your wife dies, you shouldn't blame yourself; it isn't your fault she has cancer."
Conventional level			
Stage 3: good-child orientation	Obey rules to get approval	"No one will think you're bad if you steal the drug, but your family will think you're an inhuman husband if you don't. If you let your wife die, you'll never be able to look anybody in the face again."	"It isn't just the druggist who will think you're a criminal; everyone else will, too. After you steal it, you'll feel bad thinking how you've brought dishonor on your family and yourself, and you won't be able to face anyone again."
Stage 4: law-and-order orientation	Obey laws because they maintain the social order	"If you have any sense of honor, you won't let your wife die just because you're afraid to do the only thing that will save her. You'll always feel guilty that you caused her death if you don't do your duty to her."	"You're desperate and you may not know you're doing wrong when you steal the drug. But you'll know you did wrong after you're sent to jail. You'll always feel guilty for breaking the law."
Postconventional level			
Stage 5: social-contract orientation	Moral reasoning reflects belief in democratically accepted laws	"You'll lose other people's respect, not gain it, if you don't steal. If you let your wife die, it would be out of fear, not out of reasoning it out. So you'd lose self-respect and probably the respect of others, too."	"You would lose your standing and respect in the community and violate the law. You'd lose respect for yourself if you're carried away by emotion and forget the long-range point of view."
Stage 6: universal-ethics orientation	Moral reasoning reflects individual conscience	"If you don't steal the drug and let your wife die, you'll always condemn yourself for it afterward. You wouldn't be blamed and you would have lived up to the outside rule of the law, but you wouldn't have lived up to your own standards of conscience."	"If you stole the drug, you wouldn't be blamed by other people, but you'd condemn yourself because you wouldn't have lived up to your own conscience and standards of honesty."

Sources: Adapted from Kohlberg, L. "Stage and Sequence: The Cognitive Developmental Approach to Socialization." in D. A. Goslin, *The Handbook of Socialization Theory and Research.* Chicago: Rand McNally, 1969. p. 376 (Table 6.2).

- *Stage 6 (universal ethics orientation).* At this stage, "right" is determined by universal ethical principles that all religions or moral authorities might view as compelling or fair, such as nonviolence, human dignity, freedom, and equality, whether or not they conform to existing laws. Thus, Mohandas Gandhi, Martin Luther King, Jr., and Nelson Mandela intentionally broke laws that violated universal principles, such as human dignity. Few individuals actually achieve Stage 6 (about 1 or 2 percent of those tested worldwide), and Kohlberg found it difficult to separate Stages 5 and 6. So, in time, he combined the stages (Kohlberg, 1981).

Assessing Kohlberg's theory

Kohlberg has been credited with enormous insights and contributions, but his theories have also been the focus of three major areas of criticism:

- *Moral reasoning versus behavior.* Are people who achieve higher stages on Kohlberg's scale really more moral than others, or do they just "talk a good game"? Some studies show a positive correlation between higher stages of reasoning and higher levels of moral behavior (Borba, 2001; Rest et al., 1999), but others have found that situational factors are better predictors of moral behavior (**FIGURE 10.4**) (Bandura, 1986, 1991; Bruggeman & Hart, 1996; Nayda, 2002). For example, research participants are more likely to steal when they are told the money comes from a large company rather than from individuals (Greenberg, 2002). And both men and women will tell more sexual lies during casual relationships than during close relationships (Williams, 2001).

- *Possible gender bias.* Researcher Carol Gilligan criticized Kohlberg's model because on his scale women tend to be classified at a lower level of moral reasoning than men. Gilligan suggested that this difference occurred because Kohlberg's theory emphasizes values more often held by men, such as rationality and independence, while de-emphasizing common female values, such as concern for others and belonging (Gilligan, 1977, 1990, 1993; Hoffman, 2000). Most follow-up studies of Gilligan's theory, however, have found few, if any, gender differences in level or type of moral reasoning (Hoffman, 2000; Jaffee & Hyde, 2000; Pratt, Skoe, & Arnold, 2004).

- *Cultural differences.* Cross-cultural studies confirm that children from a variety of cultures generally follow Kohlberg's model and progress sequentially from his first level, the preconventional, to his second, the conventional (Rest et al., 1999; Snarey, 1985, 1995). At the same time, studies find differences among cultures. For example, cross-cultural comparisons of responses to Heinz's moral dilemma show that Europeans and Americans tend to consider whether they like or identify with the victim in questions of morality. In contrast, Hindu Indians consider social responsibility and personal concerns to be separate issues (Miller & Bersoff, 1998). Researchers suggest that the difference reflects the Indians' broader sense of social responsibility.

"I swear I wasn't looking at smut—I was just stealing music."

Morality gap? FIGURE 10.4

What makes people who normally behave ethically willing to steal intellectual property such as music or software?

Erikson's eight stages of psychosocial development FIGURE 10.5

www.wiley.com/
college/carpenter

Stage 1
Trust versus mistrust (birth–age 1)

Infants learn to trust that their needs will be met, especially by the mother; if not, mistrust develops.

Stage 2
Autonomy versus shame and doubt (ages 1–3)

Toddlers learn to exercise will, make choices, and control themselves. Caregivers' patience and encouragement help foster a sense of autonomy versus shame and doubt.

Stage 3
Initiative versus guilt (ages 3–6)

Preschoolers learn to initiate activities and enjoy their accomplishments. Caregivers who are supportive and encouraging promote feelings of power and self-confidence versus guilt.

Stage 4
Industry versus inferiority (ages 6–12)

Elementary school-aged children develop a sense of industry and learn productive skills that their culture requires (such as reading, writing, and counting); if not, they feel inferior.

PERSONALITY DEVELOPMENT: ERIKSON'S PSYCHOSOCIAL THEORY

Like Piaget and Kohlberg, Erik Erikson (1902–1994) developed a stage theory of development. He identified eight **psychosocial stages** of social development (FIGURE 10.5), each marked by a "psychosocial" crisis or conflict related to a specific developmental task. The name given to each stage reflects a specific crisis encountered at that stage and the two possible outcomes. For example, the crisis or task of most young adults is intimacy versus isolation. This age group's developmental task is developing deep, meaningful relationships with others. Those who don't meet this developmental challenge risk social isolation. Erikson believed that the more successfully we overcome each psychosocial crisis, the better chance we have to develop in a healthy manner (Erikson, 1950).

Many psychologists agree with Erikson's general idea that psychosocial crises contribute to personality

psychosocial stages The eight developmental stages, each involving a crisis that must be successfully resolved, that individuals pass through in Erikson's theory of psychosocial development.

Stage 5
Identity versus role confusion (ages 12–20)

During a period of serious questioning and intense soul-searching, adolescents develop a coherent sense of self and their role in society. Failure to resolve this **identity crisis** may be related to a lack of a stable identity, delinquency, and difficulty in maintaining close personal relationships in later life.

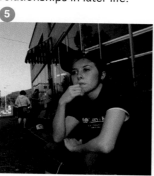

Stage 6
Intimacy versus isolation (ages 20–30)

After learning who they are and how to be independent, young adults form intimate connections with others; if not, they face isolation and consequent self-absorption.

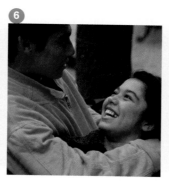

Stage 7
Generativity versus stagnation (ages 30–65)

Middle-aged adults develop concern for establishing and influencing the next generation. If this expansion and effort do not occur, an individual stagnates and is concerned solely with material possessions and personal well-being.

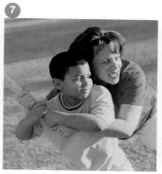

Stage 8
Ego integrity versus despair (age 65+)

Older people enter a period of reflection. They either achieve a sense of accomplishment and satisfaction with the lives they've lived and accept death or yield to despair that their lives cannot be relived.

Source: Adapted from Papalia, Olds, and Feldman, *Human Development*, 8th Edition, New York: McGraw-Hill, 2001.

development (Brendgen, Vitaro, & Bukowski, 2000; Bugental & Goodnow, 1998; Marcia, 2002). However, some critics argue that his theory oversimplifies development. Others observe that the eight stages do not apply equally to all groups. For example, in some cultures, *autonomy* is highly preferable to *shame and doubt*, but in others, the preferred resolution might be *dependence* or *merging relations* (Matsumoto, 2000).

Despite their limits, Erikson's stages have greatly contributed to the study of North American and European psychosocial development. By suggesting that development continues past adolescence, Erikson's theory has encouraged further research.

CONCEPT CHECK **STOP**

How do the securely attached, avoidant, and anxious/ambivalent attachment styles differ?

What are some cross-cultural limitations to Erikson's psychosocial theory?

What criticisms have been leveled at Kohlberg's theory of moral development?

Social, Moral, and Personality Development 265

How Sex, Gender, and Culture Affect Development

LEARNING OBJECTIVES

Identify biological sex differences in physical development.

Describe how gender differences are related to cognitive, personality, and social development.

Explain how individualistic versus collectivistic cultures shape personality development.

SEX AND GENDER INFLUENCES ON DEVELOPMENT

Imagine for a moment what your life would be like if you were a member of the other sex. Would you think differently? Would you be more or less sociable and outgoing? Would your career plans or friendship patterns change? Most people believe that whether we are male or female has a strong impact on many facets of development. But why is that? Why is it that the first question most people ask after a baby is born is, "Is it a girl or a boy?"

sex Biological maleness and femaleness, including chromosomal sex. Also, activities related to sexual behaviors, such as masturbation and intercourse.

gender Psychological and sociocultural meanings added to biological maleness or femaleness.

In this section, we will explore how our development is affected by **sex** (a biological characteristic determined at the moment of conception) and by **gender** and **gender roles** (cultural meanings that accompany biological sex).

Sex differences Physical anatomy is the most obvious biological sex difference between men and women (**FIGURE 10.6**). Men and women also differ in secondary sex characteristics, such as facial hair and breast growth; signs of reproductive capability, such as menstruation and ejaculation of sperm; and physical responses to middle age or the end of reproduction, such as *menopause* and the *male climateric*. There are also several functional and structural differences in the brains of men and women. These result partly from the influence of prenatal sex hormones on the developing fetal brain.

gender roles Societal expectations for normal and appropriate male and female behavior.

Gender differences In addition to biological sex differences, scientists have found numerous gender differences, which affect our cognitive, social, and personality development.

For example, females tend to score higher on tests of verbal skills, whereas males score higher on math and visuospatial tests (Gallagher et al., 2000; Quaiser-Pohl & Lehmann, 2002). Some researchers suggest that these differences in cognitive ability may reflect biological factors, including structural differences in the cerebral hemispheres, hormones, or the degree of hemispheric specialization.

One argument against this biological model, however, is that male-female differences in verbal ability and math scores have declined in recent years (Brown & Josephs, 1999; Halpern, 1997, 2000; Lizarraga & Ganuza, 2003). However, the gap has not been narrowed with regard to men in the highest IQ range. Men in this group still score higher on SAT math scores than women do.

Like cognitive ability, aggressive behavior also differs between the genders. From an early age, boys are more likely to engage in mock fighting and rough-and-

Body Size and Shape
The average man is 35 pounds heavier, has less body fat and is 5 inches taller than the average woman. Men tend to have broader shoulders, slimmer hips and slightly longer legs in proportion to their height.

Brain
The corpus callosum, the bridge joining the two halves of the brain, is larger in women; therefore they can integrate information from the two halves of the brain and more easily perform more than one task simultaneously.

An area of the hypothalamus causes men to have a relatively constant level of sex hormones; whereas, women have cyclic sex hormone production and menstrual cycles. Differences in the cerebral hemispheres may help explain reported sex differences in verbal and spatial skills.

Muscular System
Until puberty, boys and girls are well matched in physical strength and ability. Once hormones kick in, the average man has more muscle mass and greater upper body strength than the average woman.

Skeletal System
Men produce testosterone throughout their life span, whereas estrogen production virtually stops when a women goes through menopause. Because estrogen helps rejuvenate bones, women are more likely to have brittle bones. Women also are more prone to knee damage because a woman's wider hips may place a greater strain on the ligaments joining the thigh to the knee.

Source: Miracle, Tina S., Miracle, Andrew W., and Baumeister, R. F., *Study Guide: Human sexuality: Meeting your basic needs*, 2nd edition. (c) 2006. Reprinted by permission of Pearson Education, Inc., Upper Saddle River, NJ.

tumble play, and as adolescents and adults, they are more likely to commit aggressive crimes (Camodeca et al., 2002; Henning & Feder, 2004; Moffitt et al., 2002; Ostrov & Keating, 2004). But gender differences are clearer for physical aggression (like hitting) than for other forms of aggression. Early research suggested that females were more likely to engage in more indirect and relational forms of aggression, such as spreading rumors and ignoring or excluding someone (Bjorkqvist, 1994; Ostrov & Keating, 2004). But more recent studies have not found such clear differences (Cillessen & Mayeux, 2004; Feiring et al., 2002; Salmivalli & Kaukiainen, 2004).

Some researchers believe that biological factors cause gender differences in aggression—a nativist position. Several studies have linked the male hormone testosterone to aggressive behavior (Book, Starzyk, & Quinsey, 2002; Ramirez, 2003; Trainor, Bird, & Marler, 2004). Other studies have found that aggressive men have disturbances in their levels of serotonin, a neurotransmitter that is inversely related to aggression (Berman, Tracy, & Coccaro, 1997; Holtzworth-Munroe, 2000; Nelson & Chiavegatto, 2001). In addition, studies on identical twins have found that genetic factors account for about 50 percent of aggressive behavior (Cadoret, Leve, & Devor, 1997; Segal & Bouchard, 2000).

Other researchers take a more nurturist position. They suggest that gender differences in aggressiveness result from environmental experiences with social dominance and pressures that encourage "sex appropriate" behaviors and skills (Rowe et al., 2004) (FIGURE 10.7). TABLE 10.2 summarizes the gender differences between men and women.

Gender role development
By age two, children are well aware of gender roles. They recognize that boys "should" be strong, independent, aggressive, dominant, and achieving, whereas girls "should" be soft, dependent, passive, emotional, and "naturally" interested in children (Kimmel, 2000; Renzetti, Curran, & Kennedy-Bergen, 2006). The existence of similar gender roles in many cultures suggests that evolution and biology may play a role. However, most research emphasizes two major theories of gender role development: social learning and cognitive developmental.

Social learning theorists emphasize the power of the immediate situation and observable behaviors on gender role development. They suggest that girls learn how to be "feminine" and boys learn how to be "masculine" in two major ways: (1) They receive rewards or punishments for specific gender role behaviors and (2) they watch and imitate the behavior of others, particu-

Research-supported sex and gender differences TABLE 10.2		
Type of behavior	***More often shown by men***	***More often shown by women***
Sexual	• Begin masturbating sooner in life cycle and have higher overall occurrence rates. • Start sexual life earlier and have first orgasm through masturbation. • Are more likely to recognize their own sexual arousal. • Experience more orgasm consistency in their sexual relations.	• Begin masturbating later in life cycle and have lower overall occurrence rates. • Start sexual life later and have first orgasm from partner stimulation. • Are less likely to recognize their own sexual arousal. • Experience less orgasm consistency in their sexual relations.
Touching	• Are touched, kissed, and cuddled less by parents. • Exchange less physical contact with other men and respond more negatively to being touched. • Are more likely to initiate both casual and intimate touch with sexual partner.	• Are touched, kissed, and cuddled more by parents. • Exchange more physical contact with other women and respond more positively to being touched. • Are less likely to initiate either casual or intimate touch with sexual partner.
Friendship	• Have larger number of friends and express friendship by shared activities.	• Have smaller number of friends and express friendship by shared communication about self.
Personality	• Are more aggressive from a very early age. • Are more self-confident of future success. • Attribute success to internal factors and failures to external factors. • Achievement is task oriented; motives are mastery and competition. • Are more self-validating. • Have higher self-esteem.	• Are less aggressive from a very early age. • Are less self-confident of future success. • Attribute success to external factors and failures to internal factors. • Achievement is socially directed with emphasis on self-improvement; have higher work motives. • Are more dependent on others for validation. • Have lower self-esteem.
Cognitive Abilities	• Are slightly superior in mathematics and visuospatial skills.	• Are slightly superior in grammar, spelling, and related verbal skills.

Sources: Crooks & Baur, 2005; Masters & Johnson, 1961, 1966, 1970; Miracle, Miracle, & Baumeister, 2006.

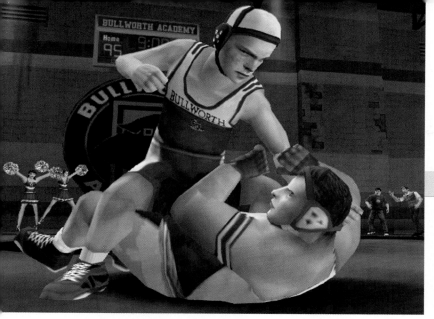

Modeling gendered behavior FIGURE 10.7

Some computer and video games include a great deal of violence and are often criticized for modeling and encouraging violence. In a new game called *Bully*, shown here, the main character ultimately takes on bullies rather than become one. The game includes plenty of fighting, but it also claims to show that actions have consequences. Do you think such an approach might encourage or discourage violent behavior?

larly the same-sex parent (Bandura, 1989, 2000; Fredricks & Eccles, 2005; Kulik, 2005). A boy who puts on his father's tie or baseball cap wins big, indulgent smiles from his parents. But what would happen if he put on his mother's nightgown or lipstick? Parents, teachers, and friends generally reward or punish behaviors according to traditional gender role expectations. Thus, a child "socially learns" what it means to be male or female.

According to the *cognitive developmental theory*, social learning is part of gender role development, but it's much more than a passive process of receiving rewards or punishments and modeling others. Instead, cognitive developmentalists argue that children actively observe, interpret, and judge the world around them (Bem, 1981, 1993; Cherney, 2005; Giles & Heyman, 2005). As children process information about the world, they also create internal rules governing correct behaviors for boys versus girls. On the basis of these rules, they form **gender schemas** (mental images) of how they should act (**FIGURE 10.8**).

Androgyny One way to overcome rigid or destructive gender role stereotypes is to express both the "masculine" and "feminine" traits found in each individual—for example, being assertive and aggressive when necessary but also gentle and nurturing. Combining characteristics in this way is known as **androgyny** [an-DRAW-juh-nee].

Researchers have found that masculine and androgynous individuals generally have higher self-esteem and creativity, are more socially competent and motivated to achieve, and exhibit better overall mental health than those with traditional feminine traits (Choi, 2004; Hittner & Daniels, 2002; Venkatesh et al., 2004).

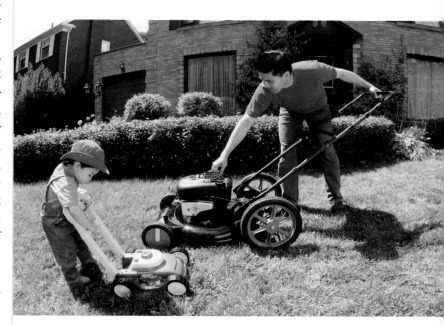

Developing gender schemas FIGURE 10.8

How would socical learning theory and congnitive developmental theory explain why children tend to choose stereo typically "appropriate" toys?

How Sex, Gender, and Culture Affect Development   269

What makes androgyny and masculinity adaptive? Research shows that traditional masculine characteristics (analytical thinking, independence) are more highly valued than traditional feminine traits (affectivity, cheerfulness) (FIGURE 10.9). For example, when college students in 14 countries were asked to describe their "current self" and their "ideal self," the ideal self-descriptions for both men and women contained more masculine than feminine qualities (Williams & Best, 1990).

Recent studies show that gender roles are becoming less rigidly defined (Kimmel, 2000; Loo & Thorpe, 1998). Asian American and Mexican American groups show some of the largest changes toward androgyny, and African Americans remain among the most androgynous of all ethnic groups (Denmark, Rabinowitz, & Sechzer, 2005; Harris, 1996; Huang & Ying, 1989; Renzetti et al., 2006).

CULTURAL INFLUENCES ON DEVELOPMENT

individualistic cultures Cultures in which the needs and goals of the individual are emphasized over the needs and goals of the group.

Our development is rooted in the concept of **self**—how we define and understand ourselves. Yet the very concept of self reflects our culture. In **individualistic cultures**, the needs and goals of the individual are emphasized over the needs and goals of the group. When asked to complete the statement "I am . . .," people from individualistic cultures tend to respond with personality traits ("I am shy"; "I am outgoing") or their occupation ("I am a teacher"; "I am a student").

In **collectivistic cultures**, however, the opposite is true. The person is defined and understood primarily by looking at his or her place in the social unit (McCrae, 2004; Montuori & Fahim, 2004; Tseng, 2004). Relatedness, connectedness, and interdependence are valued, as opposed to separateness, independence, and individualism. When asked to complete the statement "I am . . .," people from collectivistic cultures tend to

Androgyny FIGURE 10.9

For both children and adults, it is more difficult for males to express so-called female traits like nurturance and sensitivity than it is for women to adopt such traditionally male traits as assertiveness and independence (Kimmel, 2000; Leaper, 2000; Wood et al., 1997).

mention their families or nationality ("I am a daughter"; "I am Chinese").

If you are North American or Western European, you are more likely to be individualistic than collectivistic (TABLE 10.3). And you may find the concept of a self defined in terms of others almost contradictory. A core selfhood probably seems intuitively obvious to you. Recognizing that over 70 percent of the world's

collectivistic cultures Cultures in which the needs and goals of the group are emphasized over the needs and goals of the individual.

A worldwide ranking of cultures TABLE 10.3

Individualistic cultures	Intermediate cultures	Collectivistic cultures
United States	Israel	Hong Kong
Australia	Spain	Chile
Great Britain	India	Singapore
Canada	Argentina	Thailand
Netherlands	Japan	West Africa region
New Zealand	Iran	El Salvador
Italy	Jamaica	Taiwan
Belgium	Arab region	South Korea
Denmark	Brazil	Peru
France	Turkey	Costa Rica
Sweden	Uruguay	Indonesia
Ireland	Greece	Pakistan
Norway	Phillippines	Colombia
Switzerland	Mexico	Venezuela

How might these two groups differ in their physical, socioemotional, cognitive, and personality development?

population lives in collectivistic cultures, however, may improve your cultural sensitivity and prevent misunderstandings (Singelis et al., 1995). For example, North Americans generally define *sincerity* as behaving in accordance with one's inner feelings, whereas Japanese see it as behavior that conforms to a person's role expectations (carrying out one's duties) (Yamada, 1997). Can you see how Japanese behavior might appear insincere to a North American and vice versa?

CONCEPT CHECK STOP

What is the difference between sex and gender?

How do people develop gender roles?

How is the concept of "self" different in individualistic versus collectivistic cultures?

Developmental Challenges Through Adulthood

LEARNING OBJECTIVES

Describe the factors that ensure realistic expectations for marriage and long-term committed relationships.

Explain the factors that affect life satisfaction during the adult working years and retirement.

Describe the three basic concepts about death and dying that people learn to understand through the course of development.

In this section, we will explore three of the most important developmental tasks that people face during adulthood: developing a loving, committed relationship with another person; finding rewarding work and a satisfying retirement; and coping with death and dying.

COMMITTED RELATIONSHIPS: OVERCOMING UNREALISTIC EXPECTATIONS

One of the most important tasks faced during adulthood is that of establishing some kind of continuing, loving sexual relationship with another person. Yet navigating such partnerships is often very challenging. For example, nearly half of all marriages in the United States end in divorce, with serious implications for both adult and child development (Abbey & Dallos, 2004; Lengua et al., 2000; Riggio, 2004). For the adults, both spouses generally experience emotional as well as practical difficulties and are at high risk for depression and physical health problems. In many cases, these problems are present even before the marital disruption.

Realistic expectations are a key ingredient in successful relationships (Gottmann & Levenson, 2002; Waller & McLanahan, 2005). Yet many people harbor unrealistic expectations about marriage and the roles of husband and wife, opening the door to marital problems (FIGURE 10.10).

Did the Brady Bunch set unrealistic expectations?

FIGURE 10.10

Where do unrealistic marital expectations originate? Women are more likely than men to try to model their family lives after what they have seen on TV programs and to expect their partners to act like the men they have seen on TV (Morrison & Westman, 2001). Men's expectations are more likely to be driven by certain myths about marriage, such as "Men are from Mars and women are from Venus" or "Affairs are the main cause of divorce."

Are Your Relationship Expectations Realistic?

To evaluate your own expectations, answer the following questions about traits and factors common to happy marriages and committed long-term relationships (Gottman & Levenson, 2002; Gottman & Notarius, 2000; Greeff & Malherbe, 2001; Harker & Keltner, 2001; Heaton, 2002; Rosen-Grandon, Myers, & Hattie, 2004):

1. **Established "love maps"**

   Yes___ No___ *Do you believe that emotional close-ness "naturally" develops when two people have the right chemistry?*

   In successful relationships, both partners are willing to share their feelings and life goals. This sharing leads to detailed "love maps" of each other's inner emotional life and the creation of shared meaning in the relationship.

2. **Shared power and mutual support**

   Yes___ No___ *Have you unconsciously accepted the imbalance of power promoted by many TV sitcoms, or are you willing to fully share power and to respect your partner's point of view, even if you disagree?*

   The common portrayal of husbands as "head of household" and wives as the "little women" who se-cretly wield the true power may help create unrealis-tic expectations for marriage.

3. **Conflict management**

   Yes___ No___ *Do you expect to "change" your partner or to be able to resolve all your problems?*

   Successful couples work hard (through negotiation and accommodation) to solve their solvable con-flicts, to accept their unsolvable ones, and to know the difference.

4. **Similarity**

   Yes___ No___ *Do you believe that "opposites attract?"*

   Although we all know couples who are very different but are still happy, similarity (in values, beliefs, reli-gion, and so on) is one of the best predictors of long-lasting relationships (Chapter 15).

5. **Supportive social environment**

   Yes___ No___ *Do you believe that "love conquers all"?*

   Unfortunately, several environmental factors can overpower or slowly erode even the strongest love. These include age (younger couples have higher divorce rates), money and employment (divorce is higher among the poor and unemployed), parents' marriages (divorce is higher for children of divorced parents), length of courtship (longer is better), and premarital pregnancy (no pregnancy is better, and waiting a while after marriage is even better).

6. **Positive emphasis**

   Yes___ No___ *Do you believe that an intimate rela-tionship is a place where you can in-dulge your bad moods and openly crit-icize one another?*

   Think again. Positive emotions, positive mood, and positive behavior toward one's partner are vitally important to a lasting, happy relationship.

WORK AND RETIREMENT: HOW THEY AFFECT US

Throughout most of our adult lives, work defines us in fundamental ways. It affects our health, our friendships, where we live, and even our leisure activities. Too often, however, career choices are made based on dreams of high income. In a 1995 survey conducted by the Higher Education Research Institute, nearly 74 percent of college freshmen said that being "very well-off financially" was "very important" or "essential." Seventy-one percent felt the same way about raising a family. These young people understandably hope to combine both family and work roles and "live the good life," but many will find themselves in unsatisfying jobs, having to work long hours just to keep up with the rate of inflation.

Choosing an occupation is one of the most important decisions in a person's life, and the task is becoming ever more complex. The *Dictionary of Occupational Titles,* a government publication, currently lists more than 200,000 job categories. According to psychologist John Holland's **personality-job fit theory,** a match (or "good fit") between our individual personalities and our career choices is a major factor in determining job success and satisfaction. Holland's *Self-Directed Search* questionnaire scores each person on six personality types and then matches their individual interests and abilities to the job demands of various occupations (Holland, 1985, 1994; see also Brkich, Jeffs, & Carless, 2002; Kieffer, Schinka, & Curtiss, 2004; Spokane, Meir, & Catalano, 2000; Tett & Murphy, 2002) (TABLE 10.4).

Work and career are a big part of adult life and self-identity, but the large majority of men and women in the United States choose to retire sometime in their sixties. What helps people successfully navigate this important life change? According to the **activity theory of aging**, successful aging appears to be most strongly related to good health, control over one's life, social support, and participation in community services and social activities (Warr, Butcher, & Robertson, 2004; Yeh & Lo, 2004) (FIGURE 10.11).

> ■ **activity theory of aging** Successful aging is fostered by a full and active commitment to life.

The activity theory of aging has largely displaced the older notion that successful aging entails a natural and graceful withdrawal from life (Achenbaum & Bengtson, 1994; Cummings & Henry, 1961; McKee & Barber,

Are you in the right job? TABLE 10.4

Personality characteristics	Holland personality type	Matching/congruent occupation
Shy, genuine, persistent, stable, conforming, practical	1. *Realistic*: Prefers physical activities that require skill, strength, and coordination	Mechanic, drill press operator, assembly-line worker, farmer
Analytical, original, curious, independent	2. *Investigative*: Prefers activities that involve thinking, organizing, and understanding	Biologist, economist, mathematician, news reporter
Sociable, friendly, cooperative, understanding	3. *Social*: Prefers activities that involve helping and developing others	Social worker, counselor, teacher, clinical psychologist
Conforming, efficient, practical, unimaginative, inflexible	4. *Conventional*: Prefers rule-regulated, orderly, and unambiguous activities	Accountant, bank teller, file clerk, corporate manager
Imaginative, disorderly, idealistic, emotional, impractical	5. *Artistic*: Prefers ambiguous and unsystematic activities that allow creative expression	Painter, musician, writer, interior decorator
Self-confident, ambitious, energetic, domineering	6. *Enterprising*: Prefers verbal activities where there are opportunities to influence others and attain power	Lawyer, real estate agent, public relations specialist, small business manager

Reproduced by special permission of the publisher, Psychological Assessment Resources, Inc., 16204 North Florida Avenue, Lutz, Florida 33549, from the Self-Directed Search Form R by John L. Holland, Ph.D, copyright 1985. Further reproduction is prohibited without permission from PAR, Inc.

Active involvement is a key ingredient to a fulfilling old age.

disengagement theory Successful aging is characterized by mutual withdrawal between the elderly and society.

socioemotional selectivity theory A natural decline in social contact as older adults become more selective with their time.

2001; Neugarten, Havighurst, & Tobin, 1968; Rook, 2000). This **disengagement theory** has been seriously questioned and largely abandoned; we mention it because of its historical relevance and because of its connection to an influential modern theory, **socioemotional selectivity theory**. This perspective helps explain the predictable decline in social contact that almost everyone experiences as they move into their older years (Carstensen, Fung, & Charles, 2003; Lang & Carstensen, 2002) (FIGURE 10.12). According to this theory, we don't naturally withdraw from society in our later years—we just become more selective with our time. We deliberately decrease our total number of social contacts in favor of familiar people who provide emotionally meaningful interactions.

As we've seen, there are losses and stress associated with the aging process—although much less than most people think. Perhaps the greatest challenge for the elderly, at least in the United States, is the ageism they encounter. In societies that value older people as wise elders or keepers of valued traditions, the stress of aging is much less than in societies that view them as mentally slow and socially useless. In cultures like that in the United States in which youth, speed, and progress are strongly emphasized, a loss or decline in any of these qualities is deeply feared and denied (Powell, 1998).

Socioemotional selectivity FIGURE 10.12

During infancy, emotional connection is essential to our survival. During childhood, adolescence, and early adulthood, information gathering is critical and the need for emotional connection declines. During late adulthood, emotional satisfaction is again more important—we tend to invest our time in those who can be counted on in times of need.

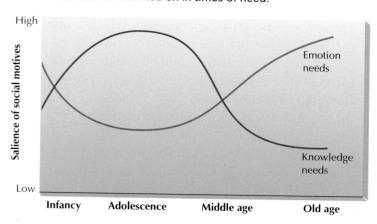

Along with the prospect of retirement, middle age often presents another challenge: the "empty nest." Many people believe that most parents experience a painful separation and some depression when the last child leaves the home. But this popular conception may be an exaggeration of the pain experienced by a few individuals and an effort to downplay positive reactions (White & Rogers, 1997; Whyte, 1992). For example, research suggests that the highest levels of marital satisfaction occur before children are born and after they leave the home (Mackey & O'Brien, 1998).

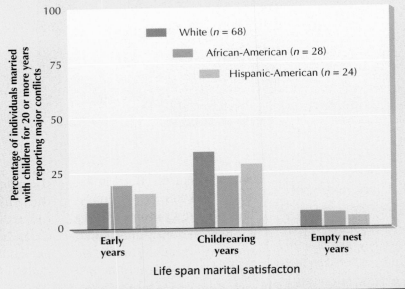

Percentage of individuals married with children for 20 or more years reporting major conflicts

White (n = 68)
African-American (n = 28)
Hispanic-American (n = 24)

Early years · Childrearing years · Empty nest years

Life span marital satisfacton

DEATH AND DYING: OUR FINAL DEVELOPMENTAL CRISIS

One unavoidable part of life is death. In this section, we will look at developmental changes in how people understand and prepare for their own deaths and for the deaths of loved others.

As adults, we understand death in terms of three basic concepts: (1) *permanence*—once a living thing dies, it cannot be brought back to life; (2) *universality*—all living things eventually die; and (3) *nonfunctionality*—all living functions, including thought, movement, and vital signs, end at death.

Research shows that permanence, the notion that death cannot be reversed, is the first and most easily understood concept (**FIGURE 10.13**). Understanding of universality comes slightly later, and by the age of 7, most children have mastered nonfunctionality and have an adultlike understanding of death.

How do children understand death?
FIGURE 10.13

Preschoolers seem to accept the fact that the dead person cannot get up again, perhaps because of their experiences with dead butterflies and beetles found while playing outside (Furman, 1990). Later, they begin to understand all that death entails and that they, too, will someday die.

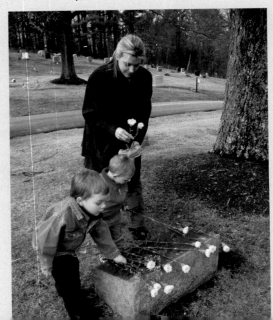

Although parents may fear that discussing death with children and adolescents will make them unduly anxious, those who are offered open, honest discussions of death have an easier time accepting it (Christ, Siegel, & Christ, 2002; Kastenbaum, 1999; Pfeffer et al., 2002).

The same is true for adults—in fact, avoiding thoughts and discussion of death and associating aging with death contribute to ageism (Atchley, 1997). Moreover, the better we understand death and the more wisely we approach it, the more fully we can live. Since the late 1990s, right-to-die and death-with-dignity advocates have brought death out in the open, and mental health professionals have suggested that understanding the psychological processes of death and dying may play a significant role in good adjustment. Cultures around the world interpret and respond to death in widely different ways: "Funerals are the occasion for avoiding people or holding parties, for fighting or having sexual orgies, for weeping or laughter, in a thousand combinations" (Metcalf & Huntington, 1991); (FIGURE 10.14).

Confronting our own death is the last major crisis we face in life. What is it like? Is there a "best" way to prepare to die? Is there such a thing as a "good" death? After spending hundreds of hours at the bedsides of the terminally ill, Elisabeth Kübler-Ross developed her stage theory of the psychological processes surrounding death (1983, 1997, 1999). She proposed that most people go through five sequential stages when facing death:

- *Denial* of the terminal condition ("This can't be true; it's a mistake!")

- *Anger* ("Why me? It isn't fair!")

- *Bargaining* ("God, if you let me live, I'll dedicate my life to you!")

- *Depression* ("I'm losing everyone and everything I hold dear.")

- *Acceptance* ("I know that death is inevitable and my time is near.")

Critics of the stage theory of dying stress that each person's death is a unique experience, depending on individual personality, life situation, age, and so on (Dunn, 2000; Wright, 2003). Others worry that popularizing such a stage theory will cause more avoidance and stereotyping of the dying ("He's just in the anger stage right now.").

Culture influences our response to death
FIGURE 10.14

In October 2006, a dairy truck driver took over a one-room Amish schoolhouse in Pennsylvania, killed and gravely injured several young girls, then shot himself. Instead of responding in rage, his Amish neighbors attended his funeral. Amish leaders urged forgiveness for the killer and called for a fund to aid his wife and three children. Rather than creating an on-site memorial, the schoolhouse was razed, to be replaced by pasture. What do you think of this response? The fact that many Americans were offended, shocked, or simply surprised by the Amish reaction illustrates how strongly culture affects our emotion, beliefs, and values.

Kübler-Ross (1983, 1997, 1999) agrees that not all people go through the same stages in the same way and regrets that anyone would use her theory as a model for a "good" death.

In spite of the potential abuses, Kübler-Ross's theory has provided valuable insights and spurred research into a long-neglected topic. **Thanatology**, the study of death and dying, has become a major topic in human development. Thanks in part to thanatology research, the dying are being helped to die with dignity by the **hospice** movement, which has created special facilities and trained staff and volunteers to provide loving support for the terminally ill and their families (McGrath, 2002; Parker-Oliver, 2002).

CONCEPT CHECK STOP

Why do people develop unrealistic expectations about marriage?

What is the socioemotional selectivity theory?

What are the stages of Kübler-Ross's theory of the psychological processes surrounding death?

Native Americans generally revere and respect the elder members of their tribe. How might viewing old age as an honor and a blessing—as opposed to a dreaded process—affect the experience of aging?

VISUAL SUMMARY

1 Social, Moral, and Personality Development

1. Nativists believe that **attachment** is innate, whereas nurturists believe that it is learned. Studies of imprinting support the biological argument for attachment. Harlow and Zimmerman's research with monkeys raised by cloth or wire "mothers" found that contact comfort might be the most important factor in attachment. The majority of infants are securely attached to their caregivers, but some exhibit either an avoidant or an anxious/ambivalent attachment style. These styles appear to carry over into adult relationships.

2. Kohlberg proposed three levels in the evolution of moral reasoning: the **preconventional level** (Stages 1 and 2), the **conventional level** (Stages 3 and 4), and the **postconventional level** (Stages 5 and 6). According to Kohlberg, Stage 4 is the highest level attained by most adolescents and adults.

3. Erikson identified eight **psychosocial stages** of development, each marked by a crisis (such as the adolescent identity crisis) related to a specific developmental task. Although some critics argue that Erikson's theory oversimplifies development and does not hold true across all cultures, many psychologists agree with Erikson's general idea that psychosocial crises do contribute to personality development.

2 How Sex, Gender, and Culture Affect Development

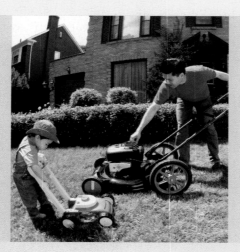

1. In addition to biological **sex** differences, scientists have found numerous **gender** differences relevant to cognitive, social, and personality development (for example, in cognitive ability and aggressive behavior) and have proposed both biological and environmental explanations for these differences. Most research emphasizes two major theories of **gender role** development: social learning and cognitive developmental. One way to overcome gender role stereotypes is to combine "masculine" and "feminine" traits (androgyny).

2. Our culture shapes how we define and understand ourselves. In **individualistic cultures**, the needs and goals of the individual are emphasized; in **collectivistic cultures**, the person is defined and understood primarily by looking at his or her place in the social unit.

3 Developmental Challenges Through Adulthood

1. Having unrealistic expectations about marriage can open the door to marital problems. Developing realistic expectations is critical to a happy marriage.

2. Choosing an occupation is an important and complex decision. According to Holland's personality-job fit theory, a match between personality and career choice is a major factor in determining job success and satisfaction. The **activity theory of aging** proposes that successful aging depends on having control over one's life, having social support, and participating in community services and social activities. The older **disengagement theory** has been largely abandoned. According to the **socioemotional selectivity theory**, people don't naturally withdraw from society in late adulthood but deliberately decrease the total number of their social contacts in favor of familiar people who provide emotionally meaningful interactions.

3. Adults understand death in terms of three basic concepts: permanence, universality, and nonfunctionality. Children most easily understand the concept of permanence; understanding of universality and nonfunctionality come later. For children, adolescents, and adults, understanding the psychological processes of death and dying are important to good adjustment. Kübler-Ross proposed that most people go through five sequential stages when facing death: denial, anger, bargaining, depression, and acceptance. But she has emphasized that her theory should not be interpreted as a model for a "good" death.

KEY TERMS

- **attachment** p. 256
- **imprinting** p. 256
- **contact comfort** p. 257
- **strange situation procedure** p. 258
- **preconventional level** p. 260
- **conventional level** p. 261
- **postconventional level** p. 261
- **psychosocial stages** p. 264

- **identity crisis** p. 265
- **sex** p. 266
- **gender** p. 266
- **gender roles** p. 266
- **gender schemas** p. 269
- **androgyny** p. 269
- **self** p. 270
- **individualistic cultures** p. 270

- **collectivistic cultures** p. 270
- **personality-job fit theory** p. 274
- **activity theory of aging** p. 274
- **disengagement theory** p. 275
- **socioemotional selectivity theory** p. 275
- **thanotology** p. 277
- **hospice** p. 277

CRITICAL AND CREATIVE THINKING QUESTIONS

1. Were Harlow and his colleagues acting ethically when they separated young rhesus monkeys from their mothers and raised them with either a wire or cloth "mother"? Why or why not?

2. According to Erikson's psychosocial theory, what developmental crisis must you now resolve in order to successfully move on to the next stage? Are there any earlier crises that you feel you may not have successfully resolved?

3. Can you think of instances where you have adopted traditional gender roles? What about times where you have expressed more androgyny?

4. According to Holland's personality-job fit theory, what occupations might be a good match for your personality?

5. Would you like to know ahead of time that you were dying? Or would you prefer to die suddenly with no warning? Briefly explain your choice.

(Check your answers in Appendix B.)

1. _____ is a strong affectional bond with special others that endures over time.

a. Bonding
c. Love
b. Attachment
d. Intimacy

2. In Ainsworth's studies on infant attachment, the _____ infants sought their mothers' comfort while also squirming to get away when the mother returned to the room.

a. anxious/ambivalent
c. insecurely attached
b. caregiver-based
d. dependent

3. Kohlberg believed his stages of moral development to be _____ .

a. universal and invariant
b. culturally bound, but invariable within a culture
c. universal, but variable within each culture
d. culturally bound and variable

4. Once an individual has accepted and complied with the rules and values of society, that person has advanced to the _____ level of moral development.

a. instrumental-exchange
c. conventional
b. social-contract
d. postconventional

5. According to Erikson, humans progress through eight stages of psychosocial development. Label the **CORRECT** sequence on the figure below for the "successful" completion of the first four stages.

Stage 1

Stage 2

Stage 3

Stage 4

6. Label the **CORRECT** sequence on the figure below for the "successful" completion of the second four stages of Erikson's eight stages of psychosocial development.

Stage 5

Stage 6

Stage 7

Stage 8

7. Your _____ is related to societal expectations for normal and appropriate male or female behavior.

a. gender identity
b. gender role
c. sexual identity
d. sexual orientation

8. On the basis of internal rules governing correct behaviors for boys versus girls, children form _____ of how they should act.

a. sexual identities
b. gender identities
c. gender schemas
d. any of these options

9. Cultures in which the needs and goals of the group are emphasized over the needs and goals of the individual are _____ .

a. collectivistic cultures
b. individualistic cultures
c. worldwide cultures
d. all of the above

10. _____ has consistently been found to be a key factor in successful relationships.

a. Realistic expectations
b. The ability to fight for equity
c. Satisfying sexual activity
d. Emotional stability

11. The _____ theory of aging suggests that successful adjustment is fostered by a full and active commitment to life.

a. activity
b. commitment
c. engagement
d. life-affirming

12. The _____ theory of aging suggests that adjustment to retirement is fostered by mutual withdrawal between the elderly and society.

a. disengagement
b. withdrawal
c. disengagement
d. death-preparation

13. Which of the following concepts is considered permanent, universal, and nonfunctional?

a. marriage
b. taxes
c. your grade on this test
d. death

14. Which of the following is **TRUE** about a child's understanding of death?

a. Preschoolers understand that death is permanent.
b. Preschoolers may not understand that death is universal.
c. Children understand that death is nonfunctional by the age of seven.
d. All of these options are true.

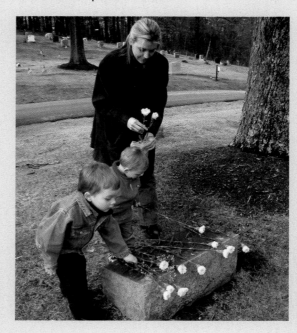

15. Which of the following is the correct sequence for Kübler-Ross's stage theory of dying?

a. denial; anger; bargaining; depression; acceptance
b. anger; denial; depression; bargaining; acceptance
c. denial; bargaining; anger; depression; acceptance
d. bargaining; denial; depression; anger; acceptance

Motivation and Emotion

In 1974, Bangladeshi economics professor Muhammad Yunus lent $27 to a group of women weavers so they could buy their own materials, rather than borrowing from a "middleman" who charged exorbitant interest. Within a year, everyone had paid him back. Yunus went on to found a bank to extend very small loans to poor people. Since 1983, his Grameen Bank has lent billions of dollars, with no collateral, to some 17 million people worldwide – 97 percent of them women. Yunus's "microcredit" model has inspired many similar efforts. In 2006, Yunus and the Grameen Bank were awarded the Nobel Peace Prize for their pioneering work.

In accepting the prize, Yunus said that he plans to use his share of the $1.4 million award to establish a company to sell food to the poor for a nominal price, and to build an eye hospital for the poor in Bangladesh. What motivates such generosity? When you read about philanthropists like Yunus, are you awestruck, joyful, happy, sad, or even teary-eyed? Why? Research on motivation and emotion attempts to answer such questions.

Motivation and emotion are inseparable. Consider this example. If you saw your loved one in the arms of another, you might experience a wide variety of emotions (jealousy, fear, sadness, and anger), and your motives would determine how you would respond. Your desire for revenge might lead you to look for another partner, but your need for love and belonging might motivate you to explain your partner's behavior and to protect your relationship. As you will see in this chapter, considerable overlap exists not only between motivation and emotion but also among motivation, emotion, and most areas of psychology.

Theories and Concepts of Motivation

BIOLOGICAL THEORIES: LOOKING FOR INTERNAL "WHYS" OF BEHAVIOR

Many theories of **motivation** focus on inborn biological processes that control behavior. Among these biologically oriented theories are instinct, drive, and arousal theories.

In the earliest days of psychology, researchers like William McDougall (1908) proposed that humans had numerous "instincts," such as repulsion, curiosity, and self-assertiveness. Other researchers later added their favorite "instincts," and by the 1920s, the list of recognized instincts had become impossibly long. One researcher found listings for over 10,000 human instincts (Bernard, 1924).

In addition, the label *instinct* led to unscientific, circular explanations—"men are aggressive because they act aggressively" or "women are maternal because they act maternally." However, in recent years, a branch of biology called sociobiology has revived the case for **instincts** when strictly defined (FIGURE 11.1).

> **motivation**
> A set of factors that activate, direct, and maintain behavior, usually toward a goal.

> **instincts** Behavioral patterns that are unlearned, always expressed in the same way, and universal in a species.

Instincts FIGURE 11.1

A Instinctual behaviors are obvious in many animals. Bears hibernate, birds build nests, and salmon swim upstream to spawn.

B Sociobiologists such as Edward O. Wilson (1975, 1978) believe that humans also have instincts, like competition or aggression, that are genetically transmitted from one generation to the next.

Drive-reduction theory FIGURE 11.2

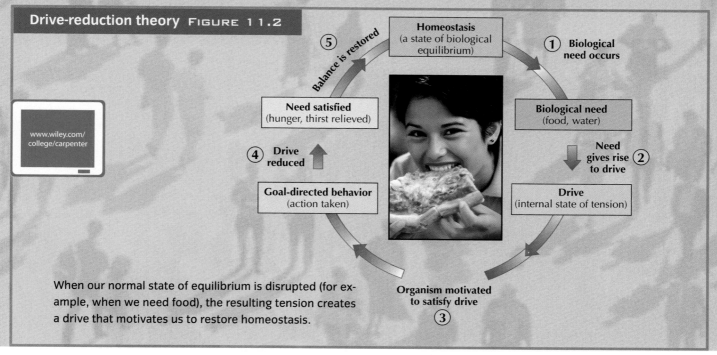

① Biological need occurs

Biological need (food, water)

② Need gives rise to drive

Drive (internal state of tension)

③ Organism motivated to satisfy drive

Goal-directed behavior (action taken)

④ Drive reduced

Need satisfied (hunger, thirst relieved)

⑤ Balance is restored

Homeostasis (a state of biological equilibrium)

www.wiley.com/college/carpenter

When our normal state of equilibrium is disrupted (for example, when we need food), the resulting tension creates a drive that motivates us to restore homeostasis.

In the 1930s, the concepts of drive and drive reduction began to replace the theory of instincts. According to **drive-reduction theory** (Hull, 1952), when biological needs (such as food, water, and oxygen) are unmet, a state of tension (known as a *drive*) is created, and the organism is motivated to reduce it. Drive-reduction theory is based largely on the biological concept of **homeostasis**, a term that literally means "standing still" (FIGURE 11.2).

■ **homeostasis** A body's tendency to maintain a relatively stable state, such as a constant internal temperature, blood sugar, oxygen level, or water balance.

In addition to our obvious biological needs, humans and other animals are innately curious and require a certain amount of novelty and complexity from the environment. This need for sensory stimulation begins shortly after birth and continues throughout the life span. Infants prefer complex versus simple visual stimuli, and adults pay more attention, and for a longer period of time, to complex and changing stimuli. Similarly, research shows that monkeys will work hard at tasks such as opening latches simply for the pleasure of satisfying their curiosity (Butler, 1954; Harlow, Harlow, & Meyer, 1950).

However, the need for arousal is not limitless. According to **arousal theory**, organisms are motivated to achieve and maintain an optimal level of arousal that maximizes their performance. Either too much or too little arousal diminishes performance (FIGURE 11.3).

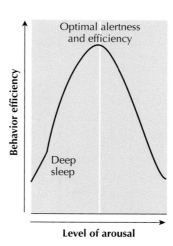

Optimal level of arousal FIGURE 11.3

Our need for stimulation (the arousal motive) suggests that behavior efficiency increases as we move from deep sleep to increased alertness. However, once we pass the maximum level of arousal, our performance declines.

Sensation Seeking

What motivates people who go bungee jumping over deep canyons or white-water rafting down dangerous rivers? According to research, these "high-sensation seekers" may be biologically "prewired" to need a higher than usual level of stimulation (Zuckerman, 1979, 1994, 2004).

To sample the kinds of questions that are asked on tests for sensation seeking, circle the choice (**A** or **B**) that BEST describes you:

1. **A** I would like a job that requires a lot of traveling.
   B I would prefer a job in one location.
2. **A** I get bored seeing the same old faces.
   B I like the comfortable familiarity of everyday friends.
3. **A** The most important goal of life is to live it to the fullest and experience as much as possible.
   B The most important goal of life is to find peace and happiness.
4. **A** I would like to try parachute jumping.
   B I would never want to try jumping out of a plane, with or without a parachute.
5. **A** I prefer people who are emotionally expressive even if they are a bit unstable.
   B I prefer people who are calm and even-tempered.

Source: Zuckerman, M. (1978, February). The search for high sensation, *Psychology Today*, pp. 38–46. Copyright © 1978 by the American Psychological Association. Reprinted by permission.

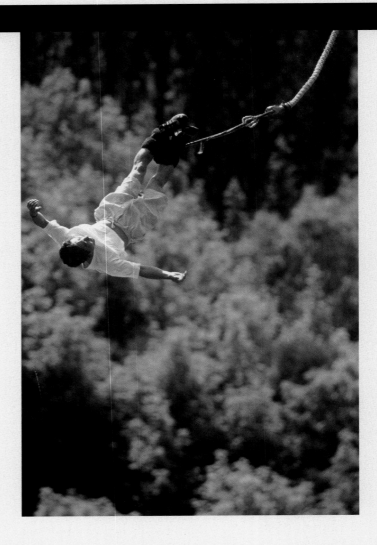

Research suggests that four distinct factors characterize sensation seeking (Diehm & Armatas, 2004; Johnson & Cropsey, 2000; Zuckerman, 2004):

1. Thrill and adventure seeking (skydiving, driving fast, or trying to beat a train)
2. Experience seeking (travel, unusual friends, drug experimentation)
3. Disinhibition ("letting loose")
4. Susceptibility to boredom (lower tolerance for repetition and sameness)

Being very high or very low in sensation seeking might cause problems in relationships with individuals who score toward the other extreme. This is true not just between partners or spouses but also between parent and child and therapist and patient. There might also be job difficulties for high-sensation seekers in routine clerical or assembly line jobs or for low-sensation seekers in highly challenging and variable occupations.

PSYCHOSOCIAL THEORIES: INCENTIVES AND COGNITIONS

Instinct and drive-reduction theories explain some motivations but not all. For example, why do we continue to eat after our biological need is completely satisfied? Or why does someone work overtime when his or her salary is sufficient to meet all basic biological needs? These questions are better answered by psychosocial theories that emphasize incentives and cognition.

Incentive theory holds that external stimuli motivate people to act to obtain desirable goals or to avoid undesirable events. People initially eat because their hunger "pushes" them, but they continue to eat because the sight of apple pie or ice cream "pulls" them.

According to *cognitive theories*, motivation is directly affected by **attributions**, or how we interpret or think about our own and others' actions. If you receive a high grade in your psychology course, for example, you can interpret that grade in several ways. You earned it because you really studied. You "lucked out." Or the textbook was exceptionally interesting and helpful (our preference!). People who attribute their successes to personal ability and effort tend to work harder toward their goals than people who attribute their successes to luck (Cheung & Rudowicz, 2003; Meltzer, 2004; Weiner, 1972, 1982).

Expectancies, or what we believe will happen, are also important to motivation (Haugen, Ommundsen, & Lund, 2004; Sirin & Rogers-Sirin, 2004) (**FIGURE 11.4**). If you anticipate that you will receive a promotion at work, you're more likely to work overtime for no pay than if you expect no promotion.

BIOPSYCHOSOCIAL THEORIES: INTERACTIONISM ONCE AGAIN

Research in psychology generally emphasizes either biological or psychosocial factors (nature or nurture). But in the final analysis, biopsychosocial factors almost always provide the best explanation. Theories of motivation are no exception. One researcher who recognized this was Abraham Maslow (1954, 1970, 1999). Maslow believed that we all have numerous needs that compete for fulfillment but that some needs are more important than others. For example, your need for food and shelter is generally more important than college

Motivated FIGURE 11.4

How might these students' attributions and expectancies affect their motivation to master a new language?

Maslow's hierarchy of needs FIGURE 11.5

As a humanistic psychologist, Maslow also believed that we all have a compelling need to "move up"— to grow, improve ourselves, and ultimately become "self-actualized."

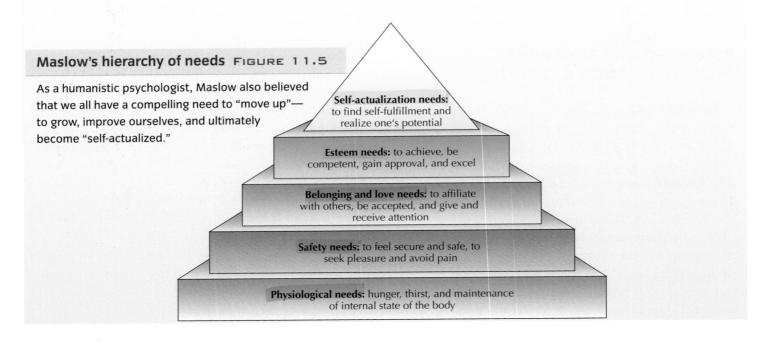

Self-actualization needs: to find self-fulfillment and realize one's potential

Esteem needs: to achieve, be competent, gain approval, and excel

Belonging and love needs: to affiliate with others, be accepted, and give and receive attention

Safety needs: to feel secure and safe, to seek pleasure and avoid pain

Physiological needs: hunger, thirst, and maintenance of internal state of the body

---

hierarchy of needs Maslow's theory of motivation that some motives (such as physiological and safety needs) must be met before going on to higher needs (such as belonging and self-actualization).

grades. Maslow's **hierarchy of needs** prioritizes needs, with survival needs at the bottom (needs that must be met before others) and social, spiritual needs at the top (FIGURE 11.5).

Maslow's hierarchy of needs seems intuitively correct—a starving person would first look for food, then love and friendship, and then self-esteem. This prioritizing and the concept of **self-actualization** are important contributions to the study of motivation (Frick, 2000; Harper, Harper, & Stills, 2003). But critics argue that parts of Maslow's theory are poorly researched and biased toward Western individualism. Critics also note that people sometimes seek to satisfy higher-level needs even when their lower-level needs have not been met (Cullen & Gotell, 2002; Hanley & Abell, 2002; Neher, 1991) (FIGURE 11.6).

Bypassing basic needs FIGURE 11.6

What higher-level needs do these people seem to be trying to fulfill? Does it appear that their lower-level needs have been met?

CONCEPT CHECK STOP

Why are modern sociobiological theories of instincts more scientifically useful than older "instinct" theories?

What four factors are associated with sensation seeking?

What criticisms have been made of Maslow's biopsychosocial (hierarchy of needs) theory?

Motivation and Behavior

LEARNING OBJECTIVES

Describe how internal (biological) and external (psychosocial) factors direct hunger and eating as well as how they are involved in serious eating disorders.

Explain why some people are more highly achievement motivated than others.

Summarize what happens to the human body during sexual activity.

Explain why providing extrinsic rewards undermines intrinsic motivation and when it does not.

Why do people put themselves in dangerous situations? Why do salmon swim upstream to spawn? Behavior results from many motives. For example, we discuss the need for sleep in Chapter 5, and we look at aggression, altruism, and interpersonal attraction in Chapter 15. Here, we will focus on three basic motives: hunger and eating, achievement, and sexuality. Then we'll turn to a discussion of how different kinds of motivation affect our intrinsic interests and performance.

HUNGER AND EATING: MULTIPLE FACTORS

What motivates hunger? Is it your growling stomach? Or is it the sight of a juicy hamburger or the smell of a freshly baked cinnamon roll?

The stomach Early hunger researchers believed that the stomach controls hunger, contracting to send hunger signals when it is empty. Today, we know that it's more complicated. As dieters who drink lots of water to keep their stomachs feeling full have been disappointed to discover, sensory input from an empty stomach is not essential for feeling hungry. In fact, humans and nonhuman animals without stomachs continue to experience hunger.

However, there is a connection between the stomach and feeling hungry. Receptors in the stomach and intestines detect levels of nutrients, and specialized pressure receptors in the stomach walls signal feelings of emptiness or **satiety** (fullness, or satiation). The stomach and other parts of the gastrointestinal tract also release chemical signals that play a role in hunger (Donini, Savina, & Cannella, 2003; Woods et al., 2000).

Biochemistry Like the stomach, the brain and other parts of the body produce numerous neurotransmitters, hormones, enzymes, and other chemicals that affect hunger and satiety (e.g., Chapelot et al., 2004; Hirosumi et al., 2002; Monteleone et al., 2003; Sawaya et al., 2001; Schwartz & Morton, 2002; Shimizu-Albergine, Ippolito, & Beavo, 2001). Research in this area is complex because of the large number of known (and unknown) bodily chemicals and the interactions among them. It's unlikely that any one chemical controls our hunger and eating. Other internal factors, such as **thermogenesis** (the heat generated in response to food ingestion), also play a role (Subramanian & Vollmer, 2002).

The brain In addition to its chemical signals, particular brain structures also influence hunger and eating. Let's look at the hypothalamus, which helps regulate eating, drinking, and body temperature.

Early research suggested that one area of the hypothalamus, the lateral hypothalamus (LH), stimulates eating, while another area, the ventromedial hypothalamus (VMH), creates feelings of satiation, signaling the animal to stop eating. When the VMH area was destroyed in rats, researchers found that the rats overate to the point of extreme obesity (**FIGURE 11.7** on the following page). In contrast, when the LH area was destroyed, the animals starved to death if they were not force-fed.

How the brain affects eating FIGURE 11.7

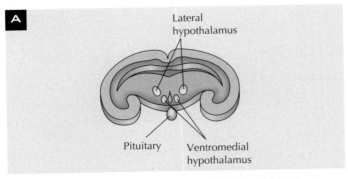

A

Lateral
hypothalamus

Pituitary   Ventromedial
hypothalamus

◀ A This diagram shows a section of a rat's brain, including the ventromedial hypothalamus (VMH) and the lateral hypothalamus (LH).

B

◀ B The rat on the right is of normal weight. In contrast, the ventromedial area of the hypothalamus of the rat on the left was destroyed, which led to the tripling of its body weight.

Later research, however, showed that the LH and VMH areas are not simple on-off switches for eating. For example, lesions (damage) to the VMH make animals picky eaters—they reject food that doesn't taste good. The lesions also increase insulin secretion, which may cause overeating (Challem et al., 2000). Today, researchers know that the hypothalamus plays an important role in hunger and eating, but it is not the brain's "eating center." In fact, hunger and eating, like virtually all behavior, are influenced by numerous neural circuits that run throughout the brain (Berthoud, 2002).

Psychosocial factors

The internal motivations for hunger we've discussed (stomach, biochemistry, the brain) are powerful. But psychosocial factors—for example, spying a dessert cart or a McDonald's billboard, or even simply noticing that it's about lunchtime—can be equally important triggers for hunger and eating.

Another important psychosocial influence on when, what, where, and why we eat is cultural conditioning. North Americans, for example, tend to eat dinner at around 6 P.M., whereas people in Spain and South America tend to eat around 10 P.M. When it comes to *what* we eat, have you ever eaten rat, dog, or horse meat? If you are a typical North American, this might sound repulsive to you, yet most Hindus would feel a similar revulsion at the thought of eating meat from cows.

Eating disorders

As you can see, hunger and eating are motivated by numerous biological, psychological, and social factors. These same biopsychosocial forces also play a role in three serious eating disorders: obesity, anorexia nervosa, and bulimia nervosa.

Obesity has reached epidemic proportions in the United States and other developed nations. Well over half of all adults in the United States meet the current criteria for clinical **obesity** (having a body weight 15 percent or more above the ideal for one's height and age). Each year, billions of dollars are spent treating serious and life-threatening medical problems related to obesity, and consumers spend billions more on largely ineffective weight-loss products and services.

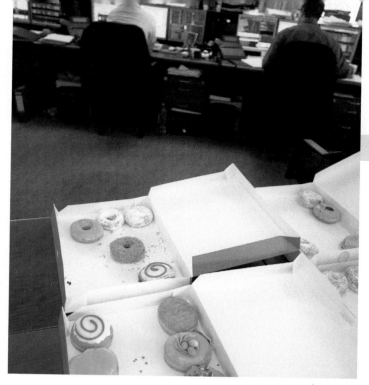

A fattening environment FIGURE 11.8

For Americans, controlling weight is a particularly difficult task. We are among the most sedentary people of all nations, and we've become accustomed to "supersized" cheeseburgers, "Big Gulp" drinks, and huge servings of dessert (Lawrence, 2002; Morgan et al., 2002; Smith, Orleans, & Jenkins, 2004). We've also learned that we should eat three meals a day (whether we're hungry or not); that "tasty" food requires lots of salt, sugar, and fat; and that food is an essential part of all social gatherings. To successfully lose (and maintain) weight, we must make permanent lifestyle changes regarding the amount and types of foods we eat and when we eat them. Can you see how our everday environments, such as in the workplace scene here, might prevent a person from making healthy lifestyle changes?

Given our culture's preference for thinness and our prejudice against fat people, why are so many Americans overweight? The simple answer is overeating and not enough exercise (FIGURE 11.8). However, we all know some people who can eat anything they want and still not add pounds. This may be a result of their ability to burn calories more effectively (thermogenesis), a higher metabolic rate, or other factors. Adoption and twin studies indicate that genes also play a role. Heritability for obesity ranges between 30 and 70 percent (Hewitt, 1977; Schmidt, 2004). Unfortunately, identifying the genes for obesity is difficult. Researchers have isolated over 2,000 genes that contribute to normal and abnormal weight (Camarena et al., 2004; Costa, Brennen, & Hochgeschwender, 2002; Devlin, Yanovski, & Wilson, 2000).

Interestingly, as obesity has reached epidemic proportions, we've seen a similar rise in two other eating disorders—**anorexia nervosa** and **bulimia nervosa**. Both disorders are serious and chronic conditions that require treatment.

More than 50 percent of women in Western industrialized countries show some signs of an eating disorder, and approximately 2 percent meet the clinical criteria for anorexia nervosa or bulimia nervosa (Porzelius, Dinsmore, & Staffelbach, 2001). These disorders also are found at all socioeconomic levels. A few men also develop eating disorders, although the incidence is rarer (Barry, Grilo, & Masheb, 2002; Coombs, 2004; Jacobi et al., 2004).

Anorexia nervosa is characterized by an overwhelming fear of becoming obese, a disturbed body image, a need for control, and the use of dangerous weight-loss measures (FIGURE 11.9 on the following page). The resulting extreme malnutrition often leads to emaciatation, osteoporosis, bone fractures, interruption of menstruation, and loss of brain tissue. A significant percentage of individuals with anorexia nervosa ultimately die of the disorder (Gordon, 2000; Werth et al., 2003).

anorexia nervosa An eating disorder characterized by a severe loss of weight resulting from self-imposed starvation and an obsessive fear of obesity.

bulimia nervosa An eating disorder involving the consumption of large quantities of food (bingeing), followed by vomiting, extreme exercise, or laxative use (purging).

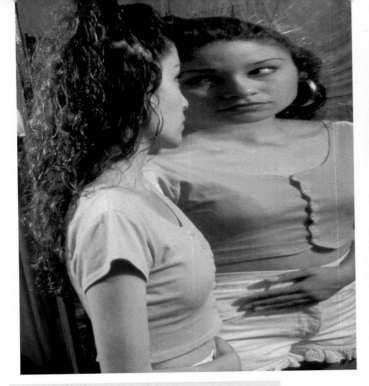

Distorted body image FIGURE 11.9

In anorexia nervosa, body image is so distorted that even a skeletal, emaciated body is perceived as fat. Many people with anorexia nervosa not only refuse to eat but also take up extreme exercise regimens—hours of cycling or running or constant walking.

The media's role in eating disorders FIGURE 11.10

Can you explain why popular movie and television stars' extreme thinness may unintentionally contribute to eating disorders?

Occasionally, the person suffering from anorexia nervosa succumbs to the desire to eat and gorges on food, then vomits or takes laxatives. However, this type of bingeing and purging is more characteristic of bulimia nervosa. Individuals with this disorder also show impulsivity in other areas, sometimes engaging in excessive shopping, alcohol abuse, or petty shoplifting (Bulik, Sullivan, & Kendler, 2002; Kane et al., 2004; Steiger et al., 2003). The vomiting associated with bulimia nervosa causes dental damage, severe damage to the throat and stomach, cardiac arrhythmias, metabolic deficiencies, and serious digestive disorders.

There are many suspected causes of anorexia nervosa and bulimia nervosa. Some theories focus on physical causes, such as hypothalamic disorders, low levels of various neurotransmitters, and genetic or hormonal disorders. Other theories emphasize psychosocial factors, such as a need for perfection, a perceived loss of control, being teased about body weight, destructive thought patterns, depression, dysfunctional families, distorted body image, and sexual abuse (e.g.,

Coombs, 2004; Jacobi et al., 2004; McCabe & Ricciardelli, 2003; Neumark-Sztainer et al., 2003).

Cultural perceptions and stereotypes about weight and eating also play important roles in eating disorders (Davis, Dionne, & Shuster, 2001; Dorian & Garfinkel, 2002). For instance, Asian and African Americans report fewer eating and dieting disorders and greater body satisfaction than do European Americans (Akan & Grilo, 1995), and Mexican students report less concern about their own weight and more acceptance of obese people than do other North American students (Crandall & Martinez, 1996).

Although social pressures for thinness certainly contribute to eating disorders (FIGURE 11.10), anorexia nervosa has also been found in nonindustrialized areas like the Caribbean island of Curaçao (Hoek et al., 2005). On that island, being overweight is socially acceptable, and the average woman is considerably heavier than the average woman in North America. However, some women there still have anorexia nervosa. This research suggests that both culture and biology help explain eating disorders.

ACHIEVEMENT: THE NEED FOR SUCCESS

Do you wonder what motivates Olympic athletes to work so hard just for a gold medal? Or what about someone like Thomas Edison, who patented more than 1,000 inventions? What drives some people to high achievement?

The key to understanding what motivates high-achieving individuals lies in what psychologist Henry Murray (1938) identified as a high need for achievement (nAch), or **achievement motivation**.

Several traits distinguish people who have high achievement motivation (McClelland, 1958, 1987, 1993; Mueller & Dweck, 1998; Wigfield & Eccles, 2000):

■ **achievement motivation**
A desire to excel, especially in competition with others.

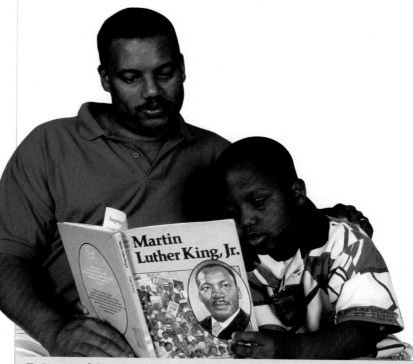

- *Preference for moderately difficult tasks.* People high in nAch (need for achievement) avoid tasks that are too easy because they offer little challenge or satisfaction. They also avoid extremely difficult tasks because the probability of success is too low.

- *Competitiveness.* High achievement–oriented people are more attracted to careers and tasks that involve competition and an opportunity to excel.

- *Preference for clear goals with competent feedback.* High achievement-oriented people tend to prefer tasks with a clear outcome and situations in which they can receive feedback on their performance. They also prefer criticism from a harsh but competent evaluator to that which comes from one who is friendlier but less competent.

- *Responsibility.* People with high nAch prefer being personally responsible for a project so that they can feel satisfied when the task is well done.

- *Persistence.* High achievement-oriented people are more likely to persist at a task when it becomes difficult. In one study, 47 percent of high nAch individuals persisted on an "unsolvable task" until time was called, compared with only 2 percent of people with low nAch.

- *More accomplished.* People who have high nAch scores do better than others on exams, earn better grades in school, and excel in their chosen professions.

Achievement orientation appears to be largely learned in early childhood, primarily through interactions with parents (**FIGURE 11.11**). Highly motivated children tend to have parents who encourage independence and frequently reward successes (Maehr & Urdan, 2000). Our cultural values also affect achievement needs (Lubinski & Benbow, 2000).

Future achiever FIGURE 11.11

A study by Richard de Charms and Gerald Moeller (1962) found a significant correlation between the achievement themes in children's literature and the industrial accomplishments of various countries.

SEXUALITY: THE WORLD'S MOST POWERFUL MOTIVE?

Obviously, there is strong motivation to engage in sexual behavior: it's essential for the survival of our species, and it's also pleasurable. But **sexuality** includes much more than reproduction. For most humans (and some other animals), a sexual relationship fulfills many needs, including the need for connection, intimacy, pleasure, and the release of sexual tension.

William Masters and Virginia Johnson (1966) were the first to conduct laboratory studies on what happens to the human body during sexual activity. They attached recording devices to male and female volunteers' and monitored or filmed their physical responses as they moved from nonarousal to orgasm and back to nonarousal. They labeled the bodily changes during this series of events a **sexual response cycle** (FIGURE 11.12).

Sexual Orientation Of course, an important part of people's sexuality is the question of to whom they are sexually attracted. What leads some people to be homosexual and others to be heterosexual? Unfortunately, the roots of **sexual orientation** are poorly understood. However, most studies suggest that genetics and biology play the dominant role (Bailey, Dunne, & Martin, 2000; Cantor et al., 2002; Hamer & Copeland, 1999; Rahman & Wilson, 2003).

■ **sexual response cycle** Master's and Johnson's description of the four-stage bodily response to sexual arousal, which consists of excitement, plateau, orgasm, and resolution.

■ **sexual orientation** Primary erotic attraction toward members of the same sex (homosexual, gay or lesbian), both sexes (bisexual), or the other sex (heterosexual).

Process Diagram

Master's and Johnson's sexual response cycle FIGURE 11.12

Note that this simplified description does not account for individual variation and should not be used to judge what's "normal."

2. During the **plateau phase**, physiological and sexual arousal continue at heightened levels. In the man, the penis becomes more engorged and erect while the testes swell and pull up closer to the body. In the woman, the clitoris pulls up under the clitoral hood and the entrance to the vagina contracts while the uterus rises slightly. This movement of the uterus causes the upper two-thirds of the vagina to balloon, or expand. As arousal reaches its peak, both sexes may experience a feeling that orgasm is imminent and inevitable.

3. The **orgasm phase** involves a highly intense and pleasurable release of tension. In the woman, muscles around the vagina squeeze the vaginal walls in and out and the uterus pulsates. Muscles at the base of the penis contract in the man, causing ejaculation, the discharge of semen or seminal fluid.

1. The **excitement phase** can last for minutes or hours. Arousal is initiated through touching, fantasy, or erotic stimuli. Heart rate and respiration increase and increased blood flow to the region causes penile or clitoral erection, and vaginal lubrication in women. In both men and women, the nipples may become erect, and both may experience a sex flush (reddening of the upper torso and face).

4. Physiological responses gradually return to normal during the **resolution phase**. After one orgasm, most men enter a **refractory period**, during which further excitement to orgasm is considered impossible. Many women (and some men), however, are capable of multiple orgasms in fairly rapid succession.

① Excitement ② Plateau ③ Orgasm ④ Resolution

Being raised by gay or lesbian (or cross-dressing) parents does *not* make a person gay. Nor is it true that gays and lesbians were seduced as children by adults of their own sex. Sons do not become gay because of domineering mothers or weak fathers, and daughters do not become lesbians because their fathers were their primary role models. Finally, having unfruitful or unhappy heterosexual experiences does not make a person gay.

Studies on identical and fraternal twins and adopted siblings found that if one identical twin was gay, 48 to 65 percent of the time so was the second twin (Kirk et al., 2000; Lynch, 2003). (Note that if the cause were totally genetic, the percentage would be 100.) The rate was 26 to 30 percent for fraternal twins and 6 to 11 percent for brothers and sisters who were adopted into one's family. Estimates of homosexuality in the general population run between 2 and 10 percent.

Research with rats and sheep hints that prenatal hormone levels may also affect fetal brain development and sexual orientation (Bagermihl, 1999). However, the effect of hormones on human fetal development is unknown. Furthermore, no well-controlled study has ever found a difference in adult hormone levels between heterosexuals and gays and lesbians (Banks & Gartrell, 1995; Le Vay, 2003).

Research has identified several widespread myths and misconceptions about homosexuality (Drescher, 2003; Fone, 2000; Lamberg, 1998; LeVay, 2003; Mitchell, 2002) (FIGURE 11.13). Although mental health authorities long ago discontinued labeling homosexuality as a mental illness, it continues to be a divisive societal issue (FIGURE 11.14). Gays, lesbians, bisexuals, and transgendered people often confront **sexual prejudice**,

sexual prejudice
Negative attitudes toward an individual because of her or his sexual orientation.

and many endure verbal and physical attacks; disrupted family and peer relationships; and high rates of anxiety, depression, and suicide (Jellison, McConnell, & Gabriel, 2004; Meyer, 2003; Mills et al., 2004; Rosenberg, 2003). Sexual prejudice is a socially reinforced phenomenon, not an individual pathology (as the older term *homophobia* implies).

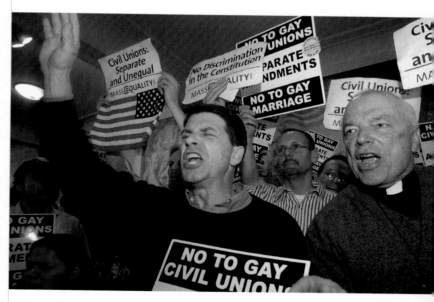

Gay marriage FIGURE 11.14

Debate, legislation, and judicial action surrounding gay marriage has brought attitudes about homosexuality into full public view.

INTRINSIC VERSUS EXTRINSIC MOTIVATION: IS ONE BETTER THAN THE OTHER?

■ extrinsic motivation
Motivation based on obvious external rewards or threats of punishment.

■ intrinsic motivation
Motivation resulting from personal enjoyment of a task or activity.

Should parents reward children for getting good grades? Many psychologists are concerned about the widespread practice of giving external, or extrinsic, rewards to motivate behavior (e.g., Deci, Koestner, & Ryan, 1999; Henderlong & Lepper, 2002; Reeve, 2005). They're concerned that providing such **extrinsic motivation** will seriously affect the individual's personal, **intrinsic motivation**. Participation in sports and hobbies, like swimming or playing guitar, is usually intrinsically mo-

tivated. Going to work is primarily extrinsically motivated.

Research has shown that people who are given extrinsic rewards (money, praise, or other incentives) for an intrinsically satisfying activity, such as watching TV, playing cards, or even engaging in sex, often lose enjoyment and interest and may decrease the time spent on the activity (Hennessey & Amabile, 1998; Kohn, 2000; Moneta & Siu, 2002).

Not all extrinsic motivation is bad (Covington & Mueller, 2001; Moneta & Siu, 2002). Extrinsic rewards are motivating if they are used to inform a person of superior performance or as a special "no strings attached" treat (Deci, 1995). In fact, they may intensify the desire to do well again. Thus, getting a raise or a gold medal can inform us and provide valuable feedback about our performance, which may increase enjoyment. But if rewards are used to control—for example, when parents give children money or privileges for achieving good grades—they inhibit intrinsic motivation (Eisenberger & Armeli, 1997; Eisenberger & Rhoades, 2002) (FIGURE 11.15).

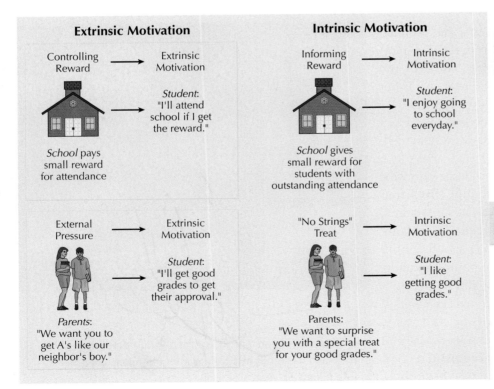

Extrinsic Motivation

Controlling Reward → Extrinsic Motivation

Student: "I'll attend school if I get the reward."

School pays small reward for attendance

External Pressure → Extrinsic Motivation

Student: "I'll get good grades to get their approval."

Parents: "We want you to get A's like our neighbor's boy."

Intrinsic Motivation

Informing Reward → Intrinsic Motivation

Student: "I enjoy going to school everyday."

School gives small reward for students with outstanding attendance

"No Strings" Treat → Intrinsic Motivation

Student: "I like getting good grades."

Parents: "We want to surprise you with a special treat for your good grades."

www.wiley.com/college/carpenter

Motivation is in the eye of the beholder FIGURE 11.15

Do rewards increase motivation, or are they seen as coercion or bribery? It depends. Note how a controlling reward and external pressure both lead to extrinsic motivation. However, an informing award and "no strings" treat produce intrinsic motivation.

Extrinsic versus Intrinsic Rewards

Have you ever noticed that for all the money and glory they receive, professional athletes often don't look like they're enjoying themselves very much? What's the problem? Why don't they appreciate how lucky they are to be able to make a living by playing games?

When we perform a task for no ulterior purpose, we use internal, personal reasons ("I like it"; "It's fun"). But when extrinsic rewards are added, the explanation shifts to external, impersonal reasons ("I did it for the money"; "I did it to please my parents"). This shift generally decreases enjoyment and hampers performance. This is as true for professional athletes as it is for anyone else.

One of the earliest experiments to demonstrate this effect was conducted with preschool children who liked to draw (Lepper, Greene, & Nisbett, 1973). As shown in the figure below, the researchers found that children who were given paper and markers and promised a reward for their drawings were subsequently less interested in drawing than children who were not given a reward or who were given an unexpected reward when they were done. Likewise, for professional athletes, what is a fun diversion for most people can easily become "just a job."

Percentage of free time spent drawing

(bar graph values: Promised certificate for drawing ≈ 8.5; Received reward after drawing ≈ 16.5; No reward ≈ 18)

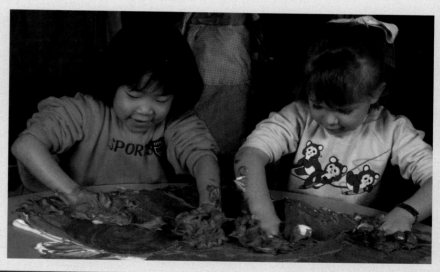

CONCEPT CHECK STOP

What are the biological, psychological, and social factors that motivate hunger and eating?

What traits characterize people with high achievement motivation?

What is sexual prejudice?

What impact does using extrinsic motivation to reward behavior have?

Theories and Concepts of Emotion

THREE COMPONENTS OF EMOTION

emotion A subjective feeling that includes arousal (heart pounding), cognitions (thoughts, values, and expectations), and expressions (frowns, smiles, and running).

Psychologists define and study **emotion** according to three basic components—physiological, cognitive, and behavioral.

The physiological (arousal) component

Internal physical changes occur in our bodies whenever we experience an emotion. Imagine walking alone on a dark street when someone jumps from behind a stack of boxes and starts running toward you. How would you respond? Like most people, you would undoubtedly interpret the situation as threatening and would run. Your predominant emotion, fear, would involve several physiological reactions, such as increased heart rate and blood pressure, perspiration, and goose bumps (pilo-erection). Such physiological reactions are controlled by certain brain structures and by the autonomic branch of the nervous system (ANS).

Our emotional experiences appear to result from important interactions between several areas of the brain, most particularly the *cerebral cortex* and *limbic system* (Langenecker et al., 2005; LeDoux, 2002; Panksepp, 2005). As we discuss in Chapter 2, the cerebral cortex, the outermost layer of the brain, serves as our body's ultimate control and information-processing center, including our ability to recognize and regulate our emotions.

The limbic system in the brain is also essential to our emotions. For example, several studies have shown that one area of the limbic system, the *amygdala*, plays a key role in emotion—especially fear.

Interestingly, emotional arousal sometimes occurs without conscious awareness. According to psychologist Joseph LeDoux (1996, 2002), when sensory inputs arrive in the *thalamus* (our brain's sensory switchboard), it sends separate messages up to the cortex (which "thinks" about the stimulus) and the amygdala (which immediately activates the body's alarm system).

As important as the brain is to emotion, it is the *autonomic nervous system* (ANS, Chapter 2) that produces the obvious signs of arousal. These largely automatic responses result from interconnections between the ANS and various glands and muscles (**FIGURE 11.16**).

The cognitive (thinking) component

> *There is nothing either good or bad,*
> *but thinking makes it so.*
> —Shakespeare, *Hamlet*

Emotional reactions are very individual: what you experience as intensely pleasurable may be boring or aversive to another. To study the cognitive (thought) component of emotions, psychologists typically use self-report techniques, such as paper-and-pencil tests, surveys, and interviews. However, people are sometimes unable or unwilling to accurately describe (or remember) their emotional states. For these reasons, our cognitions about our own and others' emotions are difficult to measure scientifically. This is why many researchers supplement participants' reports on their emotional experiences with methods that assess emotional

Sympathetic		Parasympathetic
Pupils dilated	**Eyes**	Pupils constricted
Dry	**Mouth**	Salivating
Goose bumps, perspiration	**Skin**	No goose bumps
Respiration increased	**Lungs**	Respiration normal
Increased rate	**Heart**	Decreased rate
Increased epinephrine and norepinephrine	**Adrenal glands**	Decreased epinephrine and norepinephrine
Decreased motility	**Digestion**	Increased motility

Emotion and the autonomic nervous system FIGURE 11.16

During emotional arousal, the sympathetic branch of the autonomic nervous system prepares the body for fight or flight. (The hormones epinephrine and norepinephrine keep the system under sympathetic control until the emergency is over.) The parasympathetic branch returns the body to a more relaxed state (homeostasis).

experience indirectly (for example, by measuring physiological response or behavior).

The behavioral (expressive) component

Emotional expression is a powerful form of communication, and facial expressions may be our most important form of emotional communication. Researchers have developed sensitive techniques to measure subtleties of feeling and to differentiate honest expressions from fake ones. Perhaps most interesting is the difference between the **social smile** and the **Duchenne smile** (named after French anatomist Duchenne de Boulogne, who first described it in 1862) (FIGURE 11.17). In a false, social smile, our voluntary cheek muscles are pulled back, but our eyes are unsmiling. Smiles of real pleasure, on the other hand, use the muscles not only around the cheeks but also around the eyes.

Duchenne smile FIGURE 11.17

People who show a Duchenne, or real, smile (A) and laughter elicit more positive responses from strangers and enjoy better interpersonal relationships and personal adjustment than those who use a social smile (B) (Keltner, Kring, & Bonanno, 1999; Prkachin & Silverman, 2002).

You've heard of IQ, the intelligence quotient, but what do you know about EI—emotional intelligence? Daniel Goleman's (1995, 2000) best-selling books have popularized the concept of **emotional intelligence (EI)**, based on original work by Peter Salovey and John Mayer (1990).

According to the theory, emotional intelligence involves knowing and managing one's emotions, empathizing with others, and maintaining satisfying relationships. In other words, an emotionally intelligent person successfully combines the three components of emotions (cognitive, physiological, and behavioral). Proponents of EI have suggested ways in which the close collaboration between emotion and reason may promote personal well-being and growth (Salovey et al., 2000).

Popular accounts such as Goleman's have suggested that traditional measures of human intelligence ignore a crucial range of abilities that characterize people who excel in real life: self-awareness, impulse control, persistence, zeal and self-motivation, empathy, and social deftness. Goleman also proposes that many societal problems, such as domestic abuse and youth violence, can be attributed to a low EI. Therefore, he argues, EI should be fostered in everyone.

Critics fear that a handy term like EI invites misuse, but their strongest reaction is to Goleman's proposals for teach-

ing EI. For example, Paul McHugh, director of psychiatry at Johns Hopkins University, suggests that Goleman is "presuming that someone has the key to the right emotions to be taught to children. We don't even know the right emotions to be taught to adults" (cited in Gibbs, 1995, p. 68).

FOUR MAJOR THEORIES OF EMOTION

Researchers generally agree on the three components of emotion (physiological, cognitive, and behavioral), but there is less agreement on how we become emotional. The major competing theories are the James-Lange, Cannon-Bard, facial-feedback, and Schachter's two-factor. As shown in **FIGURE 11.18**, each of these four theories emphasizes different sequences or aspects of the three elements (cognitions, arousal, and expression).

According to the **James-Lange theory** (originated by psychologist William James and later expanded by physiologist Carl Lange), emotions depend on feedback from our physiological arousal and behavioral expression (Figure 11.18A). In other words, as James wrote: "We feel sorry because we cry, angry because we strike, afraid because we tremble" (James, 1890). In short, arousal and expression cause emotion. Without arousal or expression, there is no emotion.

In contrast, the **Cannon-Bard theory** holds that all emotions are physiologically similar and that arousal, cognitions, and expression occur simultaneously. Importantly in this theory, arousal is not a necessary or even major factor in emotion. Walter Cannon (1927) and Philip Bard (1934) proposed that the thalamus sends simultaneous messages to both the ANS and the cerebral cortex (Figure 11.18B). Messages to the cortex produce the cognitive experience of emotion (such as fear), whereas messages to the autonomic nervous system produce physiological arousal and behavioral expressions (such as heart palpitations, running, widening eyes, and open mouth).

Cannon supported his position with several experiments in which animals were surgically prevented from experiencing physiological arousal. Yet these animals still showed emotion-like behaviors, such as growling and defensive postures (Cannon, Lewis, & Britton, 1927).

The third major explanation of emotion focuses on the expressive component of emotions. According to the **facial-feedback hypothesis** (Figure 11.18C), facial changes not only correlate with and intensify emotions but also cause or initiate the emotions themselves (Ceschi & Scherer, 2001; Keillor et al., 2002; Soussignan, 2002). Contractions of the various facial muscles send specific messages to the brain, identifying each basic emotion.

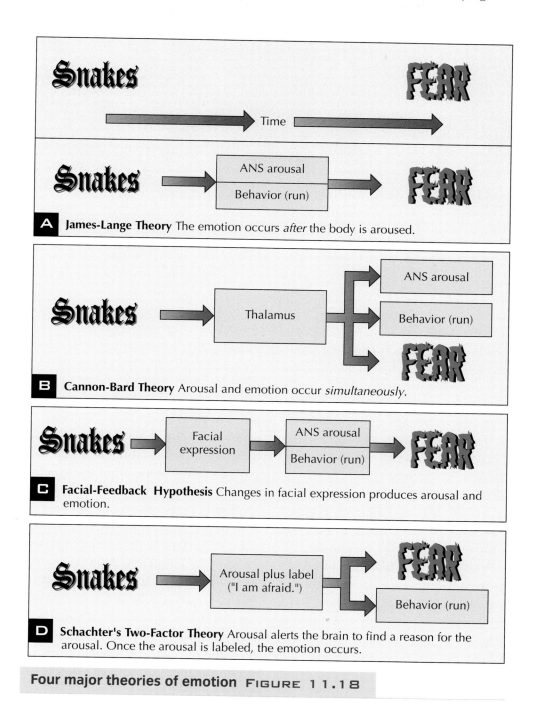

A James-Lange Theory The emotion occurs *after* the body is aroused.

B Cannon-Bard Theory Arousal and emotion occur *simultaneously*.

C Facial-Feedback Hypothesis Changes in facial expression produces arousal and emotion.

D Schachter's Two-Factor Theory Arousal alerts the brain to find a reason for the arousal. Once the arousal is labeled, the emotion occurs.

Four major theories of emotion FIGURE 11.18

Testing the facial-feedback hypothesis FIGURE 11.19

Hold a pen or pencil between your teeth with your mouth open. Spend about 30 seconds in this position. How do you feel? According to research, pleasant feelings are more likely when teeth are showing than when they are not. *Source:* Adapted from Starck, Martin, & Stepper, 1988.

The facial-feedback hypothesis also supports Charles Darwin's (1872) evolutionary theory that freely expressing an emotion intensifies it, whereas suppressing outward expression diminishes it (FIGURE 11.19). Interestingly, even watching another's facial expressions causes an automatic, reciprocal change in our own facial muscles (Dimberg & Thunberg, 1998). This automatic, matching response can occur even without the participant's attention or conscious awareness (Dimberg, Thunberg, & Elmehed, 2000).

Finally, according to psychologist Stanley **Schachter's two-factor theory** (FIGURE 11.20), we look to external rather than internal cues to understand our emotions. If we cry at a wedding, we interpret our emotion as happiness, but if we cry at a funeral, we label our emotion sadness.

Which theory is correct? Each has its limits and contributions to our understanding of emotion. For example, the James-Lange theory fails to acknowledge that physical arousal can occur without emotional experience (e.g., when we exercise). This theory also requires a distinctly different pattern of arousal for each emotion. Otherwise, how do we know whether we are sad, happy, or mad? Although brain imaging studies do show subtle differences among basic emotions (Lane et al., 1997; Levenson, 1992), most people are not aware of these slight variations. Thus, there must be other explanations for why we experience emotion.

The Cannon-Bard theory (that the cortex and autonomic nervous system receive simultaneous messages from the thalamus) has received some experimental support. Victims of spinal cord damage still experience emotions, often more intensely than before their injuries (Bermond et al., 1991). Instead of the thalamus, however, research shows that it is the limbic system, hypothalamus, and prefrontal cortex that are most important in emotional experience (Langenecker et al., 2005; Panksepp, 2005).

Research on the facial-feedback hypothesis has found that facial feedback does seem to contribute to the intensity of our subjective emotions and overall moods. Thus, if you want to change a bad mood or intensify a particularly good emotion, adopt the appropriate facial expression. In other words, "fake it 'til you make it."

Finally, Schachter's two-factor theory emphasizes the importance of cognitive processes in emotions, but his findings have been criticized. For example, as

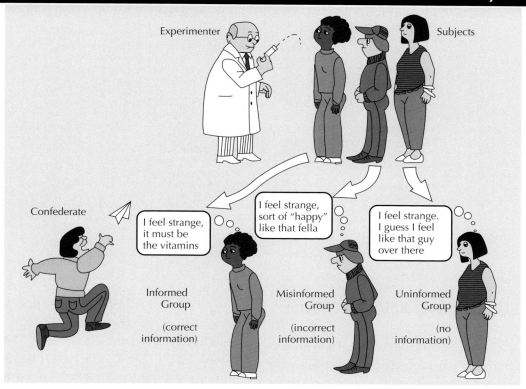

In Schachter and Singer's classic study (1962), participants were given shots of epinephrine and told it was a type of vitamin. One group of participants was correctly informed about the expected effects (hand tremors, excitement, and heart palpitations). A second group was misinformed and told to expect itching, numbness, and headache. A third group was told nothing about the possible effects.

Following the injection, each participant was placed in a room with a confederate (a "stooge" who was part of the experiment but who pretended to be a fellow volunteer) who acted either happy or unhappy.

The results showed that participants who lacked an appropriate cognitive label for their emotional arousal (the misinformed and uninformed groups) tended to look to the situation for an explanation. Thus, those placed with a happy confederate became happy, whereas those with an unhappy confederate became unhappy. Participants in the correctly informed group knew their physiological arousal was the result of the shot, so their emotions were generally unaffected by the confederate.

mentioned earlier, some neural pathways involved in emotion bypass the cortex and go directly to the limbic system. This and other evidence suggests that emotion is not simply the labeling of arousal (Dimberg, Thunberg, & Elmehed, 2000; LeDoux, 1996b, 2002; Mineka & Oehman, 2002).

The basic human emotions TABLE 11.1

Carroll Izard	Paul Ekman and Wallace Friesen	Robert Plutchik	Silvan Tomkins
Fear	Fear	Fear	Fear
Anger	Anger	Anger	Anger
Disgust	Disgust	Disgust	Disgust
Surprise	Surprise	Surprise	Surprise
Joy	Happiness	Joy	Enjoyment
Shame	—	—	Shame
Contempt	Contempt	—	Contempt
Sadness	Sadness	Sadness	—
Interest	—	Anticipation	Interest
Guilt	—	—	—
—	—	Acceptance	—
—	—	—	Distress

CULTURE, EVOLUTION, AND EMOTION

Are emotions the same across all cultures? Given the seemingly vast array of emotions within our own culture, it may surprise you to learn that some researchers believe that all our feelings can be condensed into 7 to 10 culturally universal emotions (TABLE 11.1). These researchers hold that other emotions, such as love, are simply combinations of primary emotions with variations in intensity (FIGURE 11.21).

Some research indicates that people in all cultures express and recognize the basic emotions in essentially the same way (Biehl et al., 1997; Buck, 1984; Ekman, 1993, 2004; Matsumoto, 1992, 2000; Matsumoto & Kupperbusch, 2001) (FIGURE 11.22).

From an evolutionary perspective, the idea of universal facial expressions makes adaptive sense because they signal others about our current emotional state (Ekman & Keltner, 1997). Charles Darwin first advanced the evolutionary theory of emotion in 1872. He

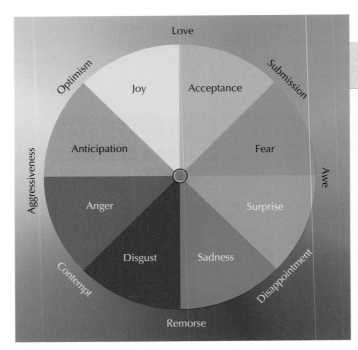

Plutchik's wheel of emotions
FIGURE 11.21

Robert Plutchik (1984, 1994, 2000) suggested that primary emotions (inner circle) combine to form secondary emotions (located outside the circle). Plutchik also found that emotions that lie next to each other are more alike than those that are farther apart.

Universal emotional expressions FIGURE 11.22

People from very different cultures display remarkably similar facial expressions when experiencing particular emotions. Furthermore, people everywhere can reliably identify at least six basic emotions: happiness, surprise, anger, sadness, fear, and disgust.

proposed that expression of emotions evolved in different species as a part of survival and natural selection. For example, fear helps animals avoid danger, whereas expressions of anger and aggression are useful when fighting for mates or resources. Modern evolutionary theory suggests that basic emotions originate in the *limbic system*, which developed earlier than higher brain areas responsible for thought (the cortex).

Studies with infants provide further support for an evolutionary basis for emotions. For example, infants only a few hours old show distinct expressions of emotion that closely match adult facial expressions (Field et al., 1982). And all infants, even those who are born deaf and blind, show similar facial expressions in similar situations (Eibl-Eibesfeldt, 1980; Feldman, 1982). In addition, a recent study showed that families may have characteristic facial expressions, shared even by family members who have been blind from birth (Peleg et al., 2006) (FIGURE 11.23). This collective evidence points to a strong biological, evolutionary basis for emotional expression and decoding.

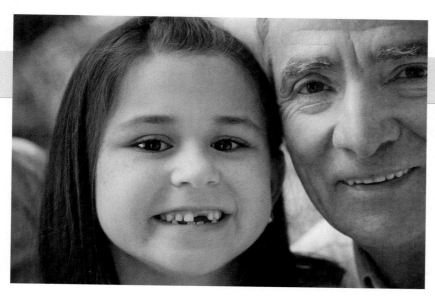

Family similarities in facial expression FIGURE 11.23

Recent research hints that some expressions may be inherited and under genetic control.

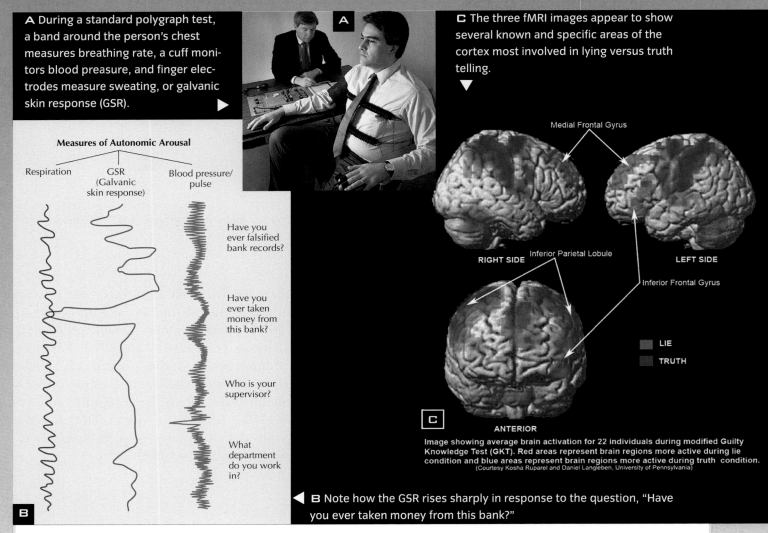

A During a standard polygraph test, a band around the person's chest measures breathing rate, a cuff monitors blood preasure, and finger electrodes measure sweating, or galvanic skin response (GSR). ▶

Measures of Autonomic Arousal

Respiration | GSR (Galvanic skin response) | Blood pressure/ pulse

Have you ever falsified bank records?

Have you ever taken money from this bank?

Who is your supervisor?

What department do you work in?

B

C The three fMRI images appear to show several known and specific areas of the cortex most involved in lying versus truth telling. ▼

Medial Frontal Gyrus

RIGHT SIDE   Inferior Parietal Lobule   LEFT SIDE

Inferior Frontal Gyrus

LIE

TRUTH

C

ANTERIOR

Image showing average brain activation for 22 individuals during modified Guilty Knowledge Test (GKT). Red areas represent brain regions more active during lie condition and blue areas represent brain regions more active during truth condition.
(Courtesy Kosha Ruparel and Daniel Langleben, University of Pennsylvania)

◀ **B** Note how the GSR rises sharply in response to the question, "Have you ever taken money from this bank?"

Polygraph Testing

Do you believe the polygraph tests are a good way to determine if someone is lying? Some people say the innocent have nothing to fear from a polygraph test. However, research suggests otherwise (DeClue, 2003; Faigman et al., 1997; Iacono & Lykken, 1997). In fact, although proponents claim that polygraph tests are 90 percent accurate or better, actual tests show error rates ranging between 25 and 75 percent.

Traditional polygraph tests are based on the theory that when people lie, they feel guilty and anxious. Special sensors supposedly detect these testable emotions by measuring sympathetic and parasympathetic nervous system responses (**A** and **B**). The problem is that lying is only loosely related to anxiety and guilt. Some people become nervous even when telling the truth, whereas others remain calm when deliberately lying. A polygraph cannot tell which emotion is being felt (nervousness, excitement, sexual arousal, etc.) or whether

a response is due to emotional arousal or something else. One study found that people could affect the outcome of a polygraph by about 50 percent simply by pressing their toes against the floor or biting their tongues (Honts & Kircher, 1994).

For these reasons, most judges and scientists have serious reservations about using polygraphs as lie detectors (DeClue, 2003). Scientific controversy and public concern led the U.S. Congress to pass a bill that severely restricts the use of polygraphs in the courts, in government, and in private industry.

In our post-9/11 world, you can see why tens of millions to hundreds of millions of dollars have been spent on new and improved lie-detection techniques (Kluger and Masters, 2006). Perhaps the most promising new technique is the use of brain scans like the function magnetic imaging (fMRI) (**C**).

Unfortunately, each of these new lie-detection techniques has serious shortcomings and unique problems. Researchers have questioned their reliability and validity, while civil libertarians and judicial scholars question their ethics and legalities.

So how do we explain cultural differences in emotions? Although we all seem to share similar facial expressions for some emotions, each culture has its own *display rules* governing how, when, and where to express emotions (Ekman, 1993, 2004; Matsumoto & Kupperbusch, 2001; Scherer & Wallbott, 1994). For instance, parents pass along their culture's specific display rules when they respond angrily to some emotions in their children, when they are sympathetic to others, and when they simply ignore an expression of emotion. Public physical contact is also governed by display rules. North Americans and Asians are less likely than people in other cultures to touch each other. For example, only the closest family and friends might hug in greeting or farewell. In contrast, Latin Americans and Middle Easterners often embrace and hold hands as a sign of casual friendship (Axtell, 1998).

CONCEPT CHECK **STOP**

HOW does the autonomic nervous system (ANS) respond to fearful stimuli?

What are the differences between the James-Lange theory and the Cannon-Bard theory of emotion?

Why do some researchers believe that the basic emotions are consistent from one culture to the next?

What is happening in this picture ?

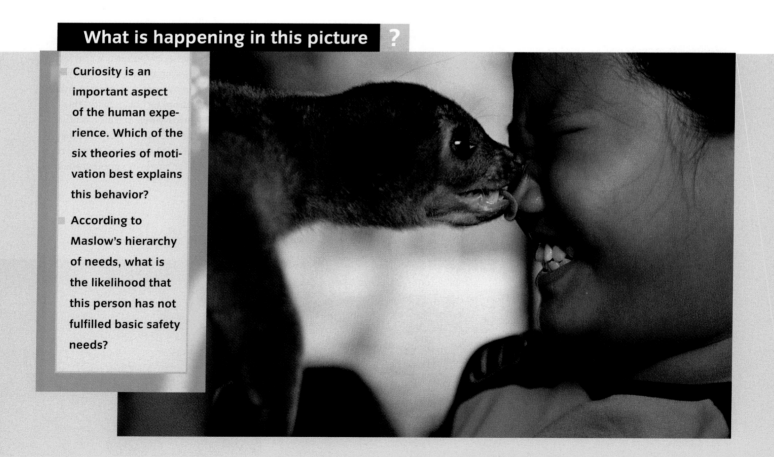

- Curiosity is an important aspect of the human experience. Which of the six theories of motivation best explains this behavior?

- According to Maslow's hierarchy of needs, what is the likelihood that this person has not fulfilled basic safety needs?

1 Theories and Concepts of Motivation

1. Because motivation and emotion are closely connected, they are often studied together.

2. The three major biological theories of **motivation** involve **instincts**, drives (produced by the body's need for **homeostasis**), and arousal, or need for novelty and complexity. For people who are high in sensation seeking, the need for stimulation is especially high.

3. Psychosocial theories emphasize the role of incentives, attributions, and expectations in cognition.

4. Maslow's **hierarchy of needs** theory takes a biopsychosocial approach. It prioritizes needs, with survival needs at the bottom and social and spiritual needs at the top. Although the theory has made important contributions, some critics argue that it is poorly researched and biased toward Western individualism.

2 Motivation and Behavior

1. Hunger is one of the strongest motivational drives, and both biological (stomach, biochemistry, the brain) and psychosocial (stimulus cues and cultural conditioning) factors affect hunger and eating. These same factors play a role in obesity, **anorexia nervosa**, and **bulimia nervosa**.

2. The key to understanding what motivates high-achieving individuals lies in a high need for achievement (nAch), or **achievement motivation**, which is learned in early childhood, primarily through interactions with parents.

3. The human motivation for sex is extremely strong. Masters and Johnson first studied and described the **sexual response cycle**, the series of physiological and sexual responses that occurs during sexual activity. Other sex research has focused on the roots of **sexual orientation**, and most studies suggest that genetics and biology play the dominant role. Sexual orientation remains a divisive issue, and gays, lesbians, bisexuals, and transgendered people often confront **sexual prejudice**.

4. Providing **extrinsic rewards** (money, praise, or other incentives) for an intrinsically satisfying activity can undermine people's enjoyment and interest (**intrinsic motivation**) for the activity. This is especially true when extrinsic motivation is used to control—for example, when parents give children money or privileges for achieving good grades.

3 Theories and Concepts of Emotion

1. **Emotion** has physiological (brain and autonomic nervous system), cognitive (thoughts), and behavioral (expressive) components.

2. The major theories of how the components of emotion interact are the James-Lange, Cannon-Bard, facial-feedback, and Schachter's two-factor. Each emphasizes different sequences or aspects of the three elements, and each has its limits and contributions to our understanding of emotion.

3. Some researchers believe that there are 7 to 10 basic, universal emotions that are shared by people of all cultures.

Some research indicates that people in all cultures express and recognize the basic emotions in essentially the same way, supporting the evolutionary theory of emotion. Studies with infants provide further support. Although we all seem to share similar facial expressions for some emotions, display rules for emotions (as well as for public physical contact, another aspect of the behavioral element of emotion) vary across cultures.

4. Polygraph tests attempt to detect lying by measuring physiological signs of guilt and anxiety. However, because lying is only loosely related to these emotions, most judges and scientists have serious reservations about using polygraphs as lie detectors.

KEY TERMS

- **motivation** p. 284
- **instincts** p. 284
- **drive-reduction theory** p. 285
- **homeostasis** p. 285
- **arousal theory** p. 285
- **incentive theory** p. 287
- **attributions** p. 287
- **expectancies** p. 287
- **hierarchy of needs** p. 288
- **self-actualization** p. 288
- **satiety** p. 289
- **thermogenesis** p. 289

- **obesity** p. 290
- **anorexia nervosa** p. 291
- **bulimia nervosa** p. 291
- **achievement motivation** p. 293
- **sexuality** p. 294
- **sexual response cycle** p. 294
- **sexual orientation** p. 294
- **excitement phase** p. 294
- **plateau phase** p. 294
- **orgasm phase** p. 294
- **resolution phase** p. 294
- **refractory period** p. 294

- **sexual prejudice** p. 295
- **extrinsic motivation** p. 296
- **intrinsic motivation** p. 296
- **emotion** p. 298
- **social smile** p. 299
- **Duchenne smile** p. 299
- **emotional intelligence (EI)** p. 300
- **James-Lange theory** p. 300
- **Cannon-Bard theory** p. 300
- **facial-feedback hypothesis** p. 301
- **Schachter's two-theory factor** p. 302

CRITICAL AND CREATIVE THINKING QUESTIONS

1. Like most Americans, you may have difficulty controlling your weight. Using information from this chapter, can you identify the factors or motives that best explain your experience?

2. How can you restructure elements of your personal, work, or school life to increase intrinsic versus extrinsic motivation?

3. If you were going out on a date with someone or applying for an important job, how might you use the four theories of emotion to increase the chances that things will go well?

4. Have you ever felt depressed after listening to a friend complain about his problems? How might the facial-feedback hypothesis explain this?

5. Why do you think people around the world experience and express the same basic emotions? What evolutionary advantages might help explain these similarities?

(Check your answers in Appendix B.)

1. *Motivation* is **BEST** defined as _____ .

   a. A set of factors that activate, direct, and maintain behavior, usually toward a goal

   b. the physiological and psychological arousal that occurs when a person really wants to achieve a goal

   c. what makes you do what you do

   d. the conscious and unconscious thoughts that focus a person's behaviors and emotions in the same direction toward a goal

2. The figure below illustrates the _____ theory, in which motivation decreases once homeostasis occurs.

   a. drive-induction

   b. heterogeneity

   c. drive-reduction

   d. biostability

3. _____ is the body's tendency to maintain a relatively stable state for internal processes.

   a. Homeostasis

   b. Heterogeneity

   c. Drive-induction

   d. Biostability

4. The _____ theory says people are "pulled" by external stimuli to act a certain way.

   a. cognitive

   b. incentive

   c. Maslow's hierarchy of needs

   d. incentive

5. The theory, illustrated in the figure below, that some motives have to be satisfied before a person can advance to fulfilling higher motives is based on _____ .

   a. Freud's psychosexual stages of development

   b. Kohlberg's moral stages of development

   c. Erikson's psychosocial stages of development

   d. Maslow's hierarchy of needs

6. The feeling of fullness you get once you have ingested enough food is called _____ .

   a. bloating

   b. a belly full

   c. a belly ache

   d. satiety

7. _____ involves the consumption of large quantities of food followed by self-induced vomiting, the use of laxatives, or extreme exercise.

   a. Anorexia nervosa

   b. The binge-purge syndrome

   c. Bulimia nervosa

   d. Pritikin dieting

8. According to your textbook, the desire to excel, especially in competition with others is known as _____ .

   a. drive

   b. instincts

   c. achievement motivation

   d. all of the above

9. Label the **CORRECT** sequence of events in Masters and Johnson's *sexual response cycle on the figure below*.

10. _____ is the term for negative attitudes toward someone based on his or her sexual orientation.

   a. Homophobia
   b. Sexual prejudice
   c. Heterosexism
   d. Sexual phobia

11. Extrinsic motivation is based on _____.

   a. the desire for rewards or threats of punishment
   b. the arousal motive
   c. the achievement motive
   d. all of these options

12. The three components of emotion are _____.

   a. cognitive, physiological, and behavioral
   b. perceiving, thinking, and acting
   c. positive, negative, and neutral
   d. active/passive, positive/negative, and direct/indirect

13. In a _____, as shown in the figure below, the cheek muscles are pulled back and the muscles around the eyes also contract.

   a. Duchenne smile
   b. Madonna smile
   c. Mona Lisa smile
   d. Da Vinci smile

14. According to _____, we look to external rather than internal cues to understand emotions.

   a. Cannon-Bard theory
   b. James-Lange theory
   c. the facial feedback hypothesis
   d. Schachter's two-factor theory

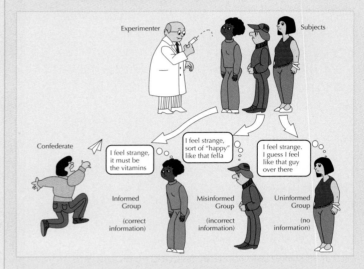

15. Researchers believe all our feelings can be condensed to _____ culturally universal emotions.

   a. 2 to 3
   b. 5 to 6
   c. 7 to 10
   d. 11 to 15

Personality

Consider the following personality description. How well does it describe you?

You have a strong need for other people to like and admire you. You tend to be critical of yourself. Although you have some personality weaknesses, you are generally able to compensate for them. At times, you have serious doubts about whether you have made the right decision or done the right thing.

—Adapted from Ulrich, Stachnik, & Stainton, 1963.

Does this sound like you? A high percentage of research participants who read a similar personality description reported that the description was "very accurate"—even after they were informed that it was a phony horoscope (Hyman, 1981). Other research shows that about three-quarters of adults read newspaper horoscopes and that many of them believe that these horoscopes were written especially for them (Halpern, 1998).

Why are such spurious personality assessments so popular? In part, it is because they seem to tap into our unique selves. However, the traits they supposedly reveal are characteristics that almost everyone shares. Do you know anyone who *doesn't* "have a strong need for other people to like and admire [them]"? The traits in horoscopes are also generally flattering, or at least neutral. In this chapter, rather than relying on these unscientific methods, we will focus on research-based methods used by psychologists to assess personality.

Unlike the pseudopsychologies offered in supermarket tabloids, newspaper horoscopes, and Chinese fortune cookies, the descriptions presented by personality researchers are based on empirical studies. In this chapter, we examine the five most prominent theories and findings in personality research and discuss the techniques that psychologists use to assess personality.

Cognitions   Behavior

Personality

Environment

Trait Theories

LEARNING OBJECTIVES

Explain how early trait theorists approached the study of personality.

Identify the "Big Five" personality traits.

Summarize the major critiques of trait theory.

Personality describes you as a person—how you are different from other people, and what patterns of behavior are typical of you. You might qualify as an "extrovert," for example, if you are talkative and outgoing most of the time. Or you might be considered "conscientious" if you are responsible and self-disciplined most of the time.

The terms we use to describe other people (and ourselves) are called **traits**. Trait theorists are interested in discovering which traits best describe people and in measuring the degree of variation in traits within individuals and among people.

personality
Relatively stable and enduring patterns of thoughts, feelings, and actions.

traits
Relatively stable and consistent characteristics that can be used to describe someone.

EARLY TRAIT THEORISTS: ALLPORT, CATTELL, AND EYSENCK

An early study of dictionary terms found almost 4,500 words that described personality traits (Allport & Odbert, 1936). Faced with this enormous list, Gordon Allport (1937) believed that the best way to understand personality was to arrange a person's unique personality traits into a hierarchy, with the most pervasive or important traits at the top.

Later psychologists reduced the list of possible personality traits using a statistical technique called **factor analysis**, in which large arrays of data are grouped into more basic units (factors). Raymond Cattell (1950, 1965, 1990) condensed the list of traits to 30 to 35 basic characteristics. Hans Eysenck (1967, 1982, 1990) reduced the list even further. He described personality as a relationship among three basic types of traits: extroversion-introversion (E), neuroticism (N), and psychoticism (P).

THE FIVE-FACTOR MODEL: FIVE BASIC PERSONALITY TRAITS

Factor analysis was also used to develop the most promising modern trait theory, the **five-factor model (FFM)** (McCrae & Costa, 1990, 1999; McCrae, Costa, Hrebickova, et al., 2004; McCrae, Costa, Martin, et al., 2004).

Combining previous research findings and the long list of possible personality traits, researchers discovered that five traits came up repeatedly, even when different tests were used.

These five major dimensions of personality are often dubbed the **Big Five**

five-factor model (FFM)
The trait theory that explains personality in terms of the "Big Five" model, which is composed of openness, conscientiousness, extroversion, agreeableness, and neuroticism.

(FIGURE 12.1). A handy way to remember the five factors is to note that the first letters of each of the five-factor model spell the word *ocean*. The Big Five are:

O *Openness.* People who rate high in this factor are original, imaginative, curious, open to new ideas, artistic, and interested in cultural pursuits. Low scorers tend to be conventional, down-to-earth, narrower in their interests, and not artistic. Interestingly, critical thinkers tend to score higher than others on this factor (Clifford, Boufal, & Kurtz, 2004).

C *Conscientiousness.* This factor ranges from responsible, self-disciplined, organized, and achieving at the high end to irresponsible, careless, impulsive, lazy, and undependable at the other.

E *Extroversion.* This factor contrasts people who are sociable, outgoing, talkative, fun loving, and affectionate at the high end with introverted individuals who tend to be withdrawn, quiet, passive, and reserved at the low end.

A *Agreeableness.* Individuals who score high in this factor are good-natured, warm, gentle, cooperative, trusting, and helpful, whereas low scorers are irritable, argumentative, ruthless, suspicious, uncooperative, and vindictive.

N *Neuroticism* (or emotional stability). People who score high in neuroticism are emotionally unstable and prone to insecurity, anxiety, guilt, worry, and moodiness. People at the other end are emotionally stable, calm, even-tempered, easygoing, and relaxed.

Dimensions of personality FIGURE 12.1

How do you think the individuals in these three photographs might score on each of the Big Five dimensions of personality?

Love and the "Big Five"

Using this figure, plot your personality profile by placing a dot on each line to indicate your degree of openness, conscientiousness, and so on. Do the same for a current, previous, or prospective love partner. How do your scores compare?

Now look at the Mate Preferences chart. David Buss and his colleagues (1989, 2003a, 2003b) surveyed more than 10,000 men and women from 37 countries and found a surprising level of agreement in the characteristics that men and women value in a mate. Moreover, most of the Big Five personality traits are found at the top of the list. Both men and women prefer dependability (conscientiousness), emotional stability (low neuroticism), pleasing disposition (agreeableness), and sociability (extroversion) to the alternatives. These findings may reflect an evolutionary advantage for people who are open, conscientious, extroverted, agreeable, and free of neuroses.

Big Five Traits	Low Scorers	High Scorers
1 Openness	Down-to-earth Uncreative Conventional Uncurious	Imaginative Creative Original Curious
2 Conscientiousness	Negligent Lazy Disorganized Late	Conscientious Hard-working Well-organized Punctual
3 Extroversion	Loner Quiet Passive Reserved	Joiner Talkative Active Affectionate
4 Agreeableness	Suspicious Critical Ruthless Irritable	Trusting Lenient Soft-hearted Good-natured
5 Neuroticism	Calm Even-tempered Comfortable Unemotional	Worried Temperamental Self-conscious Emotional

Mate selection around the world

Note how the top four traits, below, are the same for both men and women, as well as how closely their desired traits match those of the five-factor model (FFM). *Source:* Buss et al., "International Preferences in Selecting Mates." *Journal of Cross-Cultural Psychology*, 21, pp. 5–47, 1990. Sage Publications, Inc.

♂ What Men Want in a Mate	♀ What Women Want in a Mate
1. Mutual attraction — love	1. Mutual attraction — love
2. Dependable character	2. Dependable character
3. Emotional stability and maturity	3. Emotional stability and maturity
4. Pleasing disposition	4. Pleasing disposition
5. Good health	5. Education and intelligence
6. Education and intelligence	6. Sociability
7. Sociability	7. Good health
8. Desire for home and children	8. Desire for home and children
9. Refinement, neatness	9. Ambition and industriousness
10. Good looks	10. Refinement, neatness

EVALUATING TRAIT THEORIES

The five-factor model is the first to achieve the major goal of trait theory—to describe and organize personality characteristics using the fewest number of traits. Critics argue, however, that the great variation seen in personalities cannot be accounted for by only five traits and that the Big Five model fails to offer causal explanations for these traits (Friedman & Schustack, 2006; Funder, 2000; Monte & Sollod, 2003).

Critics maintain that, in general, trait theories are good at describing personality, but they have difficulty explaining why people develop these traits or why personality traits differ across cultures. For example, trait theories do not explain why people in cultures that are geographically close tend to have similar personalities or why Europeans and Americans tend to be higher in extroversion and openness to experience and lower in agreeableness than people in Asian and African cultures (Allik & McCrae, 2004).

In addition, some critics have faulted trait theories for their lack of specificity. Although trait theorists have documented a high level of personality stability after age 30 (**FIGURE 12.2**), they haven't identified which characteristics last a lifetime and which are most likely to change.

Finally, trait theorists have been criticized for ignoring the importance of situational and environmental effects on personality. In one example, psychologists Fred Rogosch and Dante Cicchetti (2004) found that abused and neglected children scored significantly lower in the traits of openness to experience, conscientiousness, and

Personality change over time FIGURE 12.2

Have you noticed how Madonna's public image and behavior have changed over time? Cross-cultural research has found that neuroticism, extroversion, and openness to experience tend to decline from adolescence to adulthood, whereas agreeableness and conscientiousness increase (McCrae, Costa, Hrebickova, et al., 2004). How would you explain these changes? Do you think they're good or bad?

agreeableness and higher in the trait of neuroticism than did children who were not maltreated. Unfortunately, these maladaptive personality traits create significant liabilities that may trouble these children throughout their lifetimes.

CONCEPT CHECK STOP

What is the purpose of factor analysis?

What dimensions of personality are central to the five-factor model?

What are some weaknesses of trait theory?

Psychoanalytic/Psychodynamic Theories

LEARNING OBJECTIVES

Identify Freud's most basic and controversial contributions to the study of personality.

Explain how Adler's, Jung's, and Horney's theories differ from Freud's thinking.

Explore the major criticisms of Freud's psychoanalytic theories.

I n contrast to trait theories that describe personality as it exists, psychoanalytic (or psychodynamic) theories of personality attempt to explain individual differences by examining how unconscious mental forces interplay with thoughts, feelings, and actions. The founding father of psychoanalytic theory is Sigmund Freud. We will examine Freud's theories in some detail and then briefly discuss three of his most influential followers.

FREUD'S PSYCHOANALYTIC THEORY: THE POWER OF THE UNCONSCIOUS

Who is the best-known figure in all of psychology? Most people immediately name Sigmund Freud, whose theories have been applied not only to psychology but also to anthropology, sociology, religion, medicine, art, and literature. Working from about 1890 until he died in 1939, Freud developed a theory of personality that has been one of the most influential—and most controversial—theories in all of science (Allen, 2006; Domhoff, 2004; Gay, 1999). Let's examine some of Freud's most basic and debatable concepts.

Freud called the mind the "psyche" and asserted that it contains three **levels of consciousness**, or awareness: the **conscious**, the **preconscious**, and the **unconscious** (FIGURE 12.3).

conscious Freud's term for thoughts or motives that a person is currently aware of or is remembering.

preconscious Freud's term for thoughts or motives that can be easily brought to mind.

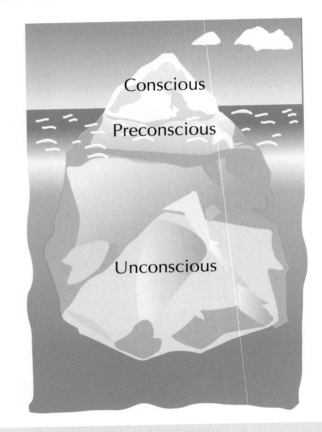

Freud's three levels of consciousness

FIGURE 12.3

Freud compared people's conscious awareness—thoughts, feelings, and actions that we are actively aware of—to the tip of an iceberg, open to easy inspection. Beneath the conscious realm, and the water's surface, is the larger preconscious. It includes mental activities that we can access with little extra effort. The large, deeply submerged base of the iceberg is like the unconscious mind, hidden from personal inspection. According to Freud, the unconscious stores our primitive, instinctual motives and anxiety-laden memories and emotions, which are prevented from entering the conscious mind.

Freud believed that most psychological disorders originate from repressed memories and instincts (sexual and aggressive) that are hidden in the unconscious. To treat these disorders, Freud developed *psychoanalysis* (Chapter 14).

In addition to proposing that the mind functions at three levels of consciousness, Freud also thought that personality was composed of three mental structures: the id, the ego, and the superego (**FIGURE 12.5**).

According to Freud, the **id** is made up of innate, biological instincts and urges. It is immature, impulsive, and irrational. The id is also totally unconscious and serves as the reservoir of mental energy. When its primitive drives build up, the id seeks immediate gratification to relieve the tension—a concept known as the **pleasure principle**.

As a child grows older, the second part of the psyche—the ego—develops. The **ego** is responsible for planning, problem solving, reasoning, and controlling the potentially destructive energy of the id. In Freud's system, the ego corresponds to the *self*—our conscious identity of ourselves as persons.

One of the ego's tasks is to channel and release the id's energy in ways that are compatible with the external world. Thus, the ego is responsible for delaying gratification when necessary. Contrary to the id's pleasure principle, the ego operates on the **reality principle** because it can understand and deal with objects and events in the "real world."

The final part of the psyche to develop is the **superego**, a set of ethical rules for behavior. The superego develops from internalized parental and societal standards. It constantly strives for perfection and is therefore as unrealistic as the id. Some Freudian followers have suggested that the superego operates on the **morality principle** because violating its rules results in feelings of guilt.

When the ego fails to satisfy both the id and the superego, anxiety slips into conscious awareness. Because anxiety is uncomfortable, people avoid it through **defense mechanisms**. For example, an alcoholic who

"Good morning, beheaded—uh, I mean beloved."

Freudian slips FIGURE 12.4

Freud believed that a small slip of the tongue (known as a "Freudian slip") can reflect unconscious feelings that we normally keep hidden.

uses his paycheck to buy drinks (a message from the id) may feel very guilty (a response from the superego). He may reduce this conflict by telling himself that he deserves a drink because he works so hard. This is an example of the defense mechanism **rationalization**.

defense mechanisms
In Freudian theory, the ego's protective method of reducing anxiety by distorting reality.

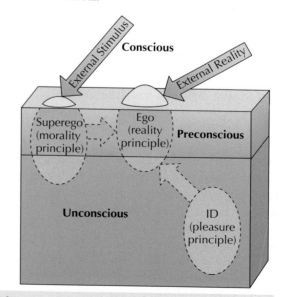

Freud's personality structure FIGURE 12.5

Sample psychological defense mechanisms TABLE 12.1

Defense mechanism	Description	Example
Repression	Preventing painful or unacceptable thoughts from entering consciousness	Forgetting the details of your parent's painful death
Sublimation	Redirecting unmet desires or unacceptable impulses into acceptable activities	Rechanneling sexual desires into school, work, art, sports, or hobbies that are constructive
Denial	Protecting oneself from an unpleasant reality by refusing to perceive it	Alcoholics refusing to admit their addiction
Rationalization	Substituting socially acceptable reasons for unacceptable ones	Justifying cheating on an exam by saying "everyone else does it"
Intellectualization	Ignoring the emotional aspects of a painful experience by focusing on abstract thoughts, words, or ideas	Emotionless discussion of your divorce while ignoring underlying pain
Projection	Transferring unacceptable thoughts, motives, or impulses to others	Becoming unreasonably jealous of your mate while denying your own attraction to others
Reaction formation	Refusing to acknowledge unacceptable urges, thoughts, or feelings by exaggerating the opposite state	Promoting a petition against adult bookstores even though you are secretly fascinated by pornography
Regression	Responding to a threatening situation in a way appropriate to an earlier age or level of development	Throwing a temper tantrum when a friend doesn't want to do what you'd like
Displacement	Redirecting impulses toward a less threatening person or object	Yelling at a coworker after being criticized by your boss

Although Freud described many kinds of defense mechanisms (TABLE 12.1), he believed that repression was the most important. **Repression** is the mechanism by which the ego prevents the most unacceptable, anxiety-provoking thoughts from entering consciousness (FIGURE 12.6).

The concept of defense mechanisms has generally withstood the test of time, and it is an accepted part of modern psychology. However, this is not the case for Freud's theory of psychosexual stages of development.

According to Freud, strong biological urges residing within the id push all children through five universal **psychosexual stages** (FIGURE 12.7). The term *psychosexual* reflects Freud's belief that children experience sexual feelings from birth (in different forms from those of adolescents or adults).

Is it bad to use defense mechanisms?

FIGURE 12.6

Although defense mechanisms do distort reality, some misrepresentation seems to be necessary for our psychological well-being (Newman, Duff, & Baumeister, 1997; Taylor & Armor, 1996). During a gruesome surgery, for example, physicians and nurses may **intellectualize** the procedure as an unconscious way of dealing with their personal anxieties. Can you see how focusing on highly objective technical aspects of the situation might help these people avoid becoming emotionally overwhelmed by the potentially tragic circumstances they often encounter?

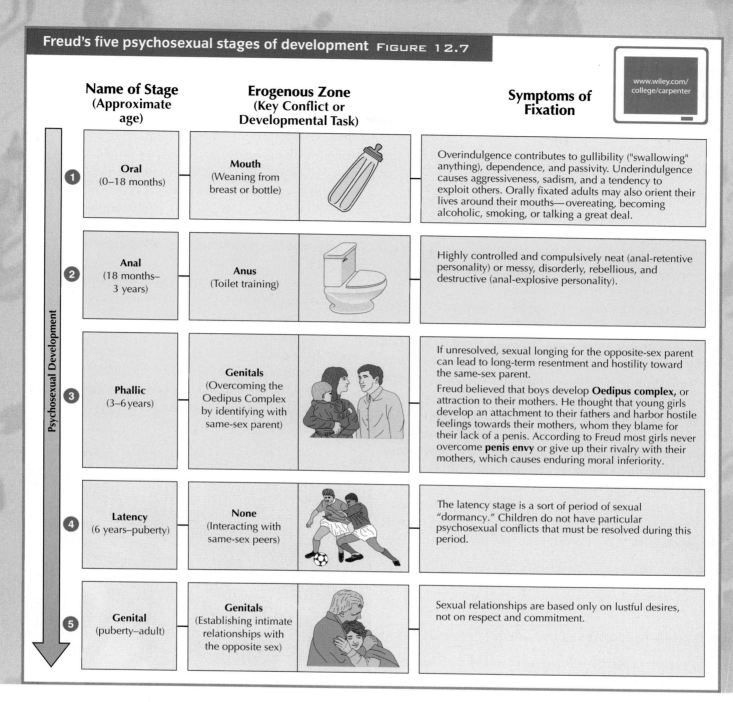

Freud's five psychosexual stages of development FIGURE 12.7

www.wiley.com/college/carpenter

Name of Stage (Approximate age)	Erogenous Zone (Key Conflict or Developmental Task)		Symptoms of Fixation
1 Oral (0–18 months)	**Mouth** (Weaning from breast or bottle)		Overindulgence contributes to gullibility ("swallowing" anything), dependence, and passivity. Underindulgence causes aggressiveness, sadism, and a tendency to exploit others. Orally fixated adults may also orient their lives around their mouths—overeating, becoming alcoholic, smoking, or talking a great deal.
2 Anal (18 months– 3 years)	**Anus** (Toilet training)		Highly controlled and compulsively neat (anal-retentive personality) or messy, disorderly, rebellious, and destructive (anal-explosive personality).
3 Phallic (3–6 years)	**Genitals** (Overcoming the Oedipus Complex by identifying with same-sex parent)		If unresolved, sexual longing for the opposite-sex parent can lead to long-term resentment and hostility toward the same-sex parent. Freud believed that boys develop **Oedipus complex,** or attraction to their mothers. He thought that young girls develop an attachment to their fathers and harbor hostile feelings towards their mothers, whom they blame for their lack of a penis. According to Freud most girls never overcome **penis envy** or give up their rivalry with their mothers, which causes enduring moral inferiority.
4 Latency (6 years–puberty)	**None** (Interacting with same-sex peers)		The latency stage is a sort of period of sexual "dormancy." Children do not have particular psychosexual conflicts that must be resolved during this period.
5 Genital (puberty–adult)	**Genitals** (Establishing intimate relationships with the opposite sex)		Sexual relationships are based only on lustful desires, not on respect and commitment.

(Left vertical label: Psychosexual Development)

psychosexual stages

In Freudian theory, the five developmental periods (oral, anal, phallic, latency, and genital) during which particular kinds of pleasures must be gratified if personality development is to proceed normally.

Freud held that if a child's needs are not met or are overindulged at one particular psychosexual stage, the child may become fixated, and a part of his or her personality will remain stuck at that stage. Furthermore, under stress, individuals may return (or regress) to a stage at which earlier needs were frustrated or overly gratified.

NEO-FREUDIAN/PSYCHODYNAMIC THEORIES: REVISING FREUD'S IDEAS

Some initial followers of Freud later rebelled and proposed theories of their own; they became known as **neo-Freudians**.

Alfred Adler (1870–1937) was the first to leave Freud's inner circle. Instead of seeing behavior as motivated by unconscious forces, he believed that it is purposeful and goal-directed. According to Adler's **individual psychology**, we are motivated by our goals in life—especially our goals of obtaining security and overcoming feelings of inferiority.

Adler believed that almost everyone suffers from an **inferiority complex**, or deep feelings of inadequacy and incompetence that arise from our feelings of helplessness as infants. According to Adler, these early feelings result in a "will-to-power" that can take one of two paths. It can either cause children to strive to develop superiority over others through dominance, aggression, or expressions of envy, or—more positively—it can cause children to develop their full potential and creativity and to gain mastery and control in their lives (Adler, 1964, 1998) (**FIGURE 12.8**).

Another early Freud follower turned dissenter, Carl Jung (pronounced "YOONG"), developed **analytical psychology**. Like Freud, Jung (1875–1961) emphasized unconscious processes, but he believed that the unconscious contains positive and spiritual motives as well as sexual and aggressive forces.

Jung also thought that we have two forms of unconscious mind: the personal unconscious and the collective unconscious. The *personal unconscious* is created from our individual experiences, whereas the *collective unconscious* is identical in each person and is inherited (Jung, 1936, 1969). The collective unconscious consists of primitive images and patterns of thought, feeling, and behavior that Jung called **archetypes** (**FIGURE 12.9**).

Because of archetypal patterns in the collective unconscious, we perceive and react in certain predictable ways. One set of archetypes refers to gender roles (Chapter 10). Jung claimed that both males and females have patterns for feminine aspects of personality (*anima*) and masculine aspects of personality (*animus*), which allow us to express both masculine and feminine personality traits and to understand the opposite sex.

An upside to feelings of inferiority? FIGURE 12.8

Adler suggested that the will-to-power could be positively expressed through social interest—identifying with others and cooperating with them for the social good. Can you explain how these volunteers might be fulfilling their will-to-power interest?

Archetypes in the collective unconscious

FIGURE 12.9

According to Jung, the collective unconscious is the ancestral memory of the human race, which explains the similarities in religion, art, symbolism, and dream imagery across cultures, such as the repeated symbol of the snake in ancient Egyptian tomb painting and early Australian aboriginal bark painting.

Like Adler and Jung, psychoanalyst Karen Horney (pronounced "HORN-eye") was an influential follower of Freud's who later came to reject major aspects of Freudian theory. She is remembered mostly for having developed a creative blend of Freudian, Adlerian, and Jungian theory, with added concepts of her own (Horney, 1939, 1945) (FIGURE 12.10).

Horney is also known for her theories of personality development. She believed that adult personality was shaped by the child's relationship to the parents—not by fixation at some stage of psychosexual development, as Freud argued. Horney believed that a child whose needs were not met by nurturing parents would experience extreme feelings of helplessness and insecurity. How people respond to this basic anxiety greatly determines emotional health.

According to Horney, everyone searches for security in one of three ways: We can move toward people (by seeking affection and acceptance from others); we can move away from people (by striving for independence, privacy, and self-reliance); or we can move against people (by trying to gain control and power over others). Emotional health requires a balance among these three styles.

Karen Horney (1885–1952) FIGURE 12.10

Horney argued that most of Freud's ideas about female personality reflected male biases and misunderstanding. She contended, for example, that Freud's concept of penis envy reflected women's feelings of cultural inferiority, not biological inferiority—*power envy*, not penis envy.

EVALUATING PSYCHOANALYTIC THEORIES: CRITICISMS AND ENDURING INFLUENCE

In this section, we look at major criticisms of Freud's psychoanalytic theories. In addition, we discuss the reasons that Freud has had enormous influence in the field of psychology. According to critics, Freud's theories are problematic for several reasons.

Today there are few Freudian purists. Instead, modern psychodynamic theorists and psychoanalysts use empirical methods and research findings to reformulate and refine traditional Freudian thinking (Shaver & Mikulincer, 2005; Westen, 1998).

But wrong as he was on many counts, Freud still ranks as one of the giants of psychology (Allen, 2006; Monte & Sollod, 2003; Weinberger & Westen, 2001). Furthermore, Freud's impact on Western intellectual history cannot be overstated. He attempted to explain dreams, religion, social groupings, family dynamics, neurosis, psychosis, humor, the arts, and literature.

It's easy to criticize Freud if you don't remember that he began his work at the start of the twentieth century and lacked the benefit of modern research findings and technology. We can only imagine how our current theories will look 100 years from now. Right or wrong, Freud has a lasting place among the pioneers in psychology (**TABLE 12.2**).

Evaluating psychoanalytic theories TABLE 12.2

Criticisms

- They prove difficult to test. From a scientific perspective, a major problem with psychoanalytic theory is that most of its concepts—such as the id or unconscious conflicts—cannot be empirically tested (Domhoff, 2004; Esterson, 2002; Friedman & Schustack, 2006).

- They overemphasize biology and unconscious forces. Modern psychologists believe that Freud did not give sufficient attention to learning and culture in shaping behavior.

- They rely on inadequate evidence. Freud based his theories almost exclusively on the subjective case histories of his adult patients. Moreover, Freud's patients represented a small and selective sample of humanity: upper-class women in Vienna (Freud's home) who had serious adjustment problems.

- They exhibit sexism. Many psychologists (beginning with Karen Horney) reject Freud's theories as derogatory toward women.

- They show a lack of cross-cultural support. The Freudian concepts that ought to be most easily supported empirically—the biological determinants of personality—are generally not borne out by cross-cultural studies.

Enduring influences

- His emphasis on the unconscious and its influence on behavior.

- The conflict among the id, ego, and superego and the resulting defense mechanisms.

- Encouraging open talk about sex in Victorian times.

- The development of psychoanalysis, an influential form of therapy.

- The sheer magnitude of his theory.

CONCEPT CHECK ⬡ **STOP**

How do the conscious, preconscious, and unconscious shape personality, in Freud's view?

What is the collective unconscious?

What is an example of sexism in Freud's psychoanalytic theory?

Humanistic Theories

Explain the importance of the self in Rogers's theory of personality.

Describe how Maslow's hierarchy of needs affects personality.

Identify three criticisms of humanistic theories.

Humanistic theories of personality emphasize each person's internal feelings, thoughts, and sense of basic worth. In contrast to Freud, humanists believe that people are naturally good (or, at worst, neutral) and that they possess a positive drive toward self-fulfillment.

According to this view, our personality and behavior depend on how we perceive and interpret the world, not on traits, unconscious impulses, or rewards and punishments. Humanistic psychology was developed largely by Carl Rogers and Abraham Maslow.

ROGERS'S THEORY: THE IMPORTANCE OF THE SELF

To psychologist Carl Rogers (1902–1987), the most important component of personality is the **self**—what a person comes to identify as "I" or "me." Today, Rogerians (followers of Rogers) use the term **self-concept** to refer to all the information and beliefs you have regarding your own nature, unique qualities, and typical behaviors.

> **self-concept** Rogers's term for all the information and beliefs that individuals have about their own nature, qualities, and behavior.

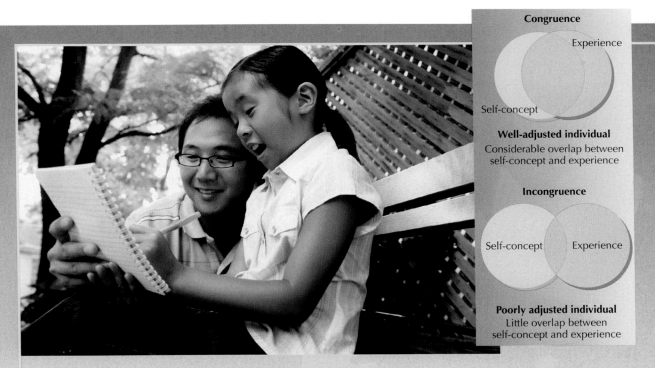

Congruence

Experience

Self-concept

Well-adjusted individual
Considerable overlap between self-concept and experience

Incongruence

Self-concept   Experience

Poorly adjusted individual
Little overlap between self-concept and experience

Congruence, Mental Health, and Self-Esteem

According to Carl Rogers, mental health and adjustment are related to the degree of congruence between a person's self-concept and life experiences. He argued that self-esteem—how we feel about ourselves—is particularly dependent on this congruence. Can you see how an artistic child would likely have higher self-esteem if her family valued art highly than if they did not?

What a Scientist Sees

Rogers believed that poor mental health and maladjustment developed from a mismatch, or incongruence, between the self-concept and actual life experiences.

Rogers believed that mental health, congruence, and self-esteem are part of our innate, biological capacities. In his view, everyone naturally approaches and values people and experiences that enhance our growth and fulfillment and avoids those that do not. Therefore, Rogers believed that we should trust our feelings to guide us toward mental health and happiness.

Then why do some people have low self-esteem and poor mental health? Rogers believed that these outcomes generally result from early childhood experiences with parents and other adults who make their love and acceptance conditional on behaving in certain ways and expressing only certain feelings.

If a child learns over time that his negative feelings and behaviors (which we all have) are totally unacceptable and unlovable, his self-concept and self-esteem may become distorted. He may always doubt the love and approval of others because they don't know "the real person hiding inside."

To help children develop to their fullest potential, adults need to create an atmosphere of **uncon**-ditional positive regard—a setting in which children realize that they will be accepted no matter what they say or do.

Some people mistakenly believe that unconditional positive regard means that we should allow people to do whatever they please. But humanists separate the value of the person from his or her behaviors. They accept the person's positive nature while discouraging destructive or hostile behaviors. Humanistic psychologists believe in guiding children to control their behavior so that they can develop a healthy self-concept and healthy relationships with others (FIGURE 12.11).

> **unconditional positive regard**
> Rogers's term for positive behavior toward a person with no contingencies attached.

MASLOW'S THEORY: THE SEARCH FOR SELF-ACTUALIZATION

Like Rogers, Abraham Maslow believed that there is a basic goodness to human nature and a natural tendency toward **self-actualization**. He saw personality as

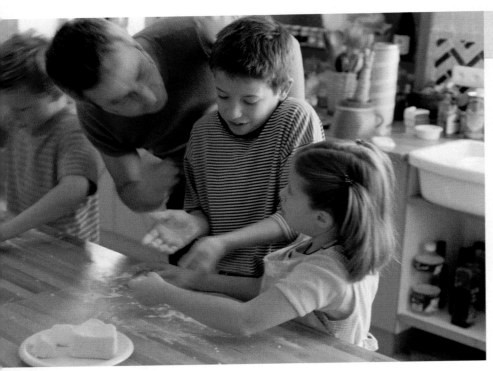

Conditional love?

FIGURE 12.11

If a child is angry and hits his younger sister, some parents might punish the child or deny his anger, saying, "Nice children don't hit their sisters; they love them!" To gain parental approval, the child has to deny his true feelings of anger, but inside he secretly suspects he is not a "nice boy" because he did hit his sister and (at that moment) did not love her. How might repeated incidents of this type have a lasting effect on someone's self-esteem? What would be a more appropriate response to the child's behavior that acknowledges that it is the behavior that is unacceptable, not the child that is unacceptable?

the quest to fulfill basic physiological needs (including safety, belonging and love, and esteem) and then move upward toward the highest level of self-actualization.

According to Maslow, self-actualization is the inborn drive to develop all one's talents and capacities. It involves understanding one's own potential, accepting oneself and others as unique individuals, and taking a problem-centered approach to life situations (Maslow, 1970). Self-actualization is an ongoing process of growth rather than an end product or accomplishment.

Maslow believed that only a few, rare individuals, such as Albert Einstein, Mohandas Gandhi, and Eleanor Roosevelt, become fully self-actualized. However, he saw self-actualization as part of every person's basic hierarchy of needs. (See Chapter 11 for more information on Maslow's theory.)

EVALUATING HUMANISTIC THEORIES: THREE MAJOR CRITICISMS

Humanistic psychology was extremely popular during the 1960s and 1970s. It was seen as a refreshing new perspective on personality after the negative determinism of the psychoanalytic approach and the mechanical nature of learning theories (Chapter 6). Although this early popularity has declined, many humanistic ideas have been incorporated into approaches to counseling and psychotherapy.

At the same time, humanistic theories have also been criticized (e.g., Funder, 2000). Three of the most important criticisms are these:

1. *Naive assumptions*. Critics suggest that the humanists are unrealistic, romantic, and even naive about human nature (**FIGURE 12.12**).

2. *Poor testability and inadequate evidence*. Like many psychoanalytic terms and concepts, humanistic

Are all people as inherently good as they say?
FIGURE 12.12

Mankind's continuing history of murders, warfare, and other acts of aggression suggests otherwise.

concepts (such as unconditional positive regard and self-actualization) are difficult to define operationally and to test scientifically.

3. *Narrowness*. Like trait theories, humanistic theories have been criticized for merely describing personality rather than explaining it. For example, where does the motivation for self-actualization come from? To say that it is an "inborn drive" doesn't satisfy those who favor using experimental research and hard data to learn about personality.

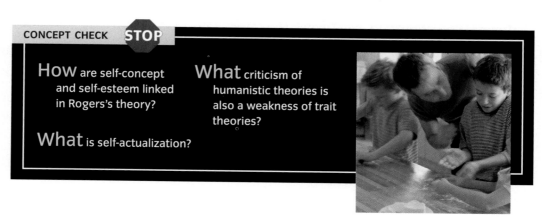

CONCEPT CHECK **STOP**

HOW are self-concept and self-esteem linked in Rogers's theory?

What is self-actualization?

What criticism of humanistic theories is also a weakness of trait theories?

Social-Cognitive Theories

LEARNING OBJECTIVES

Explain Bandura's concepts of self-efficacy and reciprocal determinism and how they affect personality.

Describe the role that Rotter's concept of locus of control plays in personality.

Summarize the attractions and criticisms of the social-cognitive perspective on personality

According to the social-cognitive perspective, each of us has a unique personality because we have individual histories of interactions with the environment (social) and because we think (cognitive) about the world and interpret what happens to us (Cervone & Shoda, 1999). Two of the most influential social-cognitive theorists are Albert Bandura and Julian Rotter.

BANDURA'S AND ROTTER'S APPROACHES: SOCIAL LEARNING PLUS COGNITIVE PROCESSES

Albert Bandura (also discussed in Chapter 6) has played a major role in reintroducing thought processes into personality theory. Cognition, or thought, is central to his concept of **self-efficacy** (Bandura, 1997, 2000, 2003).

According to Bandura, if you have a strong sense of self-efficacy, you believe you can generally succeed, regardless of past failures and current obstacles. Your self-efficacy will in turn affect which challenges you choose to accept and the effort you expend in reaching goals. However, Bandura emphasized that self-efficacy is always specific to the situation—it does not necessarily carry across situations. For example, self-defense training significantly improves women's belief that they could escape from or

self-efficacy
Bandura's term for the learned belief that one is capable of producing desired results, such as mastering new skills and achieving personal goals.

reciprocal determinism
Bandura's belief that cognitions, behaviors, and the environment interact to produce personality.

disable a potential assailant or rapist, but it does not lead them to feel more capable in all areas of their lives (Weitlauf et al., 2001). Finally, according to Bandura's concept of **reciprocal determinism**, self-efficacy beliefs will affect how others respond to you, influencing your chances for success (**FIGURE 12.13**).

Julian Rotter's theory is similar to Bandura's in that it suggests that learning experiences create **cognitive expectancies** that guide behavior and influence the environment (Rotter, 1954, 1990). According to Rotter, your behavior or personality is determined by (1) what you expect to happen following a specific action and (2) the reinforcement value attached to specific outcomes.

To understand your personality and behavior, Rotter would use personality tests

Albert Bandura's theory of reciprocal determinism
FIGURE 12.13

According to Bandura, thoughts (or cognitions), behavior, and the environment all interact to produce personality.

"We're encouraging people to become involved in their own rescue."

Locus of control and achievement
FIGURE 12.14

Research links possession of an internal locus of control with higher achievement and better mental health (Cheung & Rudowicz, 2003; Silvester et al., 2002; Sirin & Rogers-Sirin, 2004).

that measure your internal versus external **locus of control** (Chapter 3). Rotter's tests ask people to respond to statements such as, "People get ahead in this world primarily by luck and connections rather than by hard work and perseverance," and, "When someone doesn't like you, there is little you can do about it." As you may suspect, people with an external locus of control think that environment and external forces have primary control over their lives, whereas people with an internal locus of control think that they can control events in their lives through their own efforts (**FIGURE 12.14**).

EVALUATING SOCIAL-COGNITIVE THEORY: THE PLUSES AND MINUSES

The social-cognitive perspective holds several attractions. First, it emphasizes how the environment affects and is affected by individuals. Second, it offers testable, objective hypotheses and operationally defined terms, and it relies on empirical data for its basic principles. However, critics note that social-cognitive theory ignores the unconscious and emotional aspects of personality (Carducci, 1998; Westen, 1998).

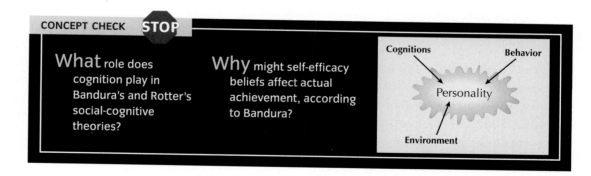

CONCEPT CHECK **STOP**

What role does cognition play in Bandura's and Rotter's social-cognitive theories?

Why might self-efficacy beliefs affect actual achievement, according to Bandura?

Biological Theories

LEARNING OBJECTIVES

Summarize the roles that brain structures and neurochemistry play in personality.

Describe how researchers study genetic influences on personality.

Describe how the biopsychosocial model integrates different theories of personality.

I n this section, we explore how inherited biological factors influence our personalities. We conclude with a discussion of how all theories of personality ultimately interact within the *biopsychosocial model*.

THREE MAJOR CONTRIBUTORS TO PERSONALITY: THE BRAIN, NEUROCHEMISTRY, AND GENETICS

Modern biological research suggests that certain brain areas may contribute to some personality traits. For instance, increased electroencephalographic (EEG) activity in the left frontal lobes of the brain is associated with sociability (or extroversion), whereas greater EEG activity in the right frontal lobes is associated with shyness (introversion) (Schmidt, 1999).

A major limitation of research on brain structures and personality is the difficulty in identifying which structures are uniquely connected with particular personality traits. Neurochemistry seems to offer more precise data on how biology influences personality (Kagan, 1998).

For example, sensation seeking (Chapter 11) has consistently been linked with levels of monoamine oxidase (MAO), an enzyme that regulates levels of neurotransmitters such as dopamine (Íbañez, Blanco, & Sáiz-Ruiz, 2002; Zuckerman, 1994, 2004). Dopamine also seems to be correlated with novelty seeking and extroversion (Depue & Collins, 1999; Laine et al., 2001).

How can neurochemistry have such effects? Studies suggest that high-sensation seekers and extroverts tend to experience less physical arousal than introverts from the same stimulus (Lissek & Powers, 2003). Extroverts' low arousal apparently motivates them to seek out situations that will elevate their arousal. Moreover, it is believed that a higher arousal threshold is genetically transmitted. In other words, personality traits like sensation seeking and extroversion may be inherited.

Finally, psychologists have recently recognized that genetic factors also have an important influence on personality (Johnson, McGue, et al., 2004; Murakami & Hayashi, 2002; Plomin & Crabbe, 2000; Sequeira et al., 2004). This relatively new area, called **behavioral genetics**, attempts to determine the extent to which behavioral differences among people are due to genetics as opposed to environment (Chapter 2).

One way to measure genetic influences is to compare similarities in personality between identical twins and fraternal twins. For example, studies of the heritability of the five-factor model personality traits suggest that genetic factors contribute about 40 to 50 percent of personality (Bouchard, 1997; Eysenck, 1967, 1991; Lensvelt-Mulders & Hettema, 2001).

In addition to twin studies, researchers compare the personalities of parents with those of their biological children and their adopted children. Studies of extroversion and neuroticism have found that parents' traits correlate moderately with those of their biological children and hardly at all with those of their adopted children (Bouchard, 1997; McCrae et al., 2000).

At the same time, researchers are careful not to overemphasize genetic influences on personality (Deckers, 2005; Funder, 2001; Maccoby, 2000). Some researchers believe that the importance of the unshared environment (aspects of the environment that differ from one individual to another, even within a family) has been overlooked (Saudino, 1997). Others fear that research on "genetic determinism"—do our genes determine who we are?—could be misused to "prove" that an ethnic or racial group is inferior, that

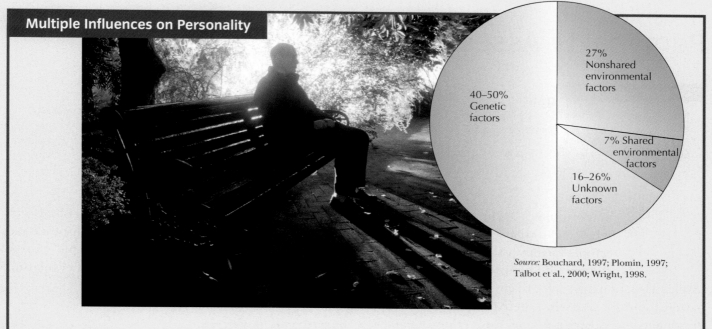

Multiple Influences on Personality

40–50% Genetic factors

27% Nonshared environmental factors

7% Shared environmental factors

16–26% Unknown factors

Source: Bouchard, 1997; Plomin, 1997; Talbot et al., 2000; Wright, 1998.

What gives a person certain personality characteristics, such as shyness, or conscientiousness, or aggressiveness? As shown in the figure here, research indicates that four major factors overlap to influence personality. These include genetic (inherited) factors; nonshared environmental factors, or how each individual's genetic factors react and adjust to his or her particular environment; shared environmental factors, involving parental patterns and shared family experiences; and error, or unidentified factors or problems with testing.

For example, Hans Eysenck (1990) believed that certain traits (like introversion and extroversion) may reflect inherited patterns of cortical arousal, as well as social learning, cognitive processes, and the environment. Can you see how someone with an introverted personality (and therefore a higher level of cortical arousal) might try to avoid excessive stimulation by seeking friends and jobs with low stimulation levels? Eysenck's work exemplifies how trait, biological, and social-cognitive theories can be combined to provide better insight into personality—the *biopsychosocial model*.

male dominance is natural, or that social progress is impossible. In short, they worry that an emphasis on genetic determinants of personality and behavior may lead people to see themselves as merely "victims" of their genes. Clearly, genetics have produced exciting and controversial results. However, more research is necessary before a cohesive biological theory of personality can be constructed.

THE BIOPSYCHOSOCIAL MODEL: PULLING THE PERSPECTIVES TOGETHER

No one personality theory explains everything we need to know about personality. Each theory offers different insights into how a person develops the distinctive set of characteristics we call "personality." That's why instead of adhering to any one theory, many psycholo-

gists believe in the biopsychosocial approach, or the idea that several factors—biological, psychological, and social—overlap in their contributions to personality (Higgins, 2004; McCrae, 2004).

CONCEPT CHECK STOP

What evidence suggests that particular brain areas contribute to some personality traits?

Why is the biopsychosocial model important to research on personality?

How can traits like sensation seeking and extroversion be related to neurochemistry?

Personality Assessment

N umerous methods have been used over the decades to assess personality. Modern personality assessments are used by clinical and counseling psychologists, psychiatrists, and others for diagnosing psychotherapy patients and for assessing their progress in therapy. Personality assessment is also used for educational and vocational counseling and to aid businesses in making hiring decisions. Personality assessments can be grouped into a few broad categories: interviews, observations, objective tests, and projective tests.

INTERVIEWS AND OBSERVATION

We all use informal "interviews" to get to know other people. When first meeting someone, we usually ask about his or her job, academic interests, family, or hobbies. Psychologists also use interviews. Unstructured interviews are often used for job and college selection and for diagnosing psychological problems. In an unstructured format, interviewers get impressions and pursue hunches or let the interviewee expand on information that promises to disclose personality characteristics. In structured interviews, the interviewer asks specific questions so that the interviewee's responses can be evaluated more objectively (and compared with others' responses).

In addition to interviews, psychologists also assess personality by directly and methodically observing behavior. The psychologist looks for examples of specific behaviors and follows a careful set of evaluation guidelines (FIGURE 12.15).

OBJECTIVE TESTS

Objective personality tests, or inventories, are the most widely used method of assessing personality, for two reasons. They can be administered to a large number of people relatively quickly and the tests can be evaluated in a standardized fashion.

Some objective tests measure one specific personality trait, such as sensation seeking (Chapter 11) or locus of control. However, psychologists in clinical, counseling, and industrial settings often wish to assess a range of personality traits. To do so, they generally use multitrait (or *multiphasic*) inventories.

objective personality tests Standardized questionnaires that require written responses, usually to multiple-choice or true-false questions.

Behavioral observation FIGURE 12.15

How might careful observation help a psychologist better understand a troubled client's personality and family dynamics?

Personality and Bumps on the Head?

In the 1800s, if you wanted to have your personality assessed, you would go to a phrenologist. Franz Gall, the founder of phrenology, believed that experts could determine personality by measuring bumps on the skull and comparing the measurements with a chart that associated different areas of the skull with particular aspects of personality. Note that terms like *sublimity* (the ability to squelch natural impulses, especially sexual) and *ideality* (the ability to live by high ideals) probably reflect traits that were important in the 1800s. What traits might be measured if we still believed in phrenology today?

The most widely studied and clinically used multitrait test is the **Minnesota Multiphasic Personality Inventory (MMPI)**—or its revision, the MMPI-2 (Butcher, 2000; Butcher & Rouse, 1996). This test consists of over 500 statements that participants respond to with *True, False,* or *Cannot Say*. The following are examples of the kinds of statements found on the MMPI:

> My stomach frequently bothers me.
> I have enemies who really wish to harm me.
> I sometimes hear things that other people can't hear.
> I would like to be a mechanic.
> I have never indulged in any unusual sex practices.

Did you notice that some of these questions are about very unusual, abnormal behavior? Although there are many "normal" questions on the full MMPI, the test is designed primarily to help clinical and counseling psychologists diagnose psychological disorders. MMPI test items are grouped into 10 clinical scales, each measuring a different disorder (TABLE 12.3). There are also a number of validity scales designed to reflect the extent to which respondents (1) distort their answers (for example, to fake psychological disturbances or to appear more psychologically healthy than they really are), (2) do not understand the items, and (3) are being uncooperative.

Personality tests like the MMPI are often confused with *career inventories* or vocational interest tests. Career counselors use these latter tests (along with aptitude and achievement tests) to help people identify occupations and careers that match their unique traits, values, and interests.

Subscales of the MMPI-2 TABLE 12.3

Clinical scales	Typical interpretations of high scores	Validity scales	Typical interpretations of high scores
1. Hypochondriasis	Numerous physical complaints	1. L (lie)	Denies common problems, projects a "saintly" or false picture
2. Depression	Seriously depressed and pessimistic		
3. Hysteria	Suggestible, immature, self-centered, demanding	2. F (confusion)	Answers are contradictory
4. Psychopathic deviate	Rebellious, nonconformist	3. K (defensiveness)	Minimizes social and emotional complaints
5. Masculinity–femininity	Interests like those of other sex		
6. Paranoia	Suspicious and resentful of others	4. ? (cannot say)	Many items left unanswered
7. Psychasthenia	Fearful, agitated, brooding		
8. Schizophrenia	Withdrawn, reclusive, bizarre thinking		
9. Hypomania	Distractible, impulsive, dramatic		
10. Social introversion	Shy, introverted, self-effacing		

PROJECTIVE TESTS

projective tests
Psychological tests that use ambiguous stimuli, such as inkblots or drawings, which allow the test taker to project his or her unconscious thoughts onto the test material.

Unlike objective tests, **projective tests** use unstructured stimuli that can be perceived in many ways. As the name implies, projective tests supposedly allow each person to project his or her own unconscious conflicts, psychological defenses, motives, and personality traits onto the test materials. Because respondents are unable (or unwilling) to express their true feelings if asked directly, the ambiguous stimuli reportedly provide an indirect "psychological X-ray" of important unconscious processes (Hogan, 2003). Two of the most widely used projective tests are the **Rorschach Inkblot Test** and the **Thematic Apperception Test (TAT)** (FIGURE 12.16).

ARE PERSONALITY MEASUREMENTS ACCURATE?

Let's evaluate the strengths and the challenges of each of the four methods of personality assessment: interviews, observation, objective tests, and projective tests.

Interviews and observations
Both interviews and observations can provide valuable insights into personality, but they are time-consuming and expensive. Furthermore, raters of personality tests frequently disagree in their evaluations of the same individuals. Interviews and observations also involve unnatural settings, and, in fact, the very presence of an observer can alter a subject's behavior.

Objective tests
Tests like the MMPI-2 provide specific, objective information about a broad range of personality traits in a relatively short period. However, they are also the subject of at least three major criticisms:

Visualizing

Projective tests FIGURE 12.16

Responses to projective tests reportedly reflect unconscious parts of the personality that "project" onto the stimuli.

A The Rorschach Inkblot Test was introduced in 1921 by Swiss psychiatrist Hermann Rorschach. With this technique, individuals are shown 10 inkblots like this, one at a time, and are asked to report what figures or objects they see in each of them. (Reproduced with permission.)

B Created by personality researcher Henry Murray in 1938, the Thematic Apperception Test (TAT) consists of a series of ambiguous black-and-white pictures that are shown to the test taker, who is asked to create a story related to each. Can you think of two different stories that a person might create for the picture of two women here? How might a psychologist interpret each story?

1. *Deliberate deception and social desirability bias.* Some items on personality inventories are easy to "see through," so respondents may intentionally, or unintentionally, fake particular personality traits. In addition, some respondents want to look good and will answer questions in ways that they perceive are socially desirable. (The validity scales of the MMPI-2 are designed to help prevent these problems.)

2. *Diagnostic difficulties.* When inventories are used for diagnosis, overlapping items sometimes make it difficult to pinpoint a diagnosis (Graham, 1991). In addition, clients with severe disorders sometimes score within the normal range, and normal clients sometimes score within the elevated range (Cronbach, 1990).

3. *Cultural bias and inappropriate use.* Some critics think that the standards for "normalcy" on objective tests fail to recognize the impact of culture. For example, Latinos generally score higher than respondents from North American and Western European cultures on the masculinity-femininity scale of the MMPI-2 (Lucio-Gómez et al., 1999).

However, this tendency reflects traditional gender roles and cultural training more than any individual personality traits.

Projective tests   Although projective tests are extremely time-consuming to administer and interpret, their proponents suggest that because they are unstructured, respondents may be more willing to talk honestly about sensitive topics. Critics point out, however, that the reliability and validity (Chapter 8) of projective tests is among the lowest of all tests of personality (Grove et al., 2002; Wood et al., 2001).

As you can see, each of these methods has its limits. Psychologists typically combine the results from various methods to create a full picture of an individual's personality.

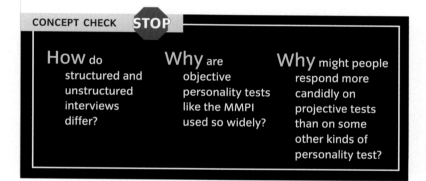

CONCEPT CHECK   STOP

How do structured and unstructured interviews differ?

Why are objective personality tests like the MMPI used so widely?

Why might people respond more candidly on projective tests than on some other kinds of personality test?

What is happening in this picture ?

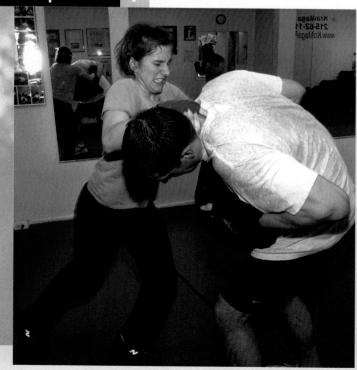

How would Bandura's social-cognitive concept of self-efficacy explain the benefits of taking a self-defense course?

How would Freud interpret this woman's aggressive behavior toward her "assailant"?

What biological factors might explain why some people choose activities that provide high levels of physiological arousal while others avoid such situations?

1 Trait Theories

1. Psychologists define **personality** as an individual's relatively stable and enduring patterns of thoughts, feelings, and actions.

2. Allport believed that the best way to understand personality was to arrange a person's unique personality **traits** into a hierarchy. Cattell and Eysenck later reduced the list of possible personality traits using factor analysis.

3. According to the **five-factor model (FFM)**, the five major dimensions of personality are openness, conscientiousness, extroversion, agreeableness, and neuroticism.

2 Psychoanalytic/ Psychodynamic Theories

1. Freud, the founder of psychodynamic theory, believed that the mind contained three levels of consciousness: the **conscious**, the **preconscious**, and the **unconscious**. He believed that most psychological disorders originate from unconscious memories and instincts. Freud also asserted that personality was composed of the id, the ego, and the superego. When the ego fails to satisfy both the id and the superego, anxiety slips into conscious awareness, which triggers **defense mechanisms**. Freud's theory of **psychosexual stages** has not withstood the test of time.

2. Neo-Freudians such as Adler, Jung, and Horney were influential followers of Freud who later came to reject major aspects of Freudian theory. Today, few Freudian purists remain, but Freud's impact on psychology and on Western intellectual history cannot be overstated.

3 Humanistic Theories

1. According to Rogers, mental health and self-esteem are related to the degree of congruence between our **self-concept** and life experiences. He argued that poor mental health results when young children do not receive **unconditional positive regard** from caregivers.

2. Maslow saw personality as the quest to fulfill basic physiological needs and to move toward the highest level of self-actualization.

4 Social-Cognitive Theories

1. Cognition is central to Bandura's concept of **self-efficacy**. According to Bandura, self-efficacy affects which challenges we choose to accept and the effort we expend in reaching goals. His concept of **reciprocal determinism** states that self-efficacy beliefs will also affect others' responses to us.

2. Rotter's theory suggests that learning experiences create cognitive expectancies that guide behavior and influence the environment. Rotter believed that having an internal versus external locus of control affects personality and achievement.

Cognitions → Personality ← Behavior
Environment →

5 Biological Theories

1. There is evidence that certain brain areas may contribute to personality. However, neurochemistry seems to offer more precise data on how biology influences personality. Research in behavioral genetics indicates that genetic factors also strongly influence personality.

2. Instead of adhering to any one theory of personality, many psychologists believe in the biopsychosocial approach—the idea that several factors overlap in their contributions to personality.

6 Personality Assessment

1. In an unstructured interview format, interviewers get impressions and pursue hunches or let the interviewee expand on information that promises to disclose personality characteristics. In structured interviews, the interviewer asks specific questions so that the interviewee's responses can be evaluated more objectively. Psychologists also assess personality by directly observing behavior.

2. **Objective personality tests** are widely used because they can be administered broadly and relatively quickly and because they can be evaluated in a standardized fashion. To assess a range of personality traits, psychologists use multitrait inventories, such as the MMPI.

3. **Projective tests** use unstructured stimuli that can be perceived in many ways. Projective tests supposedly allow each person to project his or her own unconscious conflicts, psychological defenses, motives, and personality traits onto the test materials.

KEY TERMS

CRITICAL AND CREATIVE THINKING QUESTIONS

1. How do you think you would score on each of the "Big Five" personality dimensions?

2. If scientists have so many problems with Freud, why do you think his theories are still popular with the public? Should psychologists continue to discuss his theories (and include them in textbooks)? Why or why not?

3. In what ways is Adler's individual psychology more optimistic than Freudian theory?

4. How do Bandura's and Rotter's social-cognitive theories differ from biological theories of personality?

5. Which method of personality assessment (interviews, behavioral observation, objective testing, or projective testing) do you think is likely to be most informative? Can you think of circumstances where one kind of assessment might be more effective than the others?

SELF-TEST: PERSONALITY

(Check your answers in Appendix B.)

1. A relatively stable and consistent characteristic that can be used to describe someone is known as a _____.

    a. character
    b. trait
    c. temperament
    d. personality

2. Label the list of personality traits in the Five-Factor model on the figure below.

Big Five Traits	Low Scorers	High Scorers
1	Down-to-earth Uncreative Conventional Uncurious	Imaginative Creative Original Curious
2	Negligent Lazy Disorganized Late	Conscientious Hard-working Well-organized Punctual
3	Loner Quiet Passive Reserved	Joiner Talkative Active Affectionate
4	Suspicious Critical Ruthless Irritable	Trusting Lenient Soft-hearted Good-natured
5	Calm Even-tempered Comfortable Unemotional	Worried Temperamental Self-conscious Emotional

3. Label Freud's three levels of consciousness on the figure below.

4. According to Freud, the three mental structures that form personality are the _____.

    a. unconscious, preconscious, and conscious
    b. oral, anal, and phallic
    c. Oedipus, Electra, and sexual/aggressive
    d. id, ego, and superego

5. Used excessively, defense mechanisms can be dangerous because they _____.

    a. hide true feelings
    b. become ineffective
    c. distort reality
    d. become fixated

6. Label Freud's psychosexual stages of development in correct sequence on the figure below.

Name of Stage (Approximate age)	Erogenous Zone (Key Conflict or Developmental Task)		Symptoms of Fixation
①	Mouth (Weaning from breast or bottle)		Overindulgence contributes to gullibility ("swallowing" anything), dependence, and passivity. Underindulgence causes aggressiveness, sadism, and a tendency to exploit others. Orally fixated adults may also orient their lives around their mouths—overeating, becoming alcoholic, smoking, or talking a great deal.
②	Anus (Toilet training)		Highly controlled and compulsively neat (anal-retentive personality) or messy, disorderly, rebellious, and destructive (anal-explosive personality).
③	Genitals (Overcoming the Oedipus Complex by identifying with same-sex parent)		If unresolved, sexual longing for the opposite-sex parent can lead to long-term resentment and hostility toward the same-sex parent. Freud believed that boys develop Oedipus complex, or attraction to their mothers. He thought that young girls develop an attachment to their fathers and harbor hostile feelings towards their mothers, whom they blame for their lack of a penis. According to Freud most girls never overcome penis envy or give up their rivalry with their mothers, which causes enduring moral inferiority.
④	None (Interacting with same-sex peers)		The latency stage is a sort of period of sexual "dormancy." Children do not have particular psychosexual conflicts that must be resolved during this period.
⑤	Genitals (Establishing intimate relationships with the opposite sex)		Sexual relationships are based only on lustful desires, not on respect and commitment.

7. Three of the most influential neo-Freudians were

    _____.

    a. Plato, Aristotle, and Descartes
    b. Dr. Laura, Dr. Phil, and Dr. Ruth
    c. Pluto, Mickey, and Minnie
    d. Adler, Jung, and Horney

8. Which of the following is **NOT** a criticism of Freud's psychoanalytic theory?

    a. Most of his concepts prove difficult to text.
    b. Most of his data show a lack of cross-cultural support.
    c. He under-emphasized the role of biological determinants.
    d. Some of his views were sexist.

9. *Unconditional positive regard* is a Rogerian term for
_____ .

   a. accepting any and all behavior as a positive manifestation of self-actualization
   b. positive behavior toward a person without attaching any contingencies
   c. nonjudgmental listening
   d. phenomenological congruence

10. The personality theorist who believed in the basic goodness of individuals, and their natural tendency toward self-actualization was _____ .

   a. Karen Horney
   b. Alfred Adler
   c. Abraham Maslow
   d. Carl Jung

11. According to Bandura, _____ involves a person's belief about whether he or she can successfully engage in behaviors related to personal goals.

   a. self-actualization
   b. self-esteem
   c. self-efficacy
   d. self-congruence

12. This is the study of the extent to which behavioral differences are due to genetics rather than the environment.

   a. the biobehavioral approach
   b. the genetic-environmental perspective
   c. behavioral genetics
   d. the biopsychosocial model

13. _____ appear(s) to have the largest influence (40% to 50%) on personality.

   a. Nonshared environment
   b. Shared environments
   c. Genetics
   d. Unknown factors

14. The most widely researched and clinically used self-report, personality test is the _____ .

   a. MMPI-2
   b. Rorschach Inkblot Test
   c. TAT
   d. SVII

15. The *Rorschach Inkblot Test* is an example of a(n) _____ test.

   a. aptitude
   b. projective personality
   c. the most reliable and valid personality
   d. culturally biased personality

Psychological Disorders

*M*ary's troubles first began in adolescence. She began to miss curfew, was frequently truant, and her grades declined sharply. During family counseling sessions, it was discovered that Mary also had been promiscuous and had prostituted herself several times to get drug money.... She quickly fell in love and overly idealized new friends. But when they quickly (and inevitably) disappointed her, she would angrily cast them aside.... Mary's problems, coupled with a preoccupation with inflicting pain on herself (by cutting and burning) and persistent thoughts of suicide, eventually led to her admittance to a psychiatric hospital at age 26 (Davison, Neale, & Kring, 2004, pp. 408–409).

Jim is a third-year medical student. Over the last few weeks, he has been noticing that older men appear to be frightened of him when he passes them on the street. Recently, he has become convinced that he is actually the director of the Central Intelligence Agency and that these men are secret agents of a hostile nation. Jim has found confirmatory evidence for his idea in the fact that a helicopter flies over his house every day at 8:00 a.m. and at 4:30 p.m. Surely, this surveillance is part of the plot to assassinate him (Bernheim & Lewine, 1979, p. 4).

Both Mary and Jim have severe psychological problems, and both of their stories raise interesting questions. What caused their difficulties? Was there something in their early backgrounds to explain their later behaviors? Is there something medically wrong with them? What about less-severe forms of abnormal behavior? What is the difference between being eccentric and being disordered?

In this chapter, we discuss how psychological disorders are identified, explained, and classified, and we explore six major categories of psychological disorders: anxiety disorders, mood disorders, schizophrenia, substance-related disorders, dissociative disorders, and personality disorders. Finally, we look at how gender and culture affect mental disorders.

340

Studying Psychological Disorders

LEARNING OBJECTIVES

List the four criteria for identifying abnormal behavior.

Review how views of abnormal behavior have changed through history.

Explain how the *DSM-IV-TR* is used to classify psychological disorders.

IDENTIFYING ABNORMAL BEHAVIOR: FOUR BASIC STANDARDS

abnormal behavior Patterns of emotion, thought, and action that are considered pathological (diseased or disordered) for one or more of these reasons: statistical infrequency, disability or dysfunction, personal distress, or violation of norms (Davison, Neale, & Kring, 2004).

On the continuum ranging from normal to **abnormal behavior**, people can be unusually healthy or extremely disturbed.

Mental health professionals generally agree on four criteria for abnormal behavior: statistical infrequency, disability or dysfunction, personal distress, and violation of norms (**FIGURE 13.1**). However, as we consider these criteria, remember that no single criterion is adequate for identifying all forms of abnormal behavior.

- *Statistical infrequency.* A behavior may be judged abnormal if it occurs infrequently in a given population. Statistical infrequency alone does not determine what is normal—for example, no one would classify Albert Einstein's great intelligence or Lance Armstrong's exceptional athletic ability as abnormal.

- *Disability or dysfunction.* People who suffer from psychological disorders may be unable to get along with others, hold a job, eat properly, or clean themselves. Their ability to think clearly and make rational decisions also may be impaired.

- *Personal distress.* The personal distress criterion focuses on the individual's judgment of his or her level of functioning. Yet many people with psychological disorders deny they have a problem. Also, some serious psychological disorders (such as antisocial personality disorder) cause little or no emotional discomfort. The personal distress criterion by itself is not sufficient for identifying all forms of abnormal behavior.

- *Violation of norms.* The fourth approach to identifying abnormal behavior is violation of social norms, or cultural rules that guide behavior in particular situations. A major problem with this criterion, however, is that cultural diversity can affect what people consider a violation of norms (Lopez & Guarnaccia, 2000).

EXPLAINING ABNORMALITY: FROM SUPERSTITION TO SCIENCE

What causes abnormal behavior? Historically, evil spirits and witchcraft have been blamed (Millon, 2004). Stone Age people, for example, believed that abnormal behavior stems from demonic possession; the "therapy" was to bore a hole in the skull so that the evil spirit could escape. During the European Middle Ages, a troubled person was sometimes treated with exorcism, an effort to drive the Devil out through prayer, fasting, noise-making, beating, and drinking terrible-tasting brews. During the fifteenth century, many believed that some individuals chose to consort with the Devil. Many of these supposed witches were tortured, imprisoned for life, or executed.

As the Middle Ages ended, special mental hospitals called *asylums* began to appear in Europe. Initially designed to provide quiet retreats from the world and to protect society (Millon, 2004), the asylums unfortunately became overcrowded, inhumane prisons.

Improvement came in 1792 when Philippe Pinel, a French physician in charge of a Parisian asylum, insisted that asylum inmates—whose behavior

The four criteria and a continuum of abnormal behavior FIGURE 13.1

Rather than fixed categories, both "abnormal" and "normal" behaviors exist along a continuum (Hansell & Damour, 2005).

(Rare)   (Common)

Statistical Infrequency
(e.g., believing others are plotting against you)

Normal   **Abnormal**

(Low)   (High)

Disability or Dysfunction
(e.g., being unable to go to work due to alcohol abuse)

Normal   **Abnormal**

(Low)   (High)

Personal Distress
(e.g., having thoughts of suicide)

Normal   **Abnormal**

(Rare)   (Common)

Violation of Norms
(e.g., shouting at strangers)

Normal   **Abnormal**

■ medical model
The perspective that diseases (including mental illness) have physical causes that can be diagnosed, treated, and possibly cured.

he believed to be caused by underlying physical illness—be unshackled and removed from their unlighted, unheated cells. Many inmates improved so dramatically that they could be released. Pinel's **medical model** eventually gave rise to the modern specialty of **psychiatry**.

■ psychiatry The branch of medicine that deals with the diagnosis, treatment, and prevention of mental disorders.

Unfortunately, when we label people "mentally ill," we may create new problems. One of the most outspoken critics of the medical model is psychiatrist Thomas Szasz (1960, 2000, 2004). Szasz believes that the medical model encourages people to believe that they have no responsibility for their actions. He contends that mental illness is a myth used to label individuals who are peculiar or offensive to others (Wyatt, 2004). Furthermore, labels can become self-perpetuating—that is, the person can begin to behave according to the diagnosed disorder.

Despite these potential dangers, the medical model—and the concept of mental illness—remains a founding principle of psychiatry. In contrast, psychology offers a multifaceted approach to explaining abnormal behavior.

CLASSIFYING ABNORMAL BEHAVIOR: THE DIAGNOSTIC AND STATISTICAL MANUAL IV-TR

Without a clear, reliable system for classifying the wide range of psychological disorders, scientific research on them would be almost impossible, and communication among mental health professionals would be seriously impaired. Fortunately, mental health specialists share a uniform classification system, the text revision of the fourth edition of the ***Diagnostic and Statistical Manual of Mental Disorders (DSM-IV-TR)*** (American Psychiatric Association, 2000).

Each revision of the *DSM* has expanded the list of disorders and changed the descriptions and categories to reflect both the latest in scientific research and changes in the way abnormal behaviors are viewed within our social context (First & Tasman, 2004; Smart & Smart, 1997). For example, take the terms **neurosis** and **psychosis**. In previous editions, the term *neurosis* reflected Freud's belief that all neurotic conditions arise from unconscious conflicts (Chapter 12). Now, conditions that were previously grouped under the heading *neurosis* have been formally redistributed as anxiety disorders, somatoform disorders, and dissociative disorders.

Unlike neurosis, the term *psychosis* is still listed in the *DSM-IV-TR* because it is useful for distinguishing the most severe mental disorders, such as schizophrenia and some mood disorders.

> **Diagnostic and Statistical Manual of Mental Disorders (DSM-IV-TR)**
> The classification system developed by the American Psychiatric Association used to describe abnormal behaviors; the *IV-TR* indicates that it is the text revision (*TR*) of the fourth major edition (*IV*).

The Insanity Plea—Guilty of a Crime or Mentally Ill?

On the morning of June 20, 2001, Texas mother Andrea Yates drowned her five children in the bathtub, then calmly called her husband to tell him he should come home. At Yates's trial, both the defense and prosecution agreed that Yates was mentally ill at the time of the murders, yet the jury still found her guilty and sentenced her to life in prison. (An appellate court later overturned this conviction. In 2006, Yates was found not guilty by reason of insanity and committed to a state mental hospital in which she will be held until she is no longer deemed a threat.) How could two juries come to such opposite conclusions? *Insanity* is a complicated legal term. In most states it refers to a person who cannot be held responsible for his or her actions, or is judged incompetent to manage his or her own affairs, because of mental illness. Despite high profile cases like that of Andrea Yates, it's important to keep in mind that the insanity plea is used in less than 1 percent of all cases that reach trial, and when used, it is rarely successful (Kirschner, Litwack, & Galperin, 2004; Steadman, 1993).

Seven Psychological Perspectives on Abnormal Behavior

Each of the seven major perspectives in psychology emphasizes different factors believed to contribute to abnormal behavior, but in practice they overlap. Consider the phenomenon of compulsive hoarding. Everyone sometimes makes an impulse purchase, and most people are reluctant to discard some possessions that are of questionable value. But when the acquisition of and inability to discard worthless items becomes extreme, it can interfere with basic aspects of living, such as cleaning, cooking, sleeping on a bed, and moving around one's home. This abnormal behavior is associated with several psychological disorders, but it is most commonly found in people who have obsessive-compulsive disorder, or OCD (an anxiety disorder discussed later in this chapter). How might each of the seven major perspectives explain compulsive hoarding?

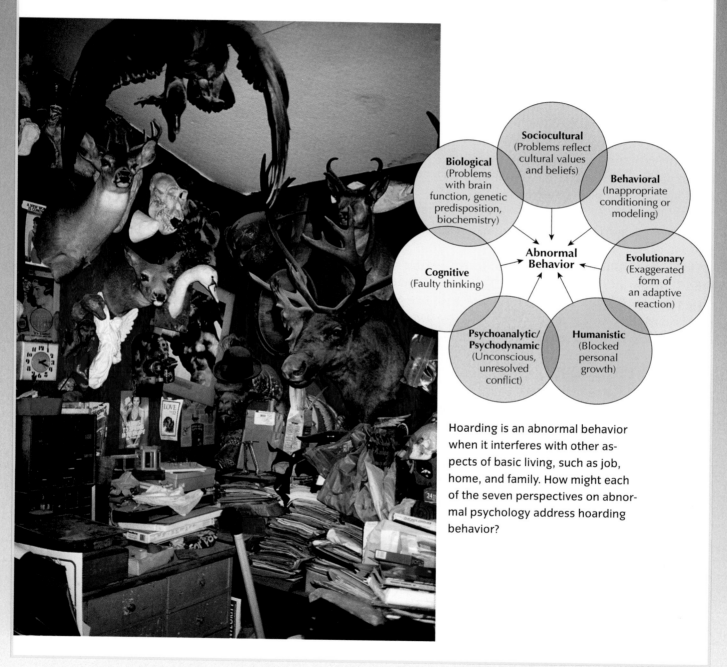

Sociocultural
(Problems reflect cultural values and beliefs)

Biological
(Problems with brain function, genetic predisposition, biochemistry)

Behavioral
(Inappropriate conditioning or modeling)

Abnormal Behavior

Cognitive
(Faulty thinking)

Evolutionary
(Exaggerated form of an adaptive reaction)

Psychoanalytic/ Psychodynamic
(Unconscious, unresolved conflict)

Humanistic
(Blocked personal growth)

Hoarding is an abnormal behavior when it interferes with other aspects of basic living, such as job, home, and family. How might each of the seven perspectives on abnormal psychology address hoarding behavior?

Understanding the DSM-IV-TR The *DSM-IV-TR* is organized according to five major dimensions called *axes*, which serve as guidelines for making decisions about symptoms (FIGURE 13.2). Axis I describes **state disorders** (the patient's current condition, or "state"), such as anxiety, substance abuse, and depression. Axis II describes **trait disorders** (enduring problems that seem to be an integral part of the self), including long-running personality disorders and mental retardation.

The other three axes are used to record important supplemental information. Medical conditions (Axis III) and psychosocial and environmental stressors (Axis IV) could contribute to moods and emotional problems. Finally, Axis V evaluates a person's overall level of functioning, on a scale from 1 (serious attempt at suicide) to 100 (happy and productive).

The *DSM-IV-TR* contains more than 200 diagnostic categories grouped into 17 subcategories (TABLE 13.1).

In this chapter, we focus on only the first 6 of the 17 categories shown in Table 13.1. Before we go on, note that the *DSM-IV-TR* classifies disorders, not people. Accordingly, we use terms such as *a person with schizophrenia*, rather than describing people as *schizophrenic*.

Evaluating the DSM-IV-TR The *DSM-IV-TR* has been praised for carefully and completely describing symptoms, standardizing diagnoses and treatments, facilitating communication, and serving as a valuable educational tool. Critics, however, suggest that it relies too heavily on the medical model and unfairly labels people (Cooper, 2004; Mitchell, 2003; Roelcke, 1997). The *DSM-IV-TR* has also been criticized for its possible cultural bias. The manual does provide a culture-specific section and a glossary of culture-bound syndromes, such as amok (Indonesia), genital retraction syndrome (Asia), and windigo psychosis (Native American cultures), which we discuss later in this chapter. However, the classification of most disorders still reflects a Western European and American perspective (Dana, 1998; Matsumoto, 2000; Smart & Smart, 1997).

Five axes of the *DSM-IV-TR* FIGURE 13.2

Each axis serves as a broad category that helps organize the wide variety of mental disorders and acts as a guideline for making decisions. (However, the *DSM* does not suggest therapies or treatment.) Reprinted with permission from the *Diagnostic and Statistical Manual of Mental Disorders,* copyright 2000, American Psychiatric Association.

Summary of main categories of mental disorders and their descriptions in *DSM-IV-TR** TABLE 13.1

1. **Anxiety Disorders:** Problems associated with severe anxiety, such as *phobias, obsessive-compulsive disorder,* and *posttraumatic stress disorder.*

2. **Mood Disorders:** Problems associated with severe disturbances of mood, such as *depression, mania,* or alternating episodes of the two *(bipolar disorder).*

3. **Schizophrenia and Other Psychotic Disorders:** A group of disorders characterized by major disturbances in perception, language and thought, emotion, and behavior.

4. **Dissociative Disorders:** Disorders in which the normal integration of consciousness, memory, or identity is suddenly and temporarily altered, such as *amnesia* and *dissociative identity disorder.*

5. **Personality Disorders:** Problems related to lifelong maladaptive personality traits, including *antisocial personality disorders.* (violation of others' rights with no sense of guilt) or *borderline personality disorders* (impulsivity and instability in mood and relationships).

6. **Substance-related Disorders:** Problems caused by alcohol, cocaine, tobacco, and other drugs.

7. **Somatoform Disorders:** Problems related to unusual preoccupation with physical health or physical symptoms with no physical cause.

8. **Factitious Disorders:** Disorders that the individual adopts to satisfy some economic or psychological need.

9. **Sexual and Gender Identity Disorders:** Problems related to unsatisfactory sexual activity, finding unusual objects or situations arousing, or gender identity problems.

10. **Eating Disorders:** Problems related to food, such as *anorexia nervosa* and *bulimia.*

11. **Sleep Disorders:** Serious disturbances of sleep, such as *insomnia* (too little sleep), *sleep terrors,* or *hypersomnia* (too much sleep).

12. **Impulse Control Disorders (not elsewhere classified):** Problems related to *kleptomania* (impulsive stealing), *pyromania* (setting of fires), and *pathological gambling.*

13. **Adjustment Disorders:** Problems involving excessive emotional reaction to specific stressors such as divorce, family discord, or economic concerns.

14. **Disorders usually first diagnosed in infancy, childhood, or early adolescence:** Problems that appear before adulthood, including mental retardation and language development disorders.

15. **Delirium, Dementia, Amnestic, and Other Cognitive Disorders:** Problems caused by known damage to the brain, including Alzheimer's disease, strokes, and physical trauma to the brain.

16. **Mental Disorders due to a general medical condition (not elsewhere classified):** Problems caused by physical deterioration of the brain due to disease, drugs, and so on.

17. **Other conditions that may be a focus of clinical attention:** Problems related to physical or sexual abuse, relational problems, occupational problems, and so forth.

* *Diagnostic Manual of Mental Disorders (DSM-IV-TR).*

Reprinted with permission from the *Diagnostic and Statistical Manual of Mental Disorders,* copyright 2000, American Psychiatric Association.

Anxiety disorders

Mood disorders

Substance-related disorders

NATIONAL GEOGRAPHIC

CONCEPT CHECK STOP

Why isn't any single criterion adequate for classifying abnormal behavior?

How does psychology diverge from the medical model?

What are the purposes of the five axes of the *DSM-IV-TR*?

Anxiety Disorders

LEARNING OBJECTIVES

Describe the symptoms of generalized anxiety disorders, panic disorder, phobias, and obsessive-compulsive disorder.

Summarize how psychological, biological, and sociocultural factors contribute to anxiety disorders.

Anxiety disorders—which are diagnosed twice as often in women as in men—are the most frequently occurring mental disorders in the general population (National Institute of Mental Health, 1999; Swartz & Margolis, 2004). Fortunately, they are also among the easiest disorders to treat and have one of the best chances for recovery (Chapter 14).

FOUR MAJOR ANXIETY DISORDERS: THE PROBLEM OF FEAR

Symptoms of anxiety, such as rapid breathing, dry mouth, and increased heart rate, plague all of us during stressful moments. But some people experience anxiety that is so intense and chronic it seriously disrupts their lives. They feel threatened, unable to cope, unhappy, and insecure in a world that seems dangerous and hostile. In this section, we consider four anxiety disorders: generalized anxiety disorder, panic disorder, phobias, and obsessive-compulsive disorder (**FIGURE 13.3**). (Posttraumatic stress disorder, another major anxiety disorder, is discussed in Chapter 3.) Although we discuss these disorders separately, people often have more than one anxiety disorder (Barlow, Esler, & Vitali, 1998).

Generalized anxiety disorder
This disorder affects twice as many women as it does men (Brawman-Mintzer & Lydiard, 1996, 1997). **Generalized anxiety disorder** is characterized by chronic, uncontrollable, and excessive fear and worry that lasts at least six months and that is not focused on any particular object

anxiety disorder A type of abnormal behavior characterized by unrealistic, irrational fear.

or situation. Because of persistent muscle tension and autonomic fear reactions, people with this disorder may develop headaches, heart palpitations, dizziness, and insomnia, making it even harder to cope with normal daily activities.

Panic disorder
Sudden, but brief, attacks of intense apprehension that cause trembling, dizziness, and difficulty breathing are symptoms of **panic disorder**. Panic attacks generally happen after frightening experiences or prolonged stress (and sometimes even after exercise). Panic disorder is diagnosed when several apparently spontaneous panic attacks lead to a persistent concern about future attacks. A common complication of panic disorder is agoraphobia (Craske, 2000; Gorman, 2000).

Phobias
Phobias involve a strong, irrational fear and avoidance of objects or situations that are usually considered harmless (fear of elevators or fear of going to the dentist, for example). Although the person recognizes that the fear is irrational, the experience is still one of overwhelming anxiety, and a full-blown panic attack may follow. The *DSM-IV-TR* divides phobic disorders into three broad categories: agoraphobia, specific phobias, and social phobias.

People with *agoraphobia* restrict their normal activities because they fear having a panic attack in crowded, enclosed, or wide-open places where they would be unable to receive help in an emergency. In severe cases, people with agoraphobia may refuse to leave the safety of their homes.

A *specific phobia* is a fear of a specific object or situation, such as needles, heights, rats, or spiders.

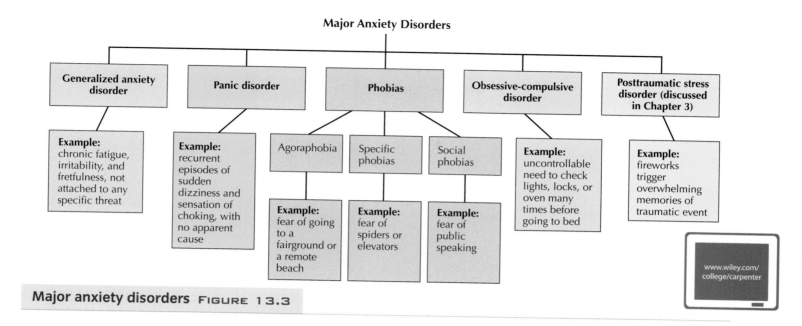

Major Anxiety Disorders

| Generalized anxiety disorder | Panic disorder | Phobias | Obsessive-compulsive disorder | Posttraumatic stress disorder (discussed in Chapter 3) |

Example: chronic fatigue, irritability, and fretfulness, not attached to any specific threat

Example: recurrent episodes of sudden dizziness and sensation of choking, with no apparent cause

Agoraphobia | Specific phobias | Social phobias

Example: fear of going to a fairground or a remote beach

Example: fear of spiders or elevators

Example: fear of public speaking

Example: uncontrollable need to check lights, locks, or oven many times before going to bed

Example: fireworks trigger overwhelming memories of traumatic event

www.wiley.com/college/carpenter

Major anxiety disorders FIGURE 13.3

Claustrophobia (fear of closed spaces) and acrophobia (fear of heights) are the specific phobias most often treated by therapists. People with specific phobias generally recognize that their fears are excessive and unreasonable, but they are unable to control their anxiety and will go to great lengths to avoid the feared stimulus (FIGURE 13.4).

People with *social phobias* are irrationally fearful of embarrassing themselves in social situations. Fear of public speaking and of eating in public are the most common social phobias. The fear of public scrutiny and potential humiliation may become so pervasive that normal life is impossible (den Boer, 2000; Swartz & Margolis, 2004).

Obsessive-compulsive disorder (OCD)

This disorder involves persistent, unwanted fearful thoughts (obsessions) or irresistible urges to perform an act or repeated ritual (compulsions), which help relieve the anxiety created by the obsession. In adults, this disorder is equally common in men and women. However, it is more prevalent among boys when the onset is in childhood (American Psychiatric Association, 2000).

Imagine what it would be like to worry so obsessively about germs that you compulsively wash your hands hundreds of times a day until they are raw and bleeding. Most sufferers of OCD realize that their actions are senseless. But when they try to stop the behavior, they experience mounting anxiety, which is relieved only by giving in to the urges.

Spider phobia FIGURE 13.4

Can you imagine how it would feel to be so frightened by a spider that you would try to jump out of a speeding car to get away from it? This is how a person suffering from a phobia might feel.

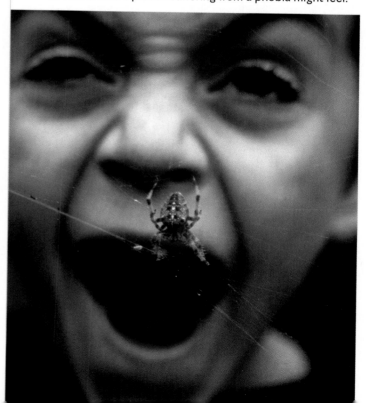

CAUSES OF ANXIETY DISORDERS

The causes of anxiety disorders are debatable. Research has focused on the roles of psychological, biological, and sociocultural processes (the *biopsychosocial model*).

Psychological contributions to anxiety disorders are primarily in the form of faulty cognitive processes and maladaptive learning.

Faulty cognitions
People with anxiety disorders have habits of thinking, or cognitive habits, that make them prone to fear. They tend to be hypervigilant—they constantly scan their environment for signs of danger and ignore signs of safety. They also tend to magnify ordinary threats and failures and to be hypersensitive to others' opinions of them (FIGURE 13.5).

Maladaptive learning
According to learning theorists, anxiety disorders generally result from conditioning and social learning (Chapter 6) (Bouton, Mineka, & Barlow, 2001; King, Clowes-Hollins, & Ollendick, 1997; Thomas & Ayres, 2004).

During classical conditioning, for example, a stimulus that is originally neutral (e.g., a harmless spider) becomes paired with a frightening event (a sudden panic attack) so that it becomes a conditioned stimulus that elicits anxiety. The person then begins to avoid spiders in order to reduce anxiety (an operant conditioning process known as negative reinforcement).

Some researchers contend that the fact that most people with phobias cannot remember a specific instance that led to their fear and that frightening experiences do not always trigger phobias suggests that conditioning may not be the only explanation. Social learning theorists propose that some phobias are the result of modeling and imitation.

Phobias may also be learned vicariously (indirectly). In one study, rhesus monkeys viewed videos showing another monkey apparently experiencing extreme fear of a toy snake, a toy rabbit, a toy crocodile, and flowers (Cook & Mineka, 1989). The "viewing" monkeys were later afraid of the snake and crocodile but not of the rabbit or flowers, suggesting that phobias have both learned and biological components.

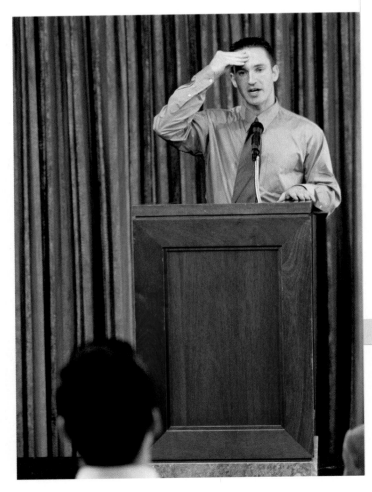

Faulty thinking patterns FIGURE 13.5

People who suffer from social phobia are excessively concerned about others' evaluations, hypersensitive to any criticism, and obsessively worried about potential mistakes. This intense self-preoccupation leads these people to perpetually believe they have failed. What changes in thinking patterns might lessen this anxiety?

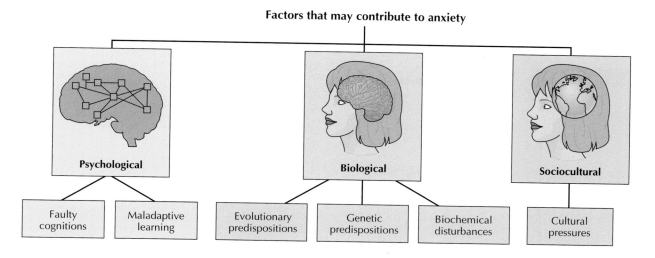

Factors that may contribute to anxiety FIGURE 13.6

Some researchers believe that phobias reflect an evolutionary predisposition to fear that which was dangerous to our ancestors (Mineka & Oehman, 2002; Rossano, 2003). Some people with panic disorder also seem to be genetically predisposed toward an overreaction of the autonomic nervous system, further supporting the argument for a biological component. In addition, stress and arousal seem to play a role in panic attacks, and drugs such as caffeine or nicotine and even hyperventilation can trigger an attack, all suggesting a biochemical disturbance.

In addition to psychological and biological components, sociocultural factors can contribute to anxiety. There has been a sharp rise in anxiety disorders in the past 50 years, particularly in Western industrialized countries. Can you see how our increasingly fast-paced lives—along with our increased mobility, decreased job security, and decreased family support—might contribute to anxiety? Unlike the dangers that humans faced in our evolutionary history, today's threats are less identifiable and immediate, which may lead some people to become hypervigilant and predisposed to anxiety disorders (FIGURE 13.6).

Further support for sociocultural influences on anxiety disorders is that anxiety disorders can have dramatically different forms in other cultures. For example, in a collectivist twist on anxiety, the Japanese have a type of social phobia called *taijin kyofusho* (TKS), which involves morbid dread of doing something to embarrass others. This disorder is quite different from the Western version of social phobia, which centers on a fear of criticism.

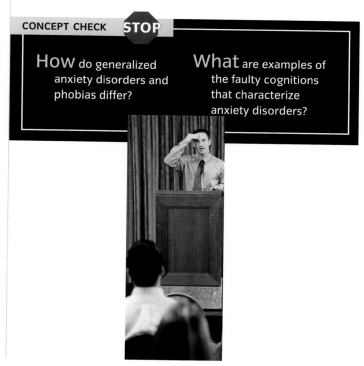

CONCEPT CHECK STOP

How do generalized anxiety disorders and phobias differ?

What are examples of the faulty cognitions that characterize anxiety disorders?

Mood Disorders

LEARNING OBJECTIVES

Explain how major depressive disorder and bipolar disorder differ.

Summarize research on the biological and psychological factors that contribute to mood disorders.

UNDERSTANDING MOOD DISORDERS: MAJOR DEPRESSIVE DISORDER AND BIPOLAR DISORDER

As the name implies, **mood disorders** (also known as affective disorders) are characterized by extreme disturbances in emotional states. There are two main types of mood disorders—major depressive disorder and bipolar disorder.

We all feel "blue" sometimes, especially following the loss of a job, end of a relationship, or death of a loved one. People suffering from **major depressive disorder**, however, may experience a lasting and continuously depressed mood without a clear trigger.

Clinically depressed people are so deeply sad and discouraged that they often have trouble sleeping, are likely to lose (or gain) weight, and may feel so fatigued that they cannot go to work or school or even comb their hair and brush their teeth. They may sleep constantly, have problems concentrating, and feel so profoundly sad and guilty that they consider suicide. Depressed individuals have a hard time thinking clearly or recognizing their own problems.

When depression is *unipolar*, the depressive episode eventually ends, and the person returns to a "normal" emotional level. People with **bipolar disorder**, however, rebound to the opposite state, known as *mania* (**FIGURE 13.7**).

During a manic episode, the person is overly excited, extremely active, and easily distracted. The person exhibits unrealistically high self-esteem, an inflated

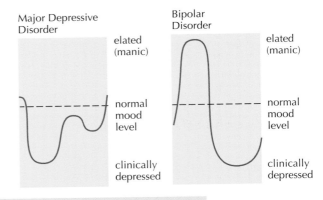

Mood disorders FIGURE 13.7

If major depressive disorders and bipolar disorders were depicted on a graph, they might look something like this.

major depressive disorder Long-lasting depressed mood that interferes with the ability to function, feel pleasure, or maintain interest in life (Swartz & Margolis, 2004).

bipolar disorder Repeated episodes of mania (unreasonable elation and hyperactivity) and depression.

sense of importance, and poor judgment (**FIGURE 13.8**). The person is hyperactive and may not sleep for days at a time yet does not become fatigued. Thinking is speeded up and can change abruptly to new topics, showing "rapid flight of ideas." Speech is also rapid ("pressured speech") and it is difficult for others to get a word in edgewise.

A manic episode may last a few days or a few months, and it generally ends abruptly. The ensuing depressive episode generally lasts three times as long as the manic episode. The lifetime risk for bipolar disorder is low—somewhere between 0.5 and 1.6 percent—but it can be one of the most debilitating and lethal disorders, with a suicide rate between 10 and 20 percent among sufferers (Goodwin et al., 2004; MacKinnon, Jamison, & DePaulo, 1997).

No limits FIGURE 13.8

Poor judgment is common during manic episodes. A person may give away valuable possessions, go on wild spending sprees, or make elaborate plans for becoming rich and famous.

CAUSES OF MOOD DISORDERS: BIOLOGICAL VERSUS PSYCHOSOCIAL FACTORS

Biological factors appear to play a significant role in both major depression and bipolar disorder. Recent research suggests that structural brain changes may contribute to these mood disorders (Almeida et al., 2003; Lyoo et al., 2004; Steffens et al., 2003). Other research points to imbalances of the neurotransmitters serotonin, norepinephrine, and dopamine (Delgado, 2004; Owens, 2004; Southwick, Vythilingam, & Charney, 2005). Drugs that alter the activity of these neurotransmitters also decrease the symptoms of depression (and are therefore called antidepressants) (Chuang, 1998).

Some research indicates that mood disorders may be inherited (Baldessarini & Hennen, 2004; Horiuchi et al., 2004; Sequeira et al., 2004). For example, when one identical twin has a mood disorder, there is about a 50 percent chance that the other twin will also develop the illness (Swartz & Margolis, 2004). However, it is important to remember that relatives generally have similar environments as well as similar genes.

Finally, the evolutionary perspective suggests that moderate depression may be a normal and healthy adaptive response to a very real loss (such as the death of a loved one). The depression helps us to step back and reassess our goals (Neese, 2000). Clinical, severe depression may just be an extreme version of this generally adaptive response.

Psychosocial theories of depression focus on environmental stressors and disturbances in the person's interpersonal relationships, thought processes, self-concept, and history of learned behaviors (Cheung, Gilbert, & Irons, 2004; Hammen, 2005; Mathews & McLeod, 2005). The psychoanalytic explanation sees depression as anger (stemming from feelings of rejection) turned inward when an important relationship is lost. The humanistic school says that depression results when a person demands perfection of him- or herself or when positive growth is blocked.

According to the **learned helplessness** theory (Seligman, 1975, 1994), depression occurs when people (and other animals) become resigned to the idea that they are helpless to escape from something painful. Learned helplessness may be particularly likely to trigger depression if the person attributes failure to causes that are internal ("my own weakness"), stable ("this weakness is long-standing and unchanging"), and global ("this weakness is a problem in lots of settings") (Chaney et al., 2004; Gotlieb & Abramson, 1999; Peterson & Vaidya, 2001).

Suicide is a major danger associated with depression. Because of the associated shame and secrecy with suicide, many fail to get or give help.

CONCEPT CHECK **STOP**

What are the key characteristics of a manic episode?

How might major depression be an exaggeration of an evolutionary adaptation?

Schizophrenia

LEARNING OBJECTIVES

Describe some common symptoms of schizophrenia.

Compare the traditional (four-group) system for classifying different types of schizophrenia with the (two-group) system that has recently emerged.

Summarize the biological and psychosocial factors that contribute to schizophrenia.

I magine that your daughter has just left for college and that you hear voices inside your head shouting, "You'll never see her again! You have been a bad mother! She'll die." Or what if you saw live animals in your refrigerator? These experiences have plagued "Mrs. T" for decades (Gershon & Rieder, 1993).

Mrs. T suffers from **schizophrenia** (skit-so-FREE-nee-uh). Schizophrenia is often so severe that it is considered a psychosis, meaning that the person is out of touch with reality. People with schizophrenia have serious problems caring for themselves, relating to others, and holding a job. In extreme cases, people with schizophrenia require institutional or custodial care.

Schizophrenia is one of the most widespread and devastating mental disorders. Approximately 1 out of every 100 people will develop schizophrenia in his or her lifetime, and approximately half of all people who are admitted to mental hospitals are diagnosed with this disorder (Gottesman, 1991; Kendler et al., 1996; Kessler et al., 1994; Regier et al., 1993). Schizophrenia usually emerges between the late teens and the mid-30s. It seems to be equally prevalent in men and women, but it's generally more severe and strikes earlier in men than in women (Salyers & Mueser, 2001).

Many people confuse schizophrenia with dissociative identity disorder, which is sometimes referred to as *split* or *multiple personality disorder*. Schizophrenia means "split mind," but when Eugen Bleuler coined the term in 1911, he was referring to the fragmenting of thought processes and emotions, not personalities (Neale, Oltmanns, & Winters, 1983). As we discuss later in this chapter, dissociative identity disorder is the rare condition of having more than one distinct personality.

SYMPTOMS OF SCHIZOPHRENIA: FIVE AREAS OF DISTURBANCE

Schizophrenia is a group of disorders characterized by a disturbance in one or more of the following areas: perception, language, thought, affect (emotions), and behavior.

Perceptual symptoms The senses of people with schizophrenia may be either enhanced or blunted. The filtering and selection processes that allow most people to concentrate on whatever they choose are impaired, and sensory stimulation is jumbled and distorted.

People with schizophrenia may also experience **hallucinations**—most commonly auditory hallucinations (hearing voices and sounds).

On rare occasions, people with schizophrenia will hurt others in response to their distorted perceptions. But a person with schizophrenia is more likely to be self-destructive and suicidal than violent toward others.

> **schizophrenia** A group of psychotic disorders involving major disturbances in perception, language, thought, emotion, and behavior. The individual withdraws from people and reality, often into a fantasy life of delusions and hallucinations.

> **hallucinations** Imaginary sensory perceptions that occur without an external stimulus.

Language and thought disturbances For people with schizophrenia, words lose their usual meanings and associations, logic is impaired, and thoughts are disorganized and bizarre. When language and thought disturbances are mild, the individual jumps from topic to topic. With more severe disturbances, the person jumbles phrases and words together (into a "word salad") or creates artificial words.

The most common—and frightening—thought disturbance experienced by people with schizophrenia is the lack of contact with reality (psychosis).

Delusions are also common in people with schizophrenia. We all experience exaggerated thoughts from time to time, such as thinking a friend is trying to avoid us, but the delusions of schizophrenia are much more extreme. For example, Jim (the med student in the chapter opener) was completely convinced that others were trying to assassinate him (a *delusion of persecution*). In *delusions of grandeur*, people believe that they are someone very important, perhaps Jesus Christ or the queen of England. In *delusions of reference*, unrelated events are given special significance, as when a person believes that a radio program is giving him or her a special message.

> **delusions** Mistaken beliefs based on misrepresentations of reality.

Emotional disturbances   Changes in emotion usually occur in people with schizophrenia. In some cases, emotions are exaggerated and fluctuate rapidly. At other times, emotions become blunted. Some people with schizophrenia have *flattened affect*—almost no emotional response of any kind.

Behavioral disturbances   Disturbances in behavior may take the form of unusual actions that have special meaning. For example, one patient massaged his head repeatedly to "clear it" of unwanted thoughts. People with schizophrenia may become *cataleptic* and assume a nearly immobile stance for an extended period.

TYPES OF SCHIZOPHRENIA: RECENT METHODS OF CLASSIFICATION

For many years, researchers divided schizophrenia into five subtypes: paranoid, catatonic, disorganized, undifferentiated, and residual (TABLE 13.2). Although these terms are still included in the *DSM-IV-TR*, critics contend that this system does not differentiate in terms of prognosis, cause, or response to treatment and that the undifferentiated type is merely a catchall for cases that are difficult to diagnose (American Psychiatric Association, 2000).

Subtypes of schizophrenia TABLE 13.2

Paranoid	Dominated by delusions (persecution and grandeur) and hallucinations (hearing voices)
Catatonic	Marked by motor disturbances (immobility or wild activity) and echo speech (repeating the speech of others)
Disorganized	Characterized by incoherent speech, flat or exaggerated emotions, and social withdrawal
Undifferentiated	Meets the criteria for schizophrenia but is not any of the above subtypes
Residual	No longer meets the full criteria for schizophrenia but still shows some symptoms

For these reasons, researchers have proposed an alternative classification system:

1. **Positive schizophrenia symptoms** involve additions to or exaggerations of normal thought processes and behaviors, including bizarre delusions and hallucinations.

2. **Negative schizophrenia symptoms** involve the loss or absence of normal thought processes and behaviors, including impaired attention, limited or toneless speech, flattened affect, and social withdrawal.

Positive symptoms are more common when schizophrenia develops rapidly, whereas negative symptoms are more often found in slow-developing schizophrenia. Positive symptoms are associated with better adjustment before the onset and a better prognosis for recovery.

In addition to these two groups, the latest *DSM* suggests adding another dimension to reflect *disorganization of behavior*, including rambling speech, erratic behavior, and inappropriate affect.

CAUSES OF SCHIZOPHRENIA: NATURE AND NURTURE THEORIES

Because schizophrenia comes in many different forms, it probably has multiple biological and psychosocial bases (Walker et al., 2004). Let's look at biological contributions first.

Prenatal viral infections, birth complications, immune responses, maternal malnutrition, and advanced paternal age all may contribute to the development of schizophrenia (Beraki et al., 2005; Cannon et al., 2002; Dalman & Allebeck, 2002; Sawa & Kamiya, 2003; Zuckerman & Weiner, 2005). However, most biological theories of schizophrenia focus on genetics, neurotransmitters, and brain abnormalities.

- *Genetics.* Although researchers are beginning to identify specific genes related to schizophrenia, most genetic studies have focused on twins and adoptions (Crow, 2004; Davalos et al., 2004; Elkin, Kalidindi, & McGuffin, 2004; Hulshoff Pol et al., 2004; Lindholm et al., 2004; Petronis, 2000). This research indicates that the risk for schizophrenia increases with genetic similarity;

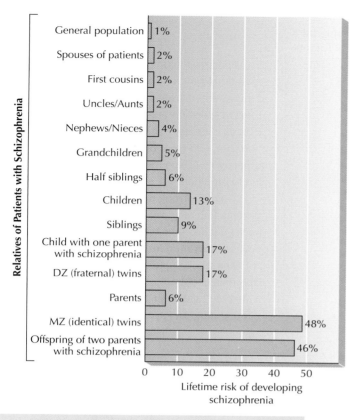

Figure shows a horizontal bar chart titled with the y-axis label "Relatives of Patients with Schizophrenia":

Relative	Lifetime risk
General population	1%
Spouses of patients	2%
First cousins	2%
Uncles/Aunts	2%
Nephews/Nieces	4%
Grandchildren	5%
Half siblings	6%
Children	13%
Siblings	9%
Child with one parent with schizophrenia	17%
DZ (fraternal) twins	17%
Parents	6%
MZ (identical) twins	48%
Offspring of two parents with schizophrenia	46%

x-axis: 0 10 20 30 40 50
Lifetime risk of developing schizophrenia

Genetics and schizophrenia FIGURE 13.9

Your lifetime risk of developing schizophrenia depends, in part, on how closely you are genetically related to someone with schizophrenia. *Source:* Gottesman, "Schizophrenia Genesis," 1991, W.H. Freeman and Company/Worth Publishers.

that is, people who share more genes with a person who has schizophrenia are more likely to develop the disorder (FIGURE 13.9).

- *Neurotransmitters.* Precisely how genetic inheritance produces schizophrenia is unclear. According to the **dopamine hypothesis**, overactivity of certain dopamine neurons in the brain causes schizophrenia (Ikemoto, 2004; Paquet et al., 2004). This hypothesis is based on two observations. First, administering amphetamines increases the amount of dopamine and can produce (or worsen) some symptoms of schizophrenia, especially in people with a genetic predisposition to the disorder. Second, drugs that reduce dopamine activity in the brain reduce or eliminate some symptoms of schizophrenia.

- *Brain abnormalities.* The third major biological theory for schizophrenia involves abnormalities in brain function and structure. Researchers have found larger cerebral ventricles (fluid-filled spaces in the brain) in some people with schizophrenia (DeLisi et al., 2004; Gaser et al., 2004).

Also, some people with chronic schizophrenia have a lower level of activity in their frontal and temporal lobes—areas that are involved in language, attention, and memory (FIGURE 13.10). Damage in these regions might explain the thought and language disturbances that characterize schizophrenia. This lower level of brain activity, and schizophrenia itself, may also result from an overall loss of gray matter (neurons in the cerebral cortex) (Gogtay et al., 2004; Hulshoff Pol et al., 2004).

Clearly, biological factors play a key role in schizophrenia. But the fact that even in identical twins—who share identical genes—the heritability of schizophrenia is only 48 percent tells us that nongenetic factors must contribute the remaining percentage. Most psychologists believe that there are at least two possible psychosocial contributors.

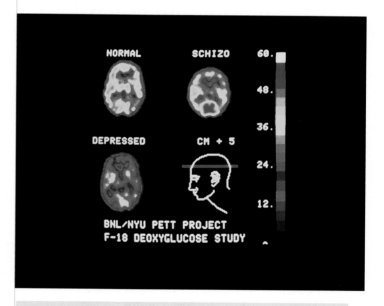

Brain activity in schizophrenia FIGURE 13.10

These positron emission tomography (PET) scans show variations in the brain activity of normal individuals, people with major depressive disorder, and individuals with schizophrenia. Warmer colors (red, yellow) indicate increased activity.

The biopsychosocial model and schizophrenia FIGURE 13.11

Biological Factors

- Malfunctioning neurotransmitters
- Metabolic and anatomical abnormalities of the brain
- Genetic predisposition
- Possible other *unknown* factors

Psychological & Social Factors

- Stress
- Family communication deviance
- Possible other *unknown* factors

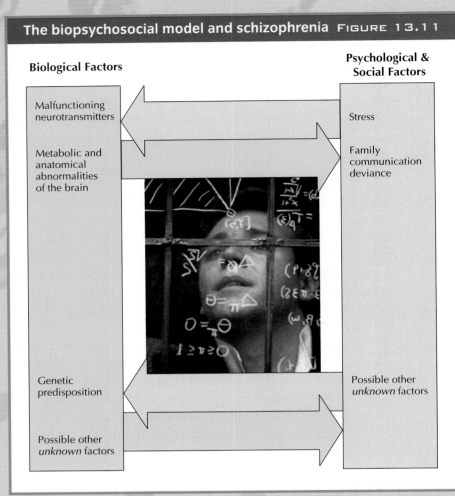

In his portrayal of the Nobel Prize-winning mathematician John Nash, actor Russell Crowe demonstrated several classic symptoms of schizophrenia, including disturbances in perception, language, thought, behavior, and emotion. For Nash, as with others with schizophrenia, no single factor led to his illness.

There is strong evidence linking schizophrenia to biological, psychological, and social factors—the biopsychosocial model. *Source:* Reprinted from Biological Psychiatry, Meltzer, "Genetics and Etiology of Schizophrenia and Bipolar Disorder," pp. 171–178, 2000, with permission from Society of Biological Psychiatry.

www.wiley.com/college/carpenter

According to the **diathesis-stress model** of schizophrenia, stress plays an essential role in triggering schizophrenic episodes in people with an inherited predisposition (or diathesis) toward the disease (Corcoran et al., 2003; Schwartz & Smith, 2004; Walker et al., 2004).

Some investigators suggest that communication disorders in family members may also be a predisposing factor for schizophrenia. Such disorders include unintelligible speech, fragmented communication, and parents' frequently sending severely contradictory messages to children. Several studies have also shown greater rates of relapse and worsening of symptoms among hospitalized patients who went home to families that were critical and hostile toward them or overly involved in their lives emotionally (Hooley & Hiller, 2001; Lefley, 2000).

How should we evaluate the different theories about the causes of schizophrenia? Critics of the dopamine hypothesis and the brain damage theory argue that those theories fit only some cases of schizophrenia. Moreover, with both theories, it is difficult to determine cause and effect. The disturbed-communication theories are also hotly debated, and research is inconclusive. Schizophrenia is probably the result of a combination of known and unknown interacting factors (**FIGURE 13.11**).

CONCEPT CHECK STOP

How do hallucinations and delusions differ?

How do positive and negative symptoms of schizophrenia differ?

What is the diathesis-stress model of schizophrenia?

Other Disorders

SUBSTANCE-RELATED DISORDERS

substance-related disorders
Abuse of or dependence on a mood- or behavior-altering drug.

The category of **substance-related disorders** is subdivided into two general groups: substance abuse and substance dependence (TABLE 13.3). When alcohol or other drug use interferes with a person's social or occupational functioning, it is called *substance abuse*. Drug use becomes *substance dependence* when it also causes physical reactions, including *tolerance* (requiring more of the drug to get the desired effect) and *withdrawal* (negative physical effects when the drug is removed).

Some people can use alcohol and other drugs and not develop a problem. Unfortunately, researchers have not been able to identify ahead of time those who can use drugs without risk of becoming abusers versus those who are likely to become abusers. Complicating diagnosis and treatment is the fact that substance-related disorders commonly coexist with other mental disorders, including anxiety disorders, mood disorders, schizophrenia, and personality disorders (Bates & Rutherford, 2003; Green et al., 2004; Teesson, Hodder, & Buhrich, 2004). This co-occurrence of disorders is called **comorbidity** (FIGURE 13.12).

Comorbidity complicates treatment

FIGURE 13.12

Can you see how comorbidity can cause serious problems? How can the appropriate cause, course, or treatment be identified for someone dealing with a combination of disorders?

DSM-IV-TR substance abuse and substance dependence TABLE 13.3

Criteria for Substance Abuse (alcohol and other drugs)	*Criteria for Substance Dependence* (alcohol and other drugs)
Maladaptive use of a substance shown by one of the following:	*Three or more of the following:*
• Failure to meet obligations	• Tolerance
• Repeated use in situations where it is physically dangerous	• Withdrawal
• Continued use despite problems caused by the substance	• Substance taken for a longer time or greater amount than intended
• Repeated substance-related legal problems	• Lack of desire or efforts to reduce or control use
	• Social, recreational, or occupational activities given up or reduced
	• Much time spent in activities to obtain the substance
	• Use continued despite knowing that psychological or physical problems are worsened by it

What causes comorbidity? Perhaps the most influential hypothesis is **self-medication**—individuals drink or use drugs to reduce their symptoms (Batel, 2000; Goswami et al., 2004; Green, 2000). Regardless of the causes, it is critical that patients, family members, and clinicians recognize and deal with comorbidity if treatment is to be effective.

DISSOCIATIVE DISORDERS: WHEN THE PERSONALITY SPLITS APART

■ **dissociative disorders**
Amnesia, fugue, or multiple personalities resulting from avoidance of painful memories or situations.

The most dramatic psychological disorders are **dissociative disorders**. There are several types of dissociative disorders, but all involve a splitting apart (a *dis*-association) of significant aspects of experience from memory or consciousness. Individuals dissociate from the core of their personality by failing to remember past experiences (*dissociative amnesia*) (**FIGURE 13.13**), by leaving home and wandering off (*dissociative fugue*), by losing their sense of reality and feeling estranged from the self (*depersonalization disorder*), or by developing completely separate personalities (*dissociative identity disorder*).

Unlike most other psychological disorders, the primary cause of dissociative disorders appears to be environmental variables, with little or no genetic influence (Waller & Ross, 1997).

The most severe dissociative disorder is **dissociative identity disorder (DID)**—previously known as multiple personality disorder—in which at least two separate and distinct personalities exist within a person at the same time. Each personality has unique memories, behaviors, and social relationships. Transition from one personality to another occurs suddenly and is often triggered by psychological stress. Usually, the original personality has no knowledge or awareness of the alternate personalities, but all of the personalities may be aware of lost periods of time. The disorder is diagnosed more among women than among men. Women with DID also tend to have more identities, averaging 15 or more, compared with men, who average 8 (American Psychiatric Association, 2000).

DID is a controversial diagnosis. Some experts suggest that many cases are faked or result from false memories and an unconscious need to please the therapist (Kihlstrom, 2005; Loftus, 1997; McNally, 2004; Stafford & Lynn, 2002). Other psychologists accept the validity of multiple personality and contend that the condition is underdiagnosed (Brown, 2001; Lipsanen et al., 2004; Spiegel & Maldonado, 1999).

Dissociation as an escape FIGURE 13.13

The major force behind all dissociative disorders is the need to escape from anxiety. Imagine witnessing a loved one's death in a horrible car accident. Can you see how your mind might cope by blocking out all memory of the event?

PERSONALITY DISORDERS: ANTISOCIAL AND BORDERLINE

What would happen if the characteristics of a personality were so inflexible and maladaptive that they significantly impaired someone's ability to function? This is what happens with **personality disorders**. Several types of personality disorders are included in this category in the *DSM-IV-TR*, but here we will focus on antisocial personality disorder and borderline personality disorder.

■ **personality disorders**
Inflexible, maladaptive personality traits that cause significant impairment of social and occupational functioning.

Antisocial personality disorder The term **antisocial personality disorder** is used interchangeably with the terms *sociopath* and *psychopath*. These labels describe behavior so far outside the ethical and legal standards of society that many consider it the most serious of all mental disorders. Unlike with anxiety, mood disorders, and schizophrenia, people with this diagnosis feel little personal distress (and may not be motivated to change). Yet their maladaptive traits generally bring considerable harm and suffering to others (Hervé et al., 2004; Kirkman, 2002; Nathan et al., 2003). Although serial killers are often seen as classic examples of antisocial personality disorder (FIGURE 13.14), many sociopaths harm people in less dramatic ways—for example, as ruthless businesspeople and crooked politicians.

The four hallmarks of antisocial personality disorder are egocentrism (preoccupation with oneself and insensitivity to the needs of others), lack of conscience, impulsive behavior, and superficial charm (American Psychiatric Association, 2000).

Unlike most adults, individuals with antisocial personality disorder act impulsively, without giving thought to the consequences. They are usually poised when confronted with their destructive behavior and feel contempt for anyone they are able to manipulate. They also change jobs and relationships suddenly, and they often have a history of truancy from school and of being expelled for destructive behavior. People with antisocial personalities can be charming and persuasive, and they have remarkably good insight into the needs and weaknesses of other people.

Twin and adoption studies suggest a possible genetic predisposition to antisocial personality disorder (Bock & Goode, 1996; Jang et al., 2003). Biological contributions are also suggested by studies that have found abnormally low autonomic activity during stress, right hemisphere abnormalities, and reduced gray matter in the frontal lobes (Kiehl et al., 2004; Raine et al., 2000).

Evidence also exists for environmental or psychological causes. Antisocial personality disorder is highly correlated with abusive parenting styles and inappropriate modeling (Farrington, 2000; Pickering, Farmer, & McGuffin, 2004). People with antisocial personality disorder often come from homes characterized by emotional deprivation, harsh and inconsistent disciplinary practices, and antisocial parental behavior. Still

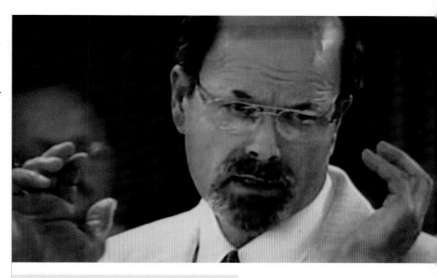

The BTK killer FIGURE 13.14

In 2005, Dennis Rader, known as the "BTK killer" for his method of binding, torturing, and killing his victims, pleaded guilty and was sentenced to life in prison for his crimes. Between 1974 and 1991, Rader murdered 10 women in and around Wichita, Kansas. After his killings, he sent taunting letters to police and local media, boasting of the crimes in graphic detail. Despite having a normal outward appearance—he was a husband, father, Cub Scout leader, and church council president—Rader's egocentrism and lack of conscience are primary characteristics of antisocial personality disorder.

other studies show a strong interaction between both heredity and environment (Paris, 2000; Rutler, 1997).

Borderline personality disorder **Borderline personality disorder (BPD)** is among the most commonly diagnosed personality disorders (Markovitz, 2004). The core features of this disorder are impulsivity and instability in mood, relationships, and self-image. Originally, the term implied that the person was on the borderline between neurosis and schizophrenia (Davison, Neale, & Kring, 2004). The modern conceptualization no longer has this connotation, but BPD remains one of the most complex and debilitating of all the personality disorders.

Mary's story of chronic, lifelong dysfunction, described in the chapter opener, illustrates the serious problems associated with this disorder. People with borderline personality disorder experience extreme difficulties in relationships. Subject to chronic feelings of depression, emptiness, and intense fear of abandonment, they also engage in destructive, impulsive behaviors, such as sexual promiscuity, drinking, gambling, and eating sprees (Chabrol et al., 2004; Trull et al., 2000). They may attempt suicide and sometimes engage in self-mutilating behavior (Bohus et al., 2004; McKay, Gavigan, & Kulchycky, 2004; Paris, 2004).

People with BPD tend to see themselves and everyone else in absolute terms—perfect or worthless (Mason & Kreger, 1998). They constantly seek reassurance from others and may quickly erupt in anger at the slightest sign of disapproval. The disorder is also typically marked by a long history of broken friendships, divorces, and lost jobs.

People with borderline personality disorder frequently have a childhood history of neglect; emotional

deprivation; and physical, sexual, or emotional abuse (Goodman & Yehuda, 2002; Helgeland & Torgersen, 2004; Schmahl et al., 2004). Borderline personality disorder also tends to run in families, and some data suggest it is a result of impaired functioning of the brain's frontal lobes and limbic system, areas that control impulsive behaviors (Schmahl et al., 2004; Tebartz van Elst et al., 2003).

Although some therapists have had success treating BPD with drug therapy and behavior therapy (Bohus et al., Markovitz, 2004), the general prognosis is not favorable. In one study, seven years after treatment, about 50 percent of the original group still had the disorder (Links, Heslegrave, & van Reekum, 1998).

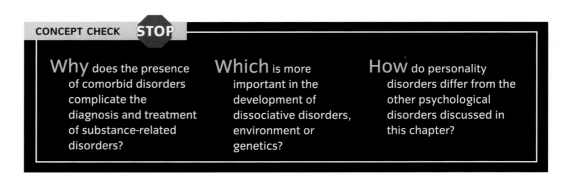

CONCEPT CHECK **STOP**

Why does the presence of comorbid disorders complicate the diagnosis and treatment of substance-related disorders?

Which is more important in the development of dissociative disorders, environment or genetics?

How do personality disorders differ from the other psychological disorders discussed in this chapter?

How Gender and Culture Affect Abnormal Behavior

LEARNING OBJECTIVES

Identify the biological, psychological, and social factors that might explain gender differences in depression.

Explain why it is difficult to directly compare mental disorders across cultures.

Explain why recognizing the difference between culture-general and culture-bound disorders and symptoms can help prevent ethnocentrism in the diagnosis and treatment of psychological disorders.

mong the Chippewa, Cree, and Montagnais-Naskapi Indians in Canada, there is a disorder called *windigo—or wiitiko—psychosis*, which is characterized by delusions and cannibalistic impulses. Believing that

they have been possessed by the spirit of a windigo, a cannibal giant with a heart and entrails of ice, victims become severely depressed (Barnouw, 1985). As the malady begins, the individual typically experiences loss of appetite, diarrhea, vomiting, and insomnia, and he or she may see people turning into beavers and other edible animals.

In later stages, the victim becomes obsessed with cannibalistic thoughts and may even attack and kill loved ones in order to devour their flesh (Berreman, 1971).

If you were a therapist, how would you treat this disorder? Does it fit neatly into any of the categories of psychological disorders that you have learned about? We began this chapter discussing the complexities and problems with defining, identifying, and classifying abnormal behavior. Before we close, we need to add two additional confounding factors: gender and culture. In this section, we explore a few of the many ways in which men and women differ in how they experience abnormal behavior. We also look at cultural variations in abnormal behavior.

GENDER AND DEPRESSION: WHY ARE WOMEN MORE DEPRESSED?

In the United States, Canada, and other countries, the rate of severe depression for women is two to three times the rate for men (Angst et al., 2002; Nolen-Hoeksema, Larson, & Grayson, 2000; Ohayon & Schatzberg, 2002; Parker & Brotchie, 2004).

Why is there such a disparity between men and women? Research explanations can be grouped under biological influences (hormones, biochemistry, and genetic predisposition), psychological processes (ruminative thought processes), and social factors (greater poverty, work-life conflicts, unhappy marriages, and sexual or physical abuse) (Cheung, Gilbert, & Irons, 2004; Garnefski et al., 2004; Kornstein, 2002; Parker & Brotchie, 2004).

According to the *biopsychosocial model,* some women inherit a genetic or hormonal predisposition toward depression. This biological predisposition combines with society's socialization processes to help reinforce behaviors—such as greater emotional expression, passivity, and dependence—that increase the chances for depression (Alloy et al., 1999; Nolen-Hoeksema, Larson, & Grayson, 2000). At the same time, focusing only on classical symptoms of depression (sadness, low energy, and feelings of helplessness) may cause large numbers of depressed men to be overlooked (FIGURE 13.15).

Depression in disguise? FIGURE 13.15

In our society, men are typically socialized to suppress their emotions and to show their distress by acting out (showing aggression), acting impulsively (driving recklessly and committing petty crimes), and engaging in substance abuse. How might such cultural pressures lead us to underestimate male depression?

CULTURE AND SCHIZOPHRENIA: DIFFERENCES AROUND THE WORLD

Peoples of different cultures experience mental disorders in a variety of ways. For example, the reported incidence of schizophrenia varies within different cultures around the world. It is unclear whether these differences result from actual differences in prevalence of the disorder or from differences in definition, diagnosis, or reporting (Kleinman & Cohen, 1997; Lefley, 2000).

The symptoms of schizophrenia also vary across cultures (Stompe et al., 2003), as do the particular stresses that may trigger its onset (FIGURE 13.16).

Finally, despite the advanced treatment facilities and methods in industrialized nations, the prognosis for people with schizophrenia is actually better in nonindustrialized societies. This may be because the core symptoms of schizophrenia (poor rapport with others, incoherent speech, etc.) make it more difficult to survive in highly industrialized countries. In addition, in most industrialized nations, families and other support groups are less likely to feel responsible for relatives and friends with schizophrenia (Brislin, 2000; Lefley, 2000).

AVOIDING ETHNOCENTRISM

Most research on psychological disorders originates and is conducted primarily in Western cultures. Such a restricted sampling can limit our understanding of disorders in general and lead to an ethnocentric view of mental disorders.

Fortunately, cross-cultural researchers have devised ways to overcome these difficulties (Draguns & Tanaka-Matsumi, 2003). For example, Robert Nishimoto (1988) has found several **culture-general symptoms** that are useful in diagnosing disorders across cultures (TABLE 13.4).

What is stressful? FIGURE 13.16

A Some stressors are culturally specific, such as feeling possessed by evil forces or being the victim of witchcraft.

B Other stressors are shared by many cultures, such as the unexpected death of a spouse or loss of a job (Al-Issa, 2000; Browne, 2001; Neria et al., 2002; Torrey & Yolken, 1998).

Twelve culture-general symptoms of mental health difficulties TABLE 13.4		
Nervous	Trouble sleeping	Low spirits
Weak all over	Personal worries	Restless
Feel apart, alone	Can't get along	Hot all over
Worry all the time	Can't do anything worthwhile	Nothing turns out right

Source: From *Understanding Culture's Influence on Behavior,* 2nd edition by Brislin. ©2000. Reprinted with permission of Wadsworth, a division of Thomson Learning. www.thomsonrights.com.

In addition, Nishimoto found several **culture-bound symptoms**. For example, the Vietnamese Chinese reported "fullness in head," the Mexican respondents had "problems with [their] memory," and the Anglo-Americans reported "shortness of breath" and "headaches." Apparently, people learn to express their problems in ways that are acceptable to others in the same culture (Brislin, 1997, 2000; Widiger & Sankis, 2000).

This division between culture-general and culture-bound symptoms also helps us understand depression. Certain symptoms of depression (such as intense sadness, poor concentration, and low energy) seem to exist across all cultures (Green, 1999). But there is also evidence of some culture-bound symptoms. For example, feelings of guilt are found more often in North America and Europe. And in China, *somatization* (the conversion of depression into bodily complaints) occurs more frequently than it does in other parts of the world (Helms & Cook, 1999).

Just as there are culture-bound and culture-general symptoms, researchers have found that mental disorders are themselves sometimes culturally bound (**FIGURE 13.17**). The earlier example of windigo psychosis, a disorder limited to a small group of Canadian Indians, illustrates just such a case.

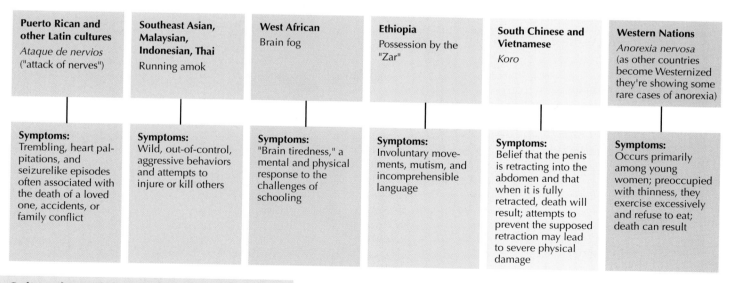

Culture-bound disorders FIGURE 13.17

Some disorders are fading as remote areas become more Westernized, whereas other disorders (such as anorexia nervosa) are spreading as other countries adopt Western values. *Source:* Carmaciu, Anderson, & Markar, 2001; Davison, Neale, & Kring, 2004; Durand & Barlow, 2003; Guarnaccia & Rogler, 1999; Matsumoto, 2000.

As you can see, culture has a strong effect on mental disorders. Studying the similarities and differences across cultures can lead to better diagnosis and understanding. It also helps mental health professionals who work with culturally diverse populations understand both culturally general and culturally bound symptoms.

CONCEPT CHECK STOP

Why might depression be frequently overlooked in men?

How does schizophrenia differ from one culture to another?

What are some examples of culture-bound disorders?

What is happening in this picture ?

Is this man's behavior abnormal? Which criteria for abnormal behavior do his piercings and tattoos meet? Which do they not?

Can you think of any behavior *you* exhibit that might be considered abnormal if your own cultural norms were not taken into account?

VISUAL SUMMARY

1 Studying Psychological Disorders

1. Criteria for **abnormal behavior** include statistical infrequency, disability or dysfunction, personal distress, and violation of norms. None of these criteria alone is adequate for classifying abnormal behavior.

2. Superstitious explanations for abnormal behavior were replaced by the **medical model**, which eventually gave rise to the modern specialty of **psychiatry**. In contrast to the medical model, psychology offers a multifaceted approach to explaining abnormal behavior.

3. The *Diagnostic and Statistical Manual of Mental Disorders*, fourth edition, text revision (*DSM-IV-TR*) is organized according to five major axes, which serve as guidelines for making decisions about symptoms.

2 Anxiety Disorders

1. Major **anxiety disorders** include generalized anxiety disorder, panic disorder, phobias (including agoraphobia, specific phobias, and social phobias), and obsessive-compulsive disorder (OCD).

2. Psychological (faulty cognitions and maladaptive learning), biological (evolutionary and genetic predispositions, biochemical disturbances), and sociocultural factors (cultural pressures toward hypervigilance) likely all contribute to anxiety.

3 Mood Disorders

1. Mood disorders are characterized by extreme disturbances in emotional states. People suffering from **major depressive disorder** may experience a lasting depressed mood without a clear trigger. In contrast, people with **bipolar disorder** alternate between periods of depression and mania (hyperactivity and poor judgment).

2. Biological factors play a significant role in mood disorders. Psychosocial theories of depression focus on environmental stressors and disturbances in interpersonal relationships, thought processes, self-concept, and learning history (including learned helplessness).

4 Schizophrenia

1. **Schizophrenia** is a group of disorders, each characterized by a disturbance in perception (including **hallucinations**), language, thought (including **delusions**), emotions, and/or behavior.

2. In the past, researchers divided schizophrenia into multiple subtypes. More recently, researchers have proposed focusing instead on positive versus negative symptoms. The latest *DSM-IV-TR* also suggests adding another dimension to reflect disorganization of behavior.

3. Most biological theories of schizophrenia focus on genetics, neurotransmitters, and brain abnormalities. Psychologists believe that there are also at least two possible psychosocial contributors: stress and communication disorders in families.

5 Other Disorders

1. **Substance-related disorders** fall into two general groups: substance abuse and substance dependence. Substance-related disorders commonly coexist with other mental disorders (comorbidity), which complicates their diagnosis and treatment.

2. **Dissociative disorders** include dissociative amnesia, dissociative fugue, depersonalization disorder, and dissociative identity disorder (DID). Environmental variables appear to be the primary cause of dissociative disorders.

3. **Personality disorders** occur when inflexible, maladaptive personality traits cause significant impairment of social and occupational functioning. The best-known type of personality disorder is antisocial personality disorder, characterized by egocentrism, lack of conscience, impulsive behavior, and superficial charm. The most common personality disorder is borderline personality disorder (BPD). Its core features are impulsivity and instability in mood, relationships, and self-image. Although some therapists have success with drug therapy and behavior therapy, prognosis is not favorable.

6 How Gender and Culture Affect Abnormal Behavior

1. Men and women differ in how they experience and express abnormal behavior. For example, in North America severe depression is much more common in women than in men. Biological, psychological, and social factors probably combine to explain this phenomenon.

2. Peoples of different cultures experience mental disorders in a variety of ways. For example, the reported incidence of schizophrenia varies within different cultures around the world, as do the disorder's symptoms, triggers, and prognosis.

3. Some symptoms of psychological disorders, as well as some disorders themselves, are culture-general, whereas others are culture-bound.

KEY TERMS

- abnormal behavior p. 342
- medical model p. 343
- psychiatry p. 343
- Diagnostic and Statistical Manual of Mental Disorders (DSM-IV-TR) p. 344
- neurosis p. 344
- psychosis p. 344
- state disorders p. 346
- trait disorders p. 346
- anxiety disorders p. 348
- generalized anxiety disorder p. 348
- panic disorder p. 348
- phobias p. 348
- obsessive-compulsive disorder (OCD) p. 349
- mood disorders p. 352
- major depressive disorders p. 352
- bipolar disorder p. 352
- learned helplessness p. 353

- schizophrenia p. 354
- hallucinations p. 354
- delusions p. 355
- positive schizophrenia symptoms p. 356
- negative schizophrenia symptoms p. 356
- dopamine hypothesis p. 357
- diathesis-stress model p. 358
- substance-related disorders p. 359
- comorbidity p. 359
- self-medication p. 360
- dissociative disorders p. 360
- dissociative identity disorder (DID) p. 360
- personality disorders p. 360
- antisocial personality disorder p. 361
- borderline personality disorder (BPD) p. 361
- culture-general symptoms p. 364
- culture-bound symptoms p. 365

CRITICAL AND CREATIVE THINKING QUESTIONS

1. Can you think of cases where each of the four criteria for abnormal behavior might *not* be suitable for classifying a person's behavior as abnormal?

2. Do you think the insanity plea, as it is currently structured, makes sense?

3. Why do you suppose that anxiety disorders are among the easiest to treat?

4. Have you ever felt depressed? How would you distinguish between "normal" depression and a major depressive disorder?

5. Culture clearly has strong effects on mental disorders. How does that influence how you think about what is normal or abnormal?

6. Most of the disorders discussed in this chapter have some evidence for a genetic predisposition. What would you tell a friend who has a family member with one of these disorders about his or her chances of developing the same disorder?

(Check answers in Appendix B.)

1. Your textbook defines *abnormal* behavior as

_____ .

a. a statistically infrequent pattern of pathological emotion, thought, or action

b. patterns of emotion, thought, and action that are considered pathological

c. a pattern of pathological emotion, thought, or action that causes personal distress, or violates social norms

d. all of these options

2. _____ is the branch of medicine that deals with the diagnosis, treatment, and prevention of mental disorders.

a. Psychology

b. Psychiatry

c. Psychobiology

d. Psychodiagnostics

3. Label the five axes of the *Diagnostic and Statistical Manual of Mental Disorders (DSM–IV–TR)* on the figure below.

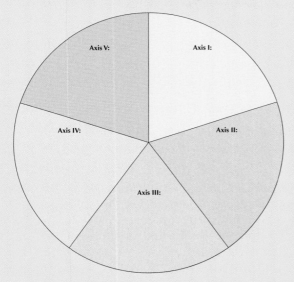

4. Anxiety disorders are _____ .

a. characterized by unrealistic, irrational fear

b. the least frequent of the mental disorders

c. twice as common in men as in women

d. all of these options

5. Label the 5 major anxiety disorders on the figure below.

6. The two main types of mood disorders are _____ .

a. major depression and bipolar disorder

b. mania and depression

c. SAD and MAD

d. learned helplessness and suicide

7. Someone who experiences repeated episodes of mania or cycles between mania and depression has a

_____ .

a. disruption of circadian rhythms

b. bipolar disorder

c. manic-depressive syndrome

d. cyclothymia disorder

8. According to the theory known as _____ , when faced with a painful situation from which there is no escape, animals and people enter a state of helplessness and resignation.

a. autonomic resignation

b. helpless resignation

c. resigned helplessness

d. learned helplessness

9. A psychotic disorder that is characterized by major disturbances in perception, language, thought, emotion, and behavior is _____ .

a. schizophrenia

b. multiple personality disorder

c. borderline psychosis

d. neurotic psychosis

10. Perceptions for which there are no appropriate external stimuli are called _____, and the most common type among people suffering from schizophrenia is

_____ .

  a. hallucinations; auditory
  b. hallucinations; visual
  c. delusions; auditory
  d. delusions; visual

11. Label the 5 subtypes of schizophrenia on the table below.

Subtypes of schizophrenia	
	Dominated by delusions (persecution and grandeur) and hallucinations (hearing voices)
	Marked by motor disturbances (immobility or wild activity) and echo speech (repeating the speech of others)
	Characterized by incoherent speech, flat or exaggerated emotions, and social withdrawal
	Meets the criteria for schizophrenia but is not any of the above subtypes
	No longer meets the full criteria for schizophrenia but still shows some symptoms

12. Failure to meet obligations may be indicative of alcohol or drug _____ whereas tolerance and withdrawal may be indicative of alcohol or drug _____ .

  a. use; abuse
  b. abuse; dependence
  c. abuse; abuse
  d. dependence; abuse

13. The disorder that is an attempt to avoid painful memories or situations, and is characterized by amnesia, fugue, or multiple personalities is _____ .

  a. dissociative disorder
  b. displacement disorder
  c. disoriented disorder
  d. identity disorder

14. Inflexible, maladaptive personality traits that cause significant impairment of social and occupational functioning is known as

_____ .

  a. nearly all mental disorders
  b. the psychotic and dissociative disorders
  c. personality disorders
  d. none of these options

15. Which of the following are examples of culture-general symptoms of mental health difficulties, useful in diagnosing disorders across cultures?

  a. trouble sleeping
  b. can't get along
  c. worry all the time
  d. all of the above

Therapy

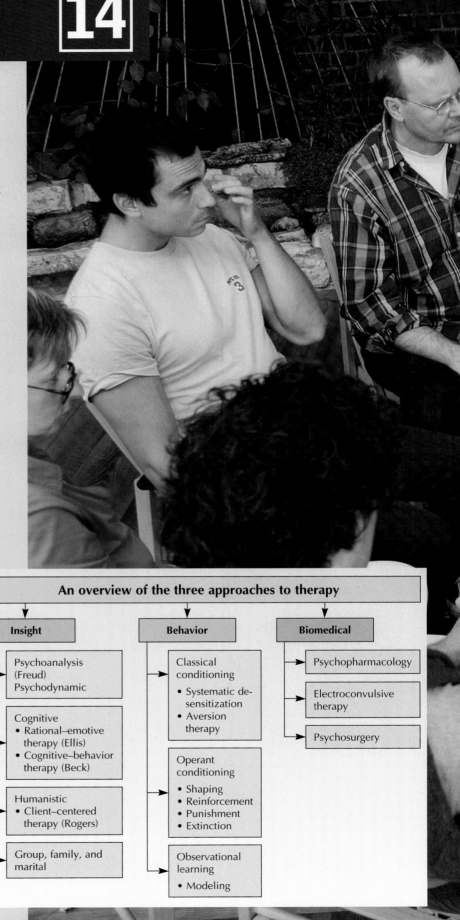

14

What images come to mind when you think of psychotherapy? Like many people, your impressions of therapy may be guided by Hollywood depictions of psychological disorders and their treatment. Since the beginning of the movie age, mentally ill people and their treatment have been the subject of some of Hollywood's most popular and influential films. But consider how people with mental illness are generally portrayed in the movies. They are either cruel, sociopathic criminals (Anthony Hopkins in *Silence of the Lambs*) or helpless, incompetent victims (Jack Nicholson in *One Flew Over the Cuckoo's Nest* and Winona Ryder in *Girl Interrupted*). Likewise, popular films about mental illness often feature mad doctors, heartless nurses, and brutal treatment methods. Although these portrayals may boost movie ticket sales, they also perpetuate harmful stereotypes.

In this chapter, we offer a balanced, factual presentation of the latest research on psychotherapy and mental illness. As you will see, modern psychotherapy can be very effective and prevent much needless suffering, not only for people with psychological disorders but also for those seeking help with everyday problems in living.

In addition to psychologists, professionals involved in psychotherapy include psychiatrists, psychiatric nurses, social workers, counselors, and clergy with training in pastoral counseling. There are numerous forms of psychotherapy. According to one expert (Kazdin, 1994), there may be over 400 approaches to treatment. To organize our discussion, we have grouped treatments into three categories: insight therapies, behavior therapies, and biomedical therapies. After exploring these approaches, we conclude with a discussion of issues that are common to all major forms of psychotherapy.

An overview of the three approaches to therapy

Insight	Behavior	Biomedical
Psychoanalysis (Freud) Psychodynamic	Classical conditioning • Systematic desensitization • Aversion therapy	Psychopharmacology
Cognitive • Rational–emotive therapy (Ellis) • Cognitive–behavior therapy (Beck)		Electroconvulsive therapy
	Operant conditioning • Shaping • Reinforcement • Punishment • Extinction	Psychosurgery
Humanistic • Client–centered therapy (Rogers)		
Group, family, and marital	Observational learning • Modeling	

Insight Therapies

We begin our discussion of professional **psychotherapy** with traditional psychoanalysis and its modern counterpart, psychodynamic therapy. Then we explore cognitive, humanistic, group, and family therapies. Although these therapies differ significantly, they're often grouped together as **insight therapies** because they seek to increase insight into clients' difficulties. The general goal is to help people gain greater control over and improvement in their thoughts, feelings, and behaviors.

psychotherapy
Techniques employed to improve psychological functioning and promote adjustment to life.

PSYCHOANALYSIS/PSYCHODYNAMIC THERAPIES: UNLOCKING THE SECRETS OF THE UNCONSCIOUS

psychoanalysis
Freudian therapy designed to bring unconscious conflicts, which usually date back to early childhood experiences, into consciousness. Also, Freud's theoretical school of thought, which emphasizes unconscious processes.

In **psychoanalysis**, a person's *psyche* (or mind) is analyzed. Traditional psychoanalysis is based on Sigmund Freud's central belief that abnormal behavior is caused by unconscious conflicts among the three parts of the psyche—the id, ego, and superego (Chapter 12).

During psychoanalysis, these conflicts are brought to consciousness (**FIGURE 14.1**). The patient comes to understand the reasons for his or her behavior and realizes that the childhood conditions under which the conflicts developed no longer exist. Once this realization (or insight) occurs, the conflicts can be resolved and the patient can develop more adaptive behavior patterns.

How can gaining insight into one's unconscious change behavior? Freud explained that becoming aware of previously hidden conflicts permits a release of tension and anxiety called **catharsis**. He observed that when his patients relived a traumatic incident, the conflict seemed to lose its power to control the person's behavior.

Modern therapy in action FIGURE 14.1

During psychoanalysis, the therapist (or psychoanalyst) works to bring unconscious conflicts, which are believed to date back to early childhood experiences, into consciousness.

Freud's infamous couch
FIGURE 14.2

The therapist's couch is a well-known fixture of traditional psychoanalysis. Freud believed that this arrangement—with the therapist out of the patient's view—makes the unconscious more accessible and helps patients relax their defenses.

According to Freud, the ego has strong defense mechanisms that block unconscious thoughts from coming to light. Thus, to gain insight into the unconscious, the ego must be "tricked" into relaxing its guard (FIGURE 14.2). With that goal, psychoanalysts employ five major methods: free association, dream analysis, analyzing resistance, analyzing transference, and interpretation.

Free association
According to Freud, when you let your mind wander and remove conscious censorship over thoughts—a process called **free association**—interesting and even bizarre connections seem to spring into awareness. Freud believed that the first thing to come to a patient's mind is often an important clue to what the person's unconscious wants to conceal. Having the patient recline on a couch, with only the ceiling to look at, is believed to encourage free association.

Dream analysis
According to Freud, defenses are lowered during sleep, and forbidden desires and unconscious conflicts can be freely expressed. Even while dreaming, however, these feelings and conflicts are recognized as being unacceptable and must be disguised as images that have deeper symbolic meaning. Thus, according to Freudian dream theory, a therapist might interpret a dream of riding a horse or driving a car (the **manifest content**) as a desire for, or concern about, sexual intercourse (the **latent content**).

Analyzing resistance
During free association or dream analysis, Freud found that patients often show **resistance**—for example, suddenly "forgetting" what they were saying or completely changing the subject. It is the therapist's job to identify these cases of resistance, help patients face their problems, and help patients learn to deal with them realistically.

Analyzing transference
During psychoanalysis, patients disclose intimate feelings and memories, and the relationship between the therapist and patient may become complex and emotionally charged. As a result, patients often apply (or transfer) some of their unresolved emotions and attitudes from past relationships onto the therapist. The therapist uses this process of **transference** to help the patient "relive" painful past relationships in a safe, therapeutic setting so that he or she can move on to healthier relationships.

Interpretation
The core of all psychoanalytic therapy is **interpretation**. During free association, dream analysis, resistance, and transference, the analyst listens closely and tries to find patterns and hidden conflicts. At the right time, the therapist explains (or interprets) the underlying meanings to the client.

As you can see, most of psychoanalysis rests on the assumption that repressed memories and unconscious conflicts actually exist. But, as noted in

Chapters 7 and 12, this assumption is questioned by modern scientists and has become the subject of a heated, ongoing debate. In addition to questioning the validity of repressed memories, critics also point to two other problems with psychoanalysis:

- *Limited applicability.* Critics argue that psychoanalysis seems to suit only a select group of highly motivated, articulate individuals with less severe disorders. Psychoanalysis is also time consuming (often lasting several years with four to five sessions a week) and expensive, and it seldom works well with severe mental disorders, such as schizophrenia, in which verbalization and rationality are significantly disrupted. Finally, critics suggest that spending years chasing unconscious conflicts from the past allows patients to escape from the responsibilities and problems of adult life.

- *Lack of scientific credibility.* Another problem with psychoanalysis is that its "insights" (and therefore its success) cannot be proved or disproved.

Partially in response to these criticisms, more streamlined forms of psychoanalysis have been developed.

In modern **psychodynamic therapy**, treatment is briefer, the patient is treated face to face (rather than reclining on a couch), and the therapist takes a more directive approach (rather than waiting for unconscious memories and desires to slowly be uncovered). Also, contemporary psychodynamic therapists focus less on unconscious, early childhood roots of problems and more on conscious processes and current problems. Such refinements have helped make psychoanalysis more available and more effective for an increasing number of people (Guthrie et al., 2004; Hilsenroth et al., 2003; Siqueland et al., 2002) (**FIGURE 14.3**).

psychodynamic therapy
A modern form of psychoanalysis that emphasizes internal conflicts, motives, and unconscious forces.

Interpersonal therapy (IPT) FIGURE 14.3

Interpersonal therapy (IPT) is an influential, brief form of psychodynamic therapy. As the name implies, interpersonal therapy focuses almost exclusively on the client's current relationships. Its goal is to relieve immediate symptoms and to help the client learn better ways to solve future interpersonal problems.

COGNITIVE THERAPIES: A FOCUS ON FAULTY THOUGHTS AND BELIEFS

Cognitive therapy assumes that faulty thought processes—beliefs that are irrational, overly demanding, or that fail to match reality—create problem behaviors and emotions (Dowd, 2004; Ellis, 1996, 2003b, 2004; Luecken, Tartaro, & Appelhans, 2004).

Like psychoanalysts, cognitive therapists believe that exploring unexamined beliefs can produce insight into the reasons for disturbed behaviors. However, instead of believing that a change in

cognitive therapy
Therapy that focuses on changing faulty thought processes and beliefs to treat problem behaviors.

Cognitive restructuring
FIGURE 14.4

By changing the way one interprets events, cognitive restructuring makes it possible to alter behavior.

behavior occurs because of insight and catharsis, cognitive therapists believe that insight into negative **self-talk** (the unrealistic things a person tells himself or herself) is most important. Through a process called **cognitive restructuring**, this insight allows clients to challenge their thoughts, change how they interpret events, and change maladaptive behaviors (**FIGURE 14.4**).

In addition, whereas psychoanalysts focus primarily on childhood family relationships, cognitive therapists assume that a broad range of events and people—both inside and outside the family—influence beliefs.

One of the best-known cognitive therapists is Albert Ellis, who developed an approach known as **rational-emotive therapy (RET)** (1961, 2003a, 2003b, 2004). Ellis calls RET an A–B–C–D approach, referring to the four steps involved in creating and dealing with maladaptive thinking: an **a**ctivating event, the person's **b**elief system, the emotional and behavioral **c**onsequences that the person experiences, and **d**isputing (or challenging) erroneous beliefs (**FIGURE 14.5**).

According to Ellis, when people demand certain "musts" ("I must get into graduate school") and "shoulds" ("He should love me") from themselves and others, they create emotional distress and behavioral

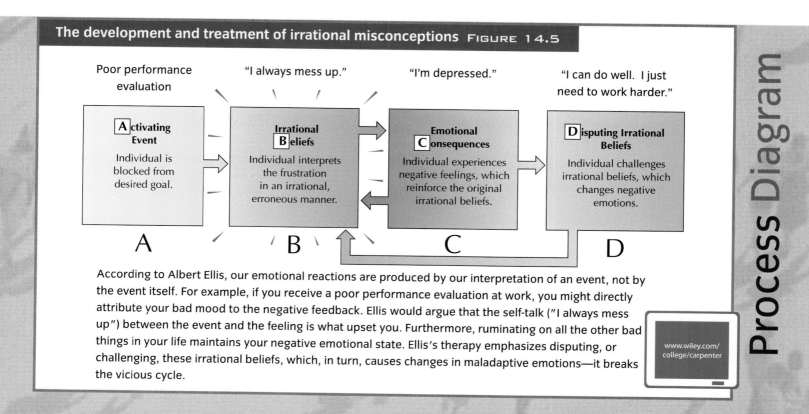

The development and treatment of irrational misconceptions FIGURE 14.5

Poor performance evaluation   "I always mess up."   "I'm depressed."   "I can do well. I just need to work harder."

Activating Event	**I**rrational **B**eliefs	**E**motional **C**onsequences	**D**isputing Irrational Beliefs
Individual is blocked from desired goal.	Individual interprets the frustration in an irrational, erroneous manner.	Individual experiences negative feelings, which reinforce the original irrational beliefs.	Individual challenges irrational beliefs, which changes negative emotions.
A	B	C	D

According to Albert Ellis, our emotional reactions are produced by our interpretation of an event, not by the event itself. For example, if you receive a poor performance evaluation at work, you might directly attribute your bad mood to the negative feedback. Ellis would argue that the self-talk ("I always mess up") between the event and the feeling is what upset you. Furthermore, ruminating on all the other bad things in your life maintains your negative emotional state. Ellis's therapy emphasizes disputing, or challenging, these irrational beliefs, which, in turn, causes changes in maladaptive emotions—it breaks the vicious cycle.

www.wiley.com/college/carpenter

Process Diagram

dysfunction (David, Schnur, & Belloiu, 2002; Ellis, 1997, 2003a, 2003b, 2004).

Ellis believes that such unrealistic, unproductive self-talk generally goes unexamined unless the client is confronted directly. In therapy, Ellis often argues with clients, cajoling and teasing them, sometimes in very blunt language. Once clients recognize their self-defeating thoughts, Ellis begins working with them on how to behave differently—to test out new beliefs and to learn better coping skills. Reflecting this increased attention to behavioral change, he renamed his therapy **rational-emotive behavior therapy (REBT)** (Crosby, 2003).

Another well-known cognitive therapist is Aaron Beck (1976, 2000). Like Ellis, Beck believes that psychological problems result from illogical thinking and destructive self-talk. But Beck seeks to directly confront and change the behaviors associated with destructive cognitions—hence the term **cognitive-behavior therapy**.

One of the most successful applications of Beck's theory has been in the treatment of depression. Beck has identified several thinking patterns that he believes are associated with depression. These are among the most important:

- *Selective perception.* Depression-prone people tend to focus selectively on negative events while ignoring positive events.

- *Overgeneralization.* Depressed people overgeneralize and draw negative conclusions about their self-worth—for example, believing that they are completely worthless because they lost a promotion or failed an exam.

- *Magnification.* Depressed people tend to exaggerate the importance of undesirable events or personal shortcomings, seeing them as catastrophic and unchangeable.

- *All-or-nothing thinking.* Depressed people see things as black-or-white: totally good or bad, right or wrong, a success or a failure.

In Beck's cognitive-behavior therapy, clients are first taught to recognize and keep track of their thoughts. Next, the therapist trains the client to develop ways to test these automatic thoughts against reality. This approach helps depressed people discover that negative attitudes are largely a product of unrealistic or faulty thought processes (**FIGURE 14.6**).

At this point, Beck introduces the second phase of therapy—persuading the client to actively pursue pleasurable activities. Depressed individuals often lose motivation, even for experiences they used to find enjoyable. Simultaneously taking an active rather than a passive role and reconnecting with enjoyable experiences help in recovering from depression.

Cognitive therapies are highly effective treatments for depression, anxiety disorders, bulimia nervosa, anger management, addiction, and even some symptoms of schizophrenia and insomnia (Beck & Rector, 2005; Ellis, 2003a, 2003b, 2004; Reeder et al., 2004; Secker et al., 2004; Turkington et al., 2004). However, both Beck and Ellis have been criticized for ignoring or denying the client's unconscious dynamics, overemphasizing rationality, and minimizing the importance of the client's past (Hammack, 2003). Other critics suggest that cognitive therapies are successful because they employ behavior techniques, not because they change the underlying cognitive structure (Bandura, 1969, 1997; Wright & Beck, 1999). Imagine that you sought treatment for depression and learned to construe events more positively and to curb your all-or-nothing thinking. Imagine that your therapist also helped you identify activities and behaviors that would promote greater fulfillment. If you found your depression lessening, would you attribute the improvement to your changing thought patterns or to changes in your overt behavior?

HUMANISTIC THERAPIES: BLOCKED PERSONAL GROWTH

Humanistic therapy assumes that people with problems are suffering from a disruption of their normal growth potential and, hence, their self-concept. When obstacles are removed, the individual is free to become the self-accepting, genuine person everyone is capable of being.

humanistic therapy
Therapy that seeks to maximize personal growth through affective restructuring (emotional readjustment).

Tracking faulty thoughts FIGURE 14.6

In cognitive-behavior therapy, clients often record their thoughts in "thought journals" so that, together with the therapist, they can compare their thoughts with reality, detecting and correcting faulty thinking.

Name: _Jordan Holley_   Date: _November 11_

DYSFUNCTIONAL THOUGHT RECORD

Directions: When you notice your mood getting worse, ask yourself, **"What's going through my mind right now?"** and as soon as possible jot down the thought or mental image in the Automatic Thought Column.

DATE/TIME	SITUATION	AUTOMATIC THOUGHT(S)	EMOTION(S)	ALTERNATIVE RESONSE	OUTCOME
	1. What actual event or stream of thoughts, or daydreams, or recollection led to the unpleasant emotion? 2. What (if any) distressing physical sensations did you have?	1. What thought(s) and/or image(s) went through your mind? 2. How much did you believe each one at the time?	1. What emotion(s) (sad, anxious, angry, etc.) did you feel at the time? 2. How intense (0-100%) was the emotion?	1. (optional) What cognitive distortion did you make? (e.g., all-or-nothing thinking, mind-reading, catastrophizing.) 2. Use questions at the bottom to compose a response to the automatic thought(s). 3. How much do you believe each response?	1. How much do you now believe each automatic thought? 2. What emotion(s) do you feel now? How intense (0-100%) is the emotion? 3. What will you do? (or did you do?)
Nov. 10, 9 p.m.	My mom called last night. When I saw her number on the caller I.D., I felt my jaw clench and my heart rate go up.	She's going to nag me again to spend the whole Thanksgiving weekend there to make me feel guilty I was 90%. this was it, so I didn't pick up.	I felt angry and frustrated Intensity = about 80%	Cognitive distortion was mind-reading or maybe all-or-nothing thinking. The evidence that the automatic thoughts were true is from my past talks with my mom. Maybe she was calling for another reason, not to make me feel guilty about Thanksgiving. The worst that could happen is that she'd make me feel guilty again... If so, I could refuse to feel that way. I could also say I didn't like the repeated pressure – it makes me guilty and sad. Maybe she'd understand and we'd break this pattern, or she'd stop nagging me so much. Now, I feel angry whenever she calls and often don't pick up the phone. If I changed my thinking, maybe I'll feel better about her calls, at least sometimes. So, I should work on not expecting the worst when she calls.	1. 50%. 2. Now I feel more optimistic, anger is decreased to about 20%. 3. Next time I will remind myself of the alternative response before picking up the phone.

Questions to help compose an alternative response: (1) What is the evidence that the automatic thought is true? Not true? (2) Is there an alternative explanation? (3) What's the worst that could happen? If it did happen, how could I cope? What's the best that could happen? What's the most realistic outcome? (4) What's the effect of my believing the automatic thought? What could be the effect of changing my thinking? (5) What should I do about it? (6) If _____ (friend's name) was in this situation and had this thought, what would I tell him/her?

One of the best-known humanistic therapists is Carl Rogers (1961, 1980), who developed an approach that encourages people to actualize their potential and to relate to others in genuine ways. His approach is referred to as **client-centered therapy** (Rogers used the term *client* because he believed the label *patient* implied that one was sick or mentally ill, rather than responsible and competent) (**FIGURE 14.7**).

Client-centered therapy, like psychoanalysis and cognitive therapies, explores thoughts and feelings as a way to obtain insight into the causes for behaviors. For Rogerian therapists, however, the focus is on providing an accepting atmosphere and encouraging healthy emotional experiences. Clients are responsible for discovering their own maladaptive patterns.

Rogerian therapists create a therapeutic relationship by focusing on four important qualities of communication: empathy, unconditional positive regard, genuineness, and active listening.

Nurturing growth FIGURE 14.7

Imagine how you feel when you are with someone who believes that you are a good person with unlimited potential, a person who believes that your "real self" is unique and valuable. These are the feelings that are nurtured in humanistic therapy.

With Your Own Eyes

Client-Centered Therapy in Action

Would you like to check your understanding of empathy, unconditional positive regard, genuineness, and active listening? Identify the techniques being used in the following excerpt (Shea, 1988, pp. 32–33).

THERAPIST (TH): What has it been like coming down to the emergency room today?

CLIENT (CL): Unsettling, to say the least. I feel very awkward here, sort of like I'm vulnerable. To be honest, I've had some horrible experiences with doctors. I don't like them.

TH: I see. Well, they scare the hell out of me, too (smiles, indicating the humor in his comment).

CL: (Chuckles) I thought you were a doctor.

TH: I am (pauses, smiles)—that's what's so scary.

CL: (Smiles and laughs)

TH: Tell me a little more about some of your unpleasant experiences with doctors, because I want to make sure I'm not doing anything that is upsetting to you. I don't want that to happen.

CL: Well, that's very nice to hear. My last doctor didn't give a hoot about what I said, and he only spoke in huge words.

In case you're wondering, this is an excerpt from an actual session—humor and informality can be an important part of the therapeutic process.

Empathy Therapists pay attention to body language and listen for subtle cues to help them understand the emotional experiences of clients. To help clients explore their feelings, the therapist uses open-ended statements such as "You found that upsetting" or "You haven't been able to decide what to do about this," rather than asking questions or offering explanations.

Unconditional positive regard Because humanists believe human nature is positive and each person is unique, clients can be respected and cherished without having to prove themselves worthy of the therapist's esteem. Humanists believe that when people receive unconditional caring from others, they become better able to value themselves in a similar way.

Genuineness Humanists believe that when therapists honestly share their thoughts and feelings with their clients, their clients will in turn develop self-trust and honest self-expression.

Active listening By reflecting, paraphrasing, and clarifying what the client says and means, the clinician communicates that he or she is genuinely interested in what the client is saying (**FIGURE 14.8**).

Supporters say that there is empirical evidence for the efficacy of client-centered therapy (e.g., Kirschenbaum & Jourdan, 2005; Teusch et al., 2001), but critics argue that outcomes such as self-actualization and self-awareness are difficult to test scientifically. In addition, research on specific therapeutic techniques such as "empathy" and "active listening" has had mixed results (Gottman, Coan, Carrere, & Swanson, 1998).

GROUP, FAMILY, AND MARITAL THERAPIES: HEALING INTERPERSONAL RELATIONSHIPS

In contrast to the therapies described so far, group, marital, and family therapies treat multiple individuals simultaneously. During sessions, therapists often apply psychoanalytic, cognitive, and humanistic techniques.

In **group therapy**, multiple people meet together to work toward therapeutic goals. Typically, a

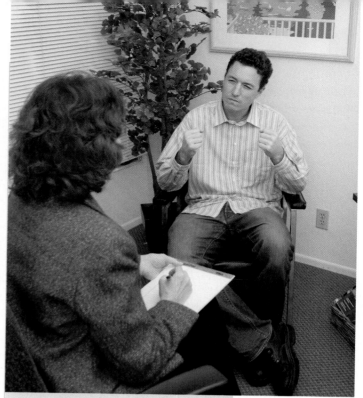

Active listening FIGURE 14.8

Noticing a client's brow furrowing and hands clenching while he is discussing his marital problems, a clinician might respond, "It sounds like you're angry with your wife and feeling pretty miserable right now." Can you see how this statement reflects the client's anger, paraphrases his complaint, and gives feedback to clarify the communication?

group of 8 to 10 people meet with a therapist on a regular basis to talk about problems in their lives.

A variation on group therapy is the **self-help group**. Unlike other group therapy approaches, a professional does not guide these groups. They are simply groups of people who share a common problem (such as alcoholism, single parenthood, or breast cancer) and who meet to give and receive support. Faith-based 12-step programs such as Alcoholics Anonymous, Narcotics Anonymous, and Spenders Anonymous are examples of self-help groups.

Although group members don't get the same level of individual attention found in one-on-one therapies, group and self-help therapies provide their own unique advantages (Davison, Neale, & Kring, 2004;

group therapy A form of therapy in which a number of people meet together to work toward therapeutic goals.

Porter, Spates, & Smitham, 2004; White & Freeman, 2000). Compared with one-on-one therapies, they are less expensive and provide a broader base of social support. Group members can learn from each other's mistakes, share insights and coping strategies, and role-play social interactions together.

Therapists often refer their patients to group therapy and self-help groups to supplement individual therapy. Research on self-help groups for alcoholism, obesity, and other disorders suggests that they can be very effective, either alone or in addition to individual psychotherapy (Bailer et al., 2004; Davidson et al., 2001; Magura et al., 2003).

Because a family or marriage is a system of interdependent parts, the problem of any one individual unavoidably affects all the others, and therapy can help everyone involved (Becvar & Becvar, 2006; Cavacuiti, 2004; Shadish & Baldwin, 2003). The line between marital (or couples) therapy and family therapy is often blurred. Here, our discussion will focus on **family therapy**, in which the primary aim is to change maladaptive family interaction patterns (FIGURE 14.9). All members of the family attend therapy sessions, though at times the therapist may see family members individually or in twos or threes.

Family therapy is also useful in treating a number of disorders and clinical problems. As we discuss in Chapter 13, schizophrenic patients are more likely to relapse if their family members

family therapy
Treatment to change maladaptive interaction patterns within a family.

Family therapy FIGURE 14.9

Many families initially come into therapy believing that one member is the cause of all their problems. However, family therapists generally find that this "identified patient" is a scapegoat for deeper disturbances. For example, instead of confronting their own problems with intimacy, a couple may focus all their attention and frustration on a delinquent child (Hanna & Brown, 1999). How could changing ways of interacting within the family system promote the health of individual family members and the family as a whole?

express emotions, attitudes, and behaviors that involve criticism, hostility, or emotional overinvolvement (Hooley & Hiller, 2001; Lefley, 2000; Quinn, Barowclough, & Tarrier, 2003). Family therapy can help family members modify their behavior toward the patient. Family therapy can also be the most favorable setting for the treatment of adolescent drug abuse (Kumpfer, Alvarado, & Whiteside, 2003; Stanton, 2004).

CONCEPT CHECK STOP

HOW does modern psychodynamic therapy differ from traditional psychoanalysis?

What are the four steps of Ellis's RET?

What is the significance of the term *client-centered therapy*?

When might family therapy be more successful than individual psychotherapy?

Behavior Therapies

LEARNING OBJECTIVES

Identify a key difference between these two classical conditioning techniques: systematic desensitization and aversion therapy.

Explore how operant conditioning can be used in therapy.

Summarize how modeling therapy works.

S ometimes having insight into a problem does not automatically solve it. In **behavior therapy**, the focus is on the problem behavior itself, rather than on any underlying causes. Although the person's feelings and interpretations are not disregarded, they are also not emphasized. The therapist diagnoses the problem by listing maladaptive behaviors that occur and adaptive behaviors that are absent. The therapist then attempts to shift the balance of the two, drawing on the learning principles of classical conditioning, operant conditioning, and observational learning (Chapter 6).

> **behavior therapy** A group of techniques based on learning principles that is used to change maladaptive behaviors.

© Sidney Harris

CLASSICAL CONDITIONING TECHNIQUES

Behavior therapists use the principles of classical conditioning to decrease maladaptive behaviors by creating new associations to replace the faulty ones. We will explore two techniques based on these principles: aversion therapy and systematic desensitization.

Aversion therapy uses principles of classical conditioning to create anxiety rather than extinguish it (**FIGURE 14.10**). People who engage in excessive drinking, for example, build up a number of pleasurable associations. Because these pleasurable associations cannot always be prevented, aversion therapy provides negative associations to compete with the pleasurable ones. Someone who wants to stop drinking, for example, could take a drug called Antabuse, which causes vomiting whenever alcohol enters the system. When the new connection between alcohol and nausea has been classically conditioned, engaging in the once-desirable behavior will cause an immediate negative response.

Making a nasty habit nastier FIGURE 14.10

A person who wants to quit smoking could collect a jar full of (smelly) cigarette butts or smoke several cigarettes at once to create an aversion to smoking. Can you see how this aversion therapy uses classical conditioning?

Virtual Reality Therapy

Rather than using mental imaging or actual physical experiences of a fearful situation, modern therapy can use the latest in computer technology—virtual reality headsets and data gloves. What kind of "virtual" experiences do you think a therapist might provide for a client with a fear of spiders?

Aversion therapy is controversial. First, some researchers question whether it is ethical to hurt someone (even when the person has given permission). The treatment also has been criticized because it does not provide lasting relief (Seligman, 1994). One reason is that (in the case of the aversion therapy for alcoholism) people understand that the nausea is produced by the Antabuse and do not generalize their learning to the alcohol.

In contrast to aversion therapy, **systematic desensitization** (Wolpe & Plaud, 1997) begins with relaxation training, followed by imagining or directly experiencing various versions of a feared object or situation while remaining deeply relaxed. The goal is to replace an anxiety response with a relaxation response when confronting the feared stimulus (Heriot & Pritchard, 2004).

Desensitization is a three-step process (**FIGURE 14.11**). First, a client is taught how to maintain a state of deep relaxation that is physiologically incompatible with an anxiety response. Next, the therapist and client construct a hierarchy, or ranked listing, of anxiety-arousing images. In the final step, the relaxed client mentally visualizes or physically experiences items in the hierarchy, starting at the bottom and working his or her way to the most anxiety-producing images at the top. If any image or situation begins to create anxiety, the client stops momentarily and returns to a state of complete relaxation. Eventually, the fear response is extinguished.

OPERANT CONDITIONING TECHNIQUES

One operant conditioning technique for eventually bringing about a desired (or target) behavior is **shaping**—providing rewards for successive approximations of the target behavior (Chapter 6). One of the most successful applications of shaping and reinforcement has been with developing language skills in children with autism. First, the child is rewarded for making any sounds; later, only for forming words and sentences.

Shaping can also help people acquire social skills and greater assertiveness. If you are painfully shy, for example, a clinician might first ask you to role-play simply saying hello to someone you find attractive. Then you might practice behaviors that gradually lead you to suggest a get-together or date. During such role-playing, or behavior rehearsal, the clinician would give you feedback and reinforcement.

Adaptive behaviors can also be taught or increased with techniques that provide immediate reinforcement in the form of tokens (Sarafino, 2000). For example, patients in an inpatient treatment facility might at first be given tokens (to be exchanged for primary rewards, such as food, treats, TV time, a private room, or outings) for merely attending group therapy sessions. Later they will be rewarded only for actually participating in the sessions. Eventually, the tokens can be discontinued when the patient receives the reinforcement of being helped by participation in the therapy sessions.

OBSERVATIONAL LEARNING TECHNIQUES

We all learn many things by observing others. Therapists use this principle in **modeling therapy**, in which clients are asked to observe and imitate appropriate models as they perform desired behav-

modeling therapy A learning technique in which the subject watches and imitates models who demonstrate desirable behaviors.

Madeleine, a college student, has been offered a lucrative summer job painting commercial buildings. She is deathly afraid of heights, but she needs the money. She goes to see a therapist, who tells her that she can be treated in a relatively brief period of time, through systematic desensitization. To understand how this works, look at the six behavioral steps depicted on the continuum below. Can you see how this technique might help Madeleine gradually overcome her fear of heights?

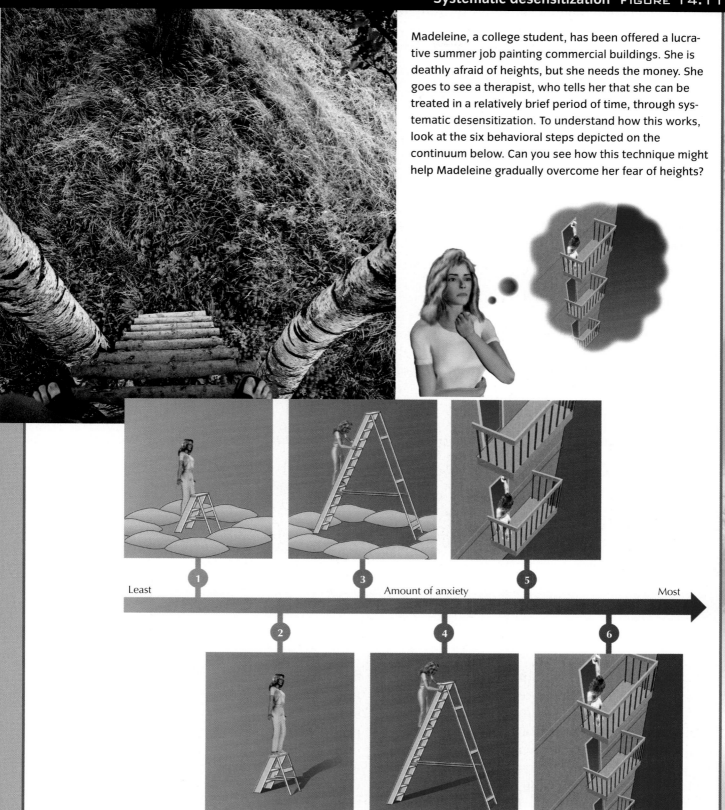

Least ① ③ Amount of anxiety ⑤ Most

② ④ ⑥

iors. For example, Albert Bandura and his colleagues (1969) asked clients with snake phobias to watch other (nonphobic) people handle snakes. After only two hours of exposure, over 92 percent of the phobic observers allowed a snake to crawl over their hands, arms, and necks.

Modeling is also part of social skills training and assertiveness training. Clients learn how to interview for a job by first watching the therapist role-play the part of the interviewee. The therapist models the appropriate language (assertively asking for a job), body posture, and so forth, and then asks the client to imitate the behavior and play the same role. Over the course of several sessions, the client gradually becomes desensitized to the anxiety of interviews and learns interview skills.

EVALUATING BEHAVIOR THERAPIES

Behavior therapy has been effective in treating various problems, including phobias, obsessive-compulsive disorder (OCD), eating disorders, autism, mental retardation, and delinquency (Heriot & Pritchard, 2004; Herrera, Johnston, & Steele, 2004; Sarafino, 2000; Wilson, 2005). Critics of behavior therapy, however, raise important questions that fall into two major categories:

- *Generalizability.* Critics argue that in the "real world" patients are not consistently reinforced, and their newly acquired behaviors may disappear. To deal with this possibility, behavior therapists work to gradually shape clients toward real-world rewards.

- *Ethics.* Critics contend that it is unethical for one person to control another's behavior. Behaviorists, however, argue that rewards and punishments already control our behaviors and that behavior therapy actually increases a person's freedom by making these controls overt and by teaching people how to change their own behavior.

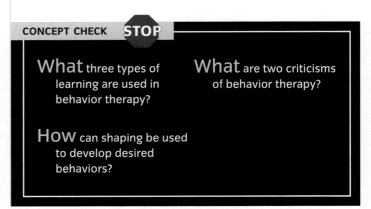

CONCEPT CHECK STOP

What three types of learning are used in behavior therapy?

What are two criticisms of behavior therapy?

How can shaping be used to develop desired behaviors?

Biomedical Therapies

LEARNING OBJECTIVES

Identify the major types of drugs used to treat psychological disorders.

Explain what happens in electroconvulsive therapy and psychosurgery.

Describe the risks associated with biomedical therapies.

Biomedical therapies are based on the premise that problem behaviors are caused, at least in part, by chemical imbalances or disturbed nervous system functioning. A physician, rather than a psychologist, must prescribe biomedical therapies, but psychologists do work with patients receiving biomedical therapies and are frequently involved in research programs to evaluate their effectiveness. In this section, we will discuss three aspects of biomedical therapies: psychopharmacology, electroconvulsive therapy (ECT), and psychosurgery.

biomedical therapy
The use of physiological interventions (drugs, electroconvulsive therapy, and psychosurgery) to reduce or alleviate symptoms of psychological disorders.

PSYCHOPHARMACOLOGY: TREATING PSYCHOLOGICAL DISORDERS WITH DRUGS

Since the 1950s, drug companies have developed an amazing variety of chemicals to treat abnormal behaviors. In some cases, discoveries from **psychopharmacology** have helped correct chemical imbalances. In other cases, drugs have been used to relieve or suppress the symptoms of psychological disturbances even when the underlying cause was not thought to be biological. As shown in TABLE 14.1, psychiatric drugs are classified into four major categories: antianxiety, antipsychotic, mood stabilizer, and antidepressant.

> **psychopharma-cology**
> The study of drug effects on the mind and behavior.

"Before Prozac, she loathed company."

Common drug treatments for psychological disorders TABLE 14.1

Type of drug	Psychological disorder	Chemical group	Generic name	Brand name
Antianxiety drugs	Anxiety disorders	Benzodiazepines	Alprazolam	Xanax
			Diazepam	Valium
		Glycerol derivatives	Meprobamate	Miltown
				Equanil
Antipsychotic drugs	Schizophrenia	Phenothiazines	Chlorpromazine	Thorazine
			Fluphenazine	Prolixin
			Thioridazine	Mellaril
		Butyrophenones	Haloperidol	Haldol
		Dibenzodiazepine	Clozapine	Clozaril
		Atypical antipsychotics	Resperidone	Risperdal
Mood stabilizer drugs	Bipolar disorder	Antimanic	Lithium carbonate	Lithonate
				Lithane
			Carbamazepine	Tegretol
Antidepressant drugs	Depressive disorders	Tricyclic antidepressants	Imipramine	Tofranil
			Amitriptyline	Elavil
		Monoamine oxidase inhibitors (MAOIs)	Phenelzine	Nardil
		Selective serotonin reuptake inhibitors (SSRIs)	Paroxetine	Paxil
			Fluoxetine	Prozac
		Atypical antidepressants	Buspirone	Buspar
			Venlafaxine	Effexor

- **Antianxiety drugs** (also known as "minor tranquilizers") lower the sympathetic activity of the brain—the crisis mode of operation—so that anxious responses are diminished or prevented and are replaced by feelings of tranquility and calmness (Blanco et al., 2003; Swartz & Margolis, 2004).

- **Antipsychotic drugs**, or neuroleptics, are used to treat schizophrenia and other acute psychotic states. They are often referred to as "major tranquilizers," creating the mistaken impression that they invariably have a strong sedating effect. However, the main effect of antipsychotic drugs is to diminish or eliminate psychotic symptoms, including hallucinations, delusions, withdrawal, and apathy. They are not designed to sedate the patient. Traditional antipsychotics work by decreasing activity at the dopamine synapses in the brain. A large majority of patients show marked improvement when treated with antipsychotic drugs.

- **Mood stabilizer drugs**, such as lithium, can help relieve manic episodes and depression for people suffering from bipolar disorder. Because lithium acts relatively slowly—it can be three or four weeks before it takes effect—its primary use is in preventing future episodes and helping to break the manic-depressive cycle.

- **Antidepressant drugs** are used to treat people with depression. There are four types of antidepressant drugs: tricyclics, monoamine oxidase inhibitors (MAOIs), selective serotonin reuptake inhibitors (SSRIs), and atypical antidepressants. Each class of drugs affects neurochemical pathways in the brain in slightly different ways, increasing or decreasing the availability of certain chemicals. SSRIs (such as Paxil and Prozac) are by far the most commonly prescribed antidepressants. The atypical antidepressants are a miscellaneous group of drugs used for patients who fail to respond to the other drugs or for people who experience side effects common to other antidepressants.

ELECTROCONVULSIVE THERAPY AND PSYCHOSURGERY

In **electroconvulsive therapy (ECT)**, also known as electroshock therapy (EST), a moderate electrical current is passed through the brain between two electrodes placed on the outside of the head. The current triggers a widespread firing of neurons, or convulsions. The convulsions produce many changes in the central and peripheral nervous systems, including activation of the autonomic nervous system, increased secretion of various hormones and neurotransmitters, and changes in the blood-brain barrier.

During the early years of ECT, some patients received hundreds of treatments (Fink, 1999), but today most receive 12 or fewer treatments. Sometimes the electrical current is applied only to the right hemisphere, which causes less interference with verbal memories and left hemisphere functioning.

Modern ECT is used primarily on patients with severe depression who do not respond to antidepressant drugs or psychotherapy and on suicidal patients because it works faster than antidepressant drugs (Birkenhäger, Renes, & Pluijms, 2004; Sylvester et al., 2000).

Although clinical studies of ECT conclude that it is effective for very severe depression (Carney et al., 2003; Prudic et al., 2004), its use remains controversial because it triggers massive changes in the brain—and also because we do not understand why it works (Baldwin & Oxlad, 2000; Cloud, 2001; Pearlman, 2002).

The most extreme, and least used, biomedical therapy is **psychosurgery**—brain surgery performed to reduce serious, debilitating psychological problems.

You may have heard of the now-outmoded form of psychosurgery called **lobotomy**. This technique origi-

> ■ **electroconvulsive therapy (ECT)** Biomedical therapy in which electrical current is passed through the brain.

> **psychosurgery** Operative procedures on the brain designed to relieve severe mental symptoms that have not responded to other forms of treatment.

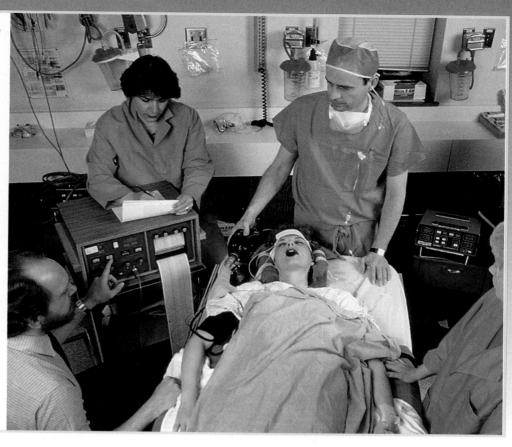

Electroconvulsive therapy (ECT)

Although electroconvulsive therapy may seem barbaric, for some severely depressed people it is their only hope for lifting the depression. Unlike portrayals of ECT in movies like *One Flew Over the Cuckoo's Nest* and *The Snake Pit*, patients show few, if any, visible reactions to the treatment owing to modern muscle-relaxant drugs that dramatically reduce muscle contractions during the seizure. Most ECT patients are also given an anesthetic to block their memories of the treatment, but some patients still find the treatment extremely uncomfortable (Johnstone, 1999). However, many others find it lifesaving.

nated in 1936, when Portuguese neurologist Egaz Moniz first treated uncontrollable psychoses by cutting the nerve fibers between the frontal lobes (where association areas for monitoring and planning behavior are found) and the thalamus and hypothalamus.

Although these surgeries did reduce emotional outbursts and aggressiveness, many patients were left with permanent brain damage. In the mid-1950s, when antipsychotic drugs came into use, psychosurgery virtually stopped. Recently, however, psychiatrists have been experimenting with a much more limited and precise surgical procedure called *cingulotomy*, in which the cingulum (a small structure in the brain's limbic system known to be involved in emotionality) is partially destroyed. On rare occasions, it is used in the treatment of severely debilitating cases of obsessive-compulsive disorder, severe depression, and chronic, intractable pain (Cohen et al., 2001; Dougherty et al., 2002; Sotres-Bayón & Pellicer, 2000).

EVALUATING BIOMEDICAL THERAPIES

Like all forms of therapy, biomedical therapies have both proponents and critics.

Pitfalls of psychopharmacology

Drug therapy poses several potential problems. First, although drugs may relieve symptoms for some people, they seldom provide "cures." In addition, some patients become physically dependent on the drugs. Also, researchers are still learning about the long-term effects and potential interactions. Furthermore, psychiatric medications can cause a variety of side effects, ranging from mild fatigue to severe impairments in memory and movement.

A final potential problem with drug treatment is that its relative inexpensiveness and generally fast results have led to its overuse in some cases. One report found that antidepressants are prescribed roughly 50 percent of the time a patient walks into a psychiatrist's office (Olfson et al., 1998).

Despite the problems associated with them, psychotherapeutic drugs have led to revolutionary changes in mental health. Before the use of drugs, some patients were destined to spend a lifetime in psychiatric institutions. Today, most patients improve enough to return to their homes and live successful lives if they continue to take their medications to prevent relapse.

Challenges to ECT and psychosurgery
As we mentioned, ECT is a controversial treatment for several reasons. However, problems with ECT may become obsolete thanks to **repetitive transcranial magnetic stimulation (rTMS)**, which delivers a brief (but powerful) electrical current through a coil of wire placed on the head. Unlike ECT, which passes a strong electrical current directly through the brain, the rTMS coil creates a strong magnetic field that is applied to certain areas in the brain. When used to treat depression, the coil is usually placed over the prefrontal cortex, a region linked to deeper parts of the brain that regulate mood. Currently, the benefits of rTMS over ECT remain uncertain.

Because all forms of psychosurgery have potentially serious or fatal side effects and complications, some critics suggest that they should be banned altogether. Furthermore, the consequences are irreversible. For these reasons, psychosurgery is considered experimental and remains a highly controversial treatment.

CONCEPT CHECK **STOP**

When are drug therapies used?

How does modern ECT differ from the therapy's early use?

What is cingulotomy?

NATIONAL GEOGRAPHIC

Therapy Essentials

Summarize the goals that are common to all major forms of psychotherapy.

Describe some key cross-cultural similarities and differences in therapy.

Explain why therapists need to be sensitive to gender issues that pertain to mental illness.

Explore some alternatives to long-term institutionalization for people with severe psychological disorders.

E arlier, we mentioned that there may be more than 400 forms of therapy. How would you choose one for yourself or someone you know? In this section, we help you to synthesize the material in this chapter and to put what you have learned about each of the major forms of therapy into a broader context.

THERAPY GOALS AND EFFECTIVENESS

All major forms of therapy are designed to help the client in five specific areas (**FIGURE 14.12**):

- *Disturbed thoughts.* Troubled individuals typically suffer some degree of confusion, destructive thought patterns, or blocked understanding of their problems. Therapists work to change these thoughts, provide new ideas or informa-tion, and guide individuals toward finding solu-tions to problems.

- *Disturbed emotions.* Therapists help clients un-derstand and control their emotions and re-lieve their emotional discomfort.

- *Disturbed behaviors.* Therapists help clients elimi-nate troublesome behaviors and guide them to-ward more effective lives.

- *Interpersonal and life situation difficulties.* Thera-pists help clients improve their relationships with others and avoid or minimize sources of stress in their lives.

- *Biomedical disturbances.* Troubled individuals sometimes suffer biological disruptions that di-rectly cause or contribute to psychological diffi-culties (for example, chemical imbalances that lead to depression). Therapists help relieve these problems, primarily with drugs.

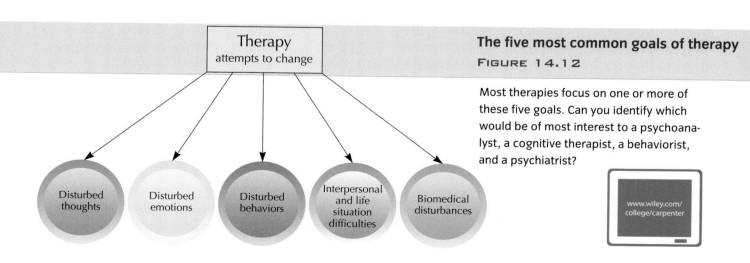

Therapy attempts to change: Disturbed thoughts · Disturbed emotions · Disturbed behaviors · Interpersonal and life situation difficulties · Biomedical disturbances

The five most common goals of therapy
FIGURE 14.12

Most therapies focus on one or more of these five goals. Can you identify which would be of most interest to a psychoana-lyst, a cognitive therapist, a behaviorist, and a psychiatrist?

www.wiley.com/college/carpenter

When stresses overwhelm our natural coping mechanisms, professional therapy can provide invaluable relief. But how do we find a good therapist for our specific needs? If you have the time (and the money) to explore options, search for a therapist who is best suited to your specific goals. However, if you are in a crisis—you have suicidal thoughts, are getting failing grades, or are the victim of abuse—get help fast. Most communities have telephone hotlines that provide counseling on a 24-hour basis. (In addition, most colleges and universities have counseling centers that provide immediate, short-term therapy to students free of charge.) If you are encouraging someone else to get therapy, you might offer to help locate a therapist and to go with him or her for the first visit. If he or she refuses help and the problem affects you, it is often a good idea to seek therapy yourself. You will gain insights and skills that will help you deal with the situation more effectively.

Although most therapists work with clients in several of these areas, the emphasis varies according to the therapist's training (psychodynamic, cognitive, humanistic, behaviorist, or biomedical). Clinicians who regularly borrow freely from various theories are said to take an **eclectic approach**.

Does therapy work? After years of controlled research and *meta-analysis* (a method of statistically combining and analyzing data from many studies), we have fairly clear evidence that it does. Forty to 90 percent of people who receive treatment are better off than people who do not. Furthermore, short-term treatments can be as effective as long-term treatments (Blanco et al., 2003; Ellis, 2004; Nordhus & Pallesen, 2003; Shadish & Baldwin, 2003; Storosum et al., 2004; Swanson, Swartz, & Elbogen, 2004).

Some therapies are more effective than others for specific problems. For example, phobias seem to respond best to systematic desensitization, and OCD can be significantly relieved with cognitive-behavior therapy accompanied by medication. Most studies that have compared medication alone versus medication plus therapy have found the combination to be more effective (e.g., Doyle & Pollack, 2004).

CULTURAL ISSUES IN THERAPY

The therapies described in this chapter are based on Western European and North American culture. Does this mean they are unique? Or do our psychotherapists do some of the same things that, say, a native healer or shaman does? Are there similarities in therapies across cultures? Conversely, are there fundamental therapeutic differences among cultures?

When we look at therapies in all cultures, we find that they have certain key features in common (Corey, 2001; Jennings & Skovholt, 1999; Matsumoto, 2000). Richard Brislin (2000) has summarized some of these features:

- *Naming the problem.* People often feel better just by knowing that others experience the same problem and that the therapist has had experience with their particular problem.

- *Qualities of the therapist.* Clients must feel that the therapist is caring, competent, approachable, and concerned with finding solutions to their problem.

- *Establishing credibility.* Word-of-mouth testimonials and status symbols (such as diplomas on the wall) establish the therapist's credibility. Among native healers, in lieu of diplomas, credibility may be established by having served as an apprentice to a revered healer.

- *Placing the problem in a familiar framework.* If the client believes that evil spirits cause psychological disorders, the therapist will direct treatment toward eliminating these spirits. Similarly, if the client believes in the importance of early childhood experiences and the unconscious mind, psychoanalysis will be the likely treatment of choice.

- *Applying techniques to bring relief.* In all cultures, therapy involves action. Either the client or the therapist must do something. Moreover, what they do must fit the client's expectations—whether it is performing a ceremony to expel demons or talking with the client about his or her thoughts and feelings.

- *A special time and place.* The fact that therapy occurs outside the client's everyday experiences seems to be an important feature of all therapies.

Although there are basic similarities in therapies across cultures, there are also important differences. In the traditional Western European and North American model, the emphasis is on the "self" and on independence and control over one's life—qualities that are highly valued in individualistic cultures. In collectivist cultures, however, the focus of therapy is on interdependence and accepting the realities of one's life (Sue & Sue, 2002).

Not only does culture affect the types of therapy that are developed. It also influences the perceptions of the therapist. What one culture considers abnormal behavior may be quite common—and even healthy—in others. For this reason, recognizing cultural differences is very important for building trust between therapists and clients and for effecting behavioral change (Snowden & Yamada, 2005; Tseng, 2004).

WOMEN AND THERAPY

Within our individualistic Western culture, men and women present different needs and problems to therapists. For example, compared with men, women are more comfortable and familiar with their emotions, have fewer negative attitudes toward therapy, and are more likely to seek psychological help (Komiya, Good, & Sherrod, 2000). Research has identified five major concerns related to women and psychotherapy (Chaikin & Prout, 2004; Holtzworth-Munroe, 2000; Whitney, Kusznir, & Dixie, 2002).

1. *Rates of diagnosis and treatment of mental disorders.* Women are diagnosed and treated for mental illness at a much higher rate than men. Is this because women are "sicker" than men as a group, or are they just more willing to admit their problems? Or perhaps the categories for illness are biased against women. More research is needed to answer this question.

Emphasizing Interdependence

In Japanese Naikan therapy, the patient sits quietly from 5:30 A.M. to 9:00 P.M. for seven days and is visited by an interviewer every 90 minutes. During this time, the patient is instructed to reflect on his or her relationships with others, with the goals of discovering personal guilt for having been ungrateful and troublesome to others and developing gratitude toward those who have helped the patient (Berry et al., 2002; Ryback, Ikemi, & Miki, 2001). The reasoning is that when these goals are attained, the patient will have a better self-image and interpersonal attitude. In what ways do the goals and methods of Naikan therapy differ from the therapies we have described in this chapter? Would this approach work with Westerners? Why or why not?

2. *Stresses of poverty.* Women are disproportionately likely to be poor. Poverty contributes to stress, which is directly related to many psychological disorders.

3. *Stresses of multiple roles.* Women today are mothers, wives, homemakers, wage earners, students, and so on. The conflicting demands of their multiple roles often create special stresses (FIGURE 14.13).

4. *Stresses of aging.* Aging brings special concerns for women. Elderly women, primarily those with age-related dementia, account for over 70 percent of the chronically mentally ill who live in nursing homes in the United States.

5. *Violence against women.* Rape, incest, and sexual harassment—which are much more likely to happen to women than to men—may lead to depression, insomnia, posttraumatic stress disorders, eating disorders, and other problems.

INSTITUTIONALIZATION

Despite Hollywood film portrayals, forced institutionalization of the mentally ill poses serious ethical problems and is generally reserved for only the most serious and life-threatening situations or when there is no reasonable, less restrictive alternative.

In emergencies, mental health professionals can authorize temporary commitment for 24 to 72 hours—time enough for laboratory tests to be performed to rule out medical illnesses that could be causing the symptoms. The patient can also receive psychological testing, medication, and short-term therapy during this time.

Today, many states have a policy of **deinstitutionalization**—discharging patients from mental hospitals as soon as possible and discouraging admissions. As an alternative to institutionalizing people in state hospitals, most clinicians suggest expanding and improving community care (FIGURE 14.14)

Meeting women's unique needs FIGURE 14.13

Therapists must be sensitive to possible connections between clients' problems and their gender. Rather than prescribing drugs to relieve depression in women, for example, it may be more appropriate for therapists to explore ways to relieve the stresses of multiple roles or poverty. Can you see how helping a single mother identify parenting resources such as play groups, parent support groups, and high-quality child care might be just as effective at relieving depression as prescribing drugs?

Outpatient support FIGURE 14.14

Community mental health (CMH) centers are a prime example of alternative treatment to institutionalization. CMH centers provide outpatient services, such as individual and group therapy and prevention programs. They also coordinate short-term inpatient care and programs for discharged mental patients, such as halfway houses and aftercare services. The downside of CMH centers and their support programs is that they are expensive. Investing in primary prevention programs (such as more intervention programs for people at high risk for mental illness) could substantially reduce these costs.

(Duckworth & Borus, 1999; Lamb, 2000). They recommend that general hospitals be equipped with special psychiatric units where acutely ill patients can receive inpatient care. For less disturbed individuals and chronically ill patients, they recommend walk-in clinics, crisis intervention services, improved residential treatment facilities, and psychosocial and vocational rehabilitation. State hospitals would be reserved for the most unmanageable patients.

CONCEPT CHECK **STOP**

What is the evidence that therapy works?

What is deinstitutionalization?

HOW do individualistic and collectivistic cultures differ in their approaches to therapy?

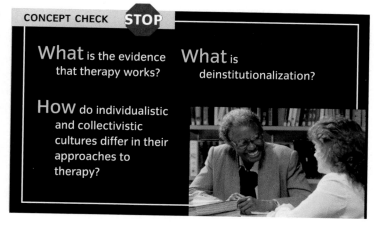

What is happening in this picture **?**

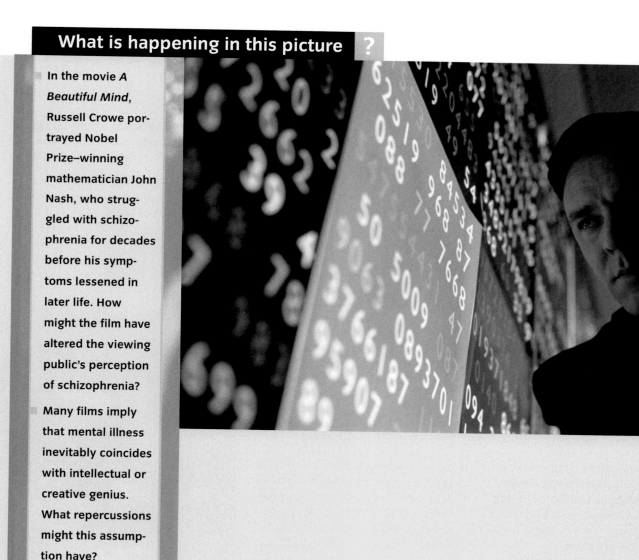

- In the movie *A Beautiful Mind*, Russell Crowe portrayed Nobel Prize–winning mathematician John Nash, who struggled with schizophrenia for decades before his symptoms lessened in later life. How might the film have altered the viewing public's perception of schizophrenia?

- Many films imply that mental illness inevitably coincides with intellectual or creative genius. What repercussions might this assumption have?

VISUAL SUMMARY

1 Insight Therapies

1. "Insight therapies" are forms of **psychotherapy** that seek to increase insight into clients' difficulties.

2. In **psychoanalysis**, the therapist seeks to identify the patient's unconscious conflicts and to help the patient resolve them through catharsis. In modern **psychodynamic therapy**, treatment is briefer and the therapist takes a more directive approach (and puts less emphasis on unconscious childhood memories) than in traditional psychoanalysis.

3. **Cognitive therapy** seeks to help clients challenge faulty thought processes and adjust maladaptive behaviors. Ellis's rational-emotive therapy (RET) and Beck's cognitive-behavior therapy are important examples of cognitive therapy.

4. **Humanistic therapy**, such as Rogers's client-centered therapy, seeks to maximize personal growth, encouraging people to actualize their potential and relate to others in genuine ways.

5. In **group therapy**, multiple people meet together to work toward therapeutic goals. A variation is the self-help group, which is not guided by a professional. Therapists often refer their patients to group therapy and self-help groups to supplement individual therapy.

6. In **family therapy**, the aim is to change maladaptive family interaction patterns. All members of the family attend therapy sessions, though at times the therapist may see family members individually or in twos or threes.

2 Behavior Therapies

1. In **behavior therapy**, the focus is on the problem behavior itself, rather than on any underlying causes. The therapist uses learning principles to change behavior.

2. Classical conditioning techniques include aversion therapy and systematic desensitization.

3. Operant conditioning techniques used to increase adaptive behaviors include shaping and reinforcement.

4. In **modeling therapy**, clients observe and imitate others who are performing the desired behaviors.

3 Biomedical Therapies

1. **Biomedical therapies** are based on the premise that chemical imbalances or disturbed nervous system functioning contribute to problem behaviors.

2. **Psychopharmacology** is the most common form of biomedical therapy. Major classes of drugs used to treat psychological disorders are antianxiety drugs, antipsychotic drugs, mood stabilizer drugs, and antidepressant drugs.

3. In **electroconvulsive therapy (ECT)**, an electrical current is passed through the brain, stimulating convulsions that produce changes in the central and peripheral nervous systems. ECT is used primarily in cases of severe depression that do not respond to other treatments.

4. The most extreme biomedical therapy is **psychosurgery**. Lobotomy, an older form of psychosurgery, is now outmoded. Recently, psychiatrists have been experimenting with a more limited and precise surgical procedure called cingulotomy.

4 Therapy Essentials

1. All major forms of therapy are designed to address disturbed thoughts, disturbed emotions, disturbed behaviors, interpersonal and life situation difficulties, and biomedical disturbances. Research indicates that, overall, therapy does work.

2. Therapies in all cultures have certain key features in common; however, there are also important differences among cultures. Therapists must recognize cultural differences in order to build trust with clients and effect behavioral change.

3. Therapists must also be sensitive to possible gender issues in therapy.

4. Forced institutionalization of the mentally ill is generally reserved for only the most serious and life-threatening situations. Many states have a policy of deinstitutionalization, and most clinicians suggest expanding and improving community care to ensure that people with mental health problems receive appropriate care.

KEY TERMS

- psychotherapy p. 374
- insight therapies p. 374
- psychoanalysis p. 374
- catharsis p. 374
- free association p. 375
- manifest content p. 375
- latent content p. 375
- resistance p. 375
- transference p. 375
- interpretation p. 375
- psychodynamic therapy p. 376
- interpersonal therapy (IPT) p. 376
- cognitive therapy p. 376
- self talk p. 377

- cognitive restructuring p. 377
- rational-emotive therapy (RET) p. 377
- rational-emotive behavior therapy (REBT) p. 378
- cognitive behavior therapy p. 378
- humanistic therapy p. 378
- client-centered therapy p. 380
- group therapy p. 381
- self-help group p. 381
- family therapy p. 382
- behavior therapy p. 383
- aversion therapy p. 383
- systematic desensitization p. 384
- shaping p. 384

- modeling therapy p. 384
- biomedical therapy p. 386
- psychopharmacology p. 387
- antianxiety drugs p. 388
- antipsychotic drugs p. 388
- mood stabilizer drugs p. 388
- antidepressant drugs p. 388
- electroconvulsive therapy (ECT) p. 388
- psychosurgery p. 388
- lobotomy p. 388
- repetitive transcranial magnetic stimulation (rTMS) p. 390
- eclectic approach p. 392
- deinstitutionalization p. 394

CRITICAL AND CREATIVE THINKING QUESTIONS

1. You undoubtedly had certain beliefs and ideas about therapy before reading this chapter. Has studying this chapter changed these beliefs and ideas? Explain.

2. Which form of therapy do you personally find most appealing? Why?

3. What do you consider the most important commonalities among the major forms of therapy described in this chapter? What are the most important differences?

4. Imagine that you were going to use the principles of cognitive-behavioral therapy to change some aspect of your own thinking and behavior. (Maybe you'd like to quit smoking, or be more organized, or overcome your fear of riding in elevators.) How would you identify faulty thinking? What could you do to change your thinking patterns and behavior?

(Check your answers in Appendix B.)

1. Psychoanalysis, cognitive, humanistic, group, and family therapy are often grouped together as _____.

   a. insight therapy
   b. behavior therapy
   c. humanistic and operant conditioning
   d. cognitive restructuring

2. The system of psychotherapy developed by Freud that seeks to bring unconscious conflicts into conscious awareness is known as _____.

   a. transference
   b. cognitive restructuring
   c. psychoanalysis
   d. the "hot seat" technique

3. This modern form of therapy emphasizes internal conflicts, motives, and unconscious forces.

   a. self-talk therapy
   b. belief-behavior therapy
   c. psychodynamic therapy
   d. thought analysis

4. The figure below illustrates the process by which the therapist and client work to change destructive ways of thinking called _____.

   a. problem-solving
   b. self-talk
   c. cognitive restructuring
   d. rational recovery

5. Beck practices _____, which attempts to change not only destructive thoughts and beliefs, but the associated behaviors as well.

   a. psycho-behavior therapy
   b. cognitive-behavior therapy
   c. thinking-acting therapy
   d. belief-behavior therapy

6. _____ therapy seeks to maximize personal growth through affective restructuring.

   a. Cognitive-emotive
   b. Emotive
   c. Humanistic
   d. Actualization

7. In Rogerian therapy, the _____ is responsible for discovering maladaptive patterns.

   a. therapist
   b. analyst
   c. patient
   d. client

8. This type of group does not have a professional leader, and members assist each other in coping with a specific problem.

   a. self-help
   b. encounter
   c. peer
   d. behavior

9. The main focus in behavior therapy is to increase _____ and decrease _____.

   a. positive thoughts and feelings; negative thoughts and feelings
   b. adaptive behaviors; maladaptive behaviors
   c. coping resources; coping deficits
   d. all of these options

10. The three steps in systematic desensitization include all **EXCEPT** _____.

   a. learning to become deeply relaxed
   b. arranging anxiety-arousing stimuli into a hierarchy from least to worst arousing
   c. practicing relaxation to anxiety-arousing stimuli, starting at the top of the hierarchy
   d. all of these options are included

11. This type of therapy involves watching and imitating appropriate models who demonstrate desirable behaviors.

   a. shaping
   b. spectatoring
   c. modeling
   d. systematic desensitization

12. Label the 4 major categories of psychiatric drugs on the chart below.

Common drug treatments for psychological disorders	
Type of drug	*Psychological disorder*
	Anxiety disorders
	Schizophrenia
	Bipolar disorder
	Depressive disorders

13. In electroconvulsive therapy (ECT), _____ .

   a. current is never applied to the left hemisphere
   b. convulsions activate the ANS, stimulate hormone and neu-
      rotransmitter release, and change the blood-brain barrier
   c. electrical current passes through the brain for up to one
      minute
   d. most patients today receive hundreds of treatments be-
      cause it is safer than in the past

14. Label the five most common areas of concern goals of therapies on the figure below.

15. Psychiatric professionals may authorize temporary commitment for assessment and treatment of a dangerous or incompetent individual for up to _____ .

   a. 12 to 24 hours
   b. 24 to 72 hours
   c. 3 to 4 days
   d. 3 to 4 weeks

Social Psychology 15

Do you remember the prisoner abuse scandal at the Iraqi Abu Ghraib prison in 2004? When degrading photos surfaced of U.S. military guards smiling as they posed next to naked prisoners at Abu Ghraib, people around the world were shocked and outraged. Politicians and military officials claimed that the abuses were the "vile acts of a few bad soldiers" and a "gross aberration for U.S. soldiers" (Ripley, 2004; Warner, 2004). But social psychologists believed otherwise. As you'll discover in this chapter, social circumstances and the demands of their roles can cause ordinary people to do cruel and abhorrent things—as well as to commit acts of astounding bravery or kindness.

Social psychologists study both the negative aspects of social behavior, such as aggression and prejudice, and its more positive aspects, such as interpersonal attraction and helping behaviors. But unlike other fields of psychology that focus on personality dynamics or individual pathology, social psychologists emphasize how groups, attitudes, social roles, and norms encourage us to behave cruelly or aggressively, as well as lovingly and altruistically. Because almost everything we do is social, the subject matter is enormous and varied.

We will approach our study by looking at components of social psychology (thoughts, feelings, and actions toward others). And we'll conclude with a discussion of how social psychology can help reduce prejudice, discrimination, and destructive behavior.

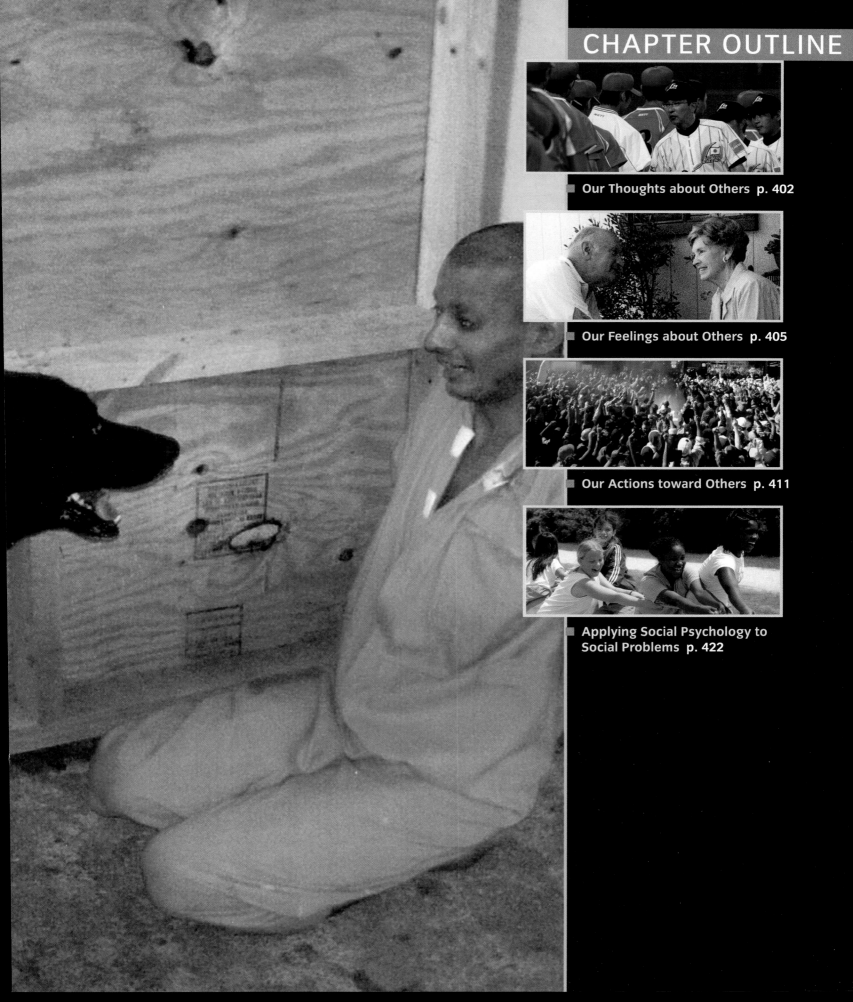

Our Thoughts about Others

LEARNING OBJECTIVES

Explain how attributions and attitudes affect the way we perceive and judge others.

Summarize the three components of attitudes.

Describe cultural differences in how people explain behavior.

ATTRIBUTION: EXPLAINING BEHAVIOR

social psychology
The study of how other people influence a person's thoughts, feelings, and actions.

attributions
Explanations for behaviors or events.

One critical aspect of **social psychology** is the search for reasons and explanations for our own and others' behavior. It's natural to want to understand and explain why people behave as they do and why events occur as they do. Many social psychologists believe that developing logical **attributions** for people's behavior makes us feel safer and more in control (Heider, 1958). To do so, most people begin with the basic question of whether a given action stems mainly from the person's internal disposition or from the external situation.

Mistaken attributions
Making the correct choice between disposition and situation is central to accurately judging why people do what they do. Unfortunately, our attributions are frequently marred by two major errors: the fundamental attribution error and the self-serving bias. Let's explore each of these.

When we consider the environmental influences on people's behavior, we generally make accurate attributions. However, given that people have enduring personality traits (Chapter 12) and a tendency to take cognitive shortcuts (Chapter 8), we more often choose dispositional attributions—that is, we blame the person. For example, if an instructor seems relaxed and talkative in class, you might conclude that she is an extroverted person. You may not realize that in situations that do not demand gregariousness, she tends to be shier. This bias is so common that it

is called the **fundamental attribution error (FAE)** (Fernández-Dols, Carrera, & Russell, 2002; Gawronski, 2003; O'Sullivan, 2003).

One reason that we tend to jump to internal, personal explanations is that human personalities and behaviors are more salient (or noticeable) than situational factors (**FIGURE 15.1**). This **saliency bias** helps explain why we tend to focus on the Abu Ghraib soldiers' behavior versus the situation itself.

When we explain our own behavior, we tend to favor internal attributions for our successes and external attributions for our failures. This **self-serving bias** is motivated by a desire to maintain positive self-esteem and a good public image (Goerke et al., 2004; Higgins & Bhatt, 2001; Ross et al., 2004). For example, students often take personal credit for doing well on an exam. If they fail the test, however, they tend to blame the instructor, the textbook, or the "tricky" questions.

fundamental attribution error (FAE)
Misjudgment of others' behavior as stemming from internal (dispositional) rather than external (situational) causes.

Culture and attributional biases
Both the fundamental attribution error and the self-serving bias may depend in part on cultural factors (Kudo & Numazaki, 2003; Matsumoto, 2000) (**FIGURE 15.2**). In highly individualistic cultures, like the United States, people are defined and understood as individual selves—largely responsible for their successes and failures. But in collectivistic cultures, like China, people are primarily defined as members of their social network—responsible for doing as others expect. Accordingly, they tend to be more aware of situational constraints on behavior, making the FAE less likely (Gilovich, Griffin, & Kahneman, 2002; Norenzayan & Nisbett, 2000; Triandis, 2001).

Saliency bias FIGURE 15.1

The saliency bias may explain why people tend to blame drug traffickers and addicts for their circumstances: The people themselves (**A**) are more concrete and conspicuous than the situational factors (**B**) that lead to drug trafficking.

Culture and attributional biases FIGURE 15.2

Westerners watching a baseball game in Japan who saw the umpire make a bad call would probably make a dispositional attribution ("He's a lousy umpire"), whereas Japanese spectators would tend to make a situational attribution ("He's under pressure"). What accounts for this difference?

The self-serving bias is also much less common in Eastern nations. In Japan, for instance, the ideal person is someone who is aware of his or her shortcomings and continually works to overcome them—not someone who thinks highly of himself or herself (Heine & Renshaw, 2002). In the East (as well as in other collectivistic cultures), self-esteem is not related to doing better than others but to fitting in with the group (Markus & Kitayama, 2003).

ATTITUDES: LEARNED PREDISPOSITIONS TOWARD OTHERS

When we observe and respond to the world around us, we are seldom completely neutral. Rather, our responses toward subjects as diverse as pizza, people, AIDS, and abortion reflect our **attitudes**. We learn our attitudes both from direct experience and from watching others.

attitudes
Learned predispositions to respond cognitively, affectively, and behaviorally to particular objects in a particular way.

NATIONAL GEOGRAPHIC

Three components of attitudes FIGURE 15.3

When social psychologists study attitudes, they measure each of the three components: cognitive, affective, and behavioral.

Social psychologists generally agree that most attitudes have three components (**FIGURE 15.3**): cognitive (thoughts and beliefs), affective (feelings), and behavioral (predisposition to act in a certain way toward an object or situation, or **attitude object**).

Cognitive dissonance theory FIGURE 15.4

When differences exist between our strongly held attitudes or between our attitudes and our behavior, we experience discomfort (dissonance) and a need to change either the attitudes or the behavior.

Although attitudes begin to form in early childhood, they are not permanent (a fact that advertisers and politicians know and exploit). One way to change attitudes is to make direct, persuasive appeals. But an even more efficient strategy is to create **cognitive dissonance** (Prinstein & Aikins, 2004; Thøgersen, 2004; Van Overwalle & Jordens, 2002). Contradictions between our attitudes and behaviors can motivate us to change our attitudes to agree with our behaviors (Festinger, 1957) (**FIGURE 15.4**).

In one of the best-known tests of cognitive dissonance theory, Leon Festinger & J. Merrill Carlsmith (1959) found that participants who received $1 for lying about a boring experimental task subsequently held more positive attitudes toward the task than those who were paid $20. Why? All participants presumably recognized the discrepancy between their attitude (the task was boring) and their behavior (saying it was interesting). In participants' minds, a $20 payment might justify the lie, resolving cognitive dissonance, but a $1 payment could not. Those paid $1 had only one way to reduce cognitive dissonance: by changing their attitude.

The experience of cognitive dissonance may depend upon a distinctly Western way of thinking about and evaluating the self. As we mentioned earlier, people in Eastern cultures tend not to define themselves in terms of their individual accomplishments. For this reason, making a bad decision may not pose the same threat to self-esteem that it would in more individualistic cultures, such as the United States (Choi & Nisbett, 2000; Markus & Kitayama, 1998, 2003).

cognitive dissonance
A feeling of discomfort caused by a discrepancy between an attitude and a behavior or between two attitudes.

CONCEPT CHECK **STOP**

Why do we tend to blame others for their misfortunes but deny responsibility for our own failures?

How can people resolve cognitive dissonance?

What role does culture play in attribution, attitudes, and cognitive dissonance?

Our Feelings about Others

LEARNING OBJECTIVES

Explain the difference between prejudice and discrimination.

Identify four explanations for why prejudice develops.

Summarize the factors that influence interpersonal attraction.

Explain how loving is different from liking.

Having explored our thoughts about others (attributions and attitudes), we now focus on our feelings about others. First we will examine the negative feelings (and thoughts and actions) associated with prejudice and discrimination. Then we'll explore the generally positive feelings of interpersonal attraction (liking and loving).

PREJUDICE AND DISCRIMINATION

Prejudice, which literally means prejudgment, creates enormous problems for its victims and also limits the perpetrator's ability to accurately judge others and to process information.

Like all attitudes, prejudice is composed of three elements: cognitive (stereotypical beliefs about people in a group), affective (emotions about the group), and behavioral (predisposition to discriminate against members of the group). Although the terms *prejudice* and *discrimination* are often used interchangeably, they are not the same. Prejudice refers to an attitude, whereas **discrimination** refers to action (Fiske, 1998). The two often coincide, but not always (**FIGURE 15.5**).

How do prejudice and discrimination originate? Why do they persist? Four commonly cited sources of prejudice are learning, mental shortcuts, economic and political competition, and displaced aggression.

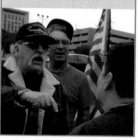

prejudice A learned, generally negative attitude directed toward specific people solely because of their membership in an identified group.

Prejudice

	Yes	No
Yes	An African American is *denied* a job because the owner of a business is prejudiced.	An African American is *denied* a job because the owner of a business fears white customers won't buy from an African American salesperson.
No	An African American is *given* a job because the owner of a business hopes to attract African-American customers.	An African American is *given* a job because he or she is the best suited for it.

Discrimination (vertical axis label)

Prejudice and discrimination FIGURE 15.5

Note how prejudice can exist without discrimination, and vice versa.

Learning   People learn prejudice the same way they learn all attitudes—through classical and operant conditioning and social learning (Chapter 6). For example, repeated exposure to stereotypical portrayals of minorities and women on television, in movies, and in books and magazines teach children that such images are correct. Hearing parents, friends, and teachers express prejudice also initiates and reinforces prejudice (Anderson & Hamilton, 2005; Bennett et al., 2004; Blaine & McElroy, 2002; Neto & Furnham, 2005). People also learn prejudice through direct experience. For example, derogating others or receiving attention for expressing racist or sexist remarks can boost one's self-esteem (Fein & Spencer, 1997; Plummer, 2001). Generalizing a single negative experience with a specific member of a group also creates prejudice (Vidmar, 1997).

Mental shortcuts   Prejudice may stem from normal attempts to simplify a complex social world (Kulik, 2005; Philipsen, 2003). Stereotypes allow people to make quick judgments about others, thereby freeing up their mental resources for other activities (**FIGURE 15.6A**). In fact, stereotypes and prejudice can occur even without a person's conscious awareness or control. This process is known as "automatic" or "implicit" bias (Devine et al., 2002; Fazio & Olson, 2003; Greenwald et al., 2002; Kawakami, Dovidio, & Dijksterhuis, 2003; Ziegert & Hanges, 2005).

People use stereotypes to classify others in terms of their membership in a group. Given that people generally classify themselves as part of the preferred group, they also create ingroups and outgroups. An **ingroup** is any category that people see themselves as belonging to; an **outgroup** is any other category (**FIGURE 15.6B**).

A Harmless stereotypes?

Even seemingly positive forms of prejudice can be destructive. For example, the stereotype that "Asian Americans are good at math" might lead Asian Americans to see few other routes to success.

B Prejudice and war

This group supports Osama Bin Laden and is demonstrating against the United States. How would you explain their loyalty to Bin Laden and prejudice against Americans?

Compared with how they see outgroup members, people tend to see ingroup members as being more attractive, having better personalities, and so on—a phenomenon known as **ingroup favoritism** (Aboud, 2003; Bennett et al., 2004). People also tend to recognize greater diversity among members of their ingroup than they do among members of outgroups (Carpenter & Radhakrisknan, 2002; Guinote & Fiske, 2003). A danger of this **outgroup homogeneity effect** is that when members of minority groups are not recognized as varied and complex individuals, it is easier to treat them in discriminatory ways. During the Vietnam War, for example, labeling Asians "gooks" made it easier for American soldiers to kill large numbers of Vietnamese civilians. As in most wars, facelessness makes it easier to kill large numbers of soldiers, as well as civilians. This type of dehumanization and facelessness also helps to perpetuate our high levels of fear and anxiety associated with terrorism (Brenner, 2005; Keen, 1991; Zimbardo, 2004).

Competition for limited resources Some theorists think that prejudice develops and is maintained because it offers significant economic and political advantages to the dominant group (Esses et al., 2001; Pettigrew, 1998) (**Figure 15.6C**).

Displaced aggression As we discuss in the next section, frustration often leads people to attack the source of frustration. As history has shown, when the cause of frustration is too powerful and capable of retaliation or is ambiguous, people often redirect their aggression toward an alternative, innocent target, or a scapegoat (**Figure 15.6D**).

Understanding the causes of prejudice is just a first step toward overcoming it. Later in this chapter, we consider several methods that psychologists have developed to reduce prejudice. In the next section, we examine the positive side of our feelings about others.

Visualizing

The cost of prejudice Figure 15.6

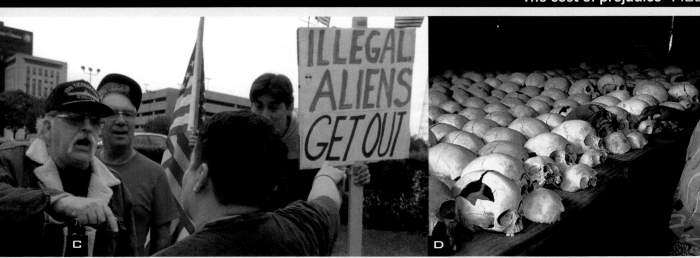

□ **Immigration and competition**

The immigration question is complex; what roles might competition for limited resources play in anti-immigrant prejudice? Prejudice can lead to atrocities.

□ **Prejudice and genocide**

These skulls are from the recent acts of genocide in Sudan, where thousands of black Africans have died from starvation, disease, and violence.

INTERPERSONAL ATTRACTION

What causes us to feel admiration, liking, friendship, intimacy, lust, or love? All of these social experiences are reflections of **interpersonal attraction**. Psychologists have found three compelling factors in interpersonal attraction: physical attractiveness, proximity, and similarity. Each influences our attraction in different ways.

> **interpersonal attraction**
> Positive feelings toward another.

Physical attractiveness (size, shape, facial characteristics, and manner of dress) is one of the most important factors in our initial liking or loving of others (Buss, 2003, 2005; Buunk et al., 2002; Li et al., 2002; Sprecher & Regan, 2002; Waynforth, 2001). Attractive individuals are seen as more poised, interesting, cooperative, achieving, sociable, independent, intelligent, healthy, and sexually warm than unattractive people (Fink & Penton-Voak,

2002; Langlois et al., 2000; Watkins & Johnston, 2000; Zebrowitz et al., 2002).

Many cultures around the world share similar standards of attractiveness, especially for women—for example, youthful appearance and facial and body symmetry (Fink et al., 2004; Gangestad & Thornhill, 2003). From an evolutionary perspective, these findings may reflect the fact that good looks generally indicate good health, sound genes, and high fertility. However, what is considered beautiful also varies from era to era and from culture to culture (**FIGURE 15.7**).

How do those of us who are not "superstar beautiful" manage to find mates? The good news is that people usually do not hold out for partners who are "ideally attractive." Instead, both men and women tend to select partners whose physical attractiveness approximately matches their own (Regan, 1998; Sprecher & Regan, 2002). Also, people use nonverbal flirting behavior to increase their attractiveness and signal interest to a potential romantic partner (Lott, 2000; Moore, 1998). Although both men and women flirt, in heterosexual couples women generally initiate courtship.

Culture and attraction FIGURE 15.7

Which of these women do you find most attractive? Can you see how your cultural background might train you to prefer one look to the others?

Proximity also promotes attraction, largely because of **mere exposure**—that is, repeated exposure increases liking (Monin, 2003; Rhodes, Halberstadt, & Brajkovich, 2001; Rossano, 2003) (FIGURE 15.8).

The major cementing factor for both liking and loving relationships is similarity in social background and values (Chen & Kenrick, 2002; Peretti & Abplanalp, 2004; Wakimoto & Fujuhara, 2004). In other words, "Birds of a feather flock together." However, what about the old saying "Opposites attract"? This adage probably refers to personality traits. That is, attraction to seemingly opposite people is most often based on the recognition that those people offer us something we lack (Dryer & Horowitz, 1997).

We can think of interpersonal attraction as a fundamental building block of how we feel about others. But how do we make sense of love? Many people find the subject to be alternately mysterious, exhilarating, comforting—and even maddening. In this section, we explore three perspectives on love: liking versus loving, romantic love, and companionate love.

Love often develops from initial feelings of friendship and liking. Using a paper-and-pencil test of liking and loving, Zick Rubin (1970) found that liking involves a favorable evaluation reflected in greater feelings of admiration and respect. Love, a more intense experience, involves caring, attachment to the other person, and intimacy (TABLE 15.1).

Couples who scored highest on Rubin's love scale spent more time looking into one another's eyes. Although both partners tended to match each other on their love scores, women liked their dating partners significantly more than they were liked in return (Rubin, 1970).

Sample items from Rubin's liking and loving test TABLE 15.1

Love scale	Liking scale
1. I feel that I can confide in _____ about virtually everything.	1. I think that _____ is unusually well adjusted.
2. I would do almost anything for _____.	2. I would highly recommend _____ for a responsible job.
3. If I could never be with _____, I would feel miserable.	3. In my opinion, _____ is an exceptionally mature person.

Source: Rubin, Z. (1970). "Measurement of romantic love," *Journal of Personality and Social Psychology, 16,* 265–273. Copyright © 1970 by the American Psychological Association. Reprinted by permission of the author.

Love Over the Life Span

Even in the most devoted couples, the intense attraction and excitement of **romantic love** generally begin to fade 6 to 30 months after the relationship begins (Hatfield & Rapson, 1996; Livingston, 1999). This is because romantic love is largely based on mystery and fantasy—people often fall in love with what they want another person to be (Fletcher & Simpson, 2000; Levine, 2001). These illusions usually fade with the realities of everyday living.

Companionate love is a strong, lasting attraction based on admiration, respect,

Years of relationship

— Romantic love
— Companionate love

trust, deep caring, and commitment. Studies of close friendships show that satisfaction grows with time as we come to recognize the value of companionship and intimacy (Kim & Hatfield, 2004). One tip for maintaining companionate love is to overlook each other's faults. People are more satisfied with relationships when they have a somewhat idealized perception of their partner (Campbell et al., 2001; Fletcher & Simpson, 2000). This makes sense in light of research on cognitive dissonance: idealizing our mates allows us to believe we have a good deal.

CONCEPT CHECK **STOP**

HOW does prejudice differ from discrimination?

What process accounts for someone saying that members of another ethnic group "all look alike"?

What are the major factors that affect our attraction to others?

Our Actions toward Others

LEARNING OBJECTIVES

Identify the factors that contribute to conformity and obedience.

Explain how groups affect behavior and decision making.

Summarize the biological and psychosocial factors believed to be involved in aggression.

Compare the egoistic model and the empathy-altruism hypothesis.

SOCIAL INFLUENCE: CONFORMITY AND OBEDIENCE

Our society and culture teach us to believe certain things, feel certain ways, and act in accordance with these beliefs and feelings. These influences are so strong and so much a part of who we are that we rarely recognize them. In this section, we discuss two kinds of social influence: conformity and obedience.

Conformity: Going along with others

Imagine that you have volunteered for a psychology experiment on perception. You are seated around a table with six other students. You are all shown a card with three lines labeled A, B, and C, as in **FIGURE 15.9**. You are then asked to select the line that is closest in length to a fourth line, X.

At first, everyone agrees on the correct line. On the third trial, however, the first participant selects line A, obviously a wrong answer. When the second, third, fourth, and fifth participants also say line A, you really start to wonder: "What's going on here? Are they blind? Or am I?"

What do you think you would do at this point in the experiment? Would you stick with your convictions and say line B, regardless of what the others have answered? Or would you go along with the group? In the original version of this experiment, conducted by Solomon Asch (1951), six of the seven participants were actually *confederates* of the experimenter (that is, they were working with the experimenter and purposely gave wrong answers). Their incorrect responses were designed to test the participant's degree of **conformity**.

> **conformity**
> The act of changing behavior as a result of real or imagined group pressure.

More than one-third of Asch's participants conformed—they agreed with the group's obviously incorrect choice. (Participants in a control group experienced no group pressure and almost always chose correctly.) Asch's study has been conducted dozens of times, in at least 17 countries, and always with similar results (Bond & Smith, 1996).

Solomon Asch's study of conformity

FIGURE 15.9

Which line in the top box is most like line X in the bottom box? Could anyone convince you otherwise?

Why would so many people conform? To the onlooker, conformity is often difficult to understand. Even the conformer sometimes has a hard time explaining his or her behavior. Let's look at three factors that drive conformity:

- *Normative social influence.* Often, people conform to group pressure out of a need for approval and acceptance by the group. **Norms** are expected behaviors that are adhered to by members of a group. Most often, norms are quite subtle and implicit. Have you ever asked what others are wearing to a party, or watched your neighbor to be sure you pick up the right fork? Such behavior reflects your desire to conform and the power of normative social influence.

- *Informational social influence.* Have you ever bought a specific product simply because of a friend's recommendation? You conform not to gain their approval (normative social influence) but because you assume they have more information than you do. Given that participants in Asch's experiment observed all the other participants give unanimous decisions on the length of the lines, they also may have conformed because they believed the others had more information.

- *Reference groups.* The third major factor in conformity is the power of **reference groups**—people we most admire, like, and want to resemble. Attractive actors and popular sports stars are paid millions of dollars to endorse products because advertisers know that we want to be as cool as Wesley Snipes or as beautiful as Jennifer Lopez. Of course, we also have more important reference groups in our lives—parents, friends, family members, teachers, religious leaders, and so on.

Norms for Personal Space

Culture and socialization have a lot to do with shaping norms for personal space. If someone invades the invisible "personal bubble" around our bodies, we generally feel very uncomfortable. People from Mediterranean, Muslim, and Latin American countries tend to maintain smaller interpersonal distances than do North Americans and Northern Europeans (Axtell, 1998; Steinhart, 1986). Children also tend to stand very close to others until they are socialized to recognize and maintain a greater personal distance. Furthermore, friends stand closer than strangers, women tend to stand closer than men, and violent prisoners prefer approximately three times the personal space of nonviolent prisoners (Axtell, 1998; Gilmour & Walkey, 1981; Roques et al., 1997).

Obedience: Following orders

As we've seen, conformity involves going along with the group. A second form of social influence, **obedience**, involves going along with a direct command, usually from someone in a position of authority.

obedience
The act of following a direct command, usually from an authority figure.

Conformity and obedience aren't always bad (FIGURE 15.10). In fact, most people conform and obey most of the time because it is in their own best interest (and everyone else's) to do so. Like most North Americans, you stand in line at the movie theatre instead of pushing ahead of others. This allows an orderly purchasing of tickets. Conformity and obedience allow social life to proceed with safety, order, and predictability.

However, on some occasions it is important not to conform or obey. We don't want teenagers (or adults) engaging in risky sex or drug use just to be part of the crowd. And we don't want soldiers (or anyone else) mindlessly following orders just because they were told to do so by an authority figure. Because recognizing and resisting destructive forms of obedience are particularly important to our society, we'll explore this material in greater depth at the end of this chapter.

Imagine that you have responded to a newspaper ad that is seeking volunteers for a study on memory. At the Yale University laboratory, an experimenter explains to you and another participant that he is studying the effects of punishment on learning and memory. You are selected to play the role of the "teacher." The experimenter leads you into a room where he straps the other participant—the "learner"—into a chair. He applies electrode paste to the learner's wrist "to avoid blisters and burns" and attaches an electrode that is connected to a shock generator.

The advantages of conformity and obedience FIGURE 15.10

These people willingly obey the firefighters who order them to evacuate a building, and many lives are saved. What would happen to our everyday functioning if people did not go along with the crowd or generally did not obey orders?

What Influences Obedience?

Milgram conducted a series of studies to discover the specific conditions that either increased or decreased obedience to authority.

Milgram's Learner
Under orders from an experimenter, would you shock a man with a known heart condition who is screaming and begging to be released? Few people believe they would. But research shows otherwise.

Milgram's Shock Generator

You are shown into an adjacent room and told to sit in front of this same shock generator, which is wired through the wall to the chair of the learner. The shock machine consists of 30 switches representing successively higher levels of shock, from 15 volts to 450 volts. Written labels appear below each group of switches, ranging from "Slight Shock" to "Danger: Severe Shock," all the way to "XXX." The experimenter explains that it is your job to teach the learner a list of word pairs and to punish any errors by administering a shock. With each wrong answer, you are to increase the shock by one level.

You begin teaching the word pairs, but the learner's responses are often wrong. Before long, you are inflicting shocks that you can only assume must be extremely painful. After you administer 150 volts, the learner begins to protest: "Get me out of here. . . . I refuse to go on."

You hesitate, and the experimenter tells you to continue. He insists that even if the learner refuses to answer, you must keep increasing the shock levels. But the other person is obviously in pain. What should you do?

Actual participants in this research—the "teachers"—suffered real conflict and distress when confronted with this problem. They sweated, trembled, stuttered, laughed nervously, and repeatedly protested that they did not want to hurt the learner. But still they obeyed.

The psychologist who designed this study, Stanley Milgram, was actually investigating not punishment and learning but obedience to authority: Would participants obey the experimenter's prompts and commands to shock another human being? In Milgram's public survey, fewer than 25 percent thought they would go beyond 150 volts. And no respondents predicted they

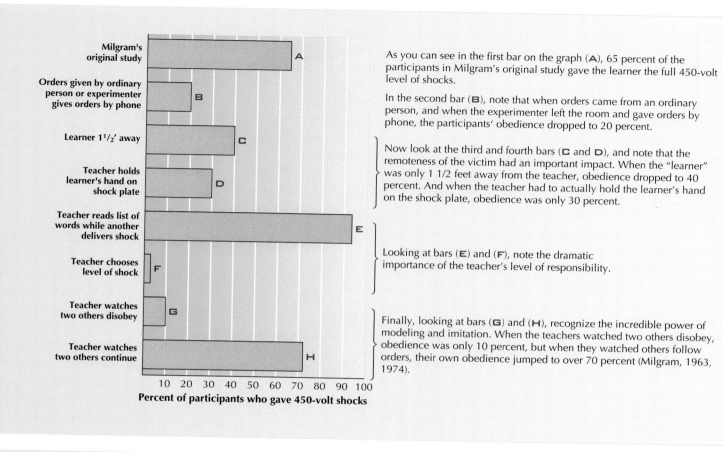

As you can see in the first bar on the graph (**A**), 65 percent of the participants in Milgram's original study gave the learner the full 450-volt level of shocks.

In the second bar (**B**), note that when orders came from an ordinary person, and when the experimenter left the room and gave orders by phone, the participants' obedience dropped to 20 percent.

Now look at the third and fourth bars (**C** and **D**), and note that the remoteness of the victim had an important impact. When the "learner" was only 1 1/2 feet away from the teacher, obedience dropped to 40 percent. And when the teacher had to actually hold the learner's hand on the shock plate, obedience was only 30 percent.

Looking at bars (**E**) and (**F**), note the dramatic importance of the teacher's level of responsibility.

Finally, looking at bars (**G**) and (**H**), recognize the incredible power of modeling and imitation. When the teachers watched two others disobey, obedience was only 10 percent, but when they watched others follow orders, their own obedience jumped to over 70 percent (Milgram, 1963, 1974).

Graph labels (top to bottom):
- Milgram's original study — A
- Orders given by ordinary person or experimenter gives orders by phone — B
- Learner 1½' away — C
- Teacher holds learner's hand on shock plate — D
- Teacher reads list of words while another delivers shock — E
- Teacher chooses level of shock — F
- Teacher watches two others disobey — G
- Teacher watches two others continue — H

x-axis: 10 20 30 40 50 60 70 80 90 100
Percent of participants who gave 450-volt shocks

would go past the 300-volt level. Yet 65 percent of the teacher-participants in this series of studies obeyed completely—going all the way to the end of the scale, even beyond the point when the "learner" (Milgram's confederate, who actually received no shocks at all) stopped responding altogether.

Even Milgram was surprised by his results. Before the study began, he polled a group of psychiatrists, and they predicted that most people would refuse to go beyond 150 volts and that less than 1 percent of those tested would "go all the way." But, as Milgram discovered, most of his participants—men and women, of all ages, and from all walks of life—administered the highest voltage. The study was replicated many times and in many other countries, with similarly high levels of obedience.

In a series of follow-up studies, Milgram found several important factors that influenced obedience:

(1) legitimacy and closeness of the authority figure, (2) remoteness of the victim, (3) assignment of responsibility, (4) modeling or imitating others. (Blass, 1991, 2000; Meeus & Raaijmakers, 1989; Snyder, 2003). These are summarized in What a Scientist Sees.

Many people are upset both by Milgram's findings and by the treatment of the participants in his research. Although deception is a necessary part of some research, the degree of deception and discomfort of participants in Milgram's research is now viewed as highly unethical. His study would never be undertaken today. On the other hand, Milgram carefully debriefed every subject after the study and followed up with the participants for several months. Most of his "teachers" reported the experience as being personally informative and valuable.

GROUP PROCESSES

Although we seldom recognize the power of group membership, social psychologists have identified several important ways that groups affect us.

Group membership

How do the roles that we play within groups affect our behavior? This question fascinated social psychologist Philip Zimbardo. In his famous study at Stanford University, 24 carefully screened, well-adjusted young college men were paid $15 a day for participating in a two-week simulation of prison life (Haney, Banks, & Zimbardo, 1978; Zimbardo, 1993).

The students were randomly assigned to the role of either prisoner or guard. Prisoners were "arrested," frisked, photographed, fingerprinted, and booked at the police station. They were then blindfolded and driven to the "Stanford Prison." There, they were given ID numbers, deloused, issued prison clothing (tight nylon caps, shapeless gowns, and no underwear), and locked in cells. Participants assigned to be guards were outfitted with official-looking uniforms, billy clubs, and whistles, and they were given complete control.

Not even Zimbardo foresaw how the study would turn out. Although some guards were nicer to the prisoners than others, they all engaged in some abuse of power. The slightest disobedience was punished with degrading tasks or the loss of "privileges" (such as eating, sleeping, and washing). As demands increased and abuses began, the prisoners became passive and depressed. Only one prisoner fought back with a hunger strike, which ended with a forced feeding by the guards.

Four prisoners had to be released within the first four days because of severe psychological reactions. The study was stopped after only six days because of the alarming psychological changes in the participants.

Although this was not a true experiment in that it lacked control groups and clear measurements of the dependent variable (Chapter 1), it offers insights into the potential effects of roles on individual behavior (**FIGURE 15.11**). According to interviews conducted after the study, the students became so absorbed in their roles that they forgot they were volunteers in a university study (Zimbardo, Ebbeson, & Maslach, 1977).

Zimbardo's study also demonstrates **deindividuation**. To be deindividuated means that you feel less self-conscious, less inhibited, and less personally responsible as a member of a group than when you're

Power corrupts FIGURE 15.11

Zimbardo's prison study showed how the demands of roles and situations could produce dramatic changes in behavior in just a few days. Can you imagine what happens to prisoners during life imprisonment, six-year sentences, or even a few nights in jail?

Lost in the crowd FIGURE 15.12

One of the most compelling explanations for deindividuation is the fact that the presence of others tends to increase arousal and feelings of anonymity. Anonymity is a powerful disinhibitor, helping to explain why vandalism seems to increase on Halloween (when people commonly wear masks) and why most crimes and riots occur at night—under the cover of darkness.

alone. Although deindividuation can sometimes be positive, it also helps explain angry mobs, rioters, gang rapes, lynchings, and hate crimes (Silke, 2003). Groups sometimes actively promote deindividuation by requiring its members to wear uniforms, for example, as a way to increase allegiance and conformity (FIGURE 15.12).

Group decision making

We have seen that the groups we belong to and the roles that we play within them influence how we think about ourselves. How do groups affect our decisions? Are two heads truly better than one?

Most people assume that group decisions are more conservative, cautious, and middle-of-the-road than individual decisions. But is this true? Initial investigations indicated that after discussing an issue, groups actually support riskier decisions than decisions they made as individuals before the discussion (Stoner, 1961).

Subsequent research on this **risky-shift** phenomenon, however, shows that some groups support riskier decisions while others support conservative decisions (Liu & Latané, 1998). How can we tell if a given group's decision will be risky or conservative? The final decision (risky or conservative) depends primarily on the dominant preexisting tendencies of the group. That is, as individuals interact and discuss their opinions, their initial positions become more exaggerated (or polarized).

Why are group decisions more exaggerated? This tendency toward **group polarization** stems from increased exposure to persuasive arguments that reinforce the group's original opinion (Liu & Latané, 1998).

group polarization
A group's movement toward either riskier or more conservative behavior, depending on the members' initial dominant tendencies.

groupthink
Faulty decision making that occurs when a highly cohesive group strives for agreement and avoids inconsistent information.

A related phenomenon is **groupthink** (FIGURE 15.13). When a group is highly cohesive (a family, a panel of military advisers, an athletic team), the members' desire for agreement may lead them to ignore important information or points of view held by outsiders or critics (Hergovich & Olbrich, 2003; Vaughn, 1996). During the discussion process, the members:

- come to believe that they are invulnerable,
- tend to develop common rationalizations and stereotypes of the outgroup, and
- exert considerable pressure on anyone who dares to offer a dissenting opinion.

Many presidential errors—from Franklin Roosevelt's failure to anticipate the attack on Pearl Harbor to Ronald Reagan's Iran-Contra scandal—have been blamed on groupthink. Groupthink may have contributed to the losses of the space shuttles *Challenger* and *Columbia,* the terrorist attack of September 11, and the war in Iraq (Ehrenreich, 2004; Janis, 1972, 1989; Landay & Kuhnehenn, 2004; Schwartz & Broder, 2003).

Process Diagram

How groupthink occurs FIGURE 15.13

A Two days before the space shuttle Columbia broke apart in 2003, a NASA engineer warned his supervisors of possible catastrophes on reentry. His message was not forwarded up the chain of command, and all seven astronauts died in the disaster (Schwartz & Broder, 2003). How might groupthink have contributed to this tragedy?

B The process of groupthink begins when group members feel a strong sense of cohesiveness and isolation from the judgments of qualified outsiders. Add a directive leader and little chance for debate, and you have the recipe for a potentially dangerous decision.

Antecedent Conditions **B**
1 A highly cohesive group of decision makers
2 Insulation of the group from outside influences
3 A directive leader
4 Lack of procedures to ensure careful consideration of the pros and cons of alternative actions
5 High stress from external threats with little hope of finding a better solution than that favored by the leader

Strong desire for group consensus—The groupthink tendency

Symptoms of Groupthink
1 Illusion of invulnerability
2 Belief in the morality of the group
3 Collective rationalizations
4 Stereotypes of outgroups
5 Self-censorship of doubts and dissenting opinions
6 Illusion of unanimity
7 Direct pressure on dissenters

Symptoms of Poor Decision Making
1 An incomplete survey of alternative courses of action
2 An incomplete survey of group objectives
3 Failure to examine risks of the preferred choice
4 Failure to reappraise rejected alternatives
5 Poor search for relevant information
6 Selective bias in processing information
7 Failure to develop contingency plans

Low probability of successful outcome

AGGRESSION

> **aggression**
> Any behavior intended to harm someone.

Why do people act aggressively? In this section, we explore both biological and psychosocial explanations for **aggression** and how aggression can be controlled or reduced.

Biological explanations
Instincts. Because aggression has such a long history and is found in all cultures, many theorists believe that humans are instinctively aggressive. Evolutionary psychologists and ethologists (scientists who study animal behavior) believe that aggression evolved because it prevents overcrowding and allows the strongest animals to win mates and reproduce (Dabbs & Dabbs, 2000; Lorenz, 1981; Rossano, 2003). Most social psychologists, however, reject the view that instincts drive aggression.

Genes. Twin studies suggest that some individuals are genetically predisposed to have hostile, irritable temperaments and to engage in aggressive acts (Miles & Carey, 1997; Segal & Bouchard, 2000; Wasserman & Wachbroit, 2000). Remember, though, that biology interacts with social experience to shape behavior.

The brain and nervous system. Electrical stimulation or the severing of specific parts of an animal's brain has a direct effect on aggression (Delgado, 1960; Delville, Mansour, & Ferris, 1996; Roberts & Nagel, 1996). Research with brain injuries and organic disorders has also identified possible aggression circuits in the brain (Davidson, Putnam, & Larson, 2000; Raine et al., 1998).

Substance abuse and other mental disorders. Substance abuse (particularly alcohol) is a major factor in most forms of aggression (Badawy, 2003; Gerra et al., 2004; Levinthal, 2006). Homicide rates are also higher among men with schizophrenia and antisocial disorders, particularly if they abuse alcohol (Raesaenen et al., 1998; Tiihonen et al., 1997).

Hormones and neurotransmitters. Several studies have linked the male hormone testosterone to aggressive behavior (Mong & Pfaff, 2003; O'Connor et al., 2002; Trainor, Bird & Marler, 2004). Violent behavior has also been linked with lowered levels of the neurotransmitters serotonin and GABA (gamma-aminobutyric acid) (Goveas, Csernansky, & Coccaro, 2004; Halperin et al., 2003).

Psychosocial explanations
Aversive stimuli. Noise, heat, pain, insults, and foul odors all can increase aggression (Anderson, 2001; Anderson et al., 2000; Twenge et al., 2001). According to the **frustration-aggression hypothesis**, developed by John Dollard and colleagues (1939), another aversive stimulus—frustration—creates anger, which for some leads to aggression.

Culture and learning. Social learning theory (Chapter 6) suggests that people raised in a culture with aggressive models will learn aggressive responses (Matsumoto, 2000). Among developed nations, for example, the United States has a high rate of violent crimes, and American children grow up with numerous models for aggression, which they tend to imitate.

Media and video games. The media can contribute to aggression in both children and adults (Anderson, 2004; Bartholow & Anderson, 2002; Kronenberger et al., 2005; Uhlmann & Swanson, 2004) (**FIGURE 15.14**). However, the link between media violence and aggression appears to be a two-way street. Laboratory studies, correlational research, and cross-cultural studies have found that exposure to TV violence increases aggressiveness and that aggressive children tend to seek out violent programs (Aluja-Fabregat & Torrubia-Beltri, 1998; Singer et al., 1998).

Video violence FIGURE 15.14

Video games such as *Doom, Mortal Kombat, Resident Evil,* and *Half-Life* all feature realistic sound effects and gory depictions of "lifelike" violence. Research shows that media can contribute to aggression in children and that aggressive children seek out violent television programs.

How can people control or eliminate aggression? Some therapists advise people to release their aggressive impulses by engaging in harmless forms of aggression, such as vigorous exercise, punching a pillow, or watching competitive sports. But studies suggest that this type of *catharsis* doesn't really help and may only intensify the feeling (Bushman, 2002; Bushman, Baumeister, & Stack, 1999).

A more effective approach is to introduce incompatible responses. Because certain emotional responses, such as empathy and humor, are incompatible with aggression, purposely making a joke or showing some sympathy for an opposing person's point of view can reduce anger and frustration (Harvey & Miceli, 1999; Kaukiainen et al., 1999; Oshima, 2000). Improving social and communication skills can also help keep aggression in check.

ALTRUISM: HELPING OTHERS

In 1964, a woman named Kitty Genovese was raped and stabbed repeatedly by an assailant outside her apartment in Queens, New York. It was about 3:00 A.M., but 38 of her neighbors watched the attack and heard her screams for help. The attack lasted for over half an hour; yet no one came to help her. Finally, one neighbor called the police, but it was too late. Kitty Genovese had died.

How could such a thing have happened? Why didn't her neighbors help? Under what conditions do people help and under what conditions do they ignore others' pleas for help?

Evolutionary theory suggests that **altruism** (or **prosocial behavior**) is an instinctual behavior that has evolved because it favors survival of one's genes (Bower, 2000; Ozinga, 2000, 2002; Rossano, 2003). By helping your child or other relative, for exam-

"Hey, there's Sara, padding her college-entrance résumé!"
© The New Yorker Collection 2001 Edward Koren from cartoonbank.com.
All Rights Reserved.

ple, you increase the odds of your genes' survival.

Other research suggests that helping others may actually be self-interest in disguise. According to this **egoistic model**, we help others only because we hope for later reciprocation, because it makes us feel virtuous, or because it helps us avoid feeling guilty (Cialdini, 2001; Williams et al., 1998).

Opposing the egoistic model is the **empathy-altruism hypothesis** (Batson, 1991, 1998; Batson & Ahmad, 2001). This perspective holds that although some altruism is motivated by selfish concerns, in many situations helping is truly selfless and motivated by concern for others (**FIGURE 15.15**).

According to the empathy-altruism hypothesis, simply seeing or hearing of another person's suffering can create empathy—a subjective grasp of that person's feelings or experiences. When we feel empathic toward another, we are motivated to help that person for his or her own sake. The ability to empathize may even be innate. Research with infants in the first few hours of life shows that they become distressed and cry at the sound of another infant's cries (Hay, 1994; Hoffman, 1993).

Many theories have been proposed to explain why people help, but few explain why we do not. How do we explain the Kitty Genovese tragedy? One of the most comprehensive explanations for helping or not helping comes from the research of Bibb Latané and John Darley (1970). They found that whether or not someone helps depends on a series of interconnected events and decisions: The potential helper must first notice what is happening, interpret the event as an emergency, accept personal responsibility for helping, decide how to help, and then actually initiate the helping behavior.

Where did the sequence break down for Kitty

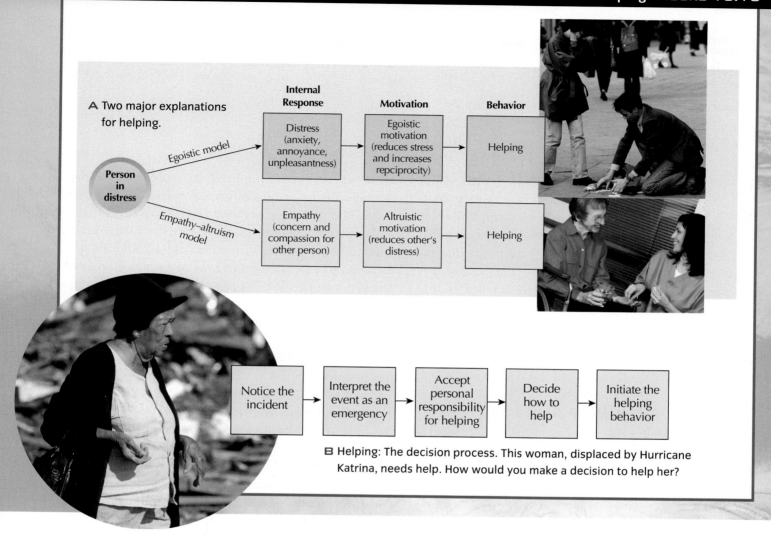

A Two major explanations for helping.

Internal Response — Motivation — Behavior

Person in distress

Egoistic model → Distress (anxiety, annoyance, unpleasantness) → Egoistic motivation (reduces stress and increases repciprocity) → Helping

Empathy–altruism model → Empathy (concern and compassion for other person) → Altruistic motivation (reduces other's distress) → Helping

Notice the incident → Interpret the event as an emergency → Accept personal responsibility for helping → Decide how to help → Initiate the helping behavior

B Helping: The decision process. This woman, displaced by Hurricane Katrina, needs help. How would you make a decision to help her?

Genovese? Kitty's neighbors obviously noticed what was happening and interpreted it as an emergency. The breakdown came at the third stage—taking personal responsibility for helping. In newspaper interviews, each of the neighbors described a great deal of anguish, but they also reported that they failed to intervene because they were certain that someone must have called the police already. Latané and Darley called this the **diffusion of responsibility** phenomenon—the dilution (or diffusion) of personal responsibility for acting by spreading it among all other group members. Ironically, if only one neighbor had seen the brutal attack and thought that there were no other witnesses, Kitty might still be alive.

CONCEPT CHECK STOP

How do norms, need for information, and reference groups produce conformity?

How does group polarization occur?

What are the most effective ways to control aggression?

What is diffusion of responsibility?

Applying Social Psychology to Social Problems

LEARNING OBJECTIVES

Describe four major approaches to reducing prejudice and discrimination.

Explain how social changes might create cognitive dissonance and eventually promote a reduction in prejudice.

Summarize the principles that explain destructive obedience to authority.

Every day we're confronted with social problems—from noisy neighbors to freeway congestion to terrorism. Unfortunately, social psychology has been more successful in describing, explaining, and predicting social problems than in changing them. However, researchers have found several helpful techniques. In this section, we first explore what scientists have discovered about methods to reduce prejudice, and then we discuss effective ways to cope with destructive forms of obedience.

REDUCING PREJUDICE AND DISCRIMINATION

What can be done to combat prejudice? Four major approaches can be used.

Cooperation and superordinate goals
One of the best ways to combat prejudice is to encourage cooperation rather than competition (Brewer, 1996; Walker & Crogan, 1998). In a classic study, Muzafer Sherif and his colleagues (1966, 1998) created strong feelings of ingroup loyalty at a summer camp by physically separating two groups of boys into different cabins and assigning different projects to each group.

Once each group developed strong feelings of group identity and allegiance, the researchers set up a series of competitive games and awarded desirable prizes to the winning teams. The groups soon began to pick fights, call each other names, and raid each other's camps.

After using competition to create prejudice between the two groups, the researchers created "mini-crises" and tasks that required expertise, labor, and cooperation from both groups, and prizes were awarded to all. The prejudice between the groups slowly began to dissipate, and by the end of camp, the earlier hostilities and ingroup favoritism had vanished. Sherif's study showed not only the importance of cooperation as opposed to competition but also the importance of **superordinate goals** (the minicrises) in reducing prejudice (Der-Karabetian, Stephenson, & Poggi, 1996).

Increased contact
A second approach to reducing prejudice is increasing contact between groups (Brown et al., 2003; London et al., 2002; Philipsen, 2003). But as you just discovered with Sherif's study of the boys at the summer camp, contact can sometimes increase prejudice. Increasing contact only works under conditions that provide for close interaction, interdependence (superordinate goals that require cooperation), and equal status. When people have positive experiences with one group, they tend to generalize to other groups (Pettigrew, 1998).

Cognitive retraining
One of the most recent strategies for prejudice reduction requires taking another's perspective or undoing associations of negative stereotypical traits (Galinsky & Ku, 2004; Galinsky & Moskowitz, 2000; Philipsen, 2003). People can also learn to be unprejudiced if they are taught to selectively pay attention to similarities between groups, rather than differences (Phillips & Ziller, 1997).

Cognitive dissonance
One of the most efficient methods of changing an attitude uses the principle of cognitive dissonance (Cook, 2000; D'Alessio & Allen, 2002). Each time we meet someone who does not

Creating cognitive dissonance FIGURE 15.16

How might social changes, such as school busing, integrated housing, and increased civil rights legislation, initially create cognitive dissonance and then eventually lead to a reduction in prejudice? What other principle of attitude change is illustrated in this photo?

conform to our prejudiced views, we experience dissonance—"I thought all gay men were effeminate. This guy is a deep-voiced professional athlete. I'm confused." To resolve the dissonance, we can maintain our stereotypes by saying, "This gay man is an exception to the rule." However, if we continue to come into contact with a variety of gay men, this "exception to the rule" defense eventually breaks down, and attitude change occurs (**FIGURE 15.16**).

OVERCOMING DESTRUCTIVE OBEDIENCE: WHEN IS IT OKAY TO SAY NO?

Obedience to authority is an important part of our lives. If we routinely refused to obey police officers, firefighters, and other official personnel, our individual safety and social world would collapse. However, there are also many times when obedience may be unnecessary and even destructive—such as obedience to someone who is abusing his or her military power or to a religious cult leader—and should be reduced.

How do we explain (and hopefully reduce) destructive obedience? Let's reexamine some of the major points discussed in this chapter while also introducing a few new ones.

Socialization   Society and culture have a tremendous influence on all our thoughts, feelings, and actions. Obedience is no exception. From very early childhood, we're taught to respect and obey our parents, teachers, and other authority figures, and without this obedience there would be social chaos. Unfortunately, this early (and lifelong) socialization often becomes so deeply ingrained that we no longer recognize it, which helps explain many instances of mindless obedience to immoral requests from people in positions of authority (such as the brutal treatment of prisoners at the Abu Ghraib prison). History is replete with instances in which atrocities were committed because people were "just following orders."

Power of the situation

Situational factors also have a strong impact on obedience. For example, the roles of police officer or public citizen, teacher or student, and parent or child all have built-in guidelines for appropriate behavior. One person is ultimately "in charge," and the other person is supposed to follow along. Because these roles are so well socialized, we mindlessly play them and find it difficult to recognize the point where they become maladaptive. As we discovered in the Zimbardo prison study, when well-adjusted and well-screened college students were suddenly given the roles of prisoners and guards, their behaviors were dramatically affected.

Groupthink

When discussing Milgram's study and other instances of destructive obedience, most people believe that they and their friends would never do such a thing. Can you see how this might be a form of groupthink, a type of faulty thinking that occurs when group members strive for agreement and avoid inconsistent information? When we proclaim that Americans would never follow the orders that some German people did during the holocaust, we're demonstrating several symptoms of groupthink—stereotypes of the outgroup, illusion of unanimity, belief in the morality of the group, and so on. We're also missing one of the most important lessons from Milgram's studies and the cross-cultural follow-ups: As philosopher Hannah Arendt has suggested, the horrifying thing about the Nazis was not that they were so deviant but that they were so "terrifyingly normal."

Foot-in-the-door

The gradual nature of many situations involving obedience may also help explain why so many people were willing to give the maximum shocks in Milgram's studies. The initial mild level of shocks may have worked as a **foot-in-the-door technique,** in which a first, small request is used as a setup for later, larger requests. Once Milgram's participants complied with the initial request, they might have felt obligated to continue (Chartrand, Pinckert, & Burger, 1999; Sabini & Silver, 1993). Can you see how this technique helps explain other forms of destructive obedience?

Relaxed moral guard

One common intellectual illusion that hinders critical thinking about obedience is the belief that only evil people do evil things, or that

evil announces itself. For example, the experimenter in Milgram's study looked and acted like a reasonable person who was simply carrying out a research project. Because he was not seen as personally corrupt and evil, the participants' normal moral guard was down, which can maximize obedience. This relaxed moral guard might similarly explain obedience to a highly respected military officer or the leader of a religious cult.

These forces toward obedience are powerful. It's important to remember that each of us must be personally alert to immoral forms of obedience. On occasion, we also need the courage to stand up and say "No!" One of the most beautiful and historically important examples of this type of bravery occurred in Alabama in 1955. Rosa Parks (**FIGURE 15.17**) boarded a bus and, as expected in those times, she obediently sat in the back section marked "Negroes." When the

Rosa Parks FIGURE 15.17

bus became crowded, the driver told her to give up her seat for a white man. Surprisingly for those days, Parks refused and was eventually forced off the bus by police and arrested. This single act of disobedience was the catalyst for the small, but growing civil rights movement. Her courageous stand inspires the rest of us to think about when it is good to obey authorities and when to disobey unethical demands from authorities.

CONCEPT CHECK STOP

What is cognitive retraining?

How might groupthink contribute to destructive obedience?

From a very early age, Andrew Golden was taught how to fire hunting rifles. At age 11, he and a friend killed four classmates and a teacher in a school shooting. What factors might have contributed to Golden's tragically aggressive act?

What if a genetic study were somehow able to prove that the boy was biologically predisposed to engage in hostile acts? Does this mean he was doomed to behave aggressively? Why or why not?

VISUAL SUMMARY

1 Our Thoughts about Others

1. **Social psychology** is the study of how other people influence our thoughts, feelings, and actions. Our **attributions** are frequently marred by the **fundamental attribution error** and the self-serving bias. Both biases may depend in part on cultural factors.

2. **Attitudes** have three components: cognitive, affective, and behavioral. An efficient strategy for changing attitudes is to create **cognitive dissonance**. Like attributional biases, the experience of cognitive dissonance may depend on culture.

2 Our Feelings about Others

1. Like all attitudes, **prejudice** involves cognitive, affective, and behavioral components. Although the terms prejudice and discrimination are often used interchangeably, they are not the same. Four commonly cited sources of prejudice are learning, mental shortcuts, economic and political competition, and displaced aggression.

2. Psychologists have found three compelling factors in **interpersonal attraction**: physical attractiveness, proximity, and similarity. Love often develops from initial feelings of friendship and liking. Romantic love is an intense but generally short-lived attraction based on mystery and fantasy, whereas companionate love is a strong, lasting attraction based on admiration, respect, trust, deep caring, and commitment.

3 Our Actions toward Others

1. Three factors drive **conformity**: normative social influence, informational social influence, and the role of reference groups. Conformity involves going along with the group, while **obedience** involves going along with a direct command, usually from someone in a position of authority. Milgram's research demonstrated the startling power of social situations to create obedience. The degree of deception and discomfort that Milgram's participants were subjected to raises serious ethical questions, and the same study would never be done today.

2. The roles that we play within groups strongly affect our behavior, as Zimbardo's Stanford Prison experiment showed. Zimbardo's study also demonstrated deindividuation. In addition, groups affect our decisions. As individuals interact and discuss their opinions, **group polarization** (a group's movement toward more extreme decisions) and **groupthink** (a group's tendency to strive for agreement and to avoid inconsistent information) tend to occur. Both processes may hinder effective decision making.

3. Several biological factors may help explain **aggression**, including instincts, genetic predisposition, aggression circuits in the brain and nervous system, mental disorders, and hormones and neurotransmitters. Researchers have also proposed several psychosocial explanations for aggression, including aversive stimuli, culture and learning, and media influences. Catharsis does not appear to help release aggressive impulses; more effective ways to control aggression are to introduce incompatible responses and to improve social and communication skills.

4. Evolutionary theory suggests that **altruism** is an evolved, instinctual behavior. Other research suggests that helping may actually be self-interest in disguise (egoistic model). The empathy-altruism hypothesis proposes that although altruism is sometimes based in selfish motivations, it is sometimes truly selfless and motivated by concern for others (empathy). Latané and Darley found that in order for helping to occur, the potential helper must notice what is happening, interpret the event as an emergency, take personal responsibility for helping, decide how to help, and then actually initiate the helping behavior.

4 Applying Social Psychology to Social Problems

1. Four major approaches can be used to combat prejudice: cooperation and superordinate goals, increased contact, cognitive retraining, and cognitive dissonance.

2. There are many times when obedience is unnecessary and destructive. Reducing destructive obedience requires an understanding of several social psychological principles, including socialization, the power of the situation, groupthink, the foot-in-the-door technique, and relaxed moral guard.

KEY TERMS

CRITICAL AND CREATIVE THINKING QUESTIONS

1. Have you ever changed a strongly held attitude? What caused the change for you?

2. Do you believe that you are free of prejudice? After reading this chapter, which of the many factors that cause prejudice do you think is most important to change?

3. How do Milgram's results—particularly the finding that the remoteness of the victim affected obedience—relate to some aspects of modern warfare?

4. What are some of the similarities between Zimbardo's prison study and the abuses at the Abu Ghraib prison in Iraq?

5. Have you ever done something in a group that you would not have done if you were alone? What happened? How did you feel? What have you learned from this chapter that might help you avoid this behavior in the future?

6. Can you think of situations when the egoistic model of altruism seems most likely correct? What about the empathy-altruism hypothesis?

(Check your answers in Appendix B.)

1. The study of how other people influence our thoughts, feelings, and actions is called ——————.

   a. sociology
   b. social science
   c. social psychology
   d. sociobehavioral psychology

2. The two major attribution mistakes people make are ——————.

   a. the fundamental attribution error and self-serving bias
   b. situational attributions and dispositional attributions
   c. the actor bias and the observer bias
   d. stereotypes and biases

3. Label the 3 components of attitudes on the figure below.

4. This theory says that contradictions between our attitudes and behavior can motivate us to change our attitudes to agree with our behavior.

   a. social learning theory
   b. cognitive dissonance theory
   c. defense mechanisms theory
   d. power of inconsistencies theory

5. —————— is a learned, generally negative, attitude toward specific people solely because of their membership in an identified group.

   a. Discrimination
   b. Stereotyping
   c. Cognitive biasing
   d. Prejudice

6. The degree of positive feelings you have toward others is called ——————.

   a. affective relations
   b. interpersonal attraction
   c. interpersonal attitudes
   d. affective connections

7. A strong and lasting attraction characterized by trust, caring, tolerance, and friendship is called ——————.

   a. companionate love
   b. intimate love
   c. passionate love
   d. all of these options

8. This is the act of changing behavior as a result of real or imagined group pressure.

   a. norm compliance
   b. obedience
   c. conformity
   d. mob rule

9. Stanley Milgram was investigating —————— in his classic teacher-learner shock study.

   a. the effects of punishment on learning
   b. the effects of reinforcement on learning
   c. obedience to authority
   d. all of these options

10. During _____, a person who feels anonymous within a group or crowd experiences an increase in arousal and a decrease in self-consciousness, inhibitions, and personal responsibility.

   a. groupthink

   b. group polarization

   c. authoritarianism

   d. deindividuation

11. Faulty decision making that is the result of a highly cohesive group striving for agreement to the point of avoiding inconsistent information is known as _____.

   a. the risky-shift

   b. group polarization

   c. groupthink

   d. destructive conformity

12. Actions that are designed to help others with no obvious benefit to the helper are collectively known as _____.

   a. empathy

   b. sympathy

   c. altruism

   d. egoism

13. Label the 2 major explanations for helping on the figure below.

14. Research suggests that one of the best ways to decrease prejudice is to encourage _____.

   a. cooperation

   b. friendly competition

   c. reciprocity of liking

   d. conformity

15. Which of the following is most **UNLIKELY** to increase obedience?

   a. Power of the situation.

   b. Foot-in-the-door.

   c. Socialization.

   d. These are all ways to increase obedience.

Appendix A
Statistics and Psychology

We are constantly bombarded by numbers: "On sale for 30 percent off," "70 percent chance of rain," "9 out of 10 doctors recommend..." The president uses numbers to try to convince us that the economy is healthy. Advertisers use numbers to convince us of the effectiveness of their products. Psychologists use statistics to support or refute psychological theories and demonstrate that certain behaviors are indeed results of specific causal factors.

When people use numbers in these ways, they are using statistics. **Statistics** is a branch of applied mathematics that uses numbers to describe and analyze information on a subject.

Statistics make it possible for psychologists to quantify the information they obtain in their studies. They can then critically analyze and evaluate this information. Statistical analysis is imperative for researchers to describe, predict, or explain behavior. For instance, Albert Bandura (1973) proposed that watching violence on television causes aggressive behavior in children. In carefully controlled experiments, he gathered numerical information and analyzed it according to specific statistical methods. The statistical analysis helped him substantiate that the aggression of his subjects and the aggressive acts they had seen on television were related, and that the relationship was not mere coincidence.

Although statistics is a branch of applied mathematics, you don't have to be a math whiz to use statistics. Simple arithmetic is all you need to do most of the calculations. For more complex statistics involving more complicated mathematics, computer programs are available for virtually every type of computer. What is more important than learning the mathematical computations, however, is developing an understanding of when and why each

type of statistic is used. The purpose of this appendix is to help you understand the significance of the statistics most commonly used.

Gathering and Organizing Data

Psychologists design their studies to facilitate gathering information about the factors they want to study. The information they obtain is known as **data** (**data** is plural; its singular is **datum**). When the data are gathered, they are generally in the form of numbers; if they aren't, they are converted to numbers. After they are gathered, the data must be organized in such a way that statistical analysis is possible. In the following section, we will examine the methods used to gather and organize information.

Variables

When studying a behavior, psychologists normally focus on one particular factor to determine whether it has an effect on the behavior. This factor is known as a **variable,** which is in effect anything that can assume more than one value (see Chapter 1). Height, weight, sex, eye color, and scores on an IQ test or a video game are all factors that can assume more than one value and are therefore variables. Some will vary between people, such as sex (you are either male *or* female but not both at the same time). Some may even vary within one person, such as scores on a video game (the same person might get 10,000 points on one try and only 800 on another). Opposed to a variable, anything that remains the same and does not vary is called a **constant.** If researchers use only females in their research, then sex is a constant, not a variable.

In nonexperimental studies, variables can be factors that are merely observed through naturalistic observation or case studies, or they can be factors about which people are questioned in a test or survey. In experimental studies, the two major types of variables are independent and dependent variables.

Independent variables are those that are manipulated by the experimenter. For example, suppose we were to conduct a study to determine whether the sex of the debater influences the outcome of a debate. In this study, one group of subjects watches a videotape of a debate between a male arguing the "pro" side and a female arguing the "con"; another group watches the same debate, but with the pro and con roles reversed. In such a study, the form of the presentation viewed by each group (whether "pro" is argued by a male or a female) is the independent variable because the experimenter manipulates the form of presentation seen by each group. Another example might be a study to determine whether a particular drug has any effect on a manual dexterity

Drawing by M. Stevens; © 1989 the New Yorker Magazine, Inc.

431

task. To study this question, we would administer the drug to one group and no drug to another. The independent variable would be the amount of drug given (some or none). The independent variable is particularly important when using **inferential statistics,** which we will discuss later.

The **dependent variable** is a factor that results from, or depends on, the independent variable. It is a measure of some outcome or, most commonly, a measure of the subjects' behavior. In the debate example, each subject's choice of the winner of the debate would be the dependent variable. In the drug experiment, the dependent variable would be each subject's score on the manual dexterity task.

Frequency Distributions

After conducting a study and obtaining measures of the variable(s) being studied, psychologists need to organize the data in a meaningful way. Table A.1 presents test scores from a statistics aptitude test collected from 50 college students. This information is called **raw data** because there is no order to the numbers. They are presented as they were collected and are therefore "raw."

The lack of order in raw data makes them difficult to study. Thus, the first step in understanding the results of an experiment is to impose some order on the raw data. There are several ways to do this. One of the simplest is to create a **frequency distribution,** which shows the number of times a score or event occurs. Although frequency distributions are helpful in several ways, the major advantages are that they allow us to see the data in an organized manner and they make it easier to represent the data on a graph.

The simplest way to make a frequency distribution is to list all the possible test scores, then tally the number of people (*N*) who received those scores. Table A.2 presents a frequency distribution using the raw data from Table A.1. As you can see, the data are now easier to read. From looking at the frequency distribution, you can see that most of the test scores lie in the middle with only a few at the very high or very low end. This was not at all evident from looking at the raw data.

This type of frequency distribution is practical when the number of possible scores is 20 or fewer. However, when there are more than 20 possible scores it can be even harder to make sense out of

the frequency distribution than the raw data. This can be seen in Table A.3, which presents the Scholastic Aptitude Test scores for 50 students. Even though there are only 50 actual scores in this table, the number of possible scores ranges from a high of 1390 to a low of 400. If we included zero frequencies there would be 100 entries in a frequency distribution of this data, making the frequency distribution much more difficult to understand than the raw data. If there are more than 20 possible scores, therefore, a **group** frequency distribution is normally used.

Frequency distribution of 50 students on Statistics Aptitude Test TABLE A.2

Score	Frequency
73	2
72	3
71	0
70	1
69	0
68	5
67	1
66	2
65	3
64	2
63	5
62	5
61	2
60	2
59	5
58	1
57	3
56	1
55	0
54	1
53	0
52	2
51	1
50	3
Total	50

Statistics Aptitude Test Scores for 50 college students TABLE A.1

73	57	63	59	50
72	66	50	67	51
63	59	65	62	65
62	72	64	73	66
61	68	62	68	63
59	61	72	63	52
59	58	57	68	57
64	56	65	59	60
50	62	68	54	63
52	62	70	60	68

Scholastic Aptitude Test scores for 50 college students TABLE A.3

1350	750	530	540	750
1120	410	780	1020	430
720	1080	1110	770	610
1130	620	510	1160	630
640	1220	920	650	870
930	660	480	940	670
1070	950	680	450	990
690	1010	800	660	500
860	520	540	880	1090
580	730	570	560	740

Group frequency distribution of Scholastic Aptitude Test scores for 50 college students

TABLE A.4

Class Interval	Frequency
1300–1390	1
1200–1290	1
1100–1190	4
1000–1090	5
900–990	5
800–890	4
700–790	7
600–690	10
500–590	9
400–490	4
Total	50

In a **group frequency distribution,** individual scores are represented as members of a group of scores or as a range of scores (see Table A.4). These groups are called **class intervals.** Grouping these scores makes it much easier to make sense out of the distribution, as you can see from the relative ease in understanding Table A.4 as compared to Table A.3. Group frequency distributions are easier to represent on a graph.

When graphing data from frequency distributions, the class intervals are represented along the **abscissa** (the horizontal or *x* axis). The frequency is represented along the **ordinate** (the vertical or *y* axis). Information can be graphed in the form of a bar graph, called a **histogram,** or in the form of a point or line graph, called a **polygon.** Figure A.1 shows a histogram presenting the data from Table A.4. Note that the class intervals are represented along the bottom line of the graph (the *x* axis) and the height of the bars indicates the frequency in each class interval. Now look at Figure A.2. The information presented here is exactly the same as that in Figure A.1 but is represented in the form of a polygon rather than a histogram. Can you see how both graphs illustrate the same information? Even though

FIGURE A.1 A histogram illustrating the information found in Table A.4.

FIGURE A.2 A polygon illustrating the information found in Table A.4.

reading information from a graph is simple, we have found that many students have never learned to read graphs. In the next section we will explain how to read a graph.

How to Read a Graph

Every graph has several major parts. The most important are the labels, the axes (the vertical and horizontal lines), and the points, lines, or bars. Find these parts in Figure A.1.

The first thing you should notice when reading a graph are the labels because they tell what data are portrayed. Usually the data consist of the descriptive statistics, or the numbers used to measure the dependent variables. For example, in Figure A.1 the horizontal axis is labeled "Scholastic Aptitude Test Scores," which is the dependent variable measure; the vertical axis is labeled "Frequency," which means the number of occurrences. If a graph is not labeled, as we sometimes see in TV commercials or magazine ads, it is useless and should be ignored. Even when a graph *is* labeled, the labels can be misleading. For example, if graph designers want to distort the information, they can elongate one of the axes. Thus, it is important to pay careful attention to the numbers as well as the words in graph labels.

Next, you should focus your attention on the bars, points, or lines on the graph. In the case of histograms like the one in Figure A.1, each bar represents the class interval. The width of the bar stands for the width of the class interval, whereas the height of the bar stands for the frequency in that interval. Look at the third bar from the left in Figure A.1. This bar represents the interval "600 to 690 SAT Scores," which has a frequency of 10. You can see that this directly corresponds to the same class interval in Table A.4, since graphs and tables are both merely alternate ways of illustrating information.

Reading point or line graphs is the same as reading a histogram. In a point graph, each point represents two numbers, one found along the horizontal axis and the other found along the vertical axis. A polygon is identical to a point graph except that it has lines connecting the points. Figure A.2 is an example of a polygon, where each point represents a class interval and is placed at the center of the interval and at the height corresponding to the frequency of that interval. To make the graph easier to read, the points are connected by straight lines.

Displaying the data in a frequency distribution or in a graph is much more useful than merely presenting raw data and can be especially helpful when researchers are trying to find relations between certain factors. However, as we explained earlier, if psychologists want to make predictions or explanations about behavior, they need to perform mathematical computations on the data.

Uses of the Various Statistics

The statistics psychologists use in a study depend on whether they are trying to describe and predict behavior or explain it. When they use statistics to describe behavior, as in reporting the average score on the Scholastic Aptitude Test, they are using **descriptive statistics.** When they use them to explain behavior, as Bandura did in his study of children modeling aggressive behavior seen on TV, they are using **inferential statistics.**

Descriptive Statistics

Descriptive statistics are the numbers used to describe the dependent variable. They can be used to describe characteristics of a **population** (an entire group, such as all people living in the United States) or a **sample** (a part of a group, such as a randomly selected group of 25 students from Cornell University). The major descriptive statistics include measures of central tendency (mean, median, and mode), measures of variation (variance and standard deviation), and correlation.

Measures of Central Tendency

Statistics indicating the center of the distribution are called **measures of central tendency** and include the mean, median, and mode. They are all scores that are typical of the center of the distribution. The **mean** is what most of us think of when we hear the word "average." The **median** is the middle score. The **mode** is the score that occurs most often.

Mean   What is your average golf score? What is the average yearly rainfall in your part of the country? What is the average reading test score in your city? When these questions ask for the average, they are really asking for the "mean." The arithmetic mean is the weighted average of all the raw scores, which is computed by totaling all the raw scores and then dividing that total by the number of scores added together. In statistical computation, the mean is represented by an "X" with a bar above it ($\overline{X}$, pronounced "X bar"), each individual raw score by an "X," and the total number of scores by an "N." For example, if we wanted to compute the $\overline{X}$ of the raw statistics test scores in Table A.1, we would sum all the X's (ΣX, with Σ meaning sum) and divide by N (number of scores). In Table A.1, the sum of all the scores is equal to 3,100 and there are 50 scores. Therefore, the mean of these scores is

$$\overline{X} = \frac{3,100}{50} = 62$$

Table A.5 illustrates how to calculate the mean for 10 IQ scores.

Computation of the mean for 10 IQ scores   TABLE A.5

IQ Scores X
143
127
116
98
85
107
106
98
104
116
$\Sigma X = 1,100$

$$\text{Mean} = \overline{X} = \frac{\Sigma X}{N} = \frac{1,100}{10} = 110$$

Median   The **median** is the middle score in the distribution once all the scores have been arranged in rank order. If N (the number of scores) is odd, then there actually is a middle score and that middle score is the median. When N is even, there are two middle scores and the median is the mean of those two scores. Table A.6 shows the computation of the median for two different sets of scores, one set with 15 scores and one with 10.

Computation of median for odd and even numbers of IQ scores   TABLE A.6

IQ	IQ
139	137
130	135
121	121
116	116
107	108 ← middle score
101	106 ← middle score
98	105
96 ← middle score	101
84	98
83	97
82	N = 10
75	N is even
75	
68	$\text{Median} = \frac{106 + 108}{2} = 107$
65	

N = 15
N is odd

Finding the mode for two different distributions
TABLE A.7

IQ	IQ
139	139
138	138
125	125
116 ←	116 ←
116 ←	116 ←
116 ←	116 ←
107	107
100	98 ←
98	98 ←
98	98 ←
Mode = most frequent score	Mode = 116 and 98
Mode = 116	

Computation of the standard deviation for 10 IQ scores TABLE A.8

IQ Scores X	$X - \bar{X}$	$(X - \bar{X})^2$
143	33	1089
127	17	289
116	6	36
98	−12	144
85	−25	625
107	−3	9
106	−4	16
98	−12	144
104	−6	36
116	6	36
$\Sigma X = 1100$		$\Sigma(X - \bar{X})^2 = 2424$

Standard Deviation = s

$$= \sqrt{\frac{\Sigma(X-\bar{X})^2}{N}} = \sqrt{\frac{2424}{10}}$$

$$= \sqrt{242.4} = 15.569$$

Mode Of all the measures of central tendency, the easiest to compute is the **mode,** which is merely the most frequent score. It is computed by finding the score that occurs most often. Whereas there is always only one mean and only one median for each distribution, there can be more than one mode. Table A.7 shows how to find the mode in a distribution with one mode (unimodal) and in a distribution with two modes (bimodal).

There are several advantages to each of these measures of central tendency, but in psychological research the mean is used most often. A book solely covering psychological statistics will provide a more thorough discussion of the relative values of these measures.

Measures of Variation

When describing a distribution, it is not sufficient merely to give the central tendency; it is also necessary to give a **measure of variation**, which is a measure of the spread of the scores. By examining the spread, we can determine whether the scores are bunched around the middle or tend to extend away from the middle. Figure A.3 shows three different distributions, all with the same mean but with different spreads of scores. You can see from this figure that, in order to describe these different distributions accurately, there must be some measures of the variation in their spread. The most widely used measure of variation is the standard deviation, which is represented by a lowercase s. The standard deviation is a standard measurement of how much the scores in a distribution deviate from the mean. The formula for the standard deviation is

$$s = \sqrt{\frac{\Sigma(X-\bar{X})^2}{N}}$$

Table A.8 illustrates how to compute the standard deviation.

Most distributions of psychological data are bell-shaped. That is, most of the scores are grouped around the mean, and the farther the scores are from the mean in either direction, the fewer the scores. Notice the bell shape of the distribution in Figure A.4. Distributions such as this are called **normal** distributions. In normal distributions, as shown in Figure A.4, approximately

FIGURE A.3 Three distributions having the same mean but a different variability.

Percent of cases under portions of the normal curve

0.13% 2.15% 13.59% 34.13% 34.13% 13.59% 2.15% 0.13%

−4 −3 −2 −1 0 +1 +2 +3 +4
Standard deviations

FIGURE A.4 The normal distribution forms a bell-shaped curve. In a normal distribution, two-thirds of the scores lie between one standard deviation above and one standard deviation below the mean.

two-thirds of the scores fall within a range that is one standard deviation below the mean to one standard deviation above the mean. For example, the Wechsler IQ tests (see Chapter 7) have a mean of 100 and a standard deviation of 15. This means that approximately two-thirds of the people taking these tests will have scores between 85 and 115.

Correlation

Suppose for a moment that you are sitting in the student union with a friend. To pass the time, you and your friend decide to play a game in which you try to guess the height of the next male who enters the union. The winner, the one whose guess is closest to the person's actual height, gets a piece of pie paid for by the loser. When it is your turn, what do you guess? If you are like most people, you will probably try to estimate the mean of all the males in the union and use that as your guess. The mean is always your best guess if you have no other information.

Now let's change the game a little and add a friend who stands outside the union and weighs the next male to enter the union. Before the male enters the union, your friend says "125 pounds." Given this new information, will you still guess the mean height? Probably not—you will probably predict *below* the mean. Why? Because you intuitively understand that there is a **correlation**, a relationship, between height and weight, with tall people usually weighing more than short people. Given that 125 pounds is less than the average weight for males, you will probably guess a less-than-average height. The statistic used to measure this type of relationship between two variables is called a correlation coefficient.

Correlation Coefficient   A correlation coefficient measures the relationship between two variables, such as height and weight or IQ and SAT scores. Given any two variables, there are three possible relationships between them: positive, negative, and zero (no relationship). A positive relationship exists when the two variables vary in the same direction (e.g., as height increases, weight normally also increases). A negative relationship occurs when the two variables vary in opposite directions (e.g., as temperatures go up, hot chocolate sales go down). There is no relationship when the two variables vary totally independently of one another (e.g., there is no relationship between peoples' height and the color of their toothbrushes). Figure A.5 illustrates these three types of correlations.

The computation and the formula for a correlation coefficient (correlation coefficient is delineated by the letter "r") are shown in Table A.9. The correlation coefficient (r) always has a value between +1 and –1 (it is never greater than +1 and it is never smaller than –1). When r is close to +1, it signifies a high positive relationship between the two variables (as one variable goes up, the other variable also goes up). When r is close to –1, it signifies a high negative relationship between the two variables (as one variable goes up, the other variable goes down). When r is 0, there is no linear relationship between the two variables being measured.

Correlation coefficients can be quite helpful in making predictions. Bear in mind, however, that predictions are just that: predictions. They will have some error as long as the correlation coefficients on which they are based are not perfect (+1 or –1).

FIGURE A.5 Three types of correlation. Positive correlation (top): As the number of days of class attendance increases, so does the number of correct exam items. Negative correlation (middle): As the number of days of class attendance increases, the number of incorrect exam items decreases. Zero correlation (bottom): The day of the month on which one is born has no relationship to the number of exam items correct.

Also, correlations cannot reveal any information regarding causation. Merely because two factors are correlated, it does not mean that one factor causes the other. Consider, for example, ice cream consumption and swimming pool use. These two variables are positively correlated with one another, in that as ice cream consumption increases, so does swimming pool use. But nobody would suggest that eating ice cream causes swimming, or vice versa. Similarly, just because Michael Jordan eats Wheaties and can do a slam dunk it does not mean that you will be able to do one if you eat the same breakfast. The only way to determine the cause of behavior is to conduct an experiment and analyze the results by using inferential statistics.

Inferential Statistics

Knowing the descriptive statistics associated with different distributions, such as the mean and standard deviation, can enable us to make comparisons between various distributions. By making these

Computation of correlation coefficient between height and weight for 10 males · TABLE A.9

Height (inches) X	X^2	Weight (pounds) Y	Y^2	XY
73	5,329	210	44,100	15,330
64	4,096	133	17,689	8,512
65	4,225	128	16,384	8,320
70	4,900	156	24,336	10,920
74	5,476	189	35,721	13,986
68	4,624	145	21,025	9,860
67	4,489	145	21,025	9,715
72	5,184	166	27,556	11,952
76	5,776	199	37,601	15,124
71	5,041	159	25,281	11,289
700	49,140	1,630	272,718	115,008

$$r = \frac{N \cdot \Sigma XY - \Sigma X \cdot \Sigma Y}{\sqrt{[N \cdot \Sigma X^2 - (\Sigma X)^2]}\ \sqrt{[N \cdot \Sigma Y^2 - (\Sigma Y)^2]}}$$

$$r = \frac{10 \cdot 115,008 - 700 \cdot 1,630}{\sqrt{[10 \cdot 49,140 - 700^2]}\ \sqrt{[10 \cdot 272,718 - 1,630^2]}}$$

$$r = 0.92$$

Reaction times in milliseconds (msec) for subjects in alcohol and no alcohol conditions and computation of t · TABLE A.10

RT (msec) Alcohol X_1	RT (msec) No Alcohol X_2
200	143
210	137
140	179
160	184
180	156
187	132
196	176
198	148
140	125
159	120
$SX_1 = 1,770$	$SX_2 = 1,500$
$N_1 = 10$	$N_2 = 10$
$\overline{X}_1 = 177$	$\overline{X}_2 = 150$
$s_1 = 24.25$	$s_2 = 21.86$

$$\Sigma_{\overline{X}_1} = \frac{s}{\sqrt{N_1 - 1}} = 8.08 \qquad \Sigma_{\overline{X}_2} = \frac{s}{\sqrt{N_2 - 1}} = 7.29$$

$$S_{\overline{X}_1 - \overline{X}_2} = \sqrt{S_{\overline{X}_1}^2 + S_{\overline{X}_2}^2} = \sqrt{8.08^2 + 7.29^2} = 10.88$$

$$t = \frac{\overline{X}_1 - \overline{X}_2}{S_{\overline{X}_1 - \overline{X}_2}} = \frac{177 - 150}{10.88} = 2.48$$

$$t = 2.48, \ p < .05$$

comparisons, we may be able to observe whether one variable is related to another or whether one variable has a causal effect on another. When we design an experiment specifically to measure causal effects between two or more variables, we use **inferential** statistics to analyze the data collected. Although there are many inferential statistics, the one we will discuss is the t-test, since it is the simplest.

t-Test Suppose we believe that drinking alcohol causes a person's reaction time to slow down. To test this hypothesis, we recruit 20 participants and separate them into two groups. We ask the participants in one group to drink a large glass of orange juice with one ounce of alcohol for every 100 pounds of body weight (e.g., a person weighing 150 pounds would get 1.5 ounces of alcohol). We ask the control group to drink an equivalent amount of orange juice with no alcohol added. Fifteen minutes after the drinks, we have each participant perform a reaction time test that consists of pushing a button as soon as a light is flashed. (The reaction time is the time between the onset of the light and the pressing of the button.) Table A.10 shows the data from this hypothetical experiment. It is clear from the data that there is definitely a difference in the reaction times of the two

groups: There is an obvious difference between the means. However, it is possible that this difference is due merely to chance. To determine whether the difference is real or due to chance, we can conduct a t-test. We have run a sample t-test in Table A.10.

The logic behind a t-test is relatively simple. In our experiment we have two samples. If each of these samples is from the same population (e.g., the population of all people, whether drunk or sober), then any difference between the samples will be due to chance. On the other hand, if the two samples are from different populations (e.g., the population of drunk people and the population of sober people), then the difference is a significant difference and not due to chance.

If there is a significant difference between the two samples, then the independent variable must have caused that difference. In our example, there is a significant difference between the alcohol and the no alcohol groups. We can tell this because p (the probability that this t value will occur by chance) is less than .05. To obtain the p, we need only look up the t value in a statistical table, which is found in any statistics book. In our example, because there is a significant difference between the groups, we can reasonably conclude that the alcohol did cause a slower reaction time.

Appendix B
Answers to Self-Tests

Chapter 1: 1d; 2d; 3c; 4a; 5b; 6, see Fig. 1.6; 7b; 8d; 9c; 10b; 11b; 12. see Fig. 1.14

Chapter 2: 1b; 2d; 3a; 4a; 5, see Fig. 2.5; 6c; 7b; 8c; 9, see Fig. 2.8; 10a; 11d; 12c; 13, see Fig. 2.16; 14, see Fig. 2.19; 15: a. left; b. right; c. right; d. left

Chapter 3: 1c; 2b; 3c; 4c; 5, See Fig. 3.4; 6c; 7c; 8b; 9d; 10d; 11c; 12b; 13, See Fig. 3.10; 14b; 15b

Chapter 4: 1c; 2c; 3, see Fig. 4.4; 4c; 5, see Fig. 4.5; 6c; 7c; 8a; 9: a. photo 1, b. photo 2; 10c; 11b; 12a; 13b; 14c; 15b

Chapter 5: 1b; 2b; 3c; 4d; 5 see Fig. 5.3; 6c; 7c; 8c; 9d; 10a; 11: a-2, b-1, c-3, d-4; 12c; 13d; 14d; 15c

Chapter 6: 1c; 2c; 3d; 4d; 5d; 6c; 7c; 8: a-2, b-1; 9a; 10c; 11d; 12a; 13c; 14a; 15d

Chapter 7: 1: a, encoding, b, retrieval, c, storage; 2: a. sensory memory, b. short-term memory (STM), c. long-term memory (LTM)]; 3d; 4a; 5: a, explicit and declarative, b, implicit and nondeclarative; 6c; 7c; 8c; 9c; 10: a. retrograde amnesia (top), b. antegrade amnesia (bottom); 11: a. decay theory, b. interference, c. motivated, d. encoding failure, e. retrieval; 12.b; 13a; 14d; 15b

Chapter 8: 1b; 2c; 3: a. preparation, b. identify given facts, c. separate relevant from irrelevant facts; 4a; 5d; 6d; 7: a. phonemes, b. morphemes, c. grammar; 8b; 9d; 10b; 11b; 12a; 13b; 14d; 15d

Chapter 9: 1c; 2c; 3: a. cross-sectional research, b. longitudinal research; 4d; 5d; 6a; 7c; 8c; 9c; 10d; 11: a. sensorimotor, b. preoperational, c. concrete operation, d, formal; 12d; 13c; 14c; 15d

Chapter 10: 1b; 2a; 3a; 4c; 5: a. trust → b. autonomy → c. initiative → d. industry; 6: a. identity → b. intimacy → c. generativity → d. ego integrity; 7b; 8c; 9a; 10a; 11a; 12a; 13d; 14d; 15a

Chapter 11: 1a; 2c; 3a; 4b; 5d; 6d; 7c; 8c; 9: a. excitement → b. plateau → c. orgasm → d. resolution; 10b; 11a; 12a; 13a; 14d; 15c

Chapter 12: 1b; 2: a. openness, b. conscientiousness, c. extroversion, d. agreeableness, e. neuroticism (OCEAN); 3: a. conscious, b. preconscious, c. unconscious; 4d; 5c; 6: a. oral → b. anal → c. phallic → d. latency → e. genital; 7d; 8c; 9b; 10c; 11c; 12c; 13c; 14a; 15b

Chapter 13: 1d; 2b; 3: Axis I. clinical disorders, Axis II. personality disorders and mental retardation, Axis III. general medical conditions, Axis IV. psychosocial and environmental problems, Axis V. global assessment of functioning; 4a; 5: a. generalized anxiety disorder, b. panic disorder, c. phobias, d. obsessive-compulsive disorder, e. post-traumatic stress disorder; 6a; 7b; 8d; 9a; 10a; 11: a. paranoid, b. catatonic, c. disorganized, d. undifferentiated, e. residual; 12b; 13a; 14c; 15d

Chapter 14: 1.a; 2c; 3c; 4c; 5b; 6c; 7d; 8a; 9b; 10c; 11c; 12: a. antianxiety drugs, b. antipsychotic drugs, c. mood stabilizer drugs, antidepressant drugs; 13b; 14: a. disturbed thoughts, b. disturbed emotions, c. disturbed behaviors, d. interpersonal and life situation difficulties, e. biomedical disturbances; 15b

Chapter 15: 1c; 2a; 3: a. cognitive element, b. affective element, c. behavioral element; 4b; 5d; 6b; 7a; 8c; 9c; 10d; 11c; 12c; 13: a. egoistic model, b. empathy-altruism; 14a; 15d

abnormal behavior Patterns of emotion, thought, and action that are considered pathological (diseased or disordered) for one or more of these reasons: statistical infrequency, disability or dysfunction, personal distress, or violation of norms (Davison, Neale, & Kring, 2004).

accommodation In Piaget's theory, the process of adjusting old schemas or developing new ones to better fit with new information.

achievement motivation A desire to excel, especially in competition with others.

activity theory of aging Successful aging is fostered by a full and active commitment to life.

aggression Any behavior intended to harm someone.

algorithm A set of steps that, if followed correctly, will eventually solve the problem.

altered state of consciousness (ASC) Mental states found during sleep, dreaming, psychoactive drug use, hypnosis, and so on.

altruism Actions designed to help others with no obvious benefit to the helper.

Alzheimer's disease (AD) Progressive mental deterioration characterized by severe memory loss.

Alzheimer's disease

anorexia nervosa An eating disorder characterized by a severe loss of weight resulting from self-imposed starvation and an obsessive fear of obesity.

anxiety disorder A type of abnormal behavior characterized by unrealistic, irrational fear.

applied research Research designed to solve practical problems.

approach–approach conflict Forced choice between two or more desirable alternatives.

approach–avoidance conflict Forced choice between two or more alternatives that have both desirable and undesirable results.

assimilation In Piaget's theory, the process of absorbing new information into existing schemas.

attachment A strong affectional bond with special others that endures over time.

Attachment

attitudes Learned predispositions to respond cognitively, affectively, and behaviorally to particular objects in a particular way.

attributions Explanations for behaviors or events.

autonomic nervous system (ANS) Subdivision of the peripheral nervous system (PNS) that controls involuntary functions. It includes the *sympathetic* nervous system and the *parasympathetic* nervous system.

avoidance–avoidance conflict Forced choice between two or more undesirable alternatives.

basic research Research conducted to advance scientific knowledge rather than for practical application.

behavior therapy A group of techniques based on learning principles that is used to change maladaptive behaviors.

behavioral genetics The study of the relative effects of heredity and environment on behavior and mental processes.

Behavior therapy

binge drinking Occurs when a man consumes five or more drinks in a row, or a woman consumes four or more drinks at a time, on at least three occasions during the previous two weeks.

biological preparedness Built-in (innate) readiness to form associations between certain stimuli and responses.

biological research Scientific studies of the brain and other parts of the nervous system.

biomedical therapy The use of physiological interventions (drugs, electroconvulsive therapy, and psychosurgery) to reduce or alleviate symptoms of psychological disorders.

biopsychosocial model A unifying theme of modern psychology that considers biological, psychological, and social processes.

bipolar disorder Repeated episodes of mania (unreasonable elation and hyperactivity) and depression.

bulimia nervosa An eating disorder involving the consumption of large quantities of food (bingeing), followed by vomiting, extreme exercise, or laxative use (purging).

central nervous system (CNS)   The brain and spinal cord.

cerebral cortex   Thin surface layer on the cerebral hemispheres that regulates most complex behavior, including receiving sensations, motor control, and higher mental processes.

chronic pain   Continuous or recurrent pain lasting six months or longer.

chunking   The act of grouping separate pieces of information into a single unit (or chunk).

circadian [ser-KAY-dee-an] rhythms   Biological changes that occur on a 24-hour cycle (in Latin, *circa* means about, and *dies* means day).

classical conditioning   Learning that occurs when a neutral stimulus (NS) becomes paired (associated) with an unconditioned stimulus (UCS) to elicit a conditioned response (CR).

coding   Process that converts a particular sensory input into a specific sensation.

cognition   Mental activities involved in acquiring, storing, retrieving, and using knowledge.

cognitive dissonance   A feeling of discomfort caused by a discrepancy between an attitude and a behavior or between two attitudes.

cognitive therapy   Therapy that focuses on changing faulty thought processes and beliefs to treat problem behaviors.

cognitive-social theory   A perspective that emphasizes the roles of thinking and social learning in behavior.

collectivistic cultures   Cultures in which the needs and goals of the group are emphasized over the needs and goals of the individual.

concrete operational stage   Piaget's third stage (roughly ages 7 to 11), in which the child can perform mental operations on concrete objects and understand reversibility and conservation, though abstract thinking is not yet present.

conditioning   The process of learning associations between environmental stimuli and behavioral responses.

conformity   The act of changing behavior as a result of real or imagined group pressure.

conscious   Freud's term for thoughts or motives that a person is currently aware of or is remembering.

consciousness   An organism's awareness of its own self and surroundings (Damasio, 1999)

conventional level   Kohlberg's second level of moral development, in which moral judgments are based on compliance with the rules and values of society.

correlational research   Scientific study in which the researcher observes or measures (without directly manipulating) two or more variables to find relationships between them.

creativity   The ability to produce valued outcomes in a novel way

critical period   A period of special sensitivity to specific types of learning that shapes the capacity for future development.

critical thinking   The process of objectively evaluating, comparing, analyzing, and synthesizing information.

cross-sectional method   Research design that measures individuals of various ages at one point in time and gives infomation about age differences.

debriefing   Informing participants after a study about the purpose of the study, the nature of the anticipated results, and any deception used.

defense mechanisms   In Freudian theory, the ego's protective method of reducing anxiety by distorting reality.

delusions   Mistaken beliefs based on misrepresentations of reality.

dependent variable (DV)   Variable that is measured; it is affected by (or dependent on) the independent variable.

descriptive research   Research methods used to observe and record behavior (without producing causal explanations).

developmental psychology   The study of age-related changes in behavior and mental processes from conception to death.

Diagnostic and Statistical Manual of Mental Disorders (DSM-IV-TR)   The classification system developed by the American Psychiatric Association used to describe abnormal behaviors; the *IV-TR* indicates that it is the text revision (*TR*) of the fourth major edition (*IV*).

disengagement theory   Successful aging is characterized by mutual withdrawal between the elderly and society.

dissociative disorder   Amnesia, fugue, or multiple personalities resulting from avoidance of painful memories or situations.

dyssomnias   Problems in the amount, timing, and quality of sleep, including insomnia, sleep apnea, and narcolepsy.

elaborative rehearsal   The process of linking new information to previously stored material.

electroconvulsive therapy (ECT)   Biomedical therapy in which electrical current is passed through the brain.

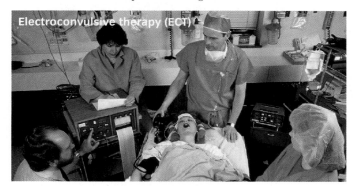
Electroconvulsive therapy (ECT)

embryonic period   The second stage of prenatal development, which begins after uterine implantation and lasts through the eighth week.

emotion   A subjective feeling that includes arousal (heart pounding), cognitions (thoughts, values, and expectations), and expressions (frowns, smiles, and running).

emotion-focused forms of coping　Coping strategies based on changing one's perceptions of stressful situations.

empirical evidence　Information acquired by direct observation and measurement using systematic scientific methods.

encoding specificity principle　Retrieval of information is improved when the conditions of recovery are similar to the conditions that existed when the information was encoded.

evolutionary/circadian theory　As a part of circadian rhythms, sleep evolved to conserve energy and to serve as protection from predators.

evolutionary psychology　A branch of psychology that studies the ways in which natural selection and adaptation can explain behavior and mental processes.

experiment　A carefully controlled scientific procedure that determines whether variables manipulated by the experimenter have a causal effect on other variables.

explicit/declarative memory　The subsystem within long-term memory that consciously stores facts, information, and personal life experiences.

external locus of control　Belief that chance or outside forces beyond one's control determine one's fate.

extrinsic motivation　Motivation based on obvious external rewards or threats of punishment.

family therapy　Treatment to change maladaptive interaction patterns within a family.

feature detectors　Specialized brain cells that respond only to certain sensory information.

Feature detectors

fetal period　The third, and final, stage of prenatal development (eight weeks to birth), which is characterized by rapid weight gain in the fetus and the fine detailing of bodily organs and systems.

five-factor model (FFM)　The trait theory that explains personality in terms of the "Big Five" model, which is composed of openness, conscientiousness, extroversion, agreeableness, and neuroticism.

formal operational stage　Piaget's fourth stage (around age 11 and beyond), which is characterized by abstract and hypothetical thinking.

fundamental attribution error (FAE)　Misjudgment of others' behavior as stemming from internal (dispositional) rather than external (situational) causes.

Piaget's formal operational stage

gate-control theory　Theory that pain sensations are processed and altered by mechanisms within the spinal cord.

gender　Psychological and sociocultural meanings added to biological maleness or femaleness.

gender roles　Societal expectations for normal and appropriate male and female behavior.

Gender roles

germinal period　The first stage of prenatal development, which begins with conception and ends with implantation in the uterus (the first two weeks).

grammar　Rules that specify how phonemes, morphemes, words, and phrases should be combined to express thoughts. These rules include syntax and semantics.

group polarization　A group's movement toward either riskier or more conservative behavior, depending on the members' initial dominant tendencies.

group therapy　A form of therapy in which a number of people meet together to work toward therapeutic goals.

groupthink　Faulty decision making that occurs when a highly cohesive group strives for agreement and avoids inconsistent information.

habituation　Tendency of the brain to ignore environmental factors that remain constant.

hallucinations　Imaginary sensory perceptions that occur without an external stimulus.

health psychology　The study of how biological, psychological, and social factors affect health and illness.

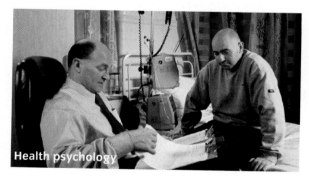

Health psychology

heuristic　A simple rule used in problem solving and decision making that does not guarantee a solution but offers a likely shortcut to it.

hierarchy of needs　Maslow's theory of motivation that some motives (such as physiological and safety needs) must be met before going on to higher needs (such as belonging and self-actualization).

homeostasis　A body's tendency to maintain a relatively stable state, such as a constant internal temperature, blood sugar, oxygen level, or water balance.

hormones Chemicals manufactured by endocrine glands and circulated in the bloodstream to produce bodily changes or maintain normal bodily function.

humanistic therapy Therapy that seeks to maximize personal growth through affective restructuring (emotional readjustment).

hypnosis A trancelike state of heightened suggestibility, deep relaxation, and intense focus.

implicit/nondeclarative memory The subsystem within long-term memory that consists of unconscious procedural skills, simple classically conditioned responses and priming.

independent variable (IV) Variable that is manipulated to determine its causal effect on the dependent variable.

individualistic cultures Cultures in which the needs and goals of the individual are emphasized over the needs and goals of the group.

informed consent A participant's agreement to take part in a study after being told what to expect.

insight Sudden understanding of a problem that implies the solution.

instinctive drift The tendency of some conditioned responses to shift (or drift) back toward innate response pattern.

instincts Behavioral patterns that are unlearned, always expressed in the same way, and universal in a species.

Instincts

intelligence The global capacity to think rationally, act purposefully, and deal effectively with the environment.

intelligence quotient (IQ) A subject's mental age divided by his or her chronological age and multiplied by 100.

internal locus of control Belief that one controls one's own fate.

interpersonal attraction Positive feelings toward another.

intrinsic motivation Motivation resulting from personal enjoyment of a task or activity.

language A form of communication using sounds and symbols combined according to specified rules.

Interpersonal attraction

latent learning Hidden learning that exists with-out behavioral signs.

law of effect Thorndike's rule that the probability of an action being repeated is strengthened when followed by a pleasant or satisfying consequence.

learning A relatively permanent change in behavior or mental processes because of practice or experience.

Learning

localization of function Specialization of various parts of the brain for particular functions.

longitudinal method Research design that measures a single individual or a group of same-aged individuals over an extended period and gives information about age changes.

Localization of function

long-term memory (LTM) This third memory stage stores information for long periods. Its capacity is limitless; its duration is relatively permanent.

long-term potentiation (LTP) Long-lasting increase in neural excitability believed to be a biological mechanism for learning and memory.

maintenance rehearsal Repeating information to maintain it in short-term memory (STM).

major depressive disorder Long-lasting depressed mood that interferes with the ability to function, feel pleasure, or maintain interest in life (Swartz & Margolis, 2004).

maturation Development governed by automatic, genetically predetermined signals.

medical model The perspective that diseases (including mental illness) have physical causes that can be diagnosed, treated, and possibly cured.

meditation A group of techniques designed to refocus attention, block out all distractions, and produce an alternate state of consciousness.

Maturation

memory An internal record or representation of some prior event or experience.

modeling therapy A learning technique in which the subject watches and imitates models who demonstrate desirable behaviors.

morpheme [MOR-feem] The smallest meaningful unit of language, formed from a combination of phonemes.

motivation A set of factors that activate, direct, and maintain behavior, usually toward a goal.

Motivation

nature-nurture controversy Ongoing dispute over the relative contributors of nature (heredity) and nurture (environment) to the development of behavior and mental processes.

neurogenesis The division and differentiation of nonneuronal cells to produce neurons.

neuron A nerve cell that receives and conducts electrical impulses from the brain.

neuroplasticity The brain's ability to reorganize and change its structure and function through the life span.

neuroscience An interdisciplinary field studying how biological processes relate to behavioral and mental processes.

neurotransmitters Chemicals that neurons release, which affect other neurons.

obedience The act of following a direct command, usually from an authority figure.

objective personality tests Standardized questionnaires that require written responses, usually to multiple-choice or true-false questions.

observational learning Learning new behavior or information by watching others (also known as social learning or modeling).

operant conditioning Learning in which voluntary responses are controlled by their consequences (also known as instrumental or Skinnerian conditioning).

opponent-process theory Theory that color perception is based on three systems of color receptors, each of which responds in an on-off fashion to opposite-color stimuli: blue-yellow, red-green, and black-white.

parasomnias Abnormal disturbances occurring during sleep, including nightmares, night terrors, sleepwalking, and sleep talking.

peripheral nervous system (PNS) All nerves and neurons connecting the CNS to the rest of the body.

personality disorders Inflexible, maladaptive personality traits that cause significant impairment of social and occupational functioning.

personality Relatively stable and enduring patterns of thoughts, feelings, and actions.

phoneme [FO-neem] The smallest basic unit of speech or sound. The English language has about 40 phonemes.

postconventional level Kohlberg's highest level of moral development, in which individuals develop personal standards for right and wrong, and they define morality in terms of abstract principles and values that apply to all situations and societies.

post-traumatic stress disorder (PTSD) An anxiety disorder following exposure to a life-threatening or other extreme event that evoked great horror or helplessness. It is characterized by flashbacks, nightmares, and impaired functioning.

preconscious Freud's term for thoughts or motives that can be easily brought to mind.

preconventional level Kohlberg's first level of moral development, in which morality is based on rewards, punishment, and the exchange of favors.

prejudice A learned, generally negative attitude directed toward specific people solely because of their membership in an identified group.

preoperational stage Piaget's second stage (roughly ages 2 to 7 years), which is characterized by the ability to employ significant language and to think symbolically, though the child lacks operations (reversible mental processes), and thinking is egocentric and animistic.

problem-focused forms of coping Coping strategies that use problem-solving strategies to decrease or eliminate the source of stress.

projective tests Psychological tests that use ambiguous stimuli, such as inkblots or drawings, which allow the test taker to project his or her unconscious thoughts onto the test material.

Parasomnias

psychiatry   The branch of medicine that deals with the diagnosis, treatment, and prevention of mental disorders.

psychoactive drugs   Chemicals that change conscious awareness, mood, or perception.

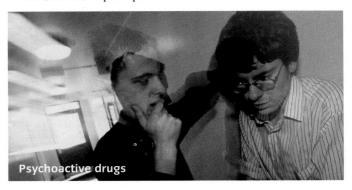
Psychoactive drugs

psychoanalysis   Freudian therapy designed to bring unconscious conflicts, which usually date back to early childhood experiences, into consciousness. Also, Freud's theoretical school of thought, which emphasizes unconscious processes.

psychodynamic therapy   A modern form of psychoanalysis that emphasizes internal conflicts, motives, and unconscious forces.

psychology   The scientific study of behavior and mental processes.

psychoneuroimmunology   [sye-koh-NEW-roh-IM-you-NOLL-oh-gee]   The interdisciplinary field that studies the effects of psychological factors on the immune system.

psychopharmacology   The study of drug effects on the mind and behavior.

psychosexual stages   In Freudian theory, the five developmental periods (oral, anal, phallic, latency, and genital) during which particular kinds of pleasures must be gratified if personality development is to proceed normally.

psychosocial stages   The eight developmental stages, each involving a crisis that must be successfully resolved, that individuals pass through in Erikson's theory of psychosocial development.

psychosurgery   Operative procedures on the brain designed to relieve severe mental symptoms that have not responded to other forms of treatment.

psychotherapy   Techniques employed to improve psychological functioning and promote adjustment to life.

punishment   A consequence that weakens a response and makes it less likely to recur.

reciprocal determinism   Bandura's belief that cognitions, behaviors, and the environment interact to produce personality.

reinforcement   A consequence that strengthens a response and makes it more likely to recur.

reliability   A measure of the consistency and stability of test scores when a test is readministered.

repair/restoration theory   Sleep serves a recuperative function, allowing organisms to repair or replenish key factors.

retrieval cue   A clue or prompt that helps stimulate recall and retrieval of a stored piece of information from long-term memory.

savant syndrome   A condition in which a person who has mental retardation exhibits exceptional skill or brilliance in some limited field.

schemas   Cognitive structures or patterns consisting of a number of organized ideas that grow and differentiate with experience.

schizophrenia   A group of psychotic disorders involving major disturbances in perception, language, thought, emotion, and behavior. The individual withdraws from people and reality, often into a fantasy life of delusions and hallucinations.

selective attention   Filtering out and attending only to important sensory messages.

self-concept   Rogers's term for all the information and beliefs that individuals have about their own nature, qualities, and behavior.

self-efficacy   Bandura's term for the learned belief that one is capable of producing desired results, such as mastering new skills and achieving personal goals.

sensorimotor stage   Piaget's first stage (birth to approximately age 2), in which schemas are developed through sensory and motor activities.

sensory adaptation   Repeated or constant stimulation decreases the number of sensory messages sent to the brain, which causes decreased sensation.

sensory memory   This first memory stage holds sensory information. It has a relatively large capacity, but duration is only a few seconds.

sensory reduction   Filtering and analyzing incoming sensations before sending a neural message to the cortex.

sex   Biological maleness and femaleness, including chromosomal sex. Also, activities related to sexual behaviors, such as masturbation and intercourse.

sexual orientation   Primary erotic attraction toward members of the same sex (homosexual, gay or lesbian), both sexes (bisexual), or the other sex (heterosexual).

sexual prejudice   Negative attitudes toward an individual because of her or his sexual orientation.

sexual response cycle   Masters and Johnson's description of the four-stage bodily response to sexual arousal, which consists of excitement, plateau, orgasm, and resolution.

shaping   Reinforcement by a series of successively improved steps leading to desired response.

short-term memory (STM)   This second memory stage temporarily stores sensory information and decides whether to send it on to long-term memory (LTM). Its capacity is limited to five to nine items, and its duration is about 30 seconds.

social psychology   The study of how other people influence a person's thoughts, feelings, and actions.

socioemotional selectivity theory A natural decline in social contact as older adults become more selective with their time.

somatic nervous system (SNS) Subdivision of the peripheral nervous system (PNS). The SNS connects the sensory receptors and controls the skeletal muscles.

standardization Establishment of the norms and uniform procedures for giving and scoring a test.

stem cells Precursor (immature) cells that give birth to new specialized cells; a stem cell holds all the information it needs to make bone, blood, brain—any part of a human body—and can also copy itself to maintain a stock of stem cells.

stereotype threat Negative stereotypes about minority groups cause some members to doubt their abilities.

stress The body's nonspecific response to any demand made on it; physical and mental arousal to situations or events that we perceive as threatening or challenging.

Stereotype threat

Stress

substance-related disorders Abuse of or dependence on a mood- or behavior-altering drug.

traits Relatively stable and consistent characteristics that can be used to describe someone.

transduction Process by which a physical stimulus is converted into neural impulses.

trichromatic theory Theory that color perception results from mixing three distinct color systems—red, green, and blue.

Type A personality Behavior characteristics that include intense ambition, competition, exaggerated time urgency, and a cynical, hostile outlook.

Type B personality Behavior characteristics consistent with a calm, patient, relaxed attitude.

unconditional positive regard Rogers's term for positive behavior toward a person with no contingencies attached.

unconscious Freud's term for thoughts or motives that lie beyond a person's normal awareness but that can be made available through psychoanalysis.

validity The ability of a test to measure what it was designed to measure.

REFERENCES

Abadinsky, H. (2004). *Drugs: An introduction* (4th ed.). Stamford, CT: Thomson Learning.

Abbey, C., & Dallos, R. (2004). The experience of the impact of divorce on sibling relationships: A qualitative study. *Clinical Child Psychology and Psychiatry, 9*(2), 241–259.

Abbott, A. (2004). Striking back. *Nature, 429*(6990), 338–339.

Aboud, F. E. (2003). The formation of in-group favoritism and out-group prejudice in young children: Are they distinct attitudes? *Developmental Psychology, 39*(1), 48–60.

About James Randi. (2002). *Detail biography.* [On-line]. Available: http://www.randi.org/jr/bio.html.

Achenbaum, W. A., & Bengtson, V. L. (1994). Re-engaging the disengagement theory of aging: On the history and assessment of theory development in gerontology. *Gerontologist, 34,* 756–763.

Adler, A. (1964). *The individual psychology of Alfred Adler.* In H. L. Ansbacher & R. R. Ansbacher (Eds.), *The individual psychology of Alfred Adler.* New York: Harper & Row.

Adler, A. (1998). *Understanding human nature.* Center City: MN: Hazelden Information Education.

Agar, M., & Reisinger, H. S. (2004). Ecstasy: Commodity or disease? *Journal of Psychoactive Drugs. 36*(2), 253–264.

Ahijevych, K., Yerardi, R., & Nedilsky, N. (2000). Descriptive outcomes of the American Lung Association of Ohio hypnotherapy smoking cessation program. *International Journal of Clinical and Experimental Hypnosis, 48*(4), 374–387.

Ainsworth, M. D. S. (1967). *Infancy in Uganda: Infant care and the growth of love.* Baltimore: Johns Hopkins University Press.

Ainsworth, M. D. S., Blehar, M., Waters, E., & Wall, S. (1978). *Patterns of attachment: Observations in the strange situation and at home.* Hillsdale, NJ: Erlbaum.

Akan, G. E., & Grilo, C. M. (1995). Sociocultural influences on eating attitudes and behaviors, body image, and psychological functioning: A comparison of African American, Asian American, and Caucasian college women. *International Journal of Eating Disorders, 18,* 181–187.

Akirav, I., & Richter-Levin, G. (1999). Biphasic modulation of hippocampal plasticity by behavioral stress and basolateral amygdala stimulation in the rat. *Journal of Neuroscience, 19,* 10530–10535.

Al'absi, M., Hugdahl, K., & Lovallo, W. R. (2002). Adrenocortical stress responses and altered working memory performance. *Psychophysiology, 39*(1), 95–99.

Al-Issa, I. (2000). Culture and mental illness in Algeria. In I. Al-Issa (Ed.), Al-Junun: *Mental illness in the Islamic world* (pp. 101–119). Madison, CT: International Universities Press.

Allen, B. (2006). *Personality theories: Development, growth, and diversity* (5th ed.). Boston: Allyn & Bacon/Longman.

Allik, J., & McCrae, R. R. (2004). Toward a geography of personality traits: Patterns of profiles across 36 cultures. *Journal of Cross-Cultural Psychology, 35*(1), 13–28.

Alloy, L. B., & Clements, C. M. (1998). Hopelessness theory of depression. *Cognitive Therapy and Research, 22,* 303–335.

Alloy, L. B., Abramson, L. Y., Whitehouse, W. G., Hogan, M. E., Tashman, N. A., Steinberg, D. L., Rose, D. T., & Donovan, P. (1999). Depressogenic cognitive styles: Predictive validity, information processing and personality characteristics, and developmental origins. *Behaviour Research and Therapy, 37,* 503–531.

Allport, G. W. (1937). *Personality: A psychological interpretation.* New York: Holt, Rinehart and Winston.

Allport, G. W., & Odbert, H. S. (1936). Traitnames: A psycho-lexical study. *Psychological Monographs: General and Applied, 47,* 1–21.

Almeida, O. P., Burton, E. J., Ferrier, N., McKeith, I. G., & O'Brien, J. T. (2003). Depression with late onset is associated with right frontal lobe atrophy. *Psychological Medicine, 33*(4), 675–681.

Aluja-Fabregat, A., & Torrubia-Beltri, R. (1998). Viewing of mass media violence, perception of violence, personality and academic achievement. *Personality and Individual Differences, 25,* 973–989.

American Cancer Society (2004). *Women and smoking.* [On-line]. Available: http://cancer.org/docroot/PED/content/PED_10_2X_Women_and_Smoking.asp?sitearea=PED.

American Heart Association (2004). *Heart disease and stroke statistics—2004 update.* [On-line]. Available: http://www.americanheart.org/presenter.jhtml?identifier=3000090.

American Medical Association. (2003). *Alcohol and other drug abuse.* [On-line]. Available: http://www.ama-assn.org/ama/pub/category/3337.html.

American Psychiatric Association. (2000). Diagnostic and statistical manual of mental disorders (4th ed. TR). Washington, DC: American Psychiatric Press.

American Psychiatric Association. (2002). *APA Let's talk facts about posttraumatic stress disorder.* [On-line]. Available: http://www.psych.org/disasterpsych/fs/ptsd.cfm.

American Psychological Association. (1984). *Behavioral research with animals.* Washington, DC: Author.

American Psychological Association. (1992). Ethical principles of psychologists and code of conduct. *American Psychologist, 47,* 1597–1611.

Amidzic, O., Riehle, H. J., Fehr, T., Wienbruch, C., & Elbert, T. (2001). Pattern of focal outbursts in chess players. *Nature, 412*(6847), 603.

Anderson, C. A. (2001). Heat and violence. *Current Directions in Psychological Science, 10*(1), 33–38.

Anderson, C. A. (2004). An update on the effects of playing violent video games. *Journal of Adolescence, 27*(1), 113–122.

Anderson, C. A., Anderson, K. B., Dorr, N., DeNeve, K. M., & Flanagan, M. (2000). Temperature and aggression. *Advances in Experimental Social Psychology, 32,* 63–133.

Anderson, C. A., & Bushman, B. J. (2001). Effects of violent video games on aggressive behavior, aggressive cognition, aggressive affect, physiological arousal, and prosocial behavior: A meta-analytic review of the scientific literature. *Psychological Science, 12*(5), 353–359.

Anderson, C. A., & Dill, K. E. (2000). Video games and aggressive thoughts, feelings, and behavior in the laboratory and in life. *Journal of Personality and Social Psychology, 78*(4), 772–790.

Anderson, C. A., & Hamilton, M. (2005). Gender role stereotyping of parents in children's picture books: The invisible father. *Sex Roles. 52*(3–4), 145–151.

Anderson, M. C., Bjork, R. A. & Bjork, E. L. (1994). Remembering can cause forgetting: Retrieval dynamics in long-term memory. *Journal of Experimental Psychology: Learning, Memory, and Cognition, 20,* 1063–1087.

Anderson, M. C., Ochsner, K. N., Kuhl, B., Cooper, J., Robertson, E., Gabrieli, S. W., Glover, G. H., & Gabrieli, J. D. E. (2004). Neural systems underlying the suppression of unwanted memories. *Science, 303*(5655), 232–235.

Andrade, T. G. C. S., & Graeff, F. G. (2001). Effect of electrolytic and neurotoxic lesions of the median raphe nucleus on anxiety and stress. *Pharmacology, Biochemistry & Behavior, 70*(1), 1–14.

Angenendt, A. (2003). Safety and security from the air traffic control services' point of view. *Human Factors & Aerospace Safety, 3*(3), 207–209.

Angst, J., Gamma, A., Gastpar, M., Lepine, J. P., Mendlewicz, J., & Tylee, A. (2002). Gender differences in depression: Epidemiological findings from the European DEPRES I and II studies. *European Archives of Psychiatry & Clinical Neuroscience, 252*(5), 201–209.

Ariznavarreta, C., Cardinali, D. P., Villanua, M. A., Granados, B., Martin, M., Chiesa, J. J., Golombek, D. A., & Tresguerres, J. A. F. (2002). Circadian rhythms in airline pilots submitted to long-haul transmeridian flights. *Aviation, Space, & Environmental Medicine, 73*(5), 445–455.

Asch, S. E. (1951). Effects of group pressure upon the modification and distortion of judgment. In H. Guetzkow (Ed.), *Groups, leadership, and men*. Pittsburgh: Carnegie Press.

Astin, J. A. (2004). Mind-body therapies for the management of pain. *Clinical Journal of Pain, 20*(1), 27–32.

Atchley, P., & Kramer, A. F. (2000). Age related changes in the control of attention in depth. *Psychology & Aging, 15*(1), 78–87.

Atchley, R. C. (1997). *Social forces and aging* (8th ed.). Belmont, CA: Wadsworth.

Atkins, C. K., Panicker, S., & Cunningham, C. L. (2005). *Laboratory animals in research and teaching: Ethics, care, and methods*. Washington, DC: American Psychological Association.

Atkinson, R. C., & Shiffrin, R. M. (1968). Human memory: A proposed system and its control processes. In K. W. Spence & J. T. Spence (Eds.), *The psychology of learning and motivation* (Vol. 2). New York: Academic Press.

Aubert, A., & Dantzer, R. (2005). The taste of sickness: Lipopolysaccharide-induced finickiness in rats. *Physiology & Behavior, 84*(3), 437–444.

Axtell, R. E. (1998). *Gestures: The do's and taboos of body language around the world*, revised and expanded ed. New York: Wiley.

Bachevalier, J., & Vargha-Khadem, F. (2005). The primate hippocampus: Ontogeny, early insult and memory. *Current Opinion in Neurobiology, 15*(2), 168–174.

Bachman, G., & Zakahi, W. R. (2000). Adult attachment and strategic relational communication: Love schemas and affinity-seeking. *Communication Reports, 13*(1), 11–19.

Badawy, A. B. (2003). Alcohol and violence and the possible role of serotonin. *Criminal Behaviour & Mental Health, 13*(1), 31–44.

Baddeley, A. D. (1992). Working memory. *Science, 255*, 556–559.

Baddeley, A. D. (1998). Recent developments in working memory. *Current Opinion in Neurobiology, 8*, 234–238.

Baddeley, A. D. (2000). Short-term and working memory. In E. Tulving & F. I. M. Craik (Eds.), *The Oxford handbook of memory* (pp. 77–92). New York: Oxford University Press.

Baer, J. (1994). Divergent thinking is not a general trait: A multi-domain training experiment. *Creativity Research Journal, 7*, 35–36.

Bagermihl, B. (1999). *Biological exuberance: Animal homosexuality and natural diversity*. New York: St Martins Press.

Bailer, U., de Zwaan, M., Leisch, F., Strnad, A., Lennkh-Wolfsberg, C., El-Giamal, N., Hornik, K., & Kasper, S. (2004). Guided self-help versus cognitive-behavioral group therapy in the treatment of bulimia nervosa. *International Journal of Eating Disorders, 35*(4), 522–537.

Bailey, J. M., Dunne, M. P., & Martin, N. G. (2000). Genetic and environmental influences on sexual orientation and its correlates in an Australian twin sample. *Journal of Personality and Social Psychology, 78*(3), 524–536.

Bailey, K. R., & Mair, R. G. (2005). Lesions of specific and nonspecific thalamic nuclei affect prefrontal cortex-dependent spects of spatial working memory. *Behavioral Neuroscience, 119*(2), 410–419.

Baillargeon, R. (2000). Reply to Bogartz, Shinskey, and Schilling; Schilling; and Cashon and Cohen. *Infancy, 1*, 447–462.

Baker, R. A. (1996). *Hidden memories*. New York: Prometheus Books.

Baker, R. A. (1998). *A view of hypnosis*. Harvard Mental Health Letter, 14, 5–6.

Baldessarini, R. J., & Hennen, J. (2004). Genetics of suicide: An overview. *Harvard Review of Psychiatry, 12*(1), 1–13.

Baldwin, S., & Oxlad, M. (2000). *Electroshock and minors: A fifty year review*. New York: Greenwood Publishing Group.

Ballesteros, S., & Manuel Reales, J. (2004). Intact haptic priming in normal aging and Alzheimer's disease: Evidence for dissociable memory systems. *Neuropsychologia, 42*(8), 1063–1070.

Bandstra, E. S., Morrow, C. E., Vogel, A. L., Fifer, R. C., Ofir, A. Y., Dausa, A. T., Xue, L., & Anthony, J. C. (2002). Longitudinal influence of prenatal cocaine exposure on child language functioning. *Neurotoxicology & Teratology, 24*(3), 297–308.

Bandura, A. (1969). *Principles of behavior modification*. New York: Holt, Rinehart and Winston.

Bandura, A. (1986). *Social foundations of thought and action: A social cognitive theory*. Englewood Cliffs, NJ: Prentice Hall.

Bandura, A. (1989). Social cognitive theory. In R. Vasta (Ed.), *Annals of child development* (Vol. 6). Greenwich, CT: JAI Press.

Bandura, A. (1991). Social cognitive theory of moral thought and action. In W. M. Kurtines &

J. L. Gewirtz (Eds.), *Handbook of moral behavior and development*: Vol. 1. Theory. Hillsdale, NJ: Erlbaum.

Bandura, A. (1997). *Self-efficacy: The exercise of control*. New York: Freeman.

Bandura, A. (2000). Exercise of human agency through collective efficacy. *Current Directions in Psychological Science, 9*(3), 75–83.

Bandura, A. (2003). On the psychosocial impact and mechanisms of spiritual modeling: Comment. *International Journal for the Psychology of Religion, 13*(3), 167–173.

Bandura, A., Ross, D., & Ross, S. (1961). Transmission of aggression through imitation of aggressive models. *Journal of Abnormal & Social Psychology, 63*, 575–582.

Bandura, A., & Walters, R. H. (1963). *Social learning and personality development*. New York: Holt, Rinehart and Winston.

Banks, A., & Gartrell, N. K. (1995). Hormones and sexual orientation: A questionable link. *Journal of Homosexuality, 28*(3–4), 247–268.

Banks, M. S., & Salapatek, P. (1983). Infant visual perception. In M. M. Haith & J. J. Campos (Eds.), *Handbook of child psychology*. New York: Wiley.

Barber, T. X. (2000). A deeper understanding of hypnosis: Its secrets, its nature, its essence. *American Journal of Clinical Hypnosis, 42*(3–4), 208–272.

Bard, C. (1934). On emotional expression after decortication with some remarks on certain theoretical views. *Psychological Review, 41*, 309–329.

Barinaga, M. (1999). Learning visualized, on the double. *Science, 286*, 1661.

Barlow, D. H., Esler, J. L., & Vitali, A. E. (1998). Psychosocial treatments for panic disorders, phobias, and generalized anxiety disorder. In P. E. Nathan & J. M. Gorman (Eds.), *A guide to treatments that work* (pp. 288–318). New York: Oxford University Press.

Barnouw, V. (1985). *Culture and personality* (4th ed.). Homewood, IL: Dorsey Press.

Barry, D. T., Grilo, C. M., & Masheb, R. M. (2002). Gender differences in patients with binge eating disorder. *International Journal of Eating Disorders, 31*(1), 63–70.

Bartholow, B. D., & Anderson, C. A. (2002). Effects of violent video games on aggressive behavior: Potential sex differences. *Journal of Experimental Social Psychology, 38*(3), 283–290.

Batel, P. (2000). Addiction and schizophrenia. *European Psychiatry, 15*, 115–122.

Bates, P., & Rutherford, J. (2003). Problem substance use and schizophrenia. *British Journal of Psychiatry, 182*(5), 455.

Batson, C. D. (1991). *The altruism question: Toward a social-psychological answer*. Hillsdale, NJ: Erlbaum.

Batson, C.D. (1998). Altruism and prosocial behavior. In D.T. Gilbert, S.T. Fiske, & G. Lindzey (Eds.). *The handbook of social psychology, Vol. 2* (4th ed.) (pp.282–316). Boston, MA: McGraw-Hill.

Batson, C. D., & Ahmad, N. (2001). Empathy-induced altruism in a prisoner's dilemma II: What if the target of empathy has defected? *European Journal of Social Psychology, 31*(1), 25–36.

Bearer, C. F., Stoler, J. M., Cook, J. D., & Carpenter, S. J. (2004–2005). Biomarkers of alcohol use in pregnancy. *Alcohol Research & Health, 28*(1), 38–43.

Beck, A. T. (1976). *Cognitive therapy and the emotional disorders.* New York: International Universities Press.

Beck, A. T. (2000). *Prisoners of hate.* New York: Harperperennial.

Beck, A., & Rector, N. A. (2005). Cognitive approaches to schizophrenia: Theory and therapy. *Annual Review of Clinical Psychology, 1,* 577–606.

Becvar, D., & Becvar, R. (2006). *Family therapy: A systemic integration* (6th ed.). Boston: Allyn & Bacon/Longman.

Begg, I. M., Needham, D. R., & Bookbinder, M. (1993). Do backward messages unconsciously affect listeners? No. *Canadian Journal of Experimental Psychology, 47,* 1–14.

Beilin, H. (1992). Piaget's enduring contribution to developmental psychology. *Developmental Psychology, 28,* 191–204.

Belsky, J., & Cassidy, J. (1994). Attachment: Theory and evidence. In M. Rutter & D. Hay (Eds.), *Development through life: A handbook for clinicians* (pp. 373–402). Oxford, England: Blackwell.

Bem, S. L. (1981). Gender schema theory: A cognitive account of sex typing. *Psychological Review, 88,* 354–364.

Bem, S. L. (1993). *The lenses of gender: Transforming the debate on sexual inequality.* New Haven, CT: Yale University Press.

Benazzi, F. (2004). Testing early-onset chronic atypical depression subtype. *Neuropsychopharmacology, 29*(2), 440–441.

Benloucif, S., Orbeta, L., Ortiz, R., Janssen, I., Finkel, S., Bleiberg, J., & Zee, P. C. (2004). Morning or evening activity improves neuropsychological performance and subjective sleep quality in older adults. *Sleep, 27*(8), 1542–1551.

Bennett, M., Barrett, M., Karakozov, R., Kipiani, G., Lyons, E., Pavlenko, V., & Riazanova, T. (2004). Young children's evaluations of the ingroup and of outgroups: A multi-national study. *Social Development, 13*(1), 124–141.

Beraki, S., Aronsson, F., Karlsson, H., Ögren, S. O., & Kristensson, K. (2005). Influenza A virus infection causes alterations in expression of synaptic regulatory genes combined with changes in cognitive and emotional behaviors in mice. *Molecular Psychiatry, 10*(3), 299–308.

Berman, M. E., Tracy, J. I., & Coccaro, E. F. (1997). The serotonin hypothesis of aggression revisited. *Clinical Psychology Review, 17*(6), 651–665.

Bermond, B., Fasotti, L., Nieuwenhuyse, B., & Schuerman, J. (1991). Spinal cord lesions, peripheral feedback, and intensities of emotional feelings. Cognition and Emotion, 5, 201–220.

Bernard, L. L. (1924). *Instinct.* New York: Holt.

Bernhardt, T., Seidler, A., & Froelich, L. (2002). Der Einfluss von psychosozialen Faktoren auf das Demenzerkrankungsrisiko [The effect of psychosocial factors on risk of dementia]. *Fortschritte der Neurologie, Psychiatrie, 70*(6), 283–288.

Bernheim, K. F., & Lewine, R. R. J. (1979). *Schizophrenia: Symptoms, causes, and treatments.* New York: Norton.

Berreman, G. (1971). *Anthropology today.* Del Mar, CA: CRM Books.

Berry, J. W., Poortinga, Y. H., Segall, M. H., & Dasen, P. R. (2002). *Cross-cultural psychology: Research and applications* (2nd ed.). New York: Cambridge University Press.

Berthoud, H. (2002). Multiple neural systems controlling food intake and body weight. *Neuroscience & Biobehavioral Reviews, 26*(4), 393–428.

Best, J. B. (1999). *Cognitive psychology* (5th ed.). Belmont, CA: Wadsworth.

Beutler, L. E., Brown, M. T., Crothers, L., Booker, K., & Seabrook, M.K. (1996). The dilemma of factitious demographic distinctions in psychological research. *Journal of Consulting and Clinical Psychology, 64,* 892–902.

Bhattacharya, S. K., & Muruganandam, A. V. (2003). Adaptogenic activity of Withania somnifera: An experimental study using a rat model of chronic stress. *Pharmacology, Biochemistry & Behavior, 75*(3), 547–555.

Biehl, M., Matsumoto, D., Ekman, P., Hearn, V., Heider, K., Kudoh, T., & Ton, V. (1997). Matsumoto and Ekman's Japanese and Caucasian facial expressions of emotion (JACFEE): Reliability data and cross-national differences. *Journal of Nonverbal Behavior, 21,* 3–21.

Birkenhäger, T. K., Renes, J., & Pluijms, E. M. (2004). One-year follow-up after successful ECT: A naturalistic study in depressed inpatients. *Journal of Clinical Psychiatry, 65*(1), 87–91.

Bjorkqvist, K. (1994). Sex differences in physical, verbal, and indirect aggression: A review of recent research. *Sex Roles, 30,* 177–188.

Blaine, B., & McElroy, J. (2002). Selling stereotypes: Weight loss infomercials, sexism, and weightism. *Sex Roles, 46*(9–10), 351–357.

Blair, H. T., Huynh, V. K., Vaz, V. T., Van, J., Patel, R. R., Hiteshi, A. K., Lee, J. E., & Tarpley, J. W. (2005). Unilateral storage of fear memories by the amygdala. *Journal of Neuroscience, 25*(16), 4198–4205.

Blair, R. J. R. (2004). The roles of orbital frontal cortex in the modulation of antisocial behavior. *Brain & Cognition, 55*(1), 198–208.

Blakemore, C., & Cooper, G. F. (1970). Development of the brain depends on the visual environment. *Nature, 228,* 477–478.

Blanco, C., Schneier, F. R., Schmidt, A., Blanco-Jerez, C. R., Marshall, R. D., Sanchez-Lacay, A., & Liebowitz, M. R. (2003). Pharmacological treatment of social anxiety disorder: A meta-analysis. *Depression & Anxiety,18*(1), 29–40.

Blass, T. (1991). Understanding behavior in the Milgram obedience experiment: The role of personality, situations, and their interactions. *Journal of Personality and Social Psychology, 60*(3), 398–413.

Blass, T. (2000). Stanley Milgram. In A. E. Kazdin (Ed.), *Encyclopedia of psychology* (Vol. 5, pp. 248–250). Washington, DC: American Psychological Association.

Bleckley, M. K., Durso, F. T., Crutchfield, J. M., Engle, R. W., & Khanna, M. M. (2003). Individual differences in working memory capacity predict visual attention allocation. *Psychonomic Bulletin & Review, 10*(4), 884–889.

Blum, K., Braverman, E. R., Holder, J. M., Lubar, J. F., Monastra, V. J., Miller, D., Lubar, J. O., Chen, T. J. H., & Comings, D. E. (2000). Reward deficiency syndrome: A biogenetic model for the diagnosis and treatment of impulsive, addictive, and compulsive behaviors. *Journal of Psychoactive Drugs, 32*(Suppl), 1–68.

Bock, G. R., & Goode, J. A. (Eds.) (1996). *Genetics of criminal and antisocial behavior.* Chichester, England: Wiley.

Bohus, M., Haaf, B., Simms, T., Limberger, M. F., Schmahl, C., Unckel, C., Lieb, K., & Linehan, M. M. (2004). Effectiveness of inpatient dialectical behavioral therapy for borderline personality disorder: A controlled trial. *Behaviour Research & Therapy, 42*(5), 487–499.

Bond, F. W., & Bunce, D. (2000). Mediators of change in emotion-focused and problem-focused worksite stress management intervention. *Journal of Occupational Health Psychology, 5,* 153–163.

Bond, M. H., & Smith, P. B. (1996). Cross-cultural social and organizational psychology. *Annual Review of Psychology, 47*, 205–235.

Bonnel, A., Mottron, L., Peretz, I., Trudel, M., Gallun, E., & Bonnel, A-M. (2003). Enhanced pitch sensitivity in individuals with autism: A signal detection analysis. *Cognitive Neuroscience, 5*(2), 226–235.

Book, A. S., Starzyk, K. B., & Qunisey, V. L. (2002). The relationship between testosterone and aggression: A meta-analysis. *Aggression & Violent Behavior, 6*(6), 579–599.

Borba, M. (2001). *Building moral intelligence: The seven essential virtues that teach kids to do the right thing.* San Francisco: Jossey-Bass.

Borbely, A. A. (1982). Circadian and sleep-dependent processes in sleep regulation. In J. Aschoff, S. Daan, & G. A. Groos (Eds.), *Vertebrate circadian rhythms* (pp. 237–242). Berlin: Springer/Verlag.

Bouchard, T. J. (2004). Genetic influence on human psychological traits: A survey. *Current Directions in Psychological Science, 13*(4), 148–151.

Bouchard, T. J., & McGue, M. (1981). Familial studies of intelligence: A review. *Science, 212*(4498), 1055–1059.

Bouchard, T. J., Jr. (1994). Genes, environment, and personality. *Science, 264*, 1700–1701.

Bouchard, T. J., Jr. (1997). The genetics of personality. In K. Blum & E. P. Noble (Eds.), *Handbook of psychiatric genetics.* Boca Raton, FL: CRC Press.

Bouchard, T. J., Jr. (1999). The search for intelligence. *Science, 284*, 922–923.

Bouchard, T. J., Jr., McGue, M., Hur, Y., & Horn, J. M. (1998). A genetic and environmental analysis of the California Psychological Inventory using adult twins reared apart and together. *European Journal of Personality, 12*, 307–320.

Boucher, L., & Dienes, Z. (2003). Two ways of learning associations. *Cognitive Science, 27*(6), 807–842.

Bourgeois-Bougrine, S., Carbon, P., Gounelle, C., Mollard, R., & Coblentz, A. (2003). Perceived fatigue for short- and long-haul flights: A survey of 739 airline pilots. *Aviation, Space, and Environmental Medicine, 74*(10), 1072–1077.

Bourne, L. E., Dominowski, R. L., & Loftus, E. F. (1979). *Cognitive processes.* Englewood Cliffs, NJ: Prentice Hall.

Bouton, M. E. (1994). Context, ambiguity, and classical conditioning. *Current Directions in Psychological Science, 2*, 49–53.

Bouton, M. E., Mineka, S., & Barlow, D. H. (2001). A modern learning theory perspective on the etiology of panic disorder. *Psychological Review, 108*(1), 4–32.

Bovbjerg, D. H. (2003). Circadian disruption and cancer: Sleep and immune regulation. *Brain, Behavior & Immunity, 17*(Suppl1), 48-S50.

Bower, B. (2000). Cooperative strangers turn a mutual profit. *Science News, 157*, 231.

Bowers, K. S., & Woody, E. Z. (1996). Hypnotic amnesia and the paradox of intentional forgetting. *Journal of Abnormal Psychology, 105*, 381–390.

Bowlby, J. (1969). *Attachment and loss: Vol. 1. Attachment.* New York: Basic Books.

Bowlby, J. (1973). *Attachment and loss: Vol. 2. Separation and anxiety.* New York: Basic Books.

Bowlby, J. (1982). Attachment and loss: Retrospect and prospect. *American Journal of Orthopsychiatry, 52*, 664–678.

Bowlby, J. (1989). *Secure attachment.* New York: Basic Books.

Bowlby, J. (2000). *Attachment.* New York: Basic Books.

Bowling, A. C., & Mackenzie, B. D. (1996). The relationship between speed of information processing and cognitive ability. *Personality & Individual Differences, 20*(6), 775–800.

Boyle, S. H., Williams, R. B., Mark, D. B., Brummett, B. H., Siegler, I. C., Helms, M. J., & Barefoot, J. C. (2004). Hostility as a predictor of survival in patients with coronary artery disease. *Psychosomatic Medicine, 66*(5), 629–632.

Bragdon, A. D., & Gamon, D. (1999). *Building left-brain power: Left-brain conditioning exercises and tips to strengthen language, math, and uniquely human skills.* Thousand Oaks, CA: Brainwaves Books.

Brandon, T. H., Collins, B. N., Juliano, L. M., & Lazev, A. B. (2000). Preventing relapse among former smokers: A comparison of minimal interventions through telephone and mail. *Journal of Consulting and Clinical Psychology, 68*(1), 103–113.

Brawman-Mintzer, O., & Lydiard, R. B. (1996). Generalized anxiety disorder: Issues in epidemiology. Journal of Clinical Psychiatry, 57, 3–8.

Brawman-Mintzer, O., & Lydiard, R. B. (1997). Biological basis of generalized anxiety disorder. *Journal of Clinical Psychiatry, 58*, 16–25.

Breland, K., & Breland, M. (1961). *The misbehavior of organisms.* American Psychologist, 16, 681–684.

Bremner, J. D., Vythilingam, M., Vermetten, E., Anderson, G., Newcomer, J. W. & Charney, D. S. (2004). Effects of Glucocorticoids on declarative memory function in major depression. *Biological Psychiatry, 55*(8), 811–815.

Brendgen, M., Vitaro, F., & Bukowski, W. M. (2000). Deviant friends and early adolescents'

emotional and behavioral adjustment. *Journal of Research on Adolescence, 10*(2), 173–189.

Brenner, S. (2005). *Vietnam war crimes.* Farmington Hills, MI: Greenhaven Press.

Brewer, J. B., Zhao, Z., Desmond, J. E., Glover, G. H., & Gabrieli, J. D. (1998). Making memories: Brain activity that predicts how well visual experience will be remembered. *Science, 281*, 1185–1187.

Brewer, M. B. (1996). When contact is not enough: Social identity and intergroup cooperation. *International Journal of Intercultural Relations, 20*(3–4), 291–303.

Brewer, W. F., & Pani, J. R. (1996). Reports of mental imagery in retrieval from long-term memory. Consciousness & Cognition: An International Journal, 5, 265–287.

Brim, O. (1999). *The MacArthur Foundation study of midlife development.* Vero Beach, FL: MacArthur Foundation.

Brislin, R. B. (2000). *Understanding culture's influence on behavior.* Ft. Worth, TX: Harcourt.

Brislin, R. W. (1997). *Understanding culture's influence on behavior* (2nd ed.). San Diego: Harcourt Brace.

Brkich, M., Jeffs, D., & Carless, S. A. (2002). A global self-report measure of person-job fit. *European Journal of Psychological Assessment, 18*(1), 43–51.

Brody, A., Olmstead, R. E., London, E. D., Farahi, J., Meyer, J. H., Grossman, P., Lee, G. S., Huang, J., Hahn, E. L., & Mandelkern, M. A. (2004). Smoking-induced ventral striatum dopamine release. *American Journal of Psychiatry, 161*(7), 1211–1218.

Broman-Fulks, J. J., Berman, M. E., Rabian, B. A., & Webster, M. J. (2004). Effects of aerobic exercise on anxiety sensitivity. Behaviour Research & Therapy, 42(2), 125–136.

Brown, D. (2001). (Mis)representations of the long-term effects of childhood sexual abuse in the courts. *Journal of Child Sexual Abuse, 9*(3–4), 79–108.

Brown, E., Deffenbacher, K., & Sturgill, K. (1977). Memory for faces and the circumstances of encounter. *Journal of Applied Psychology, 62*, 311–318.

Brown, K. T., Brown, T. N., Jackson, J., Sellers, R. M., & Manuel, W. J. (2003). Teammates on and off the field? White student athletes. *Journal of Applied Social Psychology, 33*(7), 1379–1403.

Brown, P. K., & Wald, G. (1964). Visual pigments in single rods and cones of the human retina. Direct measurements reveal mechanisms of human night and color vision. *Science, 144*, 45–52.

Brown, R. P., & Josephs, R. A. (1999). A burden of proof: Stereotype relevance and gender differences in math performance. *Journal of Personality and Social Psychology, 76*(2), 246–257.

Brown, R., & Kulik, J. (1977). Flashbulb memories. *Cognition, 5*, 73–99.

Browne, K. O. (2001). Cultural formulation of psychiatric diagnoses. *Culture, Medicine & Psychiatry, 25*(4), 411–425.

Bruck, C. S., & Allen, T. D. (2003). The relationship between big five personality traits, negative affectivity, type A behavior, and work-family conflict. *Journal of Vocational Behavior, 63*(3), 457–472.

Bruggeman, E. L., & Hart, K. J. (1996). Cheating, lying, and moral reasoning by religious and secular high school students. *Journal of Educational Research, 89*, 340–344.

Brummett, B. H., Babyak, M. A., Mark, D. B., Clapp-Channing, N. E., Siegler, I. C., & Barefoot, J. C. (2004). Prospective study of perceived stress in cardiac patients. *Annals of Behavioral Medicine, 27*(1), 22–30.

Buchanan, T. W., & Lovallo, W. R. (2001). Enhanced memory for emotional material following stress-level cortisol treatment in humans. *Psychoneuroendocrinology, 26*(3), 307–317.

Buck, R. (1984). *The communication of emotion.* New York: Guilford Press.

Buddie, A. M, & Parks, K. A. (2003). The role of the bar context and social behaviors on women's risk for aggression. *Journal of Interpersonal Violence, 18*(12), 1378–1393.

Bugental, B. B., & Goodnow, J. J. (1998). Socialization process. In W. Damon & N. Eisenberg (Eds.), *Handbook of child psychology* (5th ed.). New York: Wiley.

Bulik, C. M., Sullivan, P. F., & Kendler, K. S. (2002). Medical and psychiatric morbidity in obese women with and without binge eating. *International Journal of Eating Disorders, 32*(1), 72–78.

Bull, L. (2003). What can be done to prevent smoking in pregnancy? A literature review. *Early Child Development & Care, 173*(6), 661–667.

Burgdorf, J., Knutson, B., & Panksepp, J. (2000). Anticipation of rewarding electrical brain stimulation evokes ultrasonic vocalization in rats. *Behavioural Neuroscience, 114*(2), 320–327.

Burns, N. R., Nettelbeck, T., & Cooper, C. J. (2000). Event-related potential correlates of some human cognitive ability constructs. *Personality & Individual Differences, 29*(1), 157–168.

Burns, R. A., Johnson, K. S., Harris, B. A., Kinney, B. A., & Wright, S. E. (2004). Functional cues for position learning effects in animals. *Psychological Record, 54*(2), 233–254.

Burns, R. G., & Katovich, M. A. (2003). Examining road rage/aggressive driving: Media depiction and prevention suggestions. *Environment & Behavior, 35*(5), 621–636.

Burton, L. A., Rabin, L., Vardy, S. B., Frohlich, J., Wyatt, G., Dimitri, D., Constante, S., & Guterman, E. (2004). Gender differences in implicit and explicit memory for affective passages. *Brain & Cognition, 54*(3), 218–224.

Bushman, B. J. (2002). Does venting anger feed or extinguish the flame? Catharsis, rumination, distraction, anger and aggressive responding. Personality & Social Psychology Bulletin, 28(6), 724–731.

Bushman, B. J., Baumeister, R. F., & Stack, A. D. (1999). Catharsis, aggression, and persuasive influence: Self-fulfilling or self-defeating prophecies? *Journal of Personality and Social Psychology, 76*, 367–376.

Buss, D. M. (1989). Sex differences in human mate preferences: Evolutionary hypotheses tested in 37 cultures. *Behavioral and Brain Sciences, 12*, 1–49.

Buss, D. M. (2003). *The evolution of desire: Strategies of human mating.* New York: Basic Books.

Buss, D. M. (2005). *The handbook of evolutionary psychology.* Hoboken, NJ: Wiley.

Buss, D. M. and 40 colleagues. (1990). International preferences in selecting mates: A study of 37 cultures. *Journal of Cross-Cultural Psychology, 21*, 5–47.

Butcher, J. N. (2000). Revising psychological tests: Lessons learned from the revision of the MMPI. *Psychological Assessment, 12*(3), 263–271.

Butcher, J. N., & Rouse, S. V. (1996). Personality: Individual differences and clinical assessment. *Annual Review of Psychology, 47*, 87–111.

Butler, K. Joel, J. C., & Jeffries, M. (2000). *Clinical handbook of psychotropic drugs.* New York: Wiley.

Butler, R. A. (1954, February). Curiosity in monkeys. *Scientific American, 190*, 70–75.

Butterweck, V. (2003). Mechanism of action of St John's Wort in depression: What is known? *CNS Drugs, 17*(8), 539–562.

Buunk, B. P., Dijkstra, P., Fetchenhauer, D., & Kenrick, D. (2002). Age and gender differences in mate selection criteria for various involvement levels. *Personal Relationships, 9*(3), 271–278.

Cadoret, R. J., Leve, L. D., & Devor, E. (1997). Genetics of aggressive and violent behavior. *Psychiatric Clinics of North America, 20*, 301–322.

Camarena, B., Santiago, H., Aguilar, A., Ruvinskis, E., González-Barranco, J., & Nicolini, H. (2004). Family-based association study between the monoamine oxidase A gene and obesity: Implications for psychopharmacogenetic studies. *Neuropsychobiology, 49*(3), 126–129.

Camodeca, M., Goossens, F. A., Meerum T. M., & Schuengel, C. (2002). Bullying and victimization among school-age children: Stability

and links to proactive and reactive aggression. *Social Development, 11*(3), 332–345.

Campbell, L., Simpson, J. A., Kashy, D. A., & Fletcher, G. J. O. (2001). Ideal standards, the self, and flexibility of ideals in close relationships. *Personality & Social Psychology Bulletin, 27*(4), 447–462.

Cannon, T. D., van Erp, T. G. M., Rosso, I. M., Huttunen, M., Loenqvist, J., Pirkola, T., Salonen, O., Valanne, L., Poutanen, V., & Standertskjoeld-Nordenstam, C. (2002). Fetal hypoxia and structural brain abnormalities in schizophrenic patients, their siblings, and controls. *Archives of General Psychiatry, 59*(1), 35–41.

Cannon, W. B. (1927). The James-Lange theory of emotions: A critical examination and an alternative theory. *American Journal of Psychology, 39*, 106–124.

Cannon, W. B., Lewis, J. T., & Britton, S. W. (1927). The dispensability of the sympathetic division of the autonomic nervous system. *Boston Medical Surgery Journal, 197*, 514.

Cantor, J. M., Blanchard, R., Paterson, A. D., & Bogaert, A. F. (2002). How many gay men owe their sexual orientation to fraternal birth order? *Archives of Sexual Behavior, 31*(1), 63–71.

Carducci, B. J. (1998). *The psychology of personality: Viewpoints, research, and applications.* Pacific Grove, CA: Brooks/Cole.

Carmaciu, C. D., Anderson, C. S., & Markar, H. R. (2001). Secondary koro with unusual features in a Briton. *Transcultural Psychiatry, 38*(4), 528–533.

Carnagey, N. L., & Anderson, C. A. (2004). Violent video game exposure and aggression: A literature review. *Minerva Psichiatrica, 45*(1), 1–18.

Carney, S., Cowen, P., Geddes, J., Goodwin, G., Rogers, R., Dearness, K., Tomlin, A., Eastaugh, J., Freemantle, N., Lester, H., Harvey, A., & Scott, A. (2003). Efficacy and safety of electroconvulsive therapy in depressive disorders: A systematic review and meta-analysis. *Lancet, 361*(9360), 799–808.

Carpenter, S., & Radhakrishnan, P. (2002). The relation between allocentrism and perceptions of ingroups. *Personality & Social Psychology Bulletin, 28*(11), 1528–1537.

Carskadon, M. A., & Dement, W. C. (2002). *Adolescent sleep patterns.* New York: Cambridge University Press.

Carstensen, L. L., Fung, H. H., & Charles, S. (2003). Socioemotional selectivity theory and the regulation of emotion in the second half of life. *Motivation & Emotion, 27*(2), 103–123.

Casti, J. L. (2000). *Paradigms regained: A further exploration of the mysteries of modern science.* New York: Morrow.

Castillo, R. J. (2003). Trance, functional psychosis, and culture. *Psychiatry: Interpersonal & Biological Processes, 66*(1), 9–21.

Cattell, R. B. (1950). *Personality: A systematic, theoretical, and factual study.* New York: McGraw-Hill.

Cattell, R. B. (1963). Theory of fluid and crystallized intelligence: A critical experiment. *Journal of Educational Psychology, 54,* 1–22.

Cattell, R. B. (1965). *The scientific analysis of personality.* Baltimore: Penguin.

Cattell, R. B. (1971). *Abilities: Their structure, growth, and action.* Boston: Houghton Mifflin.

Cattell, R. B. (1990). Advances in Cattellian personality theory. In L. A. Pervin (Ed.), *Handbook of personality: Theory and research.* New York: Guilford Press.

Cavacuiti, C. A. (2004). You, me, and drugs—A love triangle: Important considerations when both members of a couple are abusing substances. *Substance Use & Misuse, 39*(4), 645–656.

Cervone, D., & Shoda, Y. (1999). Beyond traits in the study of personality coherence. *Current Directons in Psychological Science, 8*(a), 27–32.

Cesaro, P, & Ollat, H. (1997). Pain and its treatments. *European Neurology, 38,* 209–215.

Ceschi, G., & Scherer, K. R. (2001). Contrôler l'expression faciale et changer l'émotion: Une approche développementale. The role of facial expression in emotion: A developmental perspective. *Enfance, 53*(3), 257–269.

Chabrol, H., Montovany, A., Ducongé, E., Kallmeyer, A., Mullet, E., & Leichsenring, F. (2004). Factor structure of the borderline personality inventory in adolescents. *European Journal of Psychological Assessment, 20*(1), 59–65.

Chaikin, N. D., & Prout, M. F. (2004). Treating complex trauma in women within community mental health. *American Journal of Orthopsychiatry, 74*(2), 160–173.

Challem, J., Berkson, B., Smith, M. D., & Berkson, B. (2000). *Syndrome X: The complete program to prevent and reverse insulin resistance.* New York: Wiley.

Chaney, J. M., Mullins, L. L., Wagner, J. L., Hommel, K. A., Page, M. C., Doppler, & Matthew J. (2004). A longitudinal examination of causal attributions and depression symptomatology in rheumatoid arthritis. *Rehabilitation Psychology, 49*(2), 126–133.

Chang, G., Orav, J., McNamara, T. K., Tong, M-Y., & Antin, J. H. (2005). Psychosocial function after hematopoietic stem cell transplantation. *Psychosomatics: Journal of Consultation Liaison Psychiatry, 46*(1), 34–40.

Chapelot, D., Marmonier, C., Aubert, R., Gausseres, N., & Louis-Sylvestre, J. (2004). A role for glucose and insulin preprandial profiles to differentiate meals and snacks. *Physiology & Behavior, 80*(5), 721–731.

Chartrand, T., Pinckert, S., & Burger, J. M. (1999). When manipulation backfires: The effects of time delay and requester on the foot-in-the-door technique. *Journal of Applied Social Psychology, 29,* 211–221.

Chen, F., & Kenrick, D. T. (2002). Repulsion or attraction? Group membership and assumed attitude similarity. *Journal of Personality and Social Psychology, 83*(1), 111–125.

Cherney, I. D. (2005). Children's and adults' recall of sex-stereotyped toy pictures: Effects of presentation and memory task. *Infant & Child Development, 14*(1), 11–27.

Cheung, C., & Rudowicz, E. (2003). Underachievement and attributions among students attending schools stratified by student ability. *Social Psychology of Education, 6*(4), 303–323.

Cheung, M. S., Gilbert, P., & Irons, C. (2004). An exploration of shame, social rank, and rumination in relation to depression. *Personality & Individual Differences, 36*(5), 1143–1153.

Chi, I., & Chou, K. (1999). Financial strain and depressive symptoms among Hong Kong Chinese elderly: A longitudinal study. *Journal of Gerontological Social Work, 32,* 41–60.

Cho, K. (2001). Chronic "jet lag" produces temporal lobe atrophy and spatial cognitive deficits. *Nature Neuroscience, 4*(6), 567–568.

Choi, I., & Nisbett, R. E. (2000). Cultural psychology of surprise: Holistic theories and recognition of contradiction. *Journal of Personality and Social Psychology, 79*(6), 890–905.

Choi, N. (2004). Sex role group differences in specific, academic, and general self-efficacy. *Journal of Psychology: Interdisciplinary & Applied, 138*(2), 149–159.

Chomsky, N. (1968). *Language and mind.* New York: Harcourt, Brace, World.

Chomsky, N. (1980). Rules and representations. New York: Columbia University Press.

Christ, G. H., Siegel, K., & Christ, A. E. (2002). Adolescent grief: "It never really hit me...until it actually happened." *Journal of the American Medical Association, 288*(10), 1269–1278.

Christensen, D. (2000). Is snoring a diZZZease? Nighttime noises may serve as a wake-up call for future illness. *Science News, 157,* 172–173.

Christian, K. M., & Thompson, R. F. (2005). Long-term storage of an associative memory trace in the cerebellum. *Behavioral Neuroscience, 119*(2), 526–537.

Chuang, D. (1998). Cited in J. Travis, Stimulating clue hints how lithium works. *Science News, 153,* 165.

Cialdini, R. B. (2001). *Influence: Science and practice* (4th ed.). Boston: Allyn & Bacon.

Cillessen A. H. N., & Mayeux L. (2004). From censure to reinforcement: Developmental changes in the association between aggression and social status. *Child Development, 75,* 147–163.

Clark, K. B., & Clark, M. P. (1939). The development of consciousness of self and the emergence of racial identification in Negro preschool children. *Journal of Social Psychology, 10,* 591–599.

Clifford, J. S., Boufal, M. M., & Kurtz, J. E. (2004). Personality traits and critical thinking: Skills in college students empirical tests of a two-factor theory. *Assessment, 11*(2), 169–176.

Clinton, S. M., & Meador-Woodruff, J H. (2004). Thalamic dysfunction in schizophrenia: Neurochemical, neuropathological, and in vivo imaging abnormalities. *Schizophrenia Research. 69*(2–3), 237–253.

Cloud, J. (2001, February 26). New sparks over electroshock. *Time,* 60–62.

Cohen, D., Mason, K., & Farley, T. A. (2004). Beer consumption and premature mortality in Louisiana: An ecologic analysis. *Journal of Studies on Alcohol, 65*(3), 398–403.

Cohen, R. A., Paul, R., Zawacki, T. M., Moser, D. J., Sweet, L., & Wilkinson, H. (2001). Emotional and personality changes following cingulotomy. *Emotion, 1*(1), 38–50.

Cohen, S., & Williamson, G. M. (1991). Stress and infectious disease in humans. *Psychological Bulletin, 109,* 5–24.

Cohen, S., Hamrick, N., Rodriguez, M. S., Feldman, P. J., Rabin, B. S., & Manuck, S. B. (2002). Reactivity and vulnerability to stress-associated risk for upper respiratory illness. *Psychosomatic Medicine, 64*(2), 302–310.

Coley, R. L., & Chase-Lansdale, P. L. (1998). Adolescent pregnancy and parenthood: Recent evidence and future directions. *American Psychologist, 53*(2), 152–166.

Connolly, S. (2000). *LSD (just the facts).* Baltimore: Heinemann Library.

Connor, J., Norton, R., Ameratunga, S., Robinson, E., Civil, I., Dunn, R., Bailey, J., & Jackson, R. (2002). Driver sleepiness and risk of serious injury to car occupants: Population based case control study. *BMJ: British Medical Journal, 324*(7346), 1125–1128.

Connor, K. M., & Davidson, J. R. T. (2002). A placebo-controlled study of Kava kava in generalized anxiety disorder. *International Clinical Psychopharmacology, 17*(4), 185–188.

Conway, M. A., & Pleydell-Pearce, C. W. (2000). The construction of autobiographical memories in the self-memory system. *Psychological Review, 107,* 261–288.

Cook, M., & Mineka, S. (1989). Observational conditioning of fear to fear-relevant versus fear-irrelevant stimuli in rhesus monkeys. *Journal of Abnormal Psychology, 98,* 448–459.

Cook, P. F. (2000). Effects of counselors' etiology attributions on college students' procrastination. *Journal of Counseling Psychology, 47*(3), 352–361.

Coombs, R. H. (2004). *Handbook of addictive disorders.* Hoboken, NJ: Wiley.

Cooper, M. L., Shaver, P. R., & Collins, N. L. (1998). Attachment styles, emotion regulation, and adjustment in adolescence. *Journal of Personality and Social Psychology, 74*(5), 1380–1397.

Cooper, R. (2004). What is wrong with the DSM? History of Psychiatry, 15(1), 5–25.

Cooper, W. E. Jr., Pérez-Mellado, V., Vitt, L. J., & Budzinsky, B. (2002). Behavioral responses to plant toxins in two omnivorous lizard species. *Physiology & Behavior, 76*(2), 297–303.

Corcoran, C., Walker, E., Huot, R., Mittal, V., Tessner, K., Kestler, L., & Malaspina, D. (2003). The stress cascade and schizophrenia: Etiology and onset. *Schizophrenia Bulletin, 29*(4), 671–692.

Corey, G. (2001). *The art of integrative counseling.* Belmont, CA: Wadsworth.

Corey, G. (2005). *Theory and practice of counseling and psychotherapy* (7th ed.). Belmont, CA: Wadsworth.

Corkin, S. (2002). What's new with the amnesic patient H.M.? *Nature Reviews Neuroscience, 3,* 153–160.

Costa, J. L., Brennen, M. B., & Hochgeschwender, U. (2002). The human genetics of eating disorders: Lessons from the leptin/melanocortin system. *Child & Adolescent Psychiatric Clinics of North America, 11*(2), 387–397.

Courage, M. L., & Adams, R. J. (1990). Visual acuity assessment from birth to three years using the acuity card procedures: Cross-sectional and longitudinal samples. *Optometry and Vision Science, 67,* 713–718.

Covington, M. V., & Mueller, K. J. (2001). Intrinsic versus extrinsic motivation: An approach/avoidance reformulation. *Educational Psychology Review, 13*(2), 157–176.

Cowen, P. J. (2002). Cortisol, serotonin and depression: All stressed out? *British Journal of Psychiatry, 180*(2), 99–100.

Coyne, S. M., Archer, J., & Eslea, M. (2004). Cruel intentions on television and in real life: Can viewing indirect aggression increase viewers' subsequent indirect aggression? *Journal of Experimental Child Psychology, 88*(3), 234–253.

Craig, A. D., & Bushnell, M. C. (1994). The thermal grill illusion: Unmasking the burn of cold pain. *Science, 265,* 252–255.

Craik, F. I. M., & Lockhart, R. S. (1972). Levels of processing: A framework for memory research. *Journal of Verbal Learning and Verbal Behavior, 11,* 671–684.

Crair, M. C., Gillespie, D. C., & Stryker, M. P. (1998). The role of visual experience in the development of columns in cat visual cortex. *Science, 279,* 566–570.

Crandall, C. S., & Martinez, R. (1996). Culture, ideology, and antifat attitudes. *Personality and Social Psychology Bulletin, 22,* 1165–1176.

Craske, M. G. (2000). *Mastery of your anxiety and panic: Therapist guide* (3rd ed.). New York: Academic Press.

Crews, F. (1997). The verdict on Freud. *Psychological Science, 7*(2), 63–68.

Crews, F. T., Collins, M. A., Dlugos, C., Littleton, J., Wilkins, L., Neafsey, E. J., Pentney, R., Snell, L. D., Tabakoff, B., Zou, J., & Noronha, A. (2004). Alcohol-induced neurodegeneration: When, where and why? *Alcoholism: Clinical & Experimental Research, 28*(2), 350–364.

Cristofalo, V. J. (1996). Ten years later: What have we learned about human aging from studies of cell cultures? *Gerontologist, 36,* 737–741.

Cronbach, L. (1990). *Essentials of psychological testing.* New York: Harper & Row.

Crone, C. C., & Gabriel, G. (2002). Herbal and nonherbal supplements in medical-psychiatric patient populations. *Psychiatric Clinics of North America, 25*(1), 211–230.

Crooks, R. & Bauer, K. (2005). *Our sexuality* (9th ed.). Pacific Grove, CA: Brooks/Cole.

Crosby, B. (2003). Case studies in rational emotive behavior therapy with children and adolescents. *Journal of Cognitive Psychotherapy, 17*(3), 289–291.

Crow, T. J. (2004). Cerebral asymmetry and the lateralization of language: Core deficits in schizophrenia as pointers to the gene. *Current Opinion in Psychiatry, 17*(2), 97–106.

Cullen, D., & Gotell, L. (2002). From orgasms to organizations: Maslow, women's sexuality and the gendered foundations of the needs hierarchy. *Gender, Work & Organization, 9*(5), 537–555.

Cummings, E., & Henry, W. E. (1961). *Growing old: The process of disengagement.* New York: Basic Books.

Currie, S., & Wang, J. (2004). Chronic back pain and major depression in the general Canadian population. *Pain, 107*(1–2), 54–60.

Curtiss, S. (1977). *Genie: A psycholinguistic study of a modern-day "wild child."* New York: Academic Press.

D'Alessio, D., & Allen, M. (2002). Selective exposure and dissonance after decisions. *Psychological Reports, 91*(2), 527–532.

Dabbs, J. M., & Dabbs, M. G. (2000). *Heroes, rogues, and lovers: Testosterone and behavior.* New York: McGraw-Hill.

Dackis, C. A., & O'Brien, C. P. (2001). Cocaine dependence: A disease of the brain's reward centers. *Journal of Substance Abuse Treatment, 21*(3), 111–117.

Dalman, C., & Allebeck, P. (2002). Paternal age and schizophrenia: further support for an association. *American Journal of Psychiatry, 159*(9), 1591–2.

Damak, S., Rong, M., Yasumatsu, K., Kokrashvili, Z., Varadarajan, V., Zou, S., Jiang, P., Ninomiya, Y., & Margolskee, R.F. (2003). Direction of sweet and umami taste in the absence of taste receptor Tlr3. *Science, 301*(5634), 850–851.

Damasio, A. R. (1999). *The feeling of what happens: Body and emotion in the making of consciousness.* New York: Harcourt Brace.

Dana, R. H. (1998). Cultural identity assessment of culturally diverse groups. *Journal of Personality Assessment, 70*(1), 1–16.

Darmani, N. A., & Crim, J. L. (2005) Delta-9-tetrahydrocannabinol prevents emesis more potently than enhanced locomotor activity produced by chemically diverse dopamine D2/D3 receptor agonists in the least shrew (Cryptotis parva) *Pharmacology Biochemisty and Behavior, 80*:35–44.

Darwin, C. (1859). *On the origin of species.* London: Murray.

Darwin, C. (1872). *The expression of the emotions in man and animals.* London: Murray.

Davalos, D. B., Compagnon, N., Heinlein, S., & Ross, R. G. (2004). Neuropsychological deficits in children associated with increased familial risk for schizophrenia. *Schizophrenia Research, 67*(2–3), 123–130.

David, D., & Brown, R. J. (2002). Suggestibility and negative priming: Two replication studies. *International Journal of Clinical & Experimental Hypnosis, 50*(3), 215–228.

David, D., Schnur, J. E., & Belloiu, A. (2002). Another search for the "hot" cognitions: Appraisal, irrational beliefs, attributions, and their relation to emotion. *Journal of Rational-Emotive & Cognitive Behavior Therapy, 20*(2), 93–132.

David, D., Woodward, C., Esquenazi, J., & Mellman, T. A. (2004). Comparison of comorbid physical illnesses among veterans with PTSD and veterans with alcohol dependence. *Psychiatric Services, 55*(1), 82–85.

Davidson, J. R., Waisberg, J. L., Brundage, M. D., & MacLean, A. W. (2001). Nonpharmacologic group treatment of insomnia: A preliminary study with cancer survivors. *Psycho-Oncology,10*(5), 389–397.

Davies, I. (1998). A study of colour grouping in three languages: A test of the linguistic relativity hypothesis. *British Journal of Psychology, 89*, 433–452.

Davies, J. M. (1996). Dissociation, repression and reality testing in the countertransference: the controversey over memory and false memory in the psychoanalytic treatment of adult survivors of childhood sexual abuse. *Psychoanalytic Dialogues, 6*, 189–218.

Davis, C., Dionne, M., & Shuster, B. (2001). Physical and psychological correlates of appearance orientation. *Personality & Individual Differences, 30*(1), 21–30.

Davison, G.C., Neale, J.M., & Kring, A.M. (2004). *Abnormal psychology* (9th. ed.). Hoboken, NJ: Wiley.

Dawson, K. A. (2004). Temporal organization of the brain: Neurocognitive mechanisms and clinical implications. *Brain & Cognition, 54*(1), 75–94.

de Charms, R., & Moeller, G. H. (1962). Values expressed in American children's readers: 1800–1950. *Journal of Abnormal and Social Psychology, 64*(2), 136–142.

De Coteau, T. J., Hope, D. A., & Anderson, J. (2003). Anxiety, stress, and health in northern plains Native Americans. *Behavior Therapy, 34*(3), 365–380.

De Rosnay, M., & Harris, P. L. (2002). Individual differences in children's understanding of emotion: The roles of attachment and language. *Attachment & Human Development, 4*(1), 39–54.

Deary, I. J., & Stough, C. (1996). Intelligence and inspection time: Achievements, prospects, and problems. *American Psychologist, 51*, 599–608.

Deary, I. J., & Stough, C. (1997). Looking down on human intelligence. *American Psychologist, 52*, 1148–1149.

Deary, I. J., Bell, P. J., Bell, A. J., Campbell, M. L., & Fazal, N. D. (2004). Sensory discrimination and intelligence: Testing Spearman's other hypothesis. *American Journal of Psychology, 117*(1), 1–18.

Debaere, F., Wenderoth, N., Sunaert, S., Van Hecke, P., & Swinnen, S. P. (2004). Changes in brain activation during the acquisition of a new bimanual coordination task. *Neuropsychologia, 42*(7), 855–867.

DeCasper, A. J., & Fifer, W. D. (1980). Of human bonding: Newborns prefer their mother's voices. Science, 208, 1174–1176.

Deci, E. L. (1995). *Why we do what we do: The dynamics of personal autonomy*. New York: Putnam's Sons.

Deci, E. L., Koestner, R., & Ryan, R. M. (1999). A meta-analytic review of experiments examining the effects of extrinsic rewards on intrinsic motivation. *Psychological Bulletin, 125*(6), 627–668.

Deckers, L. (2005). *Motivation: Biological, psychological, and environmental* (2nd ed.). Boston: Allyn & Bacon/Longman.

DeClue, G. (2003). The polygraph and lie detection. *Journal of Psychiatry & Law, 31*(3), 361–368.

Delgado, J. M. R. (1960). Emotional behavior in animals and humans. *Psychiatric Research Report, 12*, 259–271.

Delgado, P. L. (2004). How antidepressants help depression: Mechanisms of action and clinical response. *Journal of Clinical Psychiatry, 65*, 25–30.

Delgado-Gaitan, C. (1994). Socializing young children in Mexican-American families: An intergenerational perspective. In P.M. Greenfield & R.R. Cocking (Eds.), *Cross-cultural roots of minority child development* (pp. 55–86). Hillsdale, NJ: Erlbaum.

DeLisi, L. E., Sakuma, M., Maurizio, A. M., Relja, M., & Hoff, A. L. (2004). Cerebral ventricular change over the first 10 years after the onset of schizophrenia. Psychiatry Research: *Neuroimaging, 130*(1), 57–70.

Delville, Y., Mansour, K. M., & Ferris. C. F. (1996). Testosterone facilitates aggression by modulating vasopressin receptors in the hypothalamus. *Physiology and Behavior, 60*, 25–29.

Dement, W. C., & Vaughan, C. (1999). *The promise of sleep*. New York: Delacorte Press.

den Boer, J. A. (2000). Social anxiety disorder/social phobia: Epidemiology, diagnosis, neurobiology, and treatment. *Comprehensive Psychiatry, 41*(6), 405–415.

Denmark, F. L., Rabinowitz, V. C., & Sechzer, J. A. (2005). *Engendering psychology: Women and gender revisited* (2nd ed.). Boston: Allyn and Bacon.

Dennis, W., & Dennis, M. G. (1940). Cradles and cradling customs of the Pueblo Indians. *American Anthropologist, 42*, 107–115.

Depue, R. A., & Collins, P. F. (1999). Neurobiology of the structure of personality: Dopamine, facilitation of incentive motivation, and extraversion. *Behavioral & Brain Sciences, 22*(3), 491–569.

Der-Karabetian, A., Stephenson, K., & Poggi, T. (1996). Environmental risk perception, activism and world-mindedness among samples of British and U. S. college students. *Perceptual and Motor Skills, 83*(2), 451–462.

DeStefano, D., & LeFevre, J. (2004). The role of working memory in mental arithmetic.

European Journal of Cognitive Psychology, 16(3), 353–386.

DeValois, R. L. (1965). Behavioral and electrophysiological studies of primate vision. In W. D. Neff (Ed.), *Contributions to sensory physiology (Vol. 1)*. New York: Academic Press.

Devine, P.G., Plant, E.A., Amodio, D.M., Harmon-Jones, E., & Vance, S.L. (2002). The regulation of explicit and implicit race bias: The role of motivations to respond without prejudice. *Journal of Personality and Social Psychology, 82*, 835–848.

Devlin, M. J., Yanovski, S. Z., & Wilson, G. T. (2000). Obesity: What mental health professionals need to know. *American Journal of Psychiatry, 157*(6), 854–866.

Diamond, L. M. (2004). Emerging perspectives on distinctions between romantic love and sexual desire. *Current Directions in Psychological Science, 13*(3), 116–119.

Dickens, W. T., & Flynn, J. R. (2001). Heritability estimates versus large environmental effects: The IQ paradox resolved. *Psychological Review, 108*(2), 346–369.

Dickinson, D. J., O'Connell, D. Q., & Dunn, J. S. (1996). Distributed study, cognitive study strategies and aptitude on student learning. *Psychology: A Journal of Human Behavior, 33*(3), 31–39.

Diehm, R., & Armatas, C. (2004). Surfing: An avenue for socially acceptable risk-taking, satisfying needs for sensation seeking and experience seeking. *Personality & Individual Differences, 36*(3), 663–677.

Diener, M. L., Mengelsdorf, S. C., McHale, J. L., & Frosch, C. A. (2002). Infants' behavioral strategies for emotion regulation with fathers and mothers: Associations with emotional expressions and attachment quality. *Infancy, 3*(2), 153–174.

Dieter, J. N. I., Field, T., Hernandez-Reif, M., Emory, E. K., & Redzepi, M. (2003). Stable preterm infants gain more weight and sleep less after five days of massage therapy. *Journal of Pediatric Psychology, 28*(6), 403–411.

Dietrich, A. (2003). Functional neuroanatomy of altered states of consciousness: The transient hypofrontality hypothesis. *Conciousness and Cognition, 12*, 231–256.

Dietz, T. L. (1998). An examination of violence and gender role portrayals in video games. *Sex Roles, 38*, 425–442.

DiLauro, M. D. (2004). Psychosocial factors associated with types of child maltreatment. *Child Welfare, 83*(1), 69–99.

Dimberg, U., & Thunberg, M. (1998). Rapid facial reactions to emotion facial expressions. *Scandinavian Journal of Psychology, 39*(1), 39–46.

Dimberg, U., Thunberg, M., & Elmehed, K. (2000). Unconscious facial reactions to emotional facial expressions. *Psychological Science, 11*(1), 86–89.

DiPietro, J. A. (2000). Baby and the brain: Advances in child development. *Annual Review of Public Health, 21*, 455–71.

Dobbin, A., Faulkner, S., Heaney, D., Selvaraj. S., & Gruzelier, J. (2004). Impact on health status of a hypnosis clinic in general practice. *Contemporary Hypnosis, 21*(4), 153–160.

Dodd, M. D., & MacLeod, C. M. (2004). False recognition without intentional learning. *Psychonomic Bulletin & Review, 11*(1), 137–142.

Doghramji, K. (2000, December). *Sleepless in America: Diagnosing and treating insomnia.* [On-line serial]. Available: http://psychiatry. medscape. com/Medscape/psychiatry/ClinicalMgmt/ CM.v02/public/indexCM.v02.html.

Dollard, J., Doob, L., Miller, N., Mowrer, O. H., & Sears, R. R. (1939). *Frustration and aggression.* New Haven, CT: Yale University Press.

Dols, M., Willems, B., van den Hout, M., & Bittoun, R. (2000). Smokers can learn to influence their urge to smoke. *Addictive Behaviors, 25*(1), 103–108.

Domhoff, G. W. (1999). New directions in the study of dream content using the Hall and Van de Castle coding system. *Dreaming, 9,* 115–137.

Domhoff, G. W. (2003). *The scientific study of dreams: Neural networks, cognitive development, and content analysis.* Washington, DC: American Psychological Association.

Domhoff, G. W. (2004). Why did empirical dream researchers reject Freud? A critique of historical claims by Mark Solms. *Dreaming, 14*(1), 3–17.

Domhoff, G. W. (2005). A reply to Hobson (2005). *Dreaming, 15*(1), 30–32.

Domjan, M. (2005). Pavlovian conditioning: A functional perspective. *Annual Review of Psychology, 56,* 179–206.

Dondi, M., Simion, F., & Caltran, G. (1999). Can newborns discriminate between their own cry and the cry of another newborn infant? *Developmental Psychology, 35,* 418–426.

Donini, L. M., Savina, C., & Cannella, C. (2003). Eating habits and appetite control in the elderly: The anorexia of aging. *International Psychogeriatrics, 15*(1), 73–87.

Donovan, J. J., & Radosevich, D. J. (1999). A meta-analytic review of the distribution of practice effect: Now you see it, now you don't. *Journal of Applied Psychology, 84,* 795–805.

Dorian, L., & Garfinkel, P. E. (2002). Culture and body image in Western culture. *Eating & Weight Disorders, 7*(1), 1–19.

Dougherty, D. D., Baer, L., Cosgrove, G. R., Cassem, E. H., Price, B. H., Nierenberg, A. A., Jenike, M. A., & Rauch, S. L. (2002). Prospective long-term follow-up of 44 patients who received cingulotomy for treatment-refractory obsessive-compulsive disorder. *American Journal of Psychiatry, 159*(2), 269–275.

Dowd, E. T. (2004). Cognition and the cognitive revolution in psychotherapy: Promises and advances. *Journal of Clinical Psychology, 60*(4), 415–428.

Doyle, A., & Pollack, M. H. (2004). Long-term management of panic disorder. *Journal of Clinical Psychiatry. 65*(Suppl5), 24–28.

Draguns, J., & Tanaka-Matsumi, J. (2003). Assessment of psychopathology across and within cultures: Issues and findings. *Behaviour Research & Therapy, 41*(7), 755–776.

Drescher, J. (2003). Gold or lead? Introductory remarks on conversions. *Journal of Gay & Lesbian Psychotherapy, 7*(3), 1–13.

Druckman, D., & Bjork, R. A. (Eds.) (1994). *Learning, remembering, believing: Enhancing human performance.* Washington, DC: National Academies Press.

Drummey, A. B., & Newcombe, N. S. (2002). Developmental changes in source memory. *Developmental Science, 5*(4), 502–513.

Dryer, D. C., & Horowitz, L. M. (1997). When do opposites attract? Interpersonal complementarity versus similarity. *Journal of Personality and Social Psychology, 72,* 592–603.

Duckworth, K., & Borus, J. F. (1999). Population-based psychiatry in the public sector and managed care. In A. M. Nicholi (Ed.), *The Harvard guide to psychiatry.* Cambridge, MA: Harvard University Press.

Dunn, M. (2000). *Good death guide: Everything you wanted to know but were afraid to ask.* New York: How to Books.

Durand, M. V., & Barlow, D. H. (2003). *Essentials of abnormal psychology* (3rd ed.). Belmont Park, CA: Thompson Learning.

Durham, M. D., & Dane, F. C. (1999). Juror knowledge of eyewitness behavior: Evidence for the necessity of expert testimony. *Journal of Social Behavior & Personality, 14,* 299–308.

Edery-Halpern, G., & Nachson, I. (2004). Distinctiveness in flashbulb memory: Comparative analysis of five terrorist attacks. *Memory, 12*(2), 147–157.

Edwards, B. (1999). *The new drawing on the right side of the brain.* Baltimore: J P Tarcher.

Edwards, S. (2007). *50 plus one ways to improve your study habits,* Chicago, IL: New Eductaion Encouragement Press.

Ehrenreich, B. (2004, July 15). *All together now.* [On-line]. Available: http://select.nytimes. com/gst/abstract.html?res=F00E16FA3C5E0 C768DDDAE0894DC404482.

Eibl-Eibesfeldt, I. (1980). Strategies of social interaction. In R. Plutchik & H. Kelerman (Eds.), *Emotion: Theory, research, and experience.* New York: Academic Press.

Eisen, S. A., Chantarujikapong, Sta, Xian, H., Lyons, M. J., Toomey, R., True, W. R., Scherrer, J. F., Goldberg, J., & Tsuang, M. T. (2002). Does marijuana use have residual adverse effects on self-reported health measures, sociodemographics and quality of life? A monozygotic co-twin control study in men. *Addiction, 97*(9), 1137–1144.

Eisenberger, R., & Armeli, S. (1997). Can salient reward increase creative performance without reducing intrinsic creative interest? *Journal of Personality and Social Psychology, 72,* 652–663.

Eisenberger, R., & Rhoades, L. (2002). Incremental effects of reward on creativity. *Journal of Personality and Social Psychology, 81*(4), 728–741.

Ekman, P. (1993). Facial expression and emotion. *American Psychologist, 48,* 384–392.

Ekman, P. (2004). *Emotions revealed: Recognizing faces and feelings to improve communication and emotional life.* Thousand Oaks, CA: Owl Books.

Ekman, P., & Keltner, D. (1997). Universal facial expressions of emotion: An old controversy and new findings. In U. C. Segerstrale & P. Molnar (Eds.), *Nonverbal communication: Where nature meets culture.* Mahwah, NJ: Erlbaum.

Elder, G. (1998). The life course as developmental theory. *Current Directions in Psychological Science, 69,* 1–12.

Elkin, A., Kalidindi, S., & McGuffin, P. (2004). Have schizophrenia genes been found? Current Opinion in Psychiatry, 17(2), 107–113.

Elkind, D. (1967). Egocentrism in adolescence. *Child Development, 38,* 1025–1034.

Elkind, D. (1981). *The hurried child.* Reading, MA: Addison-Wesley.

Elkind, D. (2000). A quixotic approach to issues in early childhood education. *Human Development, 43*(4–5), 279–283.

Ellis, A. (1961). *A guide to rational living.* Englewood Cliffs, NJ: Prentice-Hall.

Ellis, A. (1996). *Better, deeper, and more enduring brief therapy.* New York: Institute for Rational Emotive Therapy.

Ellis, A. (1997). Using Rational Emotive Behavior Therapy techniques to cope with disability. *Professional Psychology: Research and Practice, 28,* 17–22.

Ellis, A. (2003a). Early theories and practices of rational emotive behavior therapy and how they have been augmented and revised during the last three decades. *Journal of Rational-*

Emotive & Cognitive Behavior Therapy, 21(3–4), 219–243.

Ellis, A. (2003b). Similarities and differences between rational emotive behavior therapy and cognitive therapy. *Journal of Cognitive Psychotherapy, 17*(3), 225–240.

Ellis, A. (2004). Why rational emotive behavior therapy is the most comprehensive and effective form of behavior therapy. *Journal of Rational-Emotive & Cognitive Behavior Therapy, 22*(2), 85–92.

Engel, J. M., Jensen, M. P., & Schwartz, L. (2004). Outcome of biofeedback-assisted relaxation for pain in adults with cerebral palsy: Preliminary findings. *Applied Psychophysiology & Biofeedback, 29*(2), 135–140.

Ennis, N. E., Hobfoll, S. E., & Schroeder, K. E. E. (2000). Money doesn't talk, it swears: How economic stress and resistance resources impact inner-city women's depressive mood. *American Journal of Community Psychology, 28,* 149–173.

Erdelyi, M. H., & Applebaum, A. G. (1973). Cognitive masking: The disruptive effect of an emotional stimulus upon the perception of contiguous neutral items. *Bulletin of the Psychonomic Society, 1,* 59–61.

Eriksen, J., Jensen, M. K., Sjogren, P., Ekholm, O., & Rasmussen, N. K. (2003). Epidemiology of chronic non-malignant pain in Denmark. *Pain, 106*(3), 221–228.

Erikson, E. (1950). *Childhood and society.* New York: Norton.

Erlacher, D., & Schredl, M. (2004). Dreams reflecting waking sport activities: A comparison of sport and psychology students. *International Journal of Sport Psychology, 35*(4), 301–308.

Esses, V. M., Dovidio, J. F., Jackson, L. M., & Armstrong, T. L. (2001). The immigration dilemma: The role of perceived group competition, ethnic prejudice, and national identity. *Journal of Social Issues, 57*(3), 389–412.

Esterson, A. (2002). The myth of Freud's ostracism by the medical community in 1896–1905: Jeffrey Masson's assault on truth. *History of Psychology, 5*(2), 115–134.

Evans, J. S. B. T. (2003). In two minds: Dual-process accounts of reasoning. *Trends in Cognitive Sciences, 7*(10), 454–459.

Ewart, C. K., & Fitzgerald, S. T. (1994). Changing behaviour and promoting well-being after heart attack: A social action theory approach. *Irish Journal of Psychology, 15*(1), 219–241.

Eysenck, H. J. (1967). *The biological basis of personality.* Springfield, IL: Charles C Thomas.

Eysenck, H. J. (1982). *Personality, genetics, and behavior: Selected papers.* New York: Prager.

Eysenck, H. J. (1990). Biological dimensions of personality. In L. A. Pervin (Ed.), *Handbook of personality: Theory and research.* New York: Guilford Press.

Eysenck, H. J. (1991). *Smoking, personality, and stress: Psychosocial factors in the prevention of cancer and coronary heart disease.* New York: Springer-Verlag.

Facon, B., & Facon-Bollengier, T. (1999). Chronological age and crystallized intelligence of people with intellectual disability. *Journal of Intellectual Disability Research, 43*(6), 489–496.

Faigman, D. L., Kaye, D., Saks, M. J., & Sanders, J. (1997). *Modern scientific evidence: The law and science of expert testimony.* St. Paul, MN: West.

Fanselow, M. S., & Poulos, A. M. (2005). The neuroscience of mammalian associative learning. *Annual Review of Psychology, 56,* 207–234.

Fantz, R. L. (1956). A method for studying early visual development. *Perceptual and Motor Skills, 6,* 13–15.

Fantz, R. L. (1963). Pattern vision in newborn infants. *Science, 140,* 296–297.

Farrington, D. P. (2000). Psychosocial predictors of adult antisocial personality and adult convictions. *Behavioral Sciences & the Law, 18*(5), 605–622.

Fazio, R. H., & Olson, M. A. (2003). Implicit measures in social cognition research: Their meaning and use. *Annual Review of Psychology, 54,* 297–327.

Fein, S. & Spencer, S. J. (1997). Prejudice as self-image maintenance: Affirming the self through derogating others. *Journal of Personality and Social Psychology, 73*(1), 31–44.

Feiring, C., Deblinger, E., Hoch-Espada, A., & Haworht T. (2002). Romantic relationship aggression and attitudes in high school students: The role of gender, grade, and attachment and emotional styles. *Journal of Youth & Adolescence, 31*(5), 373–385.

Feldman, R. S. (1982). *Development of nonverbal behavior in children.* Seacaucus, NJ: Springer-Verlag.

Fernández, M. I., Bowen, G. S., Varga, L. M., Collazo, J. B., Hernandez, N., Perrino, T., & Rehbein, A. (2005). High rates of club drug use and risky sexual practices among Hispanic men who have sex with men in Miami, Florida. *Substance Use & Misuse, 40*(9–10), 1347–1362.

Fernández-Dols, J-M., Carrera, P., & Russell, J. A. (2002). Are facial displays social? Situational influences in the attribution of emotion to facial expressions. *Spanish Journal of Psychology, 5*(2), 119–124.

Festinger, L. A. (1957). *A theory of cognitive dissonance.* Palo Alto, CA: Stanford University Press.

Festinger, L. A., & Carlsmith, L. M. (1959). Cognitive consequences of forced compliance. *Journal of Abnormal and Social Psychology, 58,* 203–210.

Field, K. M., Woodson, R., Greenberg, R., & Cohen, D. (1982). Discrimination and imitation of facial expressions by neonates. *Science, 218,* 179–181.

Field, T., & Hernandez-Reif, M. (2001). Sleep problems in infants decrease following massage therapy. *Early Child Development & Care, 168,* 95–104.

Fink, B., & Penton-Voak, I. (2002). Evolutionary psychology of facial attractiveness. *Current Directions in Psychological Science, 11*(5), 154–158.

Fink, B., Manning, J. T., Neave, N., & Grammer, K. (2004). Second to fourth digit ratio and facial asymmetry. *Evolution and Human Behavior, 25*(2), 125–132.

Fink, M. (1999). *Electroshock: Restoring the mind.* London: Oxford University Press.

First, M., & Tasman, A. (2004). *DSM-IV-TR mental disorders: Diagnosis, etiology, and treatment.* Hoboken, NJ: Wiley.

Fisher, S., & Greenberg, R. P. (1996). *Freud scientifically reappraised: Testing the theories and therapy.* New York: Wiley.

Fiske, S. T. (1998). Stereotyping, prejudice, and discrimination. In D. T. Gilbert, S. T. Fiske, and G. Lindzey (Eds.), *The handbook of social psychology* (4th ed., Vol. 2, pp. 357–411). Boston: McGraw-Hill.

Flavell, J. H., Miller, P. H., & Miller, S. A. (2002). *Cognitive development* (4th ed.). Upper Saddle River, NJ: Prentice-Hall.

Fletcher, G. J. O., & Simpson, J. A. (2000). Ideal standards in close relationships: Their structure and functions. *Current Directions in Psychological Science, 9*(3), 102–105.

Flynn, J. R. (1987). Massive IQ gains in 14 nations: What IQ tests really measure. *Psychological Bulletin, 101,* 171–191.

Flynn, J. R. (2000). IQ gains and fluid g. *American Psychologist, 55*(5), 543.

Flynn, J. R. (2003). Movies about intelligence: The limitations of g. *Current Directions in Psychological Science, 12*(3), 95–99.

Folk, C. L., & Remington, R. W. (1998). Selectivity in distraction by irrelevant featural singletons: Evidence for two forms of attentional capture. *Journal of Experimental Psychology: Human Perception and Performance, 24,* 1–12.

Fone, B. R. S. (2000). *Homophobia: A history*. New York: Metropolitan Books.

Ford, T. E., Ferguson, M. A., Brooks, J. L., & Hagadone, K. M. (2004). Coping sense of humor reduces effects of stereotype threat on women's math performance. *Personality & Social Psychology Bulletin, 30*(5), 643–653.

Fraley, R. C., & Shaver, P. R. (1997). Adult attachment and the suppression of unwanted thoughts. *Journal of Personality and Social Psychology, 73*, 1080–1091.

Frank, M. J. (2005). Dynamic dopamine modulation in the basal ganglia: A neuro-computational account of cognitive deficits in medicated and nonmedicated Parkinsonism. *Journal of Cognitive Neuroscience, 17*(1), 51–72.

Frankenburg, W., Dodds, J., Archer, P., Shapiro, H., & Bresnick, B. (1992). The Denver II: A major revision and restandardization of the Denver Developmental Screening Test. *Pediatrics, 89*, 91–97.

Fredricks, J. A., & Eccles, J. S. (2005). Family socialization, gender, and sport motivation and involvement. *Journal of Sport & Exercise Psychology, 27*(1), 3–31.

Frick, W. B. (2000). Remembering Maslow: Reflections on a 1968 interview. *Journal of Humanistic Psychology, 40*(2), 128–147.

Friedman, H., & Schustack, M. (2006). *Personality: Classic theories and modern research* (3rd ed.). Boston: Allyn & Bacon/Longman.

Funder, D. C. (2000). Personality. *Annual Review of Psychology, 52*, 197–221.

Funder, D. C. (2001). The really, really fundamental attribution error. *Psychological Inquiry, 12*(1), 21–23.

Furman, E. (1990, November). Plant a potato learn about life (and death). *Young Children, 46*(1), 15–20.

Gaab, J., Rohleder, N., Nater, U. M., & Ehlert, U. (2005). Psychological determinants of the cortisol stress response: The role of anticipatory cognitive appraisal. *Psychoneuroendocrinology, 30*(6), Jul 2005, 599–610.

Gabry, K. E., Chrousos, G. P., Rice, K. C., Mostafa, R. M., Sternberg, E., Negrao, A. B., Webster, E. L., McCann, S. M., & Gold, P. W. (2002). Marked suppression of gastric ulcerogenesis and intestinal responses to stress by a novel class of drugs. *Molecular Psychiatry, 7*(5), 474–483.

Gaetz, M., Weinberg, H. Rzempoluck, E., & Jantzen, K. J. (1998). Neural network classifications and correlational analysis of EEG and MEG activity accompanying spontaneous reversals of the Necker Cube. *Cognitive Brain Research, 6*, 335–346.

Galinsky, A. D., & Moskowitz, G. B. (2000). Perspective-taking: Decreasing stereotype expression, stereotype accessibility, and in-group favoritism. *Journal of Personality and Social Psychology, 78*(4), 708–724.

Galinsky, A., D., & Ku, G. (2004). The effects of perspective-taking on prejudice: The moderating role of self-evaluation. *Personality & Social Psychology Bulletin, 30*(5), 594–604.

Gallagher, A. M., De Lisi, R., Holst, P. C., McGillicuddy-De Lisi, A. V., Morely, M., & Cahalan, C. (2000). Gender differences in advanced mathematical problem solving. *Journal of Experimental Child Psychology, 75*(3), 165–190.

Galloway, J. L. (1999, March 8). Into the heart of darkness. *U. S. News and World Report*, pp. 25–32.

Gangestad, S. W., & Thornhill, R. (2003). Facial masculinity and fluctuating asymmetry. *Evolution and Human Behavior, 24*(4), 231–241.

Garbarino, S., Beelke, M., Costa, G., Violani, C., Lucidi, F., & Ferrillo, F. (2002). Brain function and effects of shift work: Implications for clinical neuropharmacology. *Neuropsychobiology, 45*(1), 50–56.

Garcia, G. M., & Stafford, M. E. (2000). Prediction of reading by Ga and Gc specific cognitive abilities for low-SES White and Hispanic English-speaking children. *Psychology in the Schools, 37*(3), 227–235.

Garcia, J. (2003). Psychology is not an enclave. In R. Sternberg (Ed.), *Psychologists defying the crowd: Stories of those who battled the establishment and won* (pp. 67–77). Washington, DC: American Psychological Association.

Garcia, J., & Koelling, R. S. (1966). Relation of cue to consequence in avoidance learning. *Psychonomic Science, 4*, 123–124.

Gardiner, H., & Kosmitzki, G. (2005). *Lives across cultures: Cross-cultural human development* (3rd ed.). Boston: Allyn & Bacon/Longman.

Gardner, H. (1983). *Frames of mind*. New York: Basic Books.

Gardner, H. (1998). A multiplicity of intelligences. *Scientific American Presents Exploring Intelligence, 9*, 18–23.

Gardner, H. (1999, February). Who owns intelligence? *Atlantic Monthly*, pp. 67–76.

Gardner, R. A., & Gardner, B. T. (1969). Teaching sign language to a chimpanzee. *Science, 165*, 664–672.

Garnefski, N., Teerds, J., Kraaij, V., Legerstee, J., & van den Kommer, T. (2004). Cognitive emotion regulation strategies and Depressive symptoms: differences between males and females. *Personality & Individual Differences, 36*(2), 267–276.

Gaser, C., Nenadic, I., Buchsbaum, B. R., Hazeltt, E. A., & Buchsbaum, M. S. (2004). Ventricular enlargement in schizophrenia related to volume reduction of the thalamus, striatum, and superior temporal cortex. *American Journal of Psychiatry, 161*(1), 154–156.

Gawronski, B. (2003). On difficult questions and evident answers: Dispositional inference from role-constrained behavior. *Personality & Social Psychology Bulletin, 29*(11), 1459–1475.

Gay, M-C., Philippot, P., & Luminet, O. (2002). Differential effectiveness of psychological interventions for reducing osteoarthritis pain: A comparison of Erickson hypnosis and Jacobson relaxation. *European Journal of Pain, 6*, 1–16.

Gay, P. (1999, March 29). Psychoanalyst Sigmund Freud. *Time*, 65–69.

Gay, P. (2000). *Freud for historians*. Boston: Replica Books.

Geary, D. C. (2005). The motivation to control and the origin of mind: Exploring the life-mind joint point in the tree of knowledge system. *Journal of Clinical Psychology, 61*(1), 21–46.

Gemignani, A., Santarcangelo, E., Sebastiani, L., Marchese, C., Mammoliti, R., Simoni, A., & Ghelarducci, B. (2000). Changes in autonomic and EEG patterns induced by hypnotic imagination of aversive stimuli in man. *Brain Research Bulletin, 53*(1), 105–111.

Gerra, G., Angioni, L., Zaimovic, A., Moi, G., Bussandri, M., Bertacca, S., Santoro, G., Gardini, S., Caccavari, R., & Nicoli, M. A. (2004). Substance use among high-school students: Relationships with temperament, personality traits, and parental care perception. *Substance Use & Misuse, 39*(2), 345–367.

Gershon, E. S., & Rieder, R. O. (1993). Major disorders of mind and brain. *Mind and brain: Readings from Scientific American Magazine* (pp. 91–100). New York: Freeman.

Giacobbi, P. Jr., Foore, B., & Weinberg, R. S. (2004). Broken clubs and expletives: The sources of stress and coping responses of skilled and moderately skilled golfers. *Journal of Applied Sport Psychology, 16*(2), 166–182.

Gibbs, N. (1995, October 2). The EQ factor. *Time*, pp. 60–68.

Gibson, E. J., & Walk, R. D. (1960). The visual cliff. *Scientific American, 202*(2), 67–71.

Gilbert, T. (2003). Managing passive-aggressive behavior of children and youth at school and home: The angry smile. *Education & Treatment of Children, 26*(3), 305–306.

Giles, J. W., & Heyman, G. D. (2005). Young children's beliefs about the relationship between gender and aggressive behavior. *Child Development, 76*(1), 107–121.

Gilligan, C. (1977). In a different voice: Women's conception of morality. *Harvard Educational Review, 47*(4), 481–517.

Gilligan, C. (1990). Teaching Shakespeare's sister. In C. Gilligan, N. Lyons, & T. Hanmer (Eds.), *Mapping the moral domain* (pp. 73–86). Cambridge, MA: Harvard University Press.

Gilligan, C. (1993). Adolescent development reconsidered. In A. Garrod (Ed.), *Approaches to moral development: New research and emerging themes*. New York: Teachers College Press.

Gilmour, D. R., & Walkey, F. H. (1981). Identifying violent offenders using a video measure of interpersonal distance. *Journal of Consulting and Clinical Psychology, 49*, 287–291.

Gilovich, T., Griffin, D., & Kahneman, D. (2002). *Heuristics and biases: The psychology of intuitive judgment*. Cambridge, UK: Cambridge University Press.

Gitau, R., Modi, N., Gianakoulopoulos, X., Bond, C., Glover, V., & Stevenson, J. (2002). Acute effects of maternal skin-to-skin contact and massage on saliva cortisol in preterm babies. *Journal of Reproductive & Infant Psychology, 20*(2), 83–88.

Godden, D. R., & Baddeley, A. D. (1975). Context-dependent memory in two natural environments: On land and underwater. *British Journal of Psychology, 66*, 325–331.

Goebel, M. U., & Mills, P. J. (2000). Acute psychological stress and exercise and changes in peripheral leukocyte adhesion molecule expression and density. *Psychosomatic Medicine, 62*(5), 664–670.

Goerke, M., Möller, J., Schulz-Hardt, S., Napiersky, U., & Frey, D. (2004). "It's not my fault-but only I can change it": Counterfactual and prefactual thoughts of managers. *Journal of Applied Psychology, 89*(2), 279–292.

Gogtay, N., Sporn, A., Clasen, L. S., Nugent, T. F. III, Greenstein, D., Nicolson, R., Giedd, J. N., Lenane, M., Gochman, P., Evans, A., & Rapoport, J. L. (2004). Comparison of progressive cortical gray matter loss in childhood-onset schizophrenia with that in childhood-onset atypical psychoses. *Archives of General Psychiatry, 61*(1), 17–22.

Goldberg, S. (2000). *Attachment and development*. New York: Edward Arnold.

Goleman, D. (1980, February). 1,528 little geniuses and how they grew. *Psychology Today*, pp. 28–53.

Goleman, D. (1995). *Emotional intelligence: Why it can matter more than IQ*. New York: Bantam.

Goleman, D. (2000). *Working with emotional intelligence*. New York: Bantam Doubleday.

Golob, E. J., & Starr, A. (2004). Serial position effects in auditory event-related potentials during working memory retrieval. *Journal of Cognitive Neuroscience, 16*(1), 40–52.

Gonzales, P. M., Blanton, H., & Williams, K. J. (2002). The effects of stereotype threat and double-minority status on the test performance of Latino women. *Personality & Social Psychology Bulletin, 28*(5), 659–670.

Goodman, G. S., Ghetti, S., Quas, J. A., Edelstein, R. S., Alexander, K. W., Redlich, A. D., Cordon, I. M., & Jones, D. P. H. (2003). A prospective study of memory for child sexual abuse: New findings relevant to the repressed-memory controversy. *Psychological Science, 14*(2), 113–118.

Goodman, M., & Yehuda, R. (2002). The relationship between psychological trauma and borderline personality disorder. *Psychiatric Annals, 32*(6), 337–345.

Goodwin, F. K., Simon, G., Revicki, D., Hunkeler, E., Fireman, B., & Lee, E. (2004). "Pharmacotherapy and risk of suicidal behaviors among patients with bipolar disorder": Reply. *JAMA: Journal of the American Medical Association, 291*(8), 940.

Gordon, R. A. (2000). *Eating disorders: Anatomy of a social epidemic* (2nd ed.). Malden, MA: Blackwell Publishers.

Gordon, W. C. (1989). *Learning and memory*. Pacific Grove, CA: Brooks/Cole.

Gorman, J. (2000). Neuroanatomical hypothesis of panic disorder revised. *American Journal of Psychiatry, 57*, 493–505.

Goswami, S., Mattoo, S. K., Basu, D., & Singh, G. (2004). Substance-abusing schizophrenics: Do they self-medicate? *American Journal on Addictions, 13*(2), 139–150.

Gotlieb, I. H., & Abramson, L. Y. (1999). Attributional theories of emotion. In T. Dalgleish & M. Power (Eds.), *Handbook of cognition and emotion*. New York: Wiley.

Gottesman, I. I. (1991). *Schizophrenia genesis: The origins of madness*. New York: Freeman.

Gottesman, I. I., & Hanson, D. R. (2005). Human development: Biological and genetic processes. *Annual Review of Psychology, 56*, 263–286.

Gottman, J. M., & Levenson, R. W. (2002). A two-factor model for predicting when a couple will divorce: Exploratory analyses using 14-year longitudinal data. *Family Process, 41*(1), 83–96.

Gottman, J. M., & Notarius, C. I. (2000). Decade review: Observing marital interaction. *Journal of Marriage & the Family, 62*(4), 927–947.

Gottman, J. M., Coan, J., Carrere, S., & Swanson, C. (1998). Predicting happiness and stability from newlywed interactions. *Journal of Marriage and the Family, 60*, 42–48.

Goveas, J. S., Csernansky, J. G., & Coccaro, E. F. (2004). Platelet serotonin content correlates inversely with life history of aggression in personality-disordered subjects. Psychiatry Research, 126(1), 23–32.

Gracely, R. H., Farrell, M. J., & Grant, M. A. (2002). Temperature and pain perception. In H. Pashler & S. Yantis (Eds.), *Steven's handbook of experimental psychology: Vol. 1. Sensation and perception* (3rd ed.). New York: Wiley.

Graham, J. R. (1991). Comments on Duckworth's review of the Minnesota Multiphasic Personality Inventory-2. *Journal of Counseling and Development, 69*, 570–571.

Graw, P., Krauchi, K., Knoblauch, V., Wirz-Justice, A., & Cajochen, C. (2004). Circadian and wake-dependent modulation of fastest and slowest reaction times during the psychomotor vigilance task. *Physiology & Behavior, 80*(5), 695–701.

Greeff, A. P., & Malherbe, H. L. (2001). Intimacy and marital satisfaction in spouses. *Journal of Sex & Marital Therapy, 27*(3), 247–257.

Greeff, A. P., & Van Der Merwe, S. (2004). Variables associated with resilience in divorced families. *Social Indicators Research, 68*(1), 59–75.

Green, A. I. (2000). What is the relationship between schizophrenia and substance abuse? *Harvard Mental Health Letter, 17*(4), 8.

Green, A. I., Tohen, M. F., Hamer, R. M., Strakowski, S. M., Lieberman, J. A., Glick, I., Clark, W. S., & HGDH Research Group. (2004). First episode schizophrenia-related psychosis and substance use disorders: Acute response to olanzapine and haloperidol. *Schizophrenia Research, 66*(2–3), 125–135.

Green, J. W. (1999). *Cultural awareness in the human services: A multi-ethnic approach*. Needham Heights, MA: Allyn & Bacon.

Greenberg, D. L. (2004). President Bush's false "flashbulb" memory of 9/11/01. *Applied Cognitive Psychology, 18*(3), 363–370.

Greenberg, J. (2002). Who stole the money, and when? Individual and situational determinants of employee theft. *Organizational Behavior and Human Decision Processes, 89*(1), 985–1003.

Greenberg, J., Pyszczynski, T., Solomon, S., Pinel, E., Simon, L., & Jordon, K. (1993). Effects of self-esteem on vulnerability-denying defensive distortions: Further evidence of an anxiety-buffering function of self-esteem. *Journal of Experimental Social Psychology, 29*(3), 229–251.

Greene, K., Kremar, M., Walters, L. H., Rubin, D. L., Hale, J., & Hale, L. (2000). Targeting adolescent risk-taking behaviors: The contributions of egocentrism and sensation-seeking. Journal of Adolescence, 23, 439–461.

Greenwald, A. G., Banaji, M. R., Rudman, L. A., Farnham, S. D., Nosek, B. A., & Mellott, D. S. (2002). A unified theory of implicit attitudes, stereotypes, self-esteem, and self-concept. *Psychological Review, 109,* 3–25.

Grigorenko, E. L., Jarvin, L., & Sternberg, R. J. (2002). School-based tests of the triarchic theory of intelligence: Three settings, three samples, three syllabi. Contemporary *Educational Psychology, 27*(2), 167–208.

Grilly, D. (2006). *Drugs and human behavior* (5th ed.). Boston: Allyn & Bacon/Longman.

Grondin, S., Ouellet, B., & Roussel, M. (2004). Benefits and limits of explicit counting for discriminating temporal intervals. Canadian Journal of Experimental Psychology, *58*(1), 1–12.

Grossmann, K., Grossmann, K. E., Fremmer-Bombik, E., Kindler, H., Scheuerer-Englisch, H., & Zimmermann, P. (2002). The uniqueness of the child-father attachment relationship: Fathers' sensitive and challenging play as a pivotal variable in a 16-year longitudinal study. *Social Development, 11*(3), 307–331.

Grove, W. M., Barden, R. C., Garb, H. N., & Lilienfeld, S. O. (2002). Failure of Rorschach-Comprehensive-System-based testimony to be admissible under the Daubert-Joiner-Kumho standard. *Psychology, Public Policy, & Law, 8*(2), 216–234.

Gruber, U., Fegg, M., Buchmann, M., Kolb, H-J., & Hiddemann, W. (2003). The long-term psychosocial effects of haematopoetic stem cell transplantation. *European Journal of Cancer Care, 12*(3), 249–256.

Guarnaccia, P. J., & Rogler, L. H. (1999). Research on culture-bound syndromes: New directions. *American Journal of Psychiatry, 156*(9), 1322–1327.

Guidelines for the treatment of animals in behavioural research and teaching. (2005). *Animal Behaviour, 69*(1), i-vi.

Guilarte, T. R., Toscano, C. D., McGlothan, J. L., & Weaver, S. A. (2003). Environmental enrichment reverses cognitive and molecular deficits induced developmental lead exposure. *Annals of Neurology, 53*(1), 50–56.

Guilford, J. P. (1967). The nature of human intelligence. New York: McGraw-Hill.

Guinard, J. X., Zoumas-Morse, C., Dietz, J., Goldberg, S., Holz, M., Heck, E., & Amoros, A. (1996). Does consumption of beer, alcohol, and bitter substances affect bitterness perception? *Physiology and Behavior, 49,* 625–631.

Guinote, A., & Fiske, S. T. (2003). Being in the outgroup territory increases stereotypic perceptions of outgroups: Situational sources of category activation. *Group Processes & Intergroup Relations, 6*(4), 323–331.

Gunzerath, L., Faden, V., Zakhari, S., & Warren, K. (2004). National Institute on Alcohol Abuse and Alcoholism report on moderate drinking. *Alcoholism: Clinical & Experimental Research, 28*(6), 829–847.

Gustavson, C. R., & Garcia, J. (1974, August). Pulling a gag on the wily coyote. *Psychology Today,* pp. 68–72.

Guthrie, E., Margison, F., Mackay, H., Chew-Graham, C., Moorey, J., & Sibbald, B. (2004). Effectiveness of psychodynamic interpersonal therapy training for primary care counselors. *Psychotherapy Research, 14*(2), 161–175.

Haier, R. J., Chueh, D., Touchette, P., Lott, I., Buchsbaum, M. S., MacMillan, D., Sandman, C., LaCasse, L., & Sosa, E. (1995). Brain size and cerebral glucose metabolic rate in non-specific mental retardation and Down Syndrome. *Intelligence, 20,* 191–210.

Haith, M. M., & Benson, J. B. (1998). Infant cognition. In W. Damon & R. M. Lerner (Eds.), *Handbook of child psychology (Vol. 1).* New York: Wiley.

Haller, S., Radue, E. W., Erb, M., Grodd, W. & Kircher, T. (2005). Overt sentence production in event-related fMRI. *Neuropsychologia, 43*(5), 807–814.

Halperin, J. M., Schulz, K. P., McKay, K. E., Sharma, V., & Newcorn, J. H. (2003). Familial correlates of central serotonin function in children with disruptive behavior disorders. *Psychiatry Research, 119*(3), 205–216.

Halpern, D. F. (1997). Sex differences in intelligence: Implications for education. *American Psychologist, 52,* 1091–1102.

Halpern, D. F. (1998). Teaching critical thinking for transfer across domains. *American Psychologist, 53,* 449–455.

Halpern, D. F. (2000). Sex differences in cognitive abilities. Hillsdale, NJ: Erlbaum.

Hamer, D. H., & Copeland, P. (1999). *Living with our genes: Why they matter more than you think.* New York: Anchor Books.

Hamid, P. N., & Chan, W. T. (1998). Locus of control and occupational stress in Chinese professionals. *Psychological Reports, 82,* 75–79.

Hamm, J. P., Johnson, B. W., & Corballis, M. C. (2004). One good turn deserves another: An event-related brain potential study of rotated mirror-normal letter discriminations. *Neuropsychologia, 42*(6), 810–820.

Hammack, P. L. (2003). The question of cognitive therapy in a postmodern world. *Ethical Human Sciences & Services, 5*(3), 209–224.

Hammen, C. (2005). Stress and depression. *Annual Review of Clinical Psychology, 1,* 293–319.

Hammond, D. C. (2005). Neurofeedback with anxiety and affective disorders. *Child &*

Adolescent Psychiatric Clinics of North America, 14(1), 105–123.

Haney, C., Banks, C., & Zimbardo, P. (1978). Interpersonal dynamics in a simulated prison. *International Journal of Criminology and Penology, 1,* 69–97.

Hanley, S. J., & Abell, S. C. (2002). Maslow and relatedness: Creating an interpersonal model of self-actualization. *Journal of Humanistic Psychology, 42*(4), 37–56.

Hanna, S. M., & Brown, J. H. (1999). *The practice of family therapy: Key elements across models* (2nd ed.). Belmont, CA: Brooks/Cole.

Hansell, J. H., & Damour, L. K. (2005). *Abnormal psychology.* Hoboken, NJ: Wiley.

Harandi, A. A., Esfandani, A., & Shakibaei, F. (2004). The effect of hypnotherapy on procedural pain and state anxiety related to physiotherapy in women hospitalized in a burn unit. *Contemporary Hypnosis, 21*(1), 28–34.

Harker, L., & Keltner, D. (2001). Expressions of positive emotion in women's college yearbook pictures and their relationship to personality and life outcomes across adulthood. *Journal of Personality and Social Psychology. Jan80*(1), 112–124.

Harlow, H. F., & Harlow, M. K. (1966). Learning to love. *American Scientist, 54,* 244–272.

Harlow, H. F., & Zimmerman, R. R. (1959). Affectional responses in the infant monkey. *Science, 130,* 421–432.

Harlow, H. F., Harlow, M. K., & Meyer, D. R. (1950). Learning motivated by a manipulation drive. *Journal of Experimental Psychology, 40,* 228–234.

Harper, F. D., Harper, J. A., & Stills, A. B. (2003). Counseling children in crisis based on Maslow's hierarchy of basic needs. *International Journal for the Advancement of Counselling, 25*(1), 10–25.

Harris, A. C. (1996). African American and Anglo American gender identities: An empirical study. *Journal of Black Psychology, 22,* 182–194.

Harrison, E. (2005). *How meditation heals: Scientific evidence and practical applications* (2nd ed.). Berkeley, CA: Ulysses Press.

Harvey, M. G., & Miceli, N. (1999). Antisocial behavior and the continuing tragedy of the commons. *Journal of Applied Social Psychology, 29,* 109–138.

Hatfield, E., & Rapson, R. L. (1996). *Love and Sex: Cross-cultural perspectives.* Needham Heights, MA: Allyn & Bacon.

Haugen, R., Ommundsen, Y., & Lund, T. (2004). The concept of expectancy: A central factor in various personality dispositions. *Educational Psychology, 24*(1), 43–55.

Hawkins, R. S., & Hart, A. D. (2003). The use of thermal biofeedback in the treatment of pain associated with endometriosis: Preliminary findings. *Applied Psychophysiology & Biofeedback, 28*(4), 279–289.

Hawkley, L. C., & Cacioppo, J. T. (2004). Stress and the aging immune system. *Brain, Behavior & Immunity, 18*(2), 114–119.

Hay, D. F. (1994). Prosocial development. *Journal of Child Psychology and Psychiatry, 35*, 29–71.

Hayflick, L. (1977). The cellular basis for biological aging. In C. E. Finch & L. Hayflick (Eds.), *Handbook of the biology of aging* (pp. 159–186). New York: Van Nostrand Reinhold.

Hayflick, L. (1996). *How and why we age.* New York: Ballantine Books.

Hays, W. S. T. (2003). Human pheromones: Have they been demonstrated? *Behavioral Ecology & Sociobiology, 54*(2), 89–97.

Hazan, C., & Shaver, P. (1987). Romantic love conceptualized as an attachment process. *Journal of Personality and Social Psychology, 52*, 511–524.

Hazan, C., & Shaver, P. R. (1994). Attachment as an organizational framework for research on close relationships. *Psychological Inquiry, 5*, 1–22.

Healy, A. F., & McNamara, D. S. (1996). Verbal learning and memory: Does the modal model still work? *Annual Review of Psychology, 47*, 143–172.

Heaton, T. B. (2002). Factors contributing to increasing marital stability in the US. *Journal of Family Issues, 23*(3), 392–409.

Hedges, D., & Burchfield, C. (2006). *Mind, brain, and drug: An introduction to psychopharmacology.* Boston: Allyn & Bacon/Longman.

Heffelfinger, A. K., & Newcomer, J. W. (2001). Glucocorticoid effects on memory function over the human life span. *Development & Psychopathology, 13*(3), 491–513.

Heider, F. (1958). *The psychology of interpersonal relations.* New York: Wiley.

Heimann, M., & Meltzoff, A. N. (1996). Deferred imitation in 9- and 14-month-old infants. *British Journal of Developmental Psychology, 14*, 55–64.

Heine, S. J., & Renshaw, K. (2002). Interjudge agreement, self-enhancement, and liking: Cross-cultural divergences. *Personality & Social Psychology Bulletin, 28*(5), 578–587.

Helgeland, M. I., & Torgersen, S. (2004). Developmental antecedents of borderline personality disorder. *Comprehensive Psychiatry, 45*(2), 138–147.

Helms, J. E., & Cook, D. A. (1999). *Using race and culture in counseling and psychotherapy: Theory and process.* Boston: Allyn & Bacon.

Henderlong, J., & Lepper, M. R. (2002). The effects of praise on children's intrinsic motivation: A review and synthesis. *Psychological Bulletin, 128*, 774–795.

Henke, P. G. (1992). Stomach pathology and the amygdala. In J. P. Aggleton (Ed.), *The amygdala: Neurobiological aspects of emotion, memory, and mental dysfunction* (pp. 323–338). New York: Wiley.

Hennessey, B. A., & Amabile, T. M. (1998). Reward, intrinsic motivation, and creativity. *American Psychologist, 53*, 674–675.

Henning, K., & Feder, L. (2004). A comparison of men and women arrested for domestic violence: Who presents the greater threat? *Journal of Family Violence, 19*(2), 69–80.

Hergovich, A., & Olbrich, A. (2003). The impact of the Northern Ireland conflict on social identity, groupthink and integrative complexity in Great Britain. *Review of Psychology, 10*(2), 95–106.

Heriot, S. A., & Pritchard, M. (2004). Test of time: 'Reciprocal inhibition as the main basis of psychotherapeutic effects' by Joseph Wolpe (1954) [Book review]. *Clinical Child Psychology and Psychiatry, 9*(2), 297–307.

Herman, L. M., Richards, D. G., & Woltz, J. P. (1984). Comprehension of sentences by bottlenosed dolphins. *Cognition, 16*, 129–139.

Herrera, E. A., Johnston, C. A., & Steele, R. G. (2004). A comparison of cognitive and behavioral treatments for pediatric obesity. *Children's Health Care, 33*(2), 151–167.

Herrnstein, R. J., & Murray, C. (1994). *The bell curve: Intelligence and class structure in American life.* New York: Free Press.

Hervé, H., Mitchell, D., Cooper, B. S., Spidel, A., & Hare, R. D. (2004). Psychopathy and unlawful confinement: An examination of perpetrator and event characteristics. *Canadian Journal of Behavioural Science, 36*(2), 137–145.

Hewitt, J. K. (1997). The genetics of obesity: What have genetic studies told us about the environment? *Behavior Genetics, 27*, 353–358.

Heyder, K. Suchan, B., & Daum, I. (2004). Cortico-subcortical contributions to executive control. *Acta Psychologica, 115*(2–3), 271–289.

Higgins, N. C., & Bhatt, G. (2001). Culture moderates the self-serving bias: Etic and emic features of causal attributions in India and in Canada. *Social Behavior & Personality, 29*(1), 49–61.

Hilgard, E. R. (1978). Hypnosis and consciousness. *Human Nature, 1*, 42–51.

Hilgard, E. R. (1986). *Divided consciousness: Multiple controls in human thought and action* (expanded ed.). New York: Wiley-Interscience.

Hilgard, E. R. (1992). Divided consciousness and dissociation. *Consciousness and Cognition, 1*, 16–31.

Hill, E. L. (2004). Evaluating the theory of executive dysfunction in autism. *Developmental Review, 24*(2), 189–233.

Hilsenroth, K. J., Ackerman, S. J., Blagys, M. D., Baity, M. R., & Mooney, M. A. (2003). Short-term psychodynamic psychotherapy for depression: An examination of statistical, clinically significant, and technique-specific change. *Journal of Nervous & Mental Disease, 191*(6), 2003, 349–357.

Hinton, E. C., Parkinson, J. A., Holland, A. J., Arana, F. S., Roberts, A. C., & Owen, A. M. (2004). Neural contributions to the motivational control of appetite in humans. *European Journal of Neuroscience, 20*(5), 1411–1418.

Hirao-Try, Y. (2003). Hypertension and women: Gender specific differences. *Clinical Excellence for Nurse Practitioners, 7*(1–2), 4–8.

Hirosumi, J., Tuncman, G., Chang, L., Gorgun, C. Z., Uysal, K. T., Maeda, K., Karin, M., & Hotamisligil, G. S. (2002). Obesity is closely associated with insulin resistance and establishes the leading risk factor for type 2 diabetes mellitus, yet the molecular mechanisms. *Nature, 420*, 333–336.

Hirshkowitz, M., Moore, C. A., & Minhoto, G. (1997). The basics of sleep. In M. R. Pressman & W. C. Orr (Eds.), *Understanding sleep: The evaluation and treatment of sleep disorders.* Washington, DC: American Psychological Association.

Hittner, J. B., & Daniels, J.R. (2002). Gender-role orientation, creative accomplishments and cognitive styles. *Journal of Creative Behavior, 36*(1), 62–75.

Hobson, J. A. (1988). *The dreaming brain.* New York: Basic Books.

Hobson, J. A. (2002). *Dreaming: An introduction to the science of sleep.* New York: Oxford University Press.

Hobson, J. A. (2005). In bed with Mark Solms? What a nightmare! A reply to Domhoff. *Dreaming, 15*(1), 21–29.

Hobson, J. A., & McCarley, R. W. (1977). The brain as a dream state generator: An activation-synthesis hypothesis of the dream process. *American Journal of Psychiatry, 134*, 1335–1348.

Hobson, J. A., & Silvestri, L. (1999). Parasomnias. The Harvard Mental Health Letter, 15(8), 3–5.

Hoek, H. W., van Harten, P. N., Hermans, K. M. E., Katzman, M. A., Matroos, G. E., & Susser, E. S. (2005). The incidence of anorexia nervosa on Curacao. *American Journal of Psychiatry, 162*, 748–752.

Hoffman, M. L. (1993). Empathy, social cognition, and moral education. In A. Garrod (Ed.), *Approaches to moral development: New research and emerging themes.* New York: Teachers College Press.

Hoffman, M. L. (2000). *Empathy and moral development: Implications for caring and justice.* New York: Cambridge University Press.

Hogan, T.P. (2003). *Psychological testing: A practical introduction.* New York: Wiley.

Holland, J. L. (1985). *Making vocational choices: A theory of vocational personalites and work environments* (2nd ed). Englewood Cliffs, NJ: Prentice Hall.

Holland, J. L. (1994). *Self-directed search form R.* Lutz, Fl: Psychological Assessment Resources.

Holmes, T. H., & Rahe, R. H. (1967). The social readjustment rating scale. *Journal of Psychosomatic Research, 11,* 213–218.

Holt, N. L., & Dunn, J. G. H. (2004). Longitudinal idiographic analyses of appraisal and coping responses in sport. *Psychology of Sport & Exercise, 5*(2), 213–222.

Holtzworth-Munroe, A. (2000). A typology of men who are violent toward their female partners: Making sense of the heterogeneity in husband violence. *Current Directions in Psychological Science, 9*(4), 140–143.

Hong, S. (2000). Exercise and psychoneuroimmunology. *International Journal of Sport Psychology, 31*(2), 204–227.

Honts, C. R., & Kircher, J. C. (1994). Mental and physical countermeasures reduce the accuracy of polygraph tests. *Journal of Applied Psychology, 79*(2), 252–259.

Hooley, J. M., & Hiller, J. B. (2001). Family relationships and major mental disorder: Risk factors and preventive strategies. In B. R. Sarason & S. Duck (Eds.), *Personal relationships: implications for clinical and community psychology.* New York: Wiley.

Horiuchi, Y., Nakayama, K., Ishiguro, H., Ohtsuki, T., Detera-Wadleigh, S. D., Toyota, T., Yamada, K., Nankai, M., Shibuya, H., Yoshikawa, T., & Arinami, T. (2004). Possible association between a haplotype of the GABA-A receptor alpha 1 subunit gene (GABRA1) and mood disorders. *Biological Psychiatry, 55*(1), 40–45.

Horney, K. (1939). *New ways in psychoanalysis.* New York: International Universities Press.

Horney, K. (1945). *Our inner conflicts: A constructive theory of neurosis.* New York: Norton.

Hovland, C. I. (1937). The generalization of conditioned responses: II. The sensory generalization of conditioned responses with varying intensities of tone. *Journal of Genetic Psychology, 51,* 279–291.

Howe, M. L., & O'Sullivan, J. T. (1997). What children's memories tell us about recalling our childhoods: A review of storage and retrieval processes in the development of long-term retention. *Developmental Review, 17,* 148–204.

Huang, L. N., & Ying, Y. (1989). Japanese children and adolescents. In J. T. Gibbs & Ln N. Huang (Eds.), *Children of color.* San Francisco: Jossey-Bass.

Huang, M., & Hauser, R. M. (1998). Trends in Black–White test-score differentials: II. The WORDSUM Vocabulary Test. In U. Neisser (Ed.), *The rising curve: Long-term gains in IQ and related measures* (pp. 303–334). Washington, DC: American Psychological Association.

Hubel, D. H., & Wiesel, T. N. (1965). Receptive fields and the functional architecture in two nonstriate visual areas (18 and 19) of the cat. *Journal of Neurophysiology, 28,* 229–289.

Hubel, D. H., & Wiesel, T. N. (1979). Brain mechanisms of vision. *Scientific American, 241,* 150–162.

Huffman, C. J., Matthews, T. D., & Gagne, P. E. (2001). The role of part-set cuing in the recall of chess positions: Influence of chunking in memory. *North American Journal of Psychology, 3*(3), 535–542.

Hull, C. (1952). *A behavior system.* New Haven, CT: Yale University Press.

Hulshoff Pol, H. E., Brans, R. G. H., van Haren, N. E. M., Schnack, H. G., Langen, M., Baare, W. F. C., van Oel, C. J., & Kahn, R. S. (2004). Gray and white matter volume abnormalities in monozygotic and same-gender dizygotic twins discordant for schizophrenia. *Biological Psychiatry, 55*(2), 126–130.

Hvas, L. (2001). Positive aspects of menopause: A qualitative study. *Maturitas, 39*(1), 11–17.

Hyde, J.S., & Lemater, J.D. (2006). *Understanding human sexuality* (9th ed.). New York: McGraw-Hill.

Hyman, R. (1981). Cold reading: How to convince strangers that you know all about them. In K. Fraizer (Ed.), *Paranormal borderlands of science* (pp. 232–244). Buffalo, NY: Prometheus.

Hyman, R. (1996). The evidence for psychic functioning: Claims vs. reality. *Skeptical Inquirer, 20,* 24–26.

Hypericum Depression Trial Study Group (2002). Effect of Hypericum performatum (St John's wort) in major depressive disorder: A randomized controlled trial. *JAMA: Journal of the American Medical Association, 287*(14), 1807–1814.

Iacono, W. G., & Lykken, D. T. (1997). The validity of the lie detector: Two surveys of scientific opinion. *Journal of Applied Psychology, 82*(3), 426–433.

Ibáñez, A., Blanco, C., & Sáiz-Ruiz, J. (2002). Neurobiology and genetics of pathological gambling. *Psychiatric Annals, 32*(3), 181–185.

Ice, G. H., Katz-Stein, A., Himes, J., & Kane, R. L. (2004). Diurnal cycles of salivary cortisol in older adults. *Psychoneuroendocrinology, 29*(3), 355–370.

Ikemoto, K. (2004). Significance of human striatal D-neurons: Implications in neuropsychiatric functions. *Neuropsychopharmacology, 29*(4), 429–434.

Irwin, M., Mascovich, A., Gillin, J. C., Willoughby, R., Pike, J., & Smith, T. L. (1994). Partial sleep deprivation reduced natural killer cell activity in humans. *Psychosomatic Medicine, 56*(6), 493–498.

Itakura, S. (1992). Symbolic association between individuals and objects by a chimpanzee as an initiation of ownership. *Psychological Reports, 70,* 539–544.

Ivanovic, D. M., Leiva, B. P., Pérez, H. T., Olivares, M. G., Díaz, N. S., Urrutia, M. S. C., Almagià, A. F., Toro, T. D., Miller, P. T., Bosch, E. O., & Larraín, C. G. (2004). Head size and intelligence, learning, nutritional status and brain development: Head, IQ, learning, nutrition and brain. *Neuropsychologia, 42*(8), 1118–1131.

Iversen, L. (2003). Cannabis and the brain. *Brain, 126*(6), 1252–1270.

Iyer, P. (2001). *The global soul: Jet lag, shopping malls, and the search for home.* London: Vintage Books.

Jackendoff, R. (2003). Foundations of language-brain, meaning, grammar, evolution. *Applied Cognitive Psychology, 17*(1), 121–122.

Jacob, S., McClintock, M. K., Zelano, B., & Ober, C. (2002). Peternally inherited HLA alleles are associated with women's choice of male odor. *Nature Genetics, DOI: 10.1038/ng830.*

Jacobi, C., Hayward, C., de Zwaan, M., Kraemer, H. C., & Agras, W. S. (2004). Coming to terms with risk factors for eating disorders: Application of risk terminology and suggestions for a general taxonomy. *Psychological Bulletin, 130*(1), 19–65.

Jacobsen, T., & Hofmann, V. (1997). Children's attachment representations: Longitudinal relations to school behavior and academic competency in middle childhood and adolescence. *Developmental Psychology, 33,* 703–710.

Jaffee, S., & Hyde, J. S. (2000). Gender differences in moral orientation: A meta-analysis. *Psychological Bulletin, 126*(5), 703–726.

James, W. (1890). The principles of psychology (Vol. 2). New York: Holt.

Jang, K. L., Stein, M. B., Taylor, S., Asmundson, G. J. G., & Livesley, W. J. (2003). Exposure to traumatic events and experiences: Aetiological relationships with personality function. *Psychiatry Research, 120*(1), 61–69.

Janis, I. L. (1972). *Victims of groupthink: A psychological study of foreign-policy decisions and fiascoes.* Boston: Houghton Mifflin.

Janis, I. L. (1989). *Crucial decisions: Leadership in policymaking and crisis management.* New York: Free Press.

Javo, C., Ronning, J. A., Heyerdahl, S., & Rudmin, F. W. (2004). Parenting correlates of child behavior problems in a multiethnic community sample of preschool children in northern Norway. *European Child & Adolescent Psychiatry, 13*(1), 8–18.

Jellison, W. A., McConnell, A. R., & Gabriel, S. (2004). Implicit and explicit measures of sexual orientation attitudes: Ingroup preferences and related behaviors and beliefs among gay and straight men. *Personality & Social Psychology Bulletin, 30*(5), 629–642.

Jennings, L., & Skovholt, T. M. (1999). The cognitive, emotional, and relational characteristics of master therapists. *Journal of Counseling Psychology, 46*(1), 3–11.

Jiang, Y., Olson, I. R., & Chun, M. M. (2000). Organization of visual short-term memory. *Journal of Experimental Psychology: Learning, memory, and cognition, 26,* 683–702.

John, U., & Hanke, M. (2003). Tobacco- and alcohol-attributable mortality and years of potential life lost in Germany. *European Journal of Public Health, 13*(3), 275–277.

Johnson, T. J., & Cropsey, K. L. (2000). Sensation seeking and drinking game participation in heavy-drinking college students. *Addictive Behaviors, 25*(1), 109–116.

Johnson, W., Bouchard, T. J. Jr., Krueger, R. F., McGue, M., & Gottesman, I. I. (2004). Just one g: Consistent results from three test batteries. *Intelligence, 32*(1), 95–107.

Johnson, W., McGue, M., Krueger, R. F., & Bouchard, T. J. Jr. (2004). Marriage and personality: A genetic analysis. *Journal of Personality and Social Psychology. 86*(2), 285–294.

Johnstone, L. (1999). Adverse psychological effects of ECT. *Journal of Mental Health, 8*(1), 69–85.

Jolliffe, C. D., & Nicholas, M. K. (2004). Verbally reinforcing pain reports: An experimental test of the operant conditioning of chronic pain. *Pain, 107,* 167–175.

Jones, D. G., Anderson, E. R., & Galvin, K. A. (2003). Spinal cord regeneration: Moving tentatively towards new perspectives. *NeuroRehabilitation, 18*(4), 339–351.

Josephs, R. A., Newman, M. L., Brown, R. P., & Beer, J. M. (2003). Status, testosterone, and human intellectual performance: Stereotype threat as status concern. *Psychological Science, 14*(2),158–163.

Jung, C. (1969). The concept of collective unconscious. In *Collected Works* (Vol. 9, Part 1). Princeton, NJ: Princeton University Press (Original work published 1936).

Kagan, J. (1998). Biology and the child. In W. Damon & R. M. Lerner (Eds.), *Handbook of child psychology (Vol. 1).* New York: Wiley.

Kahneman, D. (2003). Experiences of collaborative research. *American Psychologist, 58*(9), 723–730.

Kane, T. A., Loxton, N. J., Staiger, P. K., & Dawe, S. (2004). Does the tendency to act impulsively underlie binge eating and alcohol use problems? An empirical investigation. *Personality & Individual Differences, 36*(1), 83–94.

Kapner, D. A. (2004). *Infofacts resources: Alcohol and other drugs on campus.* [On-line]. Available: http://www.edc.org/hec//hec/pubs/factsheets/scope.html.

Kareev, Y. (2000). Seven (indeed, plus or minus two) and the detection of correlations. *Psychological Review, 107*(2):397–402.

Karmiloff, K, & Karmiloff-Smith, A. (2002). *Pathways to language: From fetus to adolescent.* Cambridge, MA: Harvard University Press.

Karon, B. P., & Widener, A. J. (1998). Repressed memories: The real story. Professional Psychology: *Research & Practice, 29,* 482–487.

Kastenbaum, R. (1999). Dying and bereavement. In J. C. Cavanaugh & S. K. Whitbourne (Eds.), *Gerontology: An interdisciplinary perspective.* New York: Oxford University Press.

Katz, G., Knobler, H. Y., Laibel, Z., Strauss, Z., & Durst, R. (2002). Time zone change and major psychiatric morbidity: The results of a 6-year study in Jerusalem. *Comprehensive Psychiatry, 43*(1), 37–40.

Kaufman, J. C. (2002). Dissecting the golden goose: Components of studying creative writers. *Creativity Research Journal, 14*(1), 27–40.

Kaukiainen, A., Björkqvist, K., Lagerspetz, K., Österman, K., Salmivalli, C., Rothberg, S., et al. (1999). The relationships between social intelligence, empathy, and three types of aggression. *Aggressive Behavior, 25*(2), 81–89.

Kavanau, J. L. (2000). Sleep, memory maintenance, and mental disorders. Journal of Neuropsychiatry & Clinical Neurosciences, 12(2), 199–208.

Kawakami, K., Dovidio, J.F., & Dijksterhuis, A. (2003). Effect of social category priming on personal attitudes. *Psychological Science,* 14, 315–319.

Kazdin, A.E. (1994). Methodology, design, and evaluation in psychotherapy research. In A.E. Bergin & S.L. Garfield (Eds.), *Handbook of psycho- and behavior change* (4th ed.). New York: Wiley.

Keats, D. M. (1982). Cultural bases of concepts of intelligence: A Chinese versus Australian comparison. In P. Sukontasarp, N. Yongsiri, P. Intasuwan, N. Jotiban, & C. Suvannathat (Eds.), *Proceedings of the Second Asian Workshop on Child and Adolescent Development* (pp. 67–75). Bangkok: Burapasilpa Press.

Keefe, F. J., Abernethy, A. P., & Campbell, L. C. (2005). Psychological approaches to understanding and treating disease-related pain. *Annual Review of Psychology,* 56, 601–630.

Keen, S. (1991). *Faces of the enemy: Reflections of the hostile imagination.* New York: HarperCollins.

Keillor, J. M., Barrett, A. M., Crucian, G. P., Kortenkamp, S., & Heilman, K. M. (2002). Emotional experience and perception in the absence of facial feedback. *Journal of the International Neuropsychological Society,* 8, 130–135.

Keltner, D., Kring, A. M., & Bonanno, G. A. (1999). Fleeting signs of the course of life: Facial expression and personal adjustment. *Current Directions in Psychological Science, 8*(1), 18–22.

Kemps, E., Tiggemann, M., Woods, D., & Soekov, B. (2004). Reduction of food cravings through concurrent visuospatial processing. *International Journal of Eating Disorders, 36*(1), 31–40.

Kendler, K. S., Gallagher, T. J., Abelson, J. M., & Kessler, R. C. (1996). Lifetime prevalence, demographic risk factors, and diagnostic validity of nonaffective psychosis as assessed in a U. S. community sample. *Archives of General Psychiatry, 53,* 1022–1031.

Kenealy, P. M. (1997). Mood-state-dependent retrieval: The effects of induced mood on memory reconsidered. *Quarterly Journal of Experimental Psychology: Human Experimental Psychology, 50A,* 290–317.

Kessler, R. C., McGonagle, K. A., Zhao, S., Nelson, C. B., Hughes, M., Eshleman, S., Wittchen, H., & Kendler, K. S. (1994). Lifetime and 12-month prevalence of DSM-III-R psychiatric disorders in the United States. *Archives of General Psychiatry, 51,* 8–19.

Khachaturian, A. S., Corcoran, C. D., Mayer, L. S., Zandi, P. P., Breitner, J. C. S., & Cache County Study Investigators, UT, US. (2004). Apolipoprotein E e4 count affects age at onset of Alzheimer disease, but not lifetime susceptibility: The Cache County Study. *Archives of General Psychiatry, 61*(5), 518–524.

Khandai, P. (2004). Managing stress. *Social Science International, 20*(1), 100–106.

Kieffer, K. M., Schinka, J. A., & Curtiss, G. (2004). Person-environment congruence and personality domains in the prediction of job performance and work quality. *Journal of Counseling.Psychology, 51*(2), 168–177.

Kiehl, K. A., Smith, A. M., Mendrek, A., Forster, B. B., Hare, R.D., & Liddle, P. F. (2004). Temporal lobe abnormalities in semantic processing of criminal psychopaths as revealed by functional magnetic resonance imaging. *Psychiatry Research: Neuroimaging, 130*(1), 27–42.

Kiessling, J., Pichora-Fuller, M. K., Gatehouse, S., Stephen, D., Arlinger, S., Chisolm, T., Davis, A. C., Erber, N. P., Hickson, L., Holmes, A., Rosenhall, U., & von Wedel, H. (2003). Candidature for and delivery of audiological services: Special needs of older people. *International Journal of Audiology, 42*(Suppl2), 2S92–2S101.

Kihlstrom, J. F. (2004). An unbalanced balancing act: Blocked, recovered, and false memories in the laboratory and clinic. *Clinical Psychology: Science & Practice, 11*(1), 34–41.

Kihlstrom, J. F. (2005). Dissociative disorders. *Annual Review of Clinical Psychology, 1*, 227–253.

Kikusui, T., Aoyagi, A., & Kaneko, T. (2000). Spatial working memory is independent of hippocampal CA1 long-term potentiation in rats. *Behavioral Neuroscience, 114*, 700–706.

Killen, M., & Hart, D. (1999). *Morality in everyday life: Developmental perspectives.* New York: Cambridge University Press.

Kim, D-H., Moon, Y-S., Kim, H-S., Jung, J-S, Park, H-M., Suh, H-W., Kim, Y-H., & Song, D-K. (2005). Effect of Zen meditation on serum nitric oxide activity and lipid peroxidation. *Progress in Neuropsychopharmacology & Biological Psychiatry, 29*(2), 327–331.

Kim, J., & Hatfield, E. (2004). Love types and subjective well-being: A cross cultural study. *Social Behavior & Personality, 32*(2), 173–182.

Kim, J-H., Auerbach, J. M., Rodriquez-Gomez, J. A., Velasco, I., Gavin, D., Lumelsky, N., Lee, S-H., Nguyen, J., Sanchez-Pernaute, R., Bankiewicz, K., & McKay, R. (2002). Dopamine neurons derived from embryonic stem cells function in an animal model of Parkinson's disease. *Nature, 418*, 50–56.

Kim, S. U. (2004). Human neural stem cells genetically modified for brain repair in neurological disorders. *Neuropathology, 24*(3), 159–171.

Kimmel, M. S. (2000). The gendered society. London: Oxford University Press.

King, N. J., Clowes-Hollins, V., & Ollendick, T. H. (1997). The etiology of childhood dog phobia. *Behaviour Research and Therapy, 35*, 77.

Kirk, K. M., Bailey, J. M., Dunne, M. P., & Martin, N. G. (2000). Measurement models for sexual orientation in a community twin sample. *Behavior Genetics, 30*(4), 345–356.

Kirkman, C. A. (2002). Non-incarcerated psychopaths: Why we need to know more about the psychopaths who live amongst us. *Journal of Psychiatric & Mental Health Nursing, 9*(2), 155–160.

Kirsch, I., & Braffman, W. (2001). Imaginative suggestibility and hypnotizability. *Current Directions in Psychological Science, 10*(2), 57–61.

Kirsch, I., & Lynn, S. J. (1995). The altered state of hypnosis. *American Psychologist, 50*, 846–858.

Kirschenbaum, H. & Jourdan, A. (2005). The current status of Carl Rogers and the person-centered approach. *Psychotherapy: Theory, Research Practice, Training, 42*(1), 37–51.

Kirschner, S. M., Litwack, T. R., & Galperin, G. J. (2004). The defense of extreme emotional disturbance: A qualitative analysis of cases in New York county. *Psychology, Public Policy, & Law, 10*(1–2), 102–133.

Kisilevsky, B. S., Hains, S. M. J., Lee, K., Xie, X., Huang, H., Ye, H., Zhang, K., & Wang, Z. (2003). Effects of experience on fetal voice recognition. *Psychological Science, 14*(3), 220–224.

Klatzky, R. L. (1984). *Memory and awareness.* New York: Freeman.

Kleinman, A., & Cohen, A. (1997, March). Psychiatry's global challenge. *Scientific American,* 86–89.

Klohnen, E. C., & Bera, S. (1998). Behavioral and experiential patterns of avoidantly and securely attached women across adulthood: A 31-year longitudinal perspective. *Journal of Personality and Social Psychology, 74*(1), 211–223.

Kluger, J., & Masters, C. (2006, August 28). How to spot a liar. *Time,* 46–48.

Knoblauch, K., Vital-Durand, F., & Barbur, J. L. (2000). Variation of chromatic sensitivity across the life span. *Vision Research, 41*(1), 23–36.

Kobasa, S. (1979). Stressful life events, personality, and health: An inquiry into hardiness. *Journal of Personality and Social Psychology, 37*, 1–11.

Kohlberg, L. (1964). Development of moral character and moral behavior. In L. W. Hoffman & M. L. Hoffman (Eds.), *Review of child development research* (Vol. 1). New York: Sage.

Kohlberg, L. (1969). Stage and sequence: The cognitive-developmental approach to socialization. In D. A. Goslin (Ed.), *Handbook of socialization theory and research.* Chicago: Rand McNally.

Kohlberg, L. (1981). *The meaning and measurement of moral development.* Worcester, MA: Clark University Press.

Kohlberg, L. (1984). *The psychology of moral development: Essays on moral development (Vol. 2).* San Francisco: Harper & Row.

Köhler, W. (1925). *The mentality of apes.* New York: Harcourt, Brace.

Kohn, A. (2000). *Punished by rewards: The trouble with gold stars, incentive plans, A's, and other bribes.* New York: Houghton Mifflin.

Komiya, N., Good, G. E., & Sherrod, N. B. (2000). Emotional openness as a predictor of college students' attitudes toward seeking psychological help. *Journal of Counseling Psychology, 47*(1), 138–143.

Koop, C. E., Richmond, J., & Steinfeld, J. (2004). America's choice: Reducing tobacco addiction and disease. *American Journal of Public Health, 94*(2), 174–176.

Korkman, M., Kettunen, S., & Autti-Rämö, I. (2003). Neurocognitive impairment in early adolescence following prenatal alcohol exposure of varying duration. *Child Neuropsychology, 9*(2), 117–128.

Kornstein, S. G. (2002). Chronic depression in women. *Journal of Clinical Psychiatry, 63*(7), 602–609.

Kraaij, V., Arensman, E., & Spinhoven, P. (2002). Negative life events and depression in elderly persons: A meta-analysis. *Journals of Gerontology: Series B: Psychological Sciences & Social Sciences, 57B*(1), 87–94.

Kramer, A. F., Hahn, S., Irwin, D. E., & Theeuwes, J. (2000). Age differences in the control of looking behavior. *Psychological Science, 11*(3), 210–217.

Krantz, D. S., & McCeney, M. K. (2002). Effects of psychological and social factors on organic disease: A critical assessment of research on coronary heart disease. *Annual Review of Psychology, (1),* 341–369.

Krishma, G. (1999). *The dawn of a new science.* Los Angeles: Institute for Consciousness Research.

Kronenberger, W. G., Mathews, V. P., Dunn, D. W., Wang, Y., Wood, E. A., Larsen, J. J., Rembusch, M. E., Lowe, M. J., Giauque, A. L., & Lurito, J. T. (2005). Media violence exposure in aggressive and control adolescents: Differences in self- and parent-reported exposure to violence on television and in video games. *Aggressive Behavior, 31*(3), 201–216.

Kryger, M. H., Walld, R., & Manfreda, J. (2002). Diagnoses received by narcolepsy patients in the year prior to diagnosis by a sleep specialist. *Sleep: Journal of Sleep & Sleep Disorders Research, 25*(1), 36–41.

Kübler-Ross, E. (1983). *On children and death.* New York: Macmillan.

Kübler-Ross, E. (1997). *Death: The final stage of growth.* New York: Simon & Schuster.

Kübler-Ross, E. (1999). *On death and dying.* New York: Simon & Schuster.

Kudo, E., & Numazaki, M. (2003). Explicit and direct self-serving bias in Japan. Reexamination of self-serving bias for success and failure. *Journal of Cross-Cultural Psychology, 34*(5), 511–521.

Kuhn, C., Swartzwelder, S., & Wilson, W. (2003). *Buzzed: The straight facts about the most used and abused drugs from alcohol to ecstasy* (2nd ed.). New York: Norton.

Kulik, L. (2005). Intrafamiliar congruence in gender role attitudes and ethnic stereotypes: The Israeli case. *Journal of Comparative Family Studies, 36*(2), 289–303.

Kumkale, G. T., & Albarracín, D. (2004). The sleeper effect in persuasion: A meta-analytic review. *Psychological Bulletin, 130*(1), 143–172.

Kumpfer, K. L., Alvarado, R., Whiteside, H. O. (2003). Family-based interventions for substance use and misuse prevention. *Substance Use & Misuse, 38*(11–13), 1759–1787.

Kunz, D., & Hermann, W. M. (2000). Sleep-wake cycle, sleep-related disturbances, and sleep disorders: A chronobiological approach. *Comparative Psychology, 41*, 104–105.

Labiano, L. M., & Brusasca, C. (2002). Tratamientos psicológicos en la hipertensión arterial [Psychological treatments in arterial hypertension]. Interdisciplinaria, 19(1), 85–97.

Lachman, M. E. (2004). Development in midlife. *Annual Review of Psychology, 55*, 305–331.

Laine, T., Pekka, J., Ahonen, A., Raesaenen, P., & Tiihonen, J. (2001). Dopamine transporter density and novelty seeking among alcoholics. *Journal of Addictive Diseases, 20*(4), 91–96.

Lakein, A. (1998). *Give me a moment and I'll change your life: Tools for moment management.* New York: Andrews McMeel Publishing.

Lamb, H. R. (2000). Deinstitutionalization and public policy. In R. W. Menninger & J.C. Nemiah (Eds.), *American psychiatry after World War II.* Washington, DC: American Psychiatric Press.

Lamberg, L. (1998). Gay is okay with APA—Forum honors landmark 1973 events. *Journal of the American Medical Association, 280*, 497–499.

Landay, J. S., & Kuhnehenn, J. (2004, July 10). Probe blasts CIA on Iraq data. *The Philadelphia Inquirer,* p. AO1.

Landeira-Fernandez, J. (2004). Analysis of the cold-water restraint procedure in gastric ulceration and body temperature. *Physiology & Behavior, 82*(5), 827–833.

Lane, R. D., Reiman, E. M., Ahern, G. L., & Schwartz, G. E. (1997). Neuroanatomical correlates of happiness, sadness, and disgust. *American Journal of Psychiatry, 154*, 926–933.

Lang, F. R., & Carstensen, L. L. (2002). Time counts: Future time perspective, goals, and social relationships. *Psychology & Aging, 17*(1), 125–139.

Langenecker, S. A., Bieliauskas, L. A., Rapport, L. J., Zubieta, J-K., Wilde, E. A., & Berent, S. (2005). Face emotion perception and executive functioning deficits in depression. *Journal of Clinical & Experimental Neuropsychology, 27*(3), 320–333.

Langlois, J. H., Kalakanis, L., Rubenstein, A. J., Larson, A., Hallam, M., & Smoot, M. (2000). Maxims or myths of beauty? A meta-analytic and theoretical review. *Psychological Bulletin, 126*(3), 390–423.

Larzelere, R. E., & Johnson, B. (1999). Evaluations of the effects of Sweden's spanking ban on physical child abuse rates: A literature review. *Psychological Reports, 85*(2), 381–392.

Latane, B., & Darley, J. M. (1970). *The unresponsive bystander: Why doesn't he help?* New York: Appleton-Century-Crofts.

Lauc, G., Zvonar, K., Vuksic-Mihaljevic, Z., & Flögel, M. (2004). Post-awakening changes in salivary cortisol in veterans with and without PTSD. *Stress & Health: Journal of the International Society for the Investigation of Stress, 20*(2), 99–102.

Lawrence, S. A. (2002). Behavioral interventions to increase physical activity. *Journal of Human Behavior in the Social Environment, 6*(1), 25–44.

Lazar, S. W., Kerr, C. E., Wasserman, R. H., Gray, J. R., Greve, D. N., Treadway, M. T., McGarvey, M., Quinn, B. T., Dusek, J. A., Benson, H., Rauch, S. L., Moore, C. I., Fischl, B. (2005). Meditation experience is associated with increased cortical thickness. *Neuroreport, 16*(17), 1893–1897.

Lazarus, R. S. (1999). *Stress and emotion: A new synthesis.* New York: Springer.

Leaper, C. (2000). Gender, affiliation, assertion, and the interactive context of parent-child play. *Developmental Psychology, 36*(3), 381–393.

Lecrubier, Y., Clerc, G., Didi, R., & Kieser, M. (2002). Efficacy of St. John's wort extract WS 5570 in major depression: A double-blind, placebo-controlled trial. *American Journal of Psychiatry, 159*(8), 1361–1366.

LeDoux, J. (1996a). *The emotional brain: The mysterious underpinnings of emotional life.* New York: Simon & Schuster.

LeDoux, J. E. (1996b). Sensory systems and emotion: A model of affective processing. *Integrative Psychiatry, 4*, 237–243.

LeDoux, J. E. (2002). *Synaptic self: How our brains become who we are.* New York: Viking.

Lee, K. T., Mattson, S. N., & Riley, E. P. (2004). Classifying children with heavy prenatal alcohol exposure using measures of attention. Journal of the International Neuropsychological Society, 10(2), 271–277.

Lefley, H. P. (2000). Cultural perspectives on families, mental illness, and the law. *International Journal of Law and Psychiatry, 23*, 229–243.

Leichtman, M. D., & Ceci, S. J. (1995). The effects of stereotypes and suggestions on preschoolers' reports. *Developmental Psychology, 31*, 568–578.

Lengua, L. J., Wolchik, S. A., Sandler, I. N., & West, S. G. (2000). The additive and interactive effects of parenting and temperament in predicting problems of children of divorce. *Journal of Clinical Child Psychology, 29*(2), 232–244.

Lensvelt-Mulders, G., & Hettema, J. (2001). Analysis of genetic influences on the consistency and variability of the Big Five across different stressful situations. *European Journal of Personality, 15*(5), 355–371.

Leonard, B. E. (2003). *Fundamentals of psychopharmacology* (3rd ed.). New York: Wiley.

Lepper, M. R., Greene, D., & Nisbett, R. E. (1973). Undermining children's intrinsic interest with extrinsic rewards: A test of the overjustification hypothesis. *Journal of Personality and Social Psychology, 28*, 129–137.

Lerner, J. S., Gonzalez, R. M., Small, D. A., & Fischhoff, B. (2003). Effects of fear and anger on perceived risks of terrorism: A national field experiment. *Psychological Science, 14*, 144–150.

Leslie, M. (2000, July/August). The Vexing Legacy of Lewis Terman. *Stanford Magazine.* [On-line]. Available: http://www.stanford alumni.org/news/magazine/2000/julaug/articles/terman.html.

LeVay, S. (2003). Queer science: The use and abuse of research into homosexuality. *Archives of Sexual Behavior, 32*(2),187–189.

Levenson, R. W. (1992). Autonomic nervous system differences among emotions. Psychological Science, 3, 23–27.

Levine, J. R. (2001). *Why do fools fall in love: Experiencing the magic, mystery, and meaning of successful relationships.* New York: Jossey-Bass.

Levinthal, C. (2006). *Drugs, society, and criminal justice.* Boston: Allyn & Bacon.

Lewig, K. A., & Dollard, M. F. (2003). Emotional dissonance, emotional exhaustion and job satisfaction in call centre workers. *European Journal of Work & Organizational Psychology, 12*(4), 366–392.

Lewis, S. (1963). *Dear Shari.* New York: Stein & Day.

Li, N. P., Bailey, J. M., Kenrick, D. T., & Linsenmeier, J. A. W. (2002). The necessities and luxuries of mate preferences: Testing the tradeoffs. *Journal of Personality and Social Psychology, 82*(6), 947–955.

Li, S., Lindenberger, U., Hommel, B., Aschersleben, G., Prinz, W., & Baltes, P. B. (2004). Transformations in the couplings among intellectual abilities and constituent cognitive processes across the life span. *Psychological Science,15*(3), 155–163.

Lieberman, P. (1998). *Eve spoke: Human language and human evolution.* New York: Norton.

Liggett, D. R. (2000). Enhancing imagery through hypnosis: A performance aid for athletes. *American Journal of Clinical Hypnosis, 43*(2), 149–157.

Lillard, A. (1998). Ethnopsychologies: Cultural variations in theories of mind. *Psychological Bulletin, 123,* 3–32.

Linden, E. (1993). Can animals think? *Time,* pp. 54–61.

Lindholm, E., Aberg, K., Ekholm, B., Pettersson, U., Adolfsson, R., & Jazin, E. E. (2004). Reconstruction of ancestral haplotypes in a 12-generation schizophrenia pedigree. *Psychiatric Genetics, 14*(1), 1–8.

Lindsay, D. S., Allen, B. P., Chan, J. C. K., & Dahl, L. C. (2004). Eyewitness suggestibility and source similarity: Intrusions of details from one event into memory reports of another event. *Journal of Memory & Language, 50*(1), 96–111.

Links, P. S., Heslegrave, R., & van Reekum, R. (1998). Prospective follow-up of borderline personality disorder: Prognosis, prediction outcome, and Axis II comorbidity. *Canadian Journal of Psychiatry, 43,* 265–270.

Linzer, M., Gerrity, M., Douglas, J. A., McMurray, J. E., Williams, E. S., Konrad, T. R., & Society of General Medicine Career Satisfaction Study Group (2002). Physician stress: Results from the physician worklife study. *Stress & Health: Journal of the International Society for the Investigation of Stress, 18*(1), 37–42.

Lipsanen, T., Korkeila, J., Peltola, P., Järvinen, J., Langen, K., & Lauerma, H. (2004). Dissociative disorders among psychiatric patients: Comparison with a nonclinical sample. *European Psychiatry, 19*(1), 53–55.

Lissek, S., & Powers, A. S. (2003). Sensation seeking and startle modulation by physically threatening images. *Biological Psychology, 63*(2), 179–197.

Liu, H., Mantyh, P, & Basbaum, A. I. (1997). NMDA-receptor regulation of substance P release from primary afferent nociceptors. *Nature, 386,* 721–724.

Liu, J. H., & Latané, B. (1998). Extremitization of attitudes: Does thought- and discussion-induced polarization cumulate? *Basic and Applied Social Psychology, 20,* 103–110.

Livingston, J. A. (1999). Something old and something new: Love, creativity, and the enduring relationship. *Bulletin of the Menninger Clinic, 63,* 40–52.

Lizarraga, M. L. S., & Ganuza, J. M. G. (2003). Improvement of mental rotation in girls and boys. *Sex Roles, 49*(5–6), 277–286.

Loftus, E. (1982). Memory and its distortions. In A. G. Kraut (Ed.), *The G. Stanley Hall Lecture Series (Vol. 2,* pp. 123–154). Washington, DC: American Psychological Association.

Loftus, E. F. (1993). Psychologists in the eyewitness world. *American Psychologist, 48,* 550–552.

Loftus, E. F. (1997). Memory for a past that never was. *Current Directions in Psychological Science, 6*(3), 60–65.

Loftus, E. F. (2000). Remembering what never happened. In E. Tulving, et al. (Eds.), *Memory, consciousness, and the brain: The Tallinn Conference,* pp. 106–118. Philadelphia: Psychology Press/Taylor & Francis.

Loftus, E. F. (2001). Imagining the past. *Psychologist, 14*(11), 584–587.

Loftus, E., & Ketcham, K. (1994). *The myth of repressed memories: False memories and allegations of sexual abuse.* New York: St. Martin's Press.

Loftus, E., & Polage, D. C. (1999). Repressed memories: When are they real? When are they false? *Psychiatric Clinics of North America, 22,* 61–70.

London, L. H., Tierney, G., Buhin, L., Greco, D. M., & Cooper, C. J. (2002). Kids' college: Enhancing children's appreciation and acceptance of cultural diversity. *Journal of Prevention & Intervention in the Community, 24*(2), 63–78.

Loo, R., & Thorpe, K. (1998). Attitudes toward women's roles in society. *Sex Roles, 39,* 903–912.

Lopez, J. C. (2002). Brain repair: A spinal scaffold. *Nature Reviews Neuroscience, 3,* 256.

Lopez, S. R., & Guarnaccia, P. J. J. (2000). Cultural psychopathology: Uncovering the social world of mental illness. *Annual Review of Psychology, 51,* 571–598.

Lorenz, K. Z. (1937). The companion in the bird's world. *Auk, 54,* 245–273.

Lorenz, K. Z. (1981). *The foundations of ethology.* New York: Springer-Verlag.

Lott, D. A. (2000). *The new flirting game.* London: Sage.

Lu, L., Kao, S-F., Cooper, C. L., & Spector, P. E. (2000). Managerial stress, locus of control, and job strain in Taiwan and UK: A comparative study. *International Journal of Stress Management, 7*(3), 209–226.

Lu, Z. L., Williamson, S. J., & Kaufman, L. (1992). Behavioral lifetime of human auditory sensory memory predicted by physiological measures. *Science, 258,* 1668–1670.

Lubinski, D., & Benbow, C. P. (2000). States of excellence. *American Psychologist, 55*(1), 137–150.

Lubinski, D., & Dawis, R. V. (1992). Aptitudes, skills, and proficiencies. In M. D. Dunnette & L. M. Hough (Eds.), *The handbook of industrial/organizational psychology* (pp. 1–59). Palo Alto, CA: Consulting Psychologists Press.

Lucio-Gómez, E., Ampudia-Rueda, A., Durán Patiño, C., Gallegos-Mejía, L., & León Guzmán, I. (1999). La nueva versión del inventario multifásico de la personalidad de Minnesota para adolescentes mexicanos. [The new version of the Minnesota Multiphasic Personality Inventory for Mexican adolescents]. *Revista Mexicana de Psicología, 16*(2), 217–226.

Luecken, L. J., & Lemery, K. S. (2004). Early caregiving and physiological stress responses. *Clinical Psychology Review, 24*(2), 171–191.

Luecken, L. J., Tartaro, J., & Appelhans, B. (2004). Strategic coping responses and attentional biases. *Cognitive Therapy & Research, 28*(1), 23–37.

Luria, A. R. (1968). *The mind of a mnemonist: A little book about a vast memory.* New York: Basic Books.

Lutz, A., Greischar, L. L., Rawlings, N. B., Ricard, M., & Davidson, R. J. (2004). Long-term mediators self-induce high-amplitude gamma synchrony during mental practice. *Proceedings of the National Academy of Sciences, 101*(46), 16369–16373.

Lynch, P. E. (2003). An interview with Richard C. Pillard, MD. *Journal of Gay & Lesbian Psychotherapy, 7*(4), 63–70.

Lynn, S. J., Vanderhoff, H., Shindler, K., & Stafford, J. (2002). Defining hypnosis as a trance vs. cooperation: Hypnotic inductions, suggestibility, and performance standards. *American Journal of Clinical Hypnosis, 44*(3–4), 231–240.

Lyoo, I. K., Kim, M. J., Stoll, A. L., Demopulos, C. M., Parow, A. M., Dager, S. R., Friedman, S. D., Dunner, D. L., & Renshaw, P. F. (2004). Frontal lobe gray matter density decreases in Bipolar I Disorder. *Biological Psychiatry, 55*(6), 648–651.

Lytle, M. E., Bilt, J. V., Pandav, R. S., Dodge, H. H., & Ganguli, M. (2004). Exercise level and cognitive decline: The MoVIES project. *Alzheimer Disease & Associated Disorders, 18*(2), 57–64.

Lyubimov, N.N. (1992). *Electrophysiological characteristics of sensory processing and mobilization of hidden brain reserves. 2nd Russian-Swedish Symposium New Research in Neurobiology,* Moscow: Russian Academy of Science Institute of Human Brain.

Maas, J. B. (1999). *Power sleep.* New York: HarperPerennial.

Maccoby, E. E. (2000). Parenting and its effects on children: On reading and misreading behavior genetics. *Annual Review of Psychology, 51*, 1–27.

Mackey, R. A., & O'Brien, B. A. (1998). Marital conflict management: Gender and ethnic differences. *Social Work, 43*(2), 128–141.

MacKinnon, D. F., Jamison, K. R., & DePaulo, J. R. (1997). Genetics of manic-depressive illness. *Annual Review of Neuroscience, 10*, 355–373.

Macmillan, M. B. (2000). *An odd kind of fame: Stories of Phineas Gage.* Cambridge, MA: MIT Press.

Maddi, S. R. (2004). Hardiness: An operationalization of existential courage. *Journal of Humanistic Psychology, 44*(3), 279–298.

Maehr, M. L., & Urdan, T. C. (2000). *Advances in motivation and achievement: The role of context.* Greewich, CT: JAI Press.

Magura, S., Laudet, A. B., Mahmood, D., Rosenblum, A., Vogel, H. S., & Knight, E. L. (2003). Role of self-help processes in achieving abstinence among dually diagnosed persons. *Addictive Behaviors, 28*(3), 399–413.

Major, B., & O'Brien, L. T. (2005). The social psychology of stigma. *Annual Review of Psychology, 56*, 393–421.

Major, B., Spencer, S., Schmader, T., Wolfe, C., & Crocker, J. (1998). Coping with negative stereotypes about intellectual performance: The role of psychological disengagement. *Personality & Social Psychology Bulletin, 24*(1), 34–50.

Manly, J. J., Byrd, D., Touradji, P., Sanchez, D., & Stern, Y. (2004). Literacy and cognitive change among ethnically diverse elders. *International Journal of Psychology, 39*(1), 47–60.

Maquet, P., Peigneux, P., Laureys, S., Boly, M., Dang-Vu, T., Desseilles, M., & Cleeremans, A. (2003). Memory processing during human sleep as assessed functional neuroimaging. *Revue Neurologique, 159*(11), 6S27–6S29.

Marcia, J. E. (2002). Identity and psychosocial development in adulthood. *Identity, 2*(1), 7–28.

Markon, K. E., Krueger, R. F., Bouchard, T. J., & Gottesman, I. I. (2002). Normal and abnormal personality traits: Evidence for genetic and environmental relationships in the Minnesota Study of Twins Reared Apart. *Journal of Personality, 70*(5), 661–693.

Markovitz, P. J. (2004). Recent trends in the pharmacotherapy of personality disorders. *Journal of Personality Disorders, 18*(1), 99–101.

Markus, H. R., & Kitayama, S. (1998). The cultural psychology of personality. *Journal of Cross-Cultural Psychology, 29*, 63–87.

Markus, H. R., & Kitayama, S. (2003). Culture, self, and the reality of the social. *Psychological Inquiry, 14*(3–4), 277–283.

Martin, S. E., Snyder, L. B., Hamilton, M., Fleming-Milici, F., Slater, M. D., Stacy, A., Chen, M., & Grube, J. W. (2002). Alcohol advertising and youth. *Alcoholism: Clinical & Experimental Research, 26*(6), 900–906.

Martins, S. S., Mazzotti, G., & Chilcoat, H. D. (2005). Trends in ecstasy use in the United States from 1995 to 2001: Comparison with marijuana users and association with other drug use. *Experimental and Clinical Psychopharmacology, 13*(3), 244–252.

Maslow, A. H. (1954). *Motivation and personality.* New York: Harper & Row.

Maslow, A. H. (1970). *Motivation and personality (2nd ed.).* New York: Harper & Row.

Maslow, A. H. (1999). *Toward a psychology of being* (3rd ed.). New York: Wiley.

Mason, P. T., & Kreger, R. (1998). *Stop walking on eggshells: Taking your life back when someone you care about has borderline personality disorder.* New York: New Harbinger Publishers.

Masters, W. H., & Johnson, V.E. (1961). Orgasm, anatomy of the female. In A. Ellis & A. Abarbonel (Eds.), *Encyclopedia of Sexual Behavior,* Vol. 2. New York: Hawthorn.

Masters, W. H., & Johnson, V. E. (1966). *Human sexual response.* Boston: Little, Brown.

Masters, W. H., & Johnson, V.E. (1970). *Human sexual inadequacy.* Boston: Little, Brown.

Mathews, A., & MacLeod, C. (2005). Cognitive vulnerability to emotional disorders. Annual Review of Clinical Psychology, 1, 167–195.

Matlin, M. W. (2003). From menarche to menopause: Misconceptions about women's reproductive lives. Psychology Science, 45(Suppl2), 106–122.

Matlin, M. W., & Foley, H. J. (1997). *Sensation and perception* (4th ed.). Boston: Allyn and Bacon.

Matsumoto, D. (1992). More evidence for the universality of a contempt expression. *Motivation and Emotion, 16*, 363–368.

Matsumoto, D. (2000). *Culture and psychology: People around the world.* Belmont, CA: Wadsworth.

Matsumoto, D., & Kupperbusch, C. (2001). Idiocentric and allocentric differences in emotional expression, experience, and the coherence between expression and experience. *Asian Journal of Social Psychology, 4*(2), 113–131.

Matusov, E., & Hayes, R. (2000). Sociocultural critique of Piaget and Vygotsky. *New Ideas in Psychology, 18*(2–3), 215–239.

Mayers, A. G., Baldwin, D. S., Dyson, R., Middleton, R. W., & Mustapha, A. (2003). Use of St John's wort (Hypericum perforatum L) in members of a depression self-help organization: A 12-week open prospective pilot study using the HADS scale. *Primary Care Psychiatry, 9*(1), 15–20.

Maynard, A. E., & Greenfield, P. M. (2003). Implicit cognitive development in cultural tools and children: Lessons from Maya Mexico. *Cognitive Development, 18*(4), 489–510.

McAllister-Williams, R. H., & Rugg, M. D. (2002). Effects of repeated cortisol administration on brain potential correlates of episodic memory retrieval. *Psychopharmacology, 160*(1), 74–83.

McCabe, M. P., & Ricciardelli, L. A. (2003). Body image and strategies to lose weight and increase muscle among boys and girls. Health Psychology, 22(1), 39–46.

McCarley, J. S., Kramer, A. F., Colcombe, A. M., & Scialfa, C. T. (2004). Priming of pop-out in visual search: A comparison of young and old adults. *Aging, Neuropsychology, & Cognition, 11*(1), 80–88.

McCartt, A. T., Rohrbaugh, J. W., Hammer, M. C., & Fuller, S. Z. (2000). Factors associated with falling asleep at the wheel among long-distance truck drivers. *Accident Analysis and Prevention, 32*(4), 493–504.

McClelland, D. C. (1958). Risk-taking in children with high and low need for achievement. In J. W. Atkinson (Ed.), *Motives in fantasy, action, and society.* Princeton, NJ: Van Nostrand.

McClelland, D. C. (1987). Characteristics of successful entrepreneurs. *Journal of Creative Behavior, 3*, 219–233.

McClelland, D. C. (1993). Intelligence is not the best predictor of job performance. *Current Directions in Psychological Science, 2*, 5–6.

McClelland, J. L. (1995). Constructive memory and memory distortions: A parallel-distributed processing approach. In D. L. Schachter (Ed.), *Memory distortions: How minds, brains, and societies reconstruct the past* (pp. 69–90). Cambridge: Harvard University Press.

McCrae, R. R. (2004). Human nature and culture: A trait perspective. Journal of Research in Personality, 38(1), 3–14.

McCrae, R. R., & Costa, P. T. Jr. (1990). *Personality in adulthood.* New York: Guilford Press.

McCrae, R. R., & Costa, P. T. Jr. (1999). A five-factor theory of personality. In L. A. Pervin, & O. P. John (Eds.), *Handbook of personality: Theory and research.* New York: Guilford Press.

McCrae, R. R., Costa, P. T. Jr., Hrebíčková, M., Urbánek, T., Martin, T. A., Oryol, V. E., Rukavishnikov, A. A., & Senin, I. G. (2004). Age differences in personality traits across cultures: Self-report and observer perspectives. *European Journal of Personality, 18*(2), 143–157.

McCrae, R. R., Costa, P. T. Jr., Martin, T. A., Oryol, V. E., Rukavishnikov, A. A., Senin, I. G., Hrebíčková, M., & Urbánek, T. (2004). Consensual validation of personality traits across cultures. *Journal of Research in Personality, 38*(2), 179–201.

McCrae, R. R., Costa, P. T., Jr., Ostendorf, F., Angleitner, A., Hrebickova, M., Avia, M. D., Sanz, J., Sanchez-Bernardos, M. L., Kusdil, M. E., Woodfield, R., Saunders, P. R., & Smith, P. B. (2000). Nature over nurture: Temperament, personality, and life span development. *Journal of Personality and Social Psychology, 78*(1), 173–186.

McDonald, J. W., Liu, X. Z., Qu, Y., Liu, S., Mickey, S. K., Turetsky, D., Gottlieb, D. I., & Choi, D. W. (1999). Transplanted embryonic stem cells survive, differentiate, and promote recovery in injured rat spinal cord. *Nature & Medicine, 5,* 1410–1412.

McDougall, W. (1908). *Social psychology.* New York: Putnam's Sons.

McGaugh, J. L., & Roozendaal, B. (2002). Role of adrenal stress hormones in forming lasting memories in the brain. *Current Opinion in Neurobiology, 12*(2), 205–210.

McGrath, P. (2002). Qualitative findings on the experience of end-of-life care for hematological malignancies. *American Journal of Hospice & Palliative Care, 19*(2), 103–111.

McGue, M., Bouchard, T. J., Iacono, W. G., & Lykken, D. T. (1993). Behavioral genetics of cognitive ability: A life-span perspective. In R. Plomin & G. McClearn (Eds.), *Nature, nurture, and psychology.* Washington, DC: American Psychological Association.

McKay, D., Gavigan, C. A., & Kulchycky, S. (2004). Social skills and sex-role functioning in borderline personality disorder: Relationship to self-mutilating behavior. *Cognitive Behaviour Therapy, 33*(1), 27–35.

McKee, P., & Barber, C. E. (2001). Plato's theory of aging. *Journal of Aging & Identity, 6*(2), 93–104.

McKellar, P. (1972). Imagery from the standpoint of introspection. In P. W. Sheehan (Ed.), *The function and nature of imagery.* New York: Academic Press.

McKelvie, S. J., & Drumheller, A. (2001). The availability heuristic with famous names: A replication. *Perceptual & Motor Skills, 92*(2), 507–516.

McKim, W. A. (2002). Drugs and behavior: An introduction to behavioral pharmacology (5th ed). Englewood Cliffs, NJ: Prentice Hall.

McMahon, J. A. (2002). An explanation for normal and anomalous drawing ability and some implications for research on perception and imagery. *Visual Arts Research, 28*(1,Issue55), 38–52.

McNally, R. J. (2004). The science and folklore of traumatic amnesia. *Clinical Psychology: Science & Practice, 11*(1), 29–33.

Medina, J. J. (1996). *The clock of ages: Why we age.* Cambridge, MA: Cambridge University Press.

Meeus, W., & Raaijmakers, Q. (1989). Autoritätsgehorsam in Experimenten des Milgram-Typs: Eine Forschungsübersicht [Obedience to authority in Milgram-type studies: A research review]. *Zeitschrift für Sozialpsychologie, 20*(2), 70–85.

Meltzer, G. (2000). Genetics and etiology of schizophrenia and bipolar disorder. *Biological Psychiatry, 47*(3), 171–178.

Meltzer, L. (2004). Resilience and learning disabilities: Research on internal and external protective dynamics. *Learning Disabilities Research & Practice,19*(1), 1–2.

Meltzoff, A. N., & Moore, M. K. (1977). Imitation of facial and manual gestures by human neonates. *Science, 198,* 75–78.

Meltzoff, A. N., & Moore, M. K. (1985). Cognitive foundations and social functions of imitation and intermodal representation in infancy. In J. Mehler & R. Fox (Eds.), *Neonate cognition: Beyond the blooming buzzing confusion.* Hillsdale, NJ: Erlbaum.

Meltzoff, A. N., & Moore, M. K. (1994). Imitation, memory, and the representation of persons. *Infant Behavior and Development, 17,* 83–99.

Melzack, R. (1999). Pain and stress: A new perspective. In R. J. Gatchel & D. C. Turk (Eds.), *Psychosocial factors in pain: Critical perspectives.* New York: Guilford Press.

Melzack, R., & Wall, P. D. (1965). Pain mechanisms: A new theory. *Science, 150,* 971–979.

Mendrek, A., Kiehl, K. A., Smith, A. M., Irwin, D., Forster, B. B., & Liddle, P. F. (2005). Dysfunction of a distributed neural circuitry in schizophrenia patients during a working-memory performance. *Psychological Medicine, 35*(2), 187–196.

Messinger, L.M. (2001). *Georgia O'Keeffe.* London: Thames & Hudson.

Metcalf, P., & Huntington, R. (1991). *Celebrations of death: The anthropology of mortuary ritual* (2nd ed.). Cambridge, England: Cambridge University Press.

Meyer, I. H. (2003). Prejudice, social stress, and mental health in lesbian, gay, and bisexual populations: Conceptual issues and research evidence. *Psychological Bulletin, 129*(5), 674–697.

Migueles, M, & Garcia-Bajos, E. (1999). Recall, recognition, and confidence patterns in eyewitness testimony. *Applied Cognitive Psychology, 13,* 257–268.

Miles, D. R., & Carey, G. (1997). Genetic and environmental architecture on human aggression. *Journal of Personality and Social Psychology, 72,* 207–217.

Milgram, S. (1963). Behavioral study of obedience. *Journal of Abnormal and Social Psychology, 67,* 371–378.

Milgram, S. (1974). *Obedience to authority: An experimental view.* New York: Harper & Row.

Miller, G. (2004). Brain cells may pay the price for a bad night's sleep. *Science, 306*(5699), 1126.

Miller, G. A. (1956). The magical number seven, plus or minus two: Some limits on our capacity for processing information. *Psychological Review, 63,* 81–97.

Miller, J. G., & Bersoff, D. M. (1998). The role of liking in perceptions of the moral responsibility to help: A cultural perspective. *Journal of Experimental Social Psychology, 34,* 443–469.

Millon, T. (2004). *Masters of the mind: Exploring the story of mental illness from ancient times to the new millennium.* Hoboken, NJ: Wiley.

Mills, T. C., Paul, J., Stall, R., Pollack, L., Canchola, J., Chang, Y. J., Moskowitz, J. T., & Catania, J. A. (2004). Distress and depression in men who have sex with men: The urban men's health study. *American Journal of Psychiatry, 161*(2), 278–285.

Milton, J., & Wiseman, R. (1999). Does psi exist? Lack of replication of an anomalous process of information transfer. *Psychological Bulletin, 125,* 387–391.

Milton, J., & Wiseman, R. (2001). Does psi exist? Reply to Storm and Ertel 2000. *Psychological Bulletin, 127* (3), 434–438.

Mineka, S., & Oehman, A. (2002). Phobias and preparedness: The selective, automatic, and encapsulated nature of fear. *Biological Psychiatry, 51*(9), 927–937.

Mingo, C., Herman, C. J., & Jasperse, M. (2000). Women's stories: Ethnic variations in women's attitudes and experiences of menopause, hysterectomy, and hormone replacement therapy. *Journal of Women's Health and Gender Based Medicine, 9,* S27-S38.

Miracle, T. S., Miracle, A. W., & Baumeister, R. F. (2006). *Human sexuality: Meeting your basic needs* (2nd ed.). Upper Saddle River, NJ: Pearson Education.

Mita, T. H., Dermer, M., & Knight, J. (1977). Reversed facial images and the mere-exposure hypothesis. *Journal of Personality and Social Psychology, 35*(8), 597–601.

Mitchell, R. (2003). Ideological reflections on the DSM-IV-R (or pay no attention to that man behind the curtain, Dorothy!). *Child & Youth Care Forum, 32*(5), 281–298.

Mitchell, S. A. (2002). Psychodynamics, homosexuality, and the question of pathology. *Studies in Gender & Sexuality, 3*(1), 3–21.

Mitchell, T. W., Mufson, E. J., Schneider, J. A., Cochran, E. J., Nissanov, J., Han, L., et al. (2002). Parahippocampal tau pathology in healthy aging, mild cognitive impairment, and early Alzhiemer's disease. *Annals of Neurology, 51*(2), 182–189.

Mittag, O., & Maurischat, C. (2004). Die Cook-Medley Hostility Scale (Ho-Skala) im Vergleich zu den Inhaltsskalen "Zynismus", "Ärger" sowie "Typ A" aus dem MMPI-2: Zur zukünftigen Operationalisierung von Feindseligkeit [A comparison of the Cook-Medley Hostility Scale (Ho-scale) and the content scales "cynicism," "anger," and "type A" out of the MMPI-2: On the future assessment of hostility]. *Zeitschrift für Medizinische Psychologie, 13*(1), 7–12.

Moffitt, T. E., Caspi, A., Harrington, H., & Milne, B. J. (2002). Males on the life-course-persistent and adolescence-limited antisocial pathways: Follow-up at age 26 years. *Development & Psychopathology, 14*(1), 179–207.

Moneta, G. B., & Siu, C. M. Y. (2002). Trait intrinsic and extrinsic motivations, academic performance, and creativity in Hong Kong college students. *Journal of College Student Development, 43*(5), 664–683.

Mong, J. A., & Pfaff, D. W. (2003). Hormonal and genetic influences underlying arousal as it drives sex and aggression in animal and human brains. *Neurobiology of Aging, 24*(Suppl1), S83-S88.

Monin, B. (2003). The warm glow heuristic: When liking leads to familiarity. *Journal of Personality and Social Psychology, 85*(6), 1035–1048.

Monte, C. F., & Sollod, R.N., (2003). Beneath the mask: An introduction to theories of personality. Hoboken, NJ: Wiley.

Monteleone, P., Martiadis, V., Fabrazzo, M., Serritella, C., & Maj, M. (2003). Ghrelin and leptin responses to food ingestion in bulimia nervosa: Implications for binge-eating and compensatory behaviours. *Psychological Medicine, 33*(8), 1387–1394.

Montgomery, G. H., Weltz, C. R., Seltz, M., & Bovbjerg, D. H. (2002). Brief presurgery hypnosis reduces distress and pain in excisional breast biopsy patients. *International Journal of Clinical & Experimental Hypnosis, 50*(1), 17–32.

Montuori, A., & Fahim, U. (2004). Cross-cultural encounter as an opportunity for personal growth. *Journal of Humanistic Psychology. 44*(2), 243–265.

Moore, M. M. (1998). The science of sexual signaling. In G. C. Brannigan, E. R. Allgeier, & A. R. Allgeier (Eds.), *The sex scientists* (pp. 61–75). New York: Longman.

Moore, S., Grunberg, L., & Greenberg, E. (2004). Repeated downsizing contact: The effects of similar and dissimilar layoff experiences on work and well-being outcomes. *Journal of Occupational Health Psychology, 9*(3), 247–257.

Morgan, M. J., McFie, L., Fleetwood, L. H., & Robinson, J. A. (2002). Ecstasy (MDMA): Are the psychological problems associated with its use reversed by prolonged abstinence? *Psychopharmacology, 159*(3), 294–303.

Morrison, C., & Westman, A. S. (2001). Women report being more likely than men to model their relationships after what they have seen on TV. *Psychological Reports, 89*(2), 252–254.

Morrison, M. F., & Tweedy, K. (2000). Effects of estrogen on mood and cognition in aging women. *Psychiatric Annals, 30*(2), 113–119.

Moss, D. (2004). Biofeedback. Applied Psychophysiology & Biofeedback, 29(1), 75–78.

Mueller, C. M., & Dweck, C. S. (1998). Praise for intelligence can undermine children's motivation and performance. *Journal of Personality and Social Psychology, 75*, 33–52.

Murakami, K., & Hayashi, T. (2002). Interaction between mind-heart and gene. *Journal of International Society of Life Information Science, 20*(1), 122–126.

Muris, P., Merckelbach, H., Gadet, B., & Moulaert, V. (2000). Fears, worries, and scary dreams in 4- to 12-year-old children: Their content, developmental pattern, and origins. *Journal of Clinical Child Psychology, 29*(1), 43–52.

Murray, H. A. (1938). *Explorations in personality.* New York: Oxford University Press.

Myers, L. B., & Vetere, A. (2002). Adult romantic attachment styles and health-related measures. *Psychology, Health & Medicine, 7*(2), 175–180.

Myerson, J., Rank, M. R., Raines, F. Q., & Schnitzler, M. A. (1998). Race and general cognitive ability: The myth of diminishing returns to education. *Psychological Science, 9*, 139–142.

Naglieri, J. A., & Ronning, M. E. (2000). Comparison of White, African American, Hispanic, and Asian children on the Naglieri Nonverbal Ability Test. *Psychological Assessment, 12*(3), 328–334.

Nahas, G. G., Frick, H. C., Lattimer, J. K., Latour, C., & Harvey, D. (2002). Pharmacokinetics of THC in brain and testis, male gametotoxicity and premature apoptosis of spermatozoa. *Human Psychopharmacology: Clinical & Experimental., 17*(2), 103–113.

Nakajima, S., & Masaki, T. (2004). Taste aversion learning induced forced swimming in rats. *Physiology & Behavior, 80*(5), 623–628.

Nathan, R., Rollinson, L., Harvey, K., & Hill, J. (2003). The Liverpool violence assessment: An investigator-based measure of serious violence. *Criminal Behaviour & Mental Health, 13*(2), 106–120.

National Center for Health Statistics. (2004). *New report examines Americans' health behaviors.* [On-line]. Available: http://gov/nchs/pressroom/04facts/healthbehaviors.htm.

National Institute of Mental Health. (2001). *Facts about anxiety disorders.* [On-line]. Available: http://www.nimh.nih.gov/publicat/adfacts.cfm.

National Institute on Drug Abuse. (2005). *NIDA infofacts: Club drugs.* [On-line]. Available: http://www.nida.nih.gov/infofacts/club drugs.html.

National Organization on Fetal Alcohol Syndrome, 2006, What are the statistics and facts about FAS and FASD? (November 3, 2006). http://www.nofas.org/faqs.aspx?id=12.

National Research Council. (2003). *The polygraph and lie detection.* [On-line]. Available: http://fermat.nap.edu/catalog/10420.html.

National Sleep Foundation. (2006). *Myths and facts about sleep.* [On-line]. Available: http://www.sleepfoundation.org/hottopics/index.php?secid=9&id=33.

Nayda, R. (2002). Influences on registered nurses' decision-making in cases of suspected child abuse. *Child Abuse Review, 11*(3), 168–178.

Neale, J. M., Oltmanns, T. F., & Winters, K. C. (1983). Recent developments in the assessment and conceptualization of schizophrenia. *Behavioral Assessment, 5*, 33–54.

Neese, R. M. (2000). Is depression an adaptation? *Archives of General Psychiatry, 57*, 14–20.

Neher, A. (1991). Maslow's theory of motivation: A critique. *Journal of Humanistic Psychology, 31*, 89–112.

Neisser, U. (1967). *Cognitive psychology.* New York: Appleton-Century-Crofts.

Neisser, U. (1998). Introduction: Rising test scores and what they mean. In U. Neisser (Ed.), *The rising curve: Long-term gains in IQ and related measures.* Washington, DC: American Psychological Association.

Nelson, G., Chandrashekar, J., Hoon, M.A., Feng, L., Zhao, G., Ryba, N.J.P., & Zuker, C.S. (2002). An amino-acid taste receptor. *Nature, 416*, 199–202.

Nelson, R. J., & Chiavegatto, S. (2001). Molecular basis of aggression. *Trends in Neurosciences, 24*(12), 713–719.

Neria, Y., Bromet, E. J., Sievers, S., Lavelle, J., & Fochtmann, L. J. (2002). Trauma exposure and posttraumatic stress disorder in psychosis: Findings from a first-admission cohort. *Journal of Consulting & Clinical Psychology, 70*(1), 246–251.

Neto, F., & Furnham, A. (2005). Gender-role portrayals in children's television advertisements. *International Journal of Adolescence & Youth, 12*(1–2), 69–90.

Neubauer, A. C., Grabner, R. H., Freudenthaler, H. H., Beckmann, J. F., & Guthke, J. (2004). Intelligence and individual differences in becoming neurally efficient. *Acta Psychologica, 116*(1), 55–74.

Neugarten, B. L., Havighurst, R. J., & Tobin, S. S. (1968). The measurement of life satisfaction. *Journal of Gerontology, 16,* 134–143.

Neumann, K., Preibisch, C., Euler, H. A., von Gudenberg, A. W., Lanfermann, H., Gall, V., & Giraud, A-L. (2005). Cortical plasticity associated with stuttering therapy. *Journal of Fluency Disorders, 30*(1), 23–39.

Neumark-Sztainer, D., Wall, M. M., Story, M., & Perry, C. L. (2003). Correlates of unhealthy weight-control behaviors among adolescents: Implications for prevention programs. *Health Psychology, 22*(1), 88–98.

Neuringer, A., Deiss, C., & Olson, G. (2000). Reinforced variability and operant learning. *Journal of Experimental Psychology: Animal Behavior Processes, 26*(1), 98–111.

Newman, L. S., Duff, K., & Baumeister, R. (1997). A new look at defensive projection: Thought suppression, accessibility, and biased person perception. *Journal of Personality and Social Psychology, 72,* 980–1001.

Niaura, R., Todaro, J.F., Stroud, L., Spiro, A., Ward, K.D., & Weiss, S. (2002). Hostility, the metabolic syndrome, and incident cornary heart disease. *Health Psychology, 21* (6), 598–593.

Nickerson, R. (1998). Confirmation bias: A ubiquitous phenomenon in many guises. *Review of General Psychology, 2,* 175–220.

Nickerson, R. S., & Adams, M. J. (1979). Long-term memory for a common object. *Cognitive Psychology, 11,* 287–307.

Nishimoto, R. (1988). A cross-cultural analysis of psychiatric symptom expression using Langer's twenty-two item index. *Journal of Sociology and Social Welfare, 15,* 45–62.

Noble, E. P. (2000). Addiction and its reward process through polymorphisms of the D-sub-2 dopamine receptor gene: A review. *European Psychiatry, 15*(2), 79–89.

Noël, M., Désert, M., Aubrun, A., & Seron, X. (2001). Involvement of short-term memory in complex mental calculation. *Memory & Cognition, 29*(1), 34–42.

Nolen-Hoeksema, S., Larson, J., & Grayson, C. (2000). Explaining the gender difference in depressive symptoms. *Journal of Personality and Social Psychology, 77,* 1061–1072.

Nordhus, I. H., & Pallesen, S. (2003). Psychological treatment of late-life anxiety: An empirical review. *Journal of Consulting & Clinical Psychology, 71*(4), 643–651.

Norenzayan, A., & Nisbett, R. E. (2000). Culture and causal cognition. Current Directions in Psychological Science, 9(4), 132–135.

Núñez, J. P., & de Vicente, F. (2004). Unconscious learning. Conditioning to subliminal visual stimuli. *Spanish Journal of Psychology, 7*(1), 13–28.

O'Connor, D. B., Archer, J., Hair, W. M., Wu, F. C. W. (2002). Exogenous testosterone, aggression, and mood in eugonadal and hypogonadal men. *Physiology & Behavior, 75*(4), 557–566.

O'Sullivan, M. (2003). The fundamental attribution error in detecting deception: The boy-who-cried-wolf effect. *Personality & Social Psychology Bulletin, 29*(10), 1316–1327.

Oakes, M. A., & Hyman, I. E., Jr. (2001). The role of the self in false memory creation. *Journal of Aggression, Maltreatment & Trauma, 4*(2), 87–103.

Oeztuerk, L., Tufan, Y., & Gueler, F. (2002). Self-reported traffic accidents and sleepiness in a professional group of Turkish drivers. *Sleep & Hypnosis, 4*(3), 106–110.

Ohayon, M. M., & Schatzberg, A. F. (2002). Prevalence of depressive episodes with psychotic features in the general population. *American Journal of Psychiatry, 159*(11), 1855–1861.

Olds, J., & Milner, P. M. (1954). Positive reinforcement produced by electrical stimulation of septal area and other regions of rat brains. *Journal of Comparative and Physiological Psychology, 47,* 419–427.

Olfson, M., Marcus, S., Pincus, H. A., Zito, J. M., Thompson, J. W., & Zarin, D. A. (1998). Antidepressant prescribing practices of outpatient psychiatrists. *Archives of General Psychiatry, 55,* 310, 316.

Oliver, R. J. (2002). Tobacco abuse in pregnancy. *Journal of Prenatal Psychology & Health, 17*(2), 153–166.

Olson, R. K. (2004). SSSR, environment, and genes. *Scientific Studies of Reading, 8*(2), 111–124.

Oppenheimer, D. M. (2004). Spontaneous discounting of availability in frequency judgment tasks. *Psychological Science, 15*(2), 100–105.

Oshima, K. (2000). Ethnic jokes and social function in Hawaii. Humor: *International Journal of Humor Research, 13*(1), 41–57.

Ostrov, J. M., & Keating, C. F. (2004). Gender differences in preschool aggression during free play and structured interactions: An observational study. *Social Development, 13*(2), 255–277.

Overmier, J. B., & Murison, R. (2000). Anxiety and helplessness in the face of stress predisposes, precipitates, and sustains gastric ulceration. *Behavioural Brain Research, 110*(1–2), 161–174.

Owens, M. J. (2004). Selectivity of antidepressants: From the monoamine hypothesis of depression to the SSRI revolution and beyond. *Journal of Clinical Psychiatry, 65,* 5–10.

Ozinga, J. A. (2000). *Altruism.* New York: Praeger Pub.

Paice, E., Rutter, H., Wetherell, M., Winder, B., & McManus, I. C. (2002). Stressful incidents, stress and coping strategies in the pre-registration house officer year. *Medical Education, 36*(1), 56–65.

Palfai, T. P., Monti, P. M., Ostafin, B., & Hutchinson, K. (2000). Effects of nicotine deprivation on alcohol-related information processing and drinking behavior. *Journal of Abnormal Psychology, 109,* 96–105.

Panksepp, J. (2005). Affective consciousness: Core emotional feelings in animals and humans. *Consciousness & Cognition: An International Journal, 14*(1), 30–80.

Pappas, N. T., McKenry, P. C., & Catlett, B. S. (2004). Athlete aggression on the rink and off the ice: Athlete violence and aggression in hockey and interpersonal relationships. *Men & Masculinities, 6*(3), 291–312.

Paquet, F., Soucy, J. P., Stip, E., Lévesque, M., Elie, A., & Bédard, M. A. (2004). Comparison between olanzapine and haloperidol on procedural learning and the relationship with striatal D-sub-2 receptor occupancy in schizophrenia. *Journal of Neuropsychiatry & Clinical Neurosciences, 16*(1), 47–56.

Paquier, P. F., & Mariën, P. (2005). A synthesis of the role of the cerebellum in cognition. *Aphasiology, 19*(1), 3–19.

Paris, J. (2000). Childhood precursors of personality disorder. *Psychiatric Clinics of North America, 23,* 77–88.

Paris, J. (2004). Half in love with easeful death: The meaning of chronic suicidality in borderline personality disorder. *Harvard Review of Psychiatry, 12*(1), 42–48.

Parker, G. B., & Brotchie, H. L. (2004). From diathesis to dimorphism: The biology of gender differences in depression. *Journal of Nervous & Mental Disease, 192*(3), 210–216.

Parker-Oliver, D. (2002). Redefining hope for the terminally ill. *American Journal of Hospice & Palliative Care, 19*(2), 115–120.

Parrott, A., Morinan, A., Moss, M., & Scholey, A. (2004). *Understanding drugs and behavior.* Hoboken, NJ: Wiley.

Pasternak, T. & Greenlee, M. W. (2005). Working memory in primate sensory systems. *Nature Reviews Neuroscience, 6*(2), 97–107.

Patterson, F., & Linden, E. (1981). *The education of Koko.* New York: Holt, Rinehart and Winston.

Patterson, J. M., Holm, K. E., & Gurney, J. G. (2004). The impact of childhood cancer on the family: A qualitative analysis of strains, resources, and coping behaviors. *Psycho-Oncology, 13*(6), 390–407.

Patterson, P. (2002). *Penny's journal: Koko wants to have a baby.* Available: http://www.koko.org/world/journal.phtml?offset=8.

Paul, D. B., & Blumenthal, A. L. (1989). On the trail of little Albert. *The Psychological Record, 39,* 547–553.

Pearlman, C. (2002). Electroconvulsive therapy in clinical psychopharmacology. *Journal of Clinical Psychopharmacology, 22*(4), 345–346.

Peleg, G., Katzir, G., Peleg, O., Kamara, M., Brodsky, L., Hel-Or, H., et al. (2006). Hereditary family signature of facial expression. *Proceedings of the National Academy of Sciences, 103,* 15921–15926.

Pelletier, J. G., & Paré, D. (2004). Role of amygdala oscillations in the consolidation of emotional memories. *Biological Psychiatry, 55*(6), 559–562.

Penzien, D. B., Rains, J., C., & Andrasik, F. (2002). Behavioral management of recurrent headache: Three decades of experience and empiricism. *Applied Psychophysiology & Biofeedback, 27*(2), 163–181.

Peppard, P. E., Young, T., Palta, M., & Skatrud, J. (2000). Prospective study of the association between sleep-disordered breathing and hypertension. *New England Journal of Medicine, 342*(19), 1378–1384.

Peretti, P. O., & Abplanalp, R. R. Jr. (2004). Chemistry in the college dating process: Structure and function. *Social Behavior & Personality, 32*(2), 147–154.

Perry, C., Orne, M. T., London, R. W., & Orne, E. C. (1996). Rethinking per se exclusions of hypnotically elicited recall as legal testimony. *International Journal of Clinical & Experimental Hypnosis, 44*(1), 66–81.

Peterson, C., & Vaidya, R. S. (2001). Explanatory style, expectations, and depressive symptoms. *Personality & Individual Differences, 31*(7), 1217–1223.

Petronis, A. (2000). The genes for major psychosis: Aberrant sequence or regulation? *Neuropsychopharmacology, 23,* 1–12.

Pettigrew, T. F. (1998). Reactions towards the new minorities of Western Europe. *Annual Review of Sociology, 24,* 77–103.

Pfeffer, C. R., Jiang, H., Kakuma, T., Hwang, J., & Metsch, M. (2002). Group intervention for children bereaved the suicide of a relative. *Journal of the American Academy of Child & Adolescent Psychiatry, 41*(5), 505–513.

Pham, T.M., Winblad, B., Granholm, A-C., & Mohammed, A. H. (2002). Environmental influences on brain neurotrophins in rats. *Pharmacology, Biochemistry & Behavior, 73*(1), 167–175.

Philipsen, M. I. (2003). Race, the college classroom, and service learning: A practitioner's tale. *Journal of Negro Education, 72*(2), 230–240.

Phillips, S. T., & Ziller, R. C. (1997). Toward a theory and measure of the nature of nonprejudice. *Journal of Personality and Social Psychology, 72,* 420–434.

Piaget, J. (1962). *Play, Dreams and Imitation in Childhood.* New York: Norton.

Pich, E. M., Pagliusi, S. R., Tessari, M., Talabot-Ayer, D., Van Huijsduijnen, R. H., & Chiamulera, C. (1997). Common neural substrates for the addictive properties of nicotine and cocaine. *Science, 275,* 83–86.

Pickering, A., Farmer, A., & McGuffin, P. (2004). The role of personality in childhood sexual abuse. *Personality & Individual Differences, 36*(6), 1295–1303.

Pierce, J. D. Jr., Cohen, A. B., & Ulrich, P. M. (2004). Responsivity to two odorants, androstenone and amyl acetate, and the affective impact of odors on interpersonal relationships. *Journal of Comparative Psychology, 118*(1), 14–19.

Pietromonaco, P. R., & Carnelley, K. B. (1994). Gender and working models of attachment: Consequences for perception of self and romantic relationships. *Personal Relationships, 1,* 3–26.

Plomin, R. (1990). The role of inheritance in behavior. *Science, 248,* 183–188.

Plomin, R. (1997, May). Cited in B. Azar, Nature, nurture: Not mutually exclusive. *APA Monitor,* p. 32.

Plomin, R. (1999). Genetics and general cognitive ability. *Nature, 402,* C25-C29.

Plomin, R., & Crabbe, J. (2000). DNA. *Psychological Bulletin, 126,* 806–828.

Plummer, D. C. (2001). The quest for modern manhood: Masculine stereotypes, peer culture and the social significance of homophobia. *Journal of Adolescence, 24*(1), 15–23.

Plutchik, R. (1984). Emotions: A general psychoevolutionary theory. In K. R. Scherer, & P. Ekman (Eds.), *Approaches to emotion.* Hillsdale, NJ: Erlbaum.

Plutchik, R. (1994). *The psychology and biology of emotion.* New York: HarperCollins.

Plutchik, R. (2000). *Emotions in the practice of psychotherapy: Clinical implications of affect theories.* Washington, DC: American Psychological Association.

Pomponio, A. T. (2001). *Psychological consequences of terror.* New York: Wiley.

Pontius, A. A. (2005). Fastest fight/flight reaction via amygdalar visual pathway implicates simple face drawing as its marker: Neuroscientific data consistent with neuropsychological findings. *Aggression & Violent Behavior, 10*(3), 363–373.

Porter, J. F., Spates, C. R., & Smitham, S. (2004). Behavioral activation group therapy in public mental health settings: A pilot investigation. *Professional Psychology: Research & Practice, 35*(3), 297–301.

Porzelius, L. K., Dinsmore, B. D., & Staffelbach, D. (2001). Eating disorders. In M. Hersen & V. B. Van Hasselt (Eds.), *Advanced abnormal psychology* (2nd ed.). Netherlands: Klewer Academic Publishers.

Posthuma, D., de Geus, E. J. C., & Boomsma, D. I. (2001). Perceptual speed and IQ are associated through common genetic factors. *Behavior Genetics, 31*(6), 593–602.

Posthuma, D., de Geus, E. J. C., Baare, W. E. C., Hulshoff Pol, H. E., Kahn, R. S., & Boomsma, D. I. (2002). The association between brain volume and intelligence is of genetic origin. *Nature Neuroscience, 5*(2), 83–84.

Posthuma, D., Neale, M. C., Boomsma, D. I., & de Geus, E. J. C. (2001). Are smarter brains running faster? Heritability of alpha peak frequency, IQ, and their interrelation. *Behavior Genetics, 31*(6), 567–579.

Powell, D.H. (1998), *The nine myths of aging: Maximizing the quality of later life.* San Francisco: Freeman.

Powell-Hopson, D., & Hopson, D. S. (1988). Implications of doll color preferences among Black preschool children and White preschool children. *Journal of Black Psychology, 14,* 57–63.

Pratt, M. W., Skoe, E. E., & Arnold, M. (2004). Care reasoning development and family socialisation patterns in later adolescence: A longitudinal analysis. *International Journal of Behavioral Development, 28*(2), 139–147.

Preuss, U. W., Zetzsche, T., Jäger, M., Groll, C., Frodl, T., Bottlender, R., Leinsinger, G., Hegerl, U., Hahn, K., Möller, H. J., & Meisenzahl, E. M. (2005). Thalamic volume in first-episode and chronic schizophrenic subjects: A volumetric MRI study. *Schizophrenia Research, 73*(1), 91–101.

Primavera, L. H., & Herron, W. G. (1996). The effect of viewing television violence on aggression. *International Journal of Instructional Media, 23,* 91–104.

Pring, L., & Hermelin, B. (2002). Numbers and letters: Exploring an autistic savant's unpractised ability. *Neurocase, 8*(4), 330–337.

Prinstein, M. J., & Aikins, J. W. (2004). Cognitive moderators of the longitudinal association between peer rejection and adolescent depressive symptoms. *Journal of Abnormal Child Psychology, 32*(2), 147–158.

Prkachin, K. M., & Silverman, B. E. (2002). Hostility and facial expression in young men and women: Is social regulation more important than negative affect? *Health Psychology, 21*(1), 33–39.

Prudic, J., Olfson, M., Marcus, S. C., Fuller, R. B., & Sackeim, H. A. (2004). Effectiveness of electroconvulsive therapy in community settings. *Biological Psychiatry, 55*(3), 301–312.

Pullum, G. K. (1991). *The great Eskimo vocabulary hoax and other irreverent essays on the study of language.* Chicago: University of Chicago Press.

Quaiser-Pohl, C., & Lehmann, W. (2002). Girls' spatial abilities: Charting the contributions of experiences and attitudes in different academic groups. *British Journal of Educational Psychology, 72*(2), 245–260.

Quinn, J., Barrowclough, C., & Tarrier, N. (2003). The Family Questionnaire (FQ): A scale for measuring symptom appraisal in relatives of schizophrenic patients. *Acta Psychiatrica Scandinavica, 108*(4), 290–296.

Raesaenen, P. M., Tiihonen, J., Isohanni, M., Rantakallio, P., Lehtonen, J., & Moring, J. (1998). Schizophrenia, alcohol abuse, and violent behavior: A 26-year follow up study of an unselected birth cohort. *Schizophrenia Bulletin, 24,* 437–441.

Rahman, Q., & Wilson, G. D. (2003). Born gay? The psychobiology of human sexual orientation. *Personality & Individual Differences. 34*(8), 1337–1382.

Raine, A., Lencz, T., Bihrle, S., LaCasse, L., & Colletti, P. (2000). Reduced prefrontal gray matter volume and reduced autonomic activity in antisocial personality disorder. *Archives of General Psychiatry, 57,* 119–127.

Raine, A., Meloy, J.R., Bihrle, S., Stoddard, J., LaCasse, L., & Buchsbaum, M.S. (1998). Reduced prefrontal and increased subcortical brain functioning assessed using positron emission tomography in predatory and effective murderers. *Behavioral Sciences & the Law, 16*(3), 319–332.

Ramirez, G., Zemba, D., & Geiselman, R. E. (1996). Judges' cautionary instructions on eyewitness testimony. *American Journal of Forensic Psychology, 14,* 31–66.

Ramirez, J. M. (2003). Hormones and aggression in childhood and adolescence. *Aggression & Violent Behavior, 8*(6), 621–644.

Randi, J. (1997). *An encyclopedia of claims, frauds, and hoaxes of the occult and supernatural: James Randi's decidedly skeptical definitions of alternate realities.* New York: St Martin's Press.

Rauhut, A. S., Thomas, B. L., & Ayres, J. J. B. (2001). Treatments that weaken Pavlovian conditioned fear and thwart its renewal in rats: Implications for treating human phobias. *Journal of Experimental Psychology: Animal Behavior Processes, 27*(2), 99–114.

Ray, W. J. (2003). *Methods: Toward a science of behavior and experience.* (7th ed.). Wadsworth: Belmont, CA.

Rechtschaffen, A., & Siegel, J.M. (2000). Sleep and Dreaming. In E. R. Kandel, J. H. Schwartz, & T. M. Jessel (Eds.), *Principles of Neuroscience* (4th ed., pp. 936–947). New York: McGraw-Hill.

Reed, L. J., Lasserson, D., Marsden, P., Bright, P., Stanhope, N., & Kopelman, M. D. (2005). Correlations of regional cerebral metabolism with memory performance and executive function in patients with herpes encephalitis or frontal lobe lesions. *Neuropsychology, 19*(5), 555–565.

Reeder, C., Newton, E., Frangou, S., & Wykes, T. (2004). Which executive skills should we target to affect social functioning and symptom change? A study of a cognitive remediation therapy program. *Schizophrenia Bulletin, 30*(1), 87–100.

Reeve, J. (2005). *Understanding motivation and emotion* (4th ed.). Hoboken, NJ: Wiley.

Regan, P. (1998). What if you can't get what you want? Willingness to compromise ideal mate selection standards as a function of sex, mate value, and relationship context. *Personality and Social Psychology Bulletin, 24,* 1294–1303.

Regier, D. A., Narrow, W. E., Rae, D. S., Manderscheid, R. W., Locke, B. Z., & Goodwin, F. K. (1993). The de facto US mental and addictive disorders service system. *Archives of General Psychiatry, 50,* 85–93.

Reich, D. A. (2004). What you expect is not always what you get: The roles of extremity, optimism, and pessimism in the behavioral confirmation process. *Journal of Experimental Social Psychology, 40*(2), 199–215.

Reifman, A. (2000). Revisiting the Bell Curve. *Psychology, 11,* 21–29.

Reis, C., Ahmed, A. T., Amowitz, L. L., Kushner, A. L., Elahi, M., & Iacopino, V. (2004). Physician participation in human rights abuses in southern Iraq. *JAMA: Journal of the American Medical Association, 291*(12), 1480–1486.

Renzetti, C., Curran, D., & Kennedy-Bergen, R. (2006). *Understanding diversity.* Boston: Allyn & Bacon/Longman.

Resing, W. C., & Nijland, M. I. (2002). Worden kinderen intelligenter? Een kwart eeuw onderzoek met de Leidse Diagnostische Test [Are children becoming more intelligent? Twenty-five years' research using the Leiden Diagnostic Test.] *Kind en Adolescent, 23*(1), 42–49.

Ressler, K., & Davis, M. (2003). Genetics of childhood disorders: L. Learning and memory, part 3: Fear conditioning. *Journal of the American Academy of Child & Adolescent Psychiatry, 42*(5), 612–615.

Rest, J., Narvaez, D., Bebeau, M., & Thoma, S. (1999). A neo-Kohlbergian approach: The DIT and schema theory. *Educational Psychology Review, 11*(4), 291–324.

Rhodes, G., Halberstadt, J., & Brajkovich, G. (2001). Generalization of mere exposure effects to averaged composite faces. *Social Cognition,19*(1), 57–70.

Richardson, G. A., Ryan, C., Willford, J., Day, N. L., & Goldschmidt, L. (2002). Prenatal alcohol and marijuana exposure: Effects on neuropsychological outcomes at 10 years. *Neurotoxicology & Teratology, 24*(3), 311–320.

Ridley, R. M., Baker, H. F., Cummings, R. M., Green, M. E., & Leow-Dyke, A. (2005). Mild topographical memory impairment following crossed unilateral lesions of the mediodorsal thalamic nucleus and the inferotemporal cortex. *Behavioral Neuroscience, 119*(2), 518–525.

Riemann, D., & Voderholzer, U. (2003). Primary insomnia: a risk factor to develop depression? *Journal of Affective Disorders, 76*(1–3), 255–259.

Riggio, H. R. (2004). Parental marital conflict and divorce, parent-child relationships, social support, and relationship anxiety in young adulthood. *Personal Relationships, 11*(1), 99–114.

Ripley, A. (2004, June 31). Redefining torture. *Time, 163,* p. 29.

Roberts, M. (2006, March). Idol dreams. *Schizophrenia Digest,* 30–33.

Roberts, W. W., & Nagel, J. (1996). First-order projections activated by stimulation of hypothalamic sites eliciting attack and flight in rats. *Behavioral Neuroscience, 110,* 509–527.

Robinson, F. P. (1946). *Effective study.* New York: Harper & Row

Robinson, F. P. (1970). *Effective study* (4th ed.). New York: Harper & Row.

Roelcke, V. (1997). Biologizing social facts: An early 20th century debate on Kraepelin's concepts of culture, neurasthenia, and degeneration. *Culture, Medicine, and Psychiatry, 21,* 383–403.

Rogers, C. R. (1961). *On becoming a person*. Boston: Houghton Mifflin.

Rogers, C. R. (1980). *A way of being*. Boston: Houghton Mifflin.

Rogers, T. T., Lambon Ralph, M. A., Garrard, P., Bozeat, S., McClelland, J. L., Hodges, J. R., & Patterson, K. (2004). Structure and deterioration of semantic memory: A neuropsychological and computational investigation. *Psychological Review, 111*(1), 205–235.

Rogosch, F. A., & Cicchetti, D. (2004). Child maltreatment and emergent personality organization: Perspectives from the five-factor model. *Journal of Abnormal Child Psychology, 32*(2), 123–145.

Rönnberg, J., Rudner, M., & Ingvar, M. (2004). Neural correlates of working memory for sign language. *Cognitive Brain Research, 20*(2), 165–182.

Rook, K. S. (2000). The evolution of social relationships in later adulthood. In S. H. Qualls & N. Abeles (Eds.), *Psychology and the aging revolution*. Washington, DC: American Psychological Association.

Roques, P. Lambin, M., Jeunier, B., Strayer, F. (1997). Multivariate analysis of personal space in a primary school classroom. *Enfance, 4*, 451–468.

Rosch, E. (1978). Principles of organization. In E. Rosch & H. L. Lloyd (Eds.), *Cognition and categorization*. Hillsdale, NJ: Erlbaum.

Rosch, E. H. (1973). Natural categories. *Cognitive Psychology, 4*, 328–350.

Rose, C. R., & Konnerth, A. (2002). Exciting glial oscillations. *Nature Neuroscience, 4*, 773–774.

Rosenberg, M. (2003). Recognizing gay, lesbian, and transgender teens in a child and adolescent psychiatry practice. *Journal of the American Academy of Child & Adolescent Psychiatry, 42*(12), 1517–1521.

Rosen-Grandon, J. R., Myers, J. E., & Hattie, J. A. (2004). The relationship between marital characteristics, marital interaction processes, and marital satisfaction. *Journal of Counseling & Development, 82*(1), 58–68.

Rosenzweig, E. S., Barnes, C. A., & McNaughton, B. L. (2002). Making room for new memories. *Nature Neuroscience, 5*(1), 6–8.

Rosenzweig, M. R., & Bennett, E. L. (1996). Psychobiology of plasticity: Effects of training and experience on brain and behavior. *Behavioural Brain Research, 78*(1), 57–65.

Rosenzweig, M. R., Bennett, E. L., & Diamond, M. C. (1972). Brain changes in response to experience. *Scientific American, 226*, 22–29.

Ross, H., Smith, J., Spielmacher, C., & Recchia, H. (2004). Shading the truth: Self-serving biases in children's reports of sibling conflicts. *Merrill-Palmer Quarterly, 50*(1), 61–85.

Rossano, M. J. (2003). *Evolutionary psychology: The science of human behavior and evolution*. Hoboken, NJ: Wiley.

Roth, M. D., Whittaker, K., Salehi, K., Tashkin, D. P., & Baldwin, G. C. (2004). Mechanisms for impaired effector function in alveolar macrophages from marijuana and cocaine smokers. *Journal of Neuroimmunology, 147*(1–2), 82–86.

Rotter, J. B. (1954). *Social learning and clinical psychology*. Englewood Cliffs, NJ: Prentice Hall.

Rotter, J. B. (1990). Internal versus external control of reinforcement: A case history of a variable. *American Psychologist, 45*, 489–493.

Rowe, R., Maughan, B., Worthman, C. M., Costello, E. J., & Angold, A. (2004). Testosterone, antisocial behavior, and social dominance in boys: Pubertal development and biosocial interaction. *Biological Psychiatry, 55*(5), 546–552.

Roy, T. S., Seidler, F. J., & Slotkin, T. A. (2002). Prenatal nicotine exposure evokes alterations of cell structure in hippocampus and somatosensory cortex. *Journal of Pharmacology & Experimental Therapeutics, 300*(1), 124–133.

Rozencwajg, P., Cherfi, M., Ferrandez, A. M., Lautrey, J., Lemoine, C., & Loarer, E. (2005). Age related differences in the strategies used by middle aged adults to solve a block design task. *International Journal of Aging & Human Development, 60*(2), 159–182.

Rubin, Z. (1970). Measurement of romantic love. *Journal of Personality and Social Psychology, 16*, 265–273.

Rudman, L. A., & Fairchild, K. (2004). Reactions to Counterstereotypic Behavior: The Role of Backlash in Cultural Stereotype Maintenance. *Journal of Personality and Social Psychology, 87*(2), 157–176.

Rumbaugh, D. M., von Glasersfeld, E. C., Warner, H., Pisani, P., & Gill, T. V. (1974). Lana (chimpanzee) learning language: A progress report. *Brain & Language, 1*(2), 205–212.

Rumpel, S., LeDoux, J., Zador, A., & Malinow, R. (2005). Postsynaptic receptor trafficking underlying a form of associative learning. *Science, 308*(5718), 83–88.

Rutler, M. L. (1997). Nature–nurture integration: The example of antisocial behavior. *American Psychologist, 52*, 390–398.

Ryback, D., Ikemi, A., & Miki, Y. (2001). Japanese psychology in crisis: Thinking inside the (empty) box. *Journal of Humanistic Psychology, 41*(4), 124–136.

Rybacki, J.J., & Long, J.W. (1999). *The essential guide to prescription drugs 1999*. New York: Harper Perennial.

Rymer, R. (1993). *Genie: An abused child's first flight from silence*. New York: HarperCollins.

Saab, C. Y., & Willis, W. D. (2003). The cerebellum: Organization, functions and its role in nociception. *Brain Research Reviews, 42*(1), 85–95.

Saadeh, W., Rizzo, C. P., & Roberts, D. G. (2002). Spanking. *Clinical Pediatrics, 41*(2), 87–88.

Sabini, J., & Silver, M. (1993). Critical thinking and obedience to authority. In J. Chaffee (Ed.), *Critical thinking* (2nd ed., pp. 367–376). Palo Alto, CA: Houghton Mifflin.

Salmivalli, C., & Kaukiainen, A. (2004). "Female aggression" revisited: Variable- and person-centered approaches to studying gender differences in different types of aggression. *Aggressive Behavior, 30*(2),158–163.

Salovey, P., Bedell, B. T., Detweiler, J. B., & Mayer, J. D. (2000). Current directions in emotional intelligence research. In M. Lewis & J.M. Haviland (Eds.), *Handbook of emotions* (2nd ed., pp. 504–520). New York: Guilford.

Salovey, P., & Mayer, J. D. (1990). Emotional intelligence. *Imagination,Cognition, and Personality, 9*, 185–211.

Saltus, R. (2000, June 22). *Brain cells are coaxed into repair duty*. Boston Globe, A18.

Salyers, M.P., & Mueser, K.T. (2001). Schizophrenia. In M. Hersen & V.B. Van Hasselt (Eds.), *Advanced abnormal psychology* (2nd ed.). New York: Klewer Academic/Plenum.

Sampselle, C. M., Harris, V. Harlow, S. D., & Sowers, M. F. (2002). Midlife development and menopause in African American and Caucasian women. *Health Care for Women International, 23*(4), 351–363.

Sanderson, C. A. (2004). *Health psychology*. Hoboken, NJ: Wiley.

Sangha, S., McComb, C., Scheibenstock, A., Johannes, C., & Lukowiak, K. (2002). The effects of continuous versus partial reinforcement schedules on associative learning, memory, and extinction in Lymnaea stagnalis. *Journal of Experimental Biology, 205*, 1171–1178.

Sangwan, S. (2001). Ecological factors as related to I.Q. of children. *Psycho-Lingua, 31*(2), 89–92.

Sansone, R. A., Levengood, J. V., & Sellbom, M. (2004). Psychological aspects of fibromyalgia research vs. clinician impressions. *Journal of Psychosomatic Research, 56*(2), 185–188.

Sarafino, E. P. (2000). *Behavior modification: Understanding principles of behavior change*. Springfield, MA: Mayfield.

Sarafino, E. P. (2005). *Health psychology: Biopsychosocial interactions* (5th ed.). Hoboken, NJ: Wiley.

Sarnthein, J., Morel, A., von Stein, A., & Jeanmonod, D. (2005). Thalamocortical theta coherence in neurological patients at rest and during a working memory task. *International Journal of Psychophysiology, 57*(2), 87–96.

Saudino, K. J. (1997). Moving beyond the heritability question: New directions in behavioral genetic studies of personality. *Current Directions in Psychological Science, 6*, 86–90.

Savage-Rumbaugh, E. S. (1990). Language acquisition in a nonhuman species: Implications for the innateness debate. *Developmental Psychobiology, 23*, 599–620.

Sawa, A., & Kamiya, A. (2003). Elucidating the pathogenesis of schizophrenia: DISC-1 gene may predispose to neurodevelopmental changes underlying schizophrenia. *BMJ: British Medical Journal, 327*(7416), 632–633.

Sawaya, A. L., Fuss, P. J., Dallal, G. E., Tsay, R., McCrory, M. A., Young, V., & Roberts, S. B. (2001). Meal palatability, substrate oxidation and blood glucose in young and older men. *Physiology & Behavior, 72*(1–2), 5–12.

Saxe, L., & Ben-Shakhar, G. (1999). Admissibility of polygraph tests: The application of scientific standards post-Daubert. *Psychology, Public Policy, & Law, 5*(1), 203–223.

Schachter, S., & Singer, J. E. (1962). Cognitive, social, and physiological determinants of emotional state. *Psychological Review, 69*, 379–399.

Schaie, K. W. (1994). The life course of adult intellectual development. *American Psychologist, 49*, 304–313.

Scherer, K. R., & Wallbott, H. G. (1994). Evidence for universal and cultural variation of differential emotion response patterning. *Journal of Personality and Social Psychology, 66*(2), 310–328.

Schiffer, F, Zaidel, E, Bogen, J, & Chasan-Taber, S. (1998). Different psychological status in the two hemispheres of two split-brain patients. *Neuropsychiatry, Neuropsychology, & Behavioral Neurology, 11*, 151–156.

Schmahl, C. G., Vermetten, E., Elzinga, B. M., & Bremner, J. D. (2004). A positron emission tomography study of memories of childhood abuse in borderline personality disorder. *Biological Psychiatry, 55*(7), 759–765.

Schmidt, L.A. (1999). Frontal brain electrical activity in shyness and sociability. *Psychological Science, 10* (4), 316–320.

Schmidt, U. (2004). Undue influence of weight on self-evaluation: A population-based twin study of gender differences. *International Journal of Eating Disorders, 35*(2), 133–135.

Schnurr, P.P, & Green, B. L. (2004). *Trauma and health: Physical health consequences of exposure to extreme stress*. Washington, DC: American Psychological Association.

Schwartz, J. & Broder, J.M. (2003, February 13). Loss of the shuttle: The overview; Engineer warned about dire impact of liftoff damage. *New York Times*, p. A1.

Schwartz, M. W., & Morton, G. J. (2002). Obesity: Keeping hunger at bay. *Nature, 418*, 595–597.

Schwartz, R. C., & Smith, S. D. (2004). Suicidality and psychosis: The predictive potential of symptomatology and insight into illness. *Journal of Psychiatric Research, 38*(2), 185–191.

Scribner, S. (1977). Modes of thinking and ways of speaking: Culture and logic reconsidered. In P. N. Johnson-Laird & P. C. Wason (Eds.), *Thinking: Readings in cognitive science* (pp. 324–339). New York: Cambridge University Press.

Scully, J. A., Tosi, H., & Banning, K. (2000). Life event checklists: Revisiting the Social Readjustment Rating Scale after 30 years. *Educational & Psychological Measurement, 60*(6), 864–876.

Sebre, S., Sprugevica, I., Novotni, A., Bonevski, D., Pakalniskiene, V., Popescu, D., Turchina, T., Friedrich, W., & Lewis, O. (2004). Cross-cultural comparisons of child-reported emotional and physical abuse: Rates, risk factors and psychosocial symptoms. *Child Abuse & Neglect, 28*(1), 113–127.

Secker, D. L., Kazantzis, N., Pachana, N. A., & Kazantzis, N. (2004). Cognitive behavior therapy for older adults: Practical guidelines for adapting therapy structure. *Journal of Rational-Emotive & Cognitive Behavior Therapy, 22*(2), 93–109.

Segal, N. L., & Bouchard, T. J. (2000). *Entwined lives: Twins and what they tell us about human behavior.* New York: Plumsock.

Segerstrale, U. (2000). *Defenders of the truth: The battle for science in the sociobiology debate and beyond.* London: Oxford University Press.

Segerstrom, S., C., & Miller, G. E. (2004). Psychological stress and the human immune system: A meta-analytic study of 30 years of inquiry. *Psychological Bulletin, 130*(4), 601–630.

Seligman, M. E. P. (1975) *Helplessness: On depression, development, and death.* San Francisco: Freeman.

Seligman, M. E. P. (1994). *What you can change and what you can't.* New York: Alfred A. Knopf.

Selye, H. (1936). A syndrome produced by diverse nocuous agents. *Nature, 138*, 32.

Selye, H. (1974). *Stress without distress.* New York: Harper & Row.

Sequeira, A., Mamdani, F., Lalovic, A., Anguelova, M., Lesage, A., Seguin, M., Chawky, N., Desautels, A., & Turecki, G. (2004). Alpha 2A adrenergic receptor gene and suicide. *Psychiatry Research, 125*(2), 87–93.

Shadish, W. R., & Baldwin, S. A. (2003). Meta-analysis of MFT interventions. *Journal of Marital & Family Therapy, 29*(4), 547–570.

Shaver, P. R., & Mikulincer, M. (2005). Attachment theory and research: Resurrection of the psychodynamic approach to personality. *Journal of Research in Personality, 39*(1), 22–45.

Shea, S.C. (1988). *Psychiatric interviewing: The art of understanding.* Philadelphia, PA: Saunders.

Shen, B-J., McCreary, C. P., & Myers, H. F. (2004). Independent and mediated contributions of personality, coping, social support, and depressive symptoms to physical functioning outcome among patients in cardiac rehabilitation. *Journal of Behavioral Medicine, 27*(1), 39–62.

Sher, K. J., Grekin, E. R., & Williams, N. A. (2005). The development of alcohol use disorders. *Annual Review of Clinical Psychology, 1*, 493–523.

Sherif, M. (1966). *In common predicament: Social psychology of intergroup conflict and cooperation.* Boston: Houghton Mifflin.

Sherif, M. (1998). Experiments in group conflict. In J. M. Jenkins, K. Oatley, & N. L. Stein (Eds.), *Human emotions: A reader* (pp. 245–252). Malden, MA: Blackwell.

Sherwin, C. M., Christiansen, S. B., Duncan, I. J., Erhard, H. W., Lay, D. C., Jr., Mench, J. A., O'Connor, C. E., & Petherick, J. C. (2003). Guidelines for the ethical use of animals in applied ethology studies. *Applied Animal Behaviour Science, 81*(3), 291–305.

Shi, Y., Devadas, S., Greeneltch, K. M., Yin, D., Mufson, R. A., & Zhou, J. (2003). Stressed to death: Implication of lymphocyte apoptosis for psychoneuroimmunology. *Brain, Behavior & Immunity.* 7(Suppl1), S18-S26.

Shimizu-Albergine, M., Ippolito, D. L., & Beavo, J. A. (2001). Downregulation of fasting-induced cAMP response element-mediated gene induction by leptin in neuropeptide Y neurons of the arcuate nucleus. *Journal of Neuroscience, 21*(4), 1238–1246.

Shors, T. J. (2004). Learning during stressful times. *Learning and Memory, 11*, 137–144.

Sias, P. M., Heath, R. G., Perry, T., Silva, D., & Fix, B. (2004). Narratives of workplace friendship deterioration. *Journal of Social & Personal Relationships, 21*(3), 321–340.

Siegel, J. M. (2000, January). Narcolepsy. *Scientific American*, pp. 76–81.

Silke, A. (2003). Deindividuation, anonymity, and violence: Findings from Northern Ireland. *Journal of Social Psychology, 143*(4), 493–499.

Silvester, J., Anderson-Gough, F. M., Anderson, N. R., & Mohamed, A. R. (2002). Locus of control, attributions and impression management in the selection interview. *Journal of Occupational & Organizational Psychology, 75*(1), 59–76.

Simcock, G., & Hayne, H. (2002). Breaking the barrier? Children fail to translate their preverbal memories into language. *Psychological Science, 13*(3), 225–231.

Singelis, T. M., Triandis, H. C., Bhawuk, D. S., & Gelfand, M. (1995). Horizontal and vertical dimensions of individualism and collectivism: A theoretical and measurement refinement. *Cross-Cultural Research, 29*, 240–275.

Singer, M. I., Slovak, K., Frierson, T., & York, P. (1998). Viewing preferences, symptoms of psychological trauma, and violent behaviors among children who watch television. *Journal of the American Academy of Child and Adolescent Psychiatry, 37*, 1041–1048.

Sinha, R., Talih, M., Malison, R., Cooney, N., Anderson, G. M., & Kreek, M. J. (2003). Hypothalamic-pituitary-adrenal axis and sympatho-adreno-medullary responses during stress-induced and drug cue-induced cocaine craving states. *Psychopharmacology, 170*(1), 62–72.

Siqueland, L., Crits-Christoph, P., Gallop, R., Barber, J. P., Griffin, M. L., Thase, M. E., Daley, D., Frank, A., Gastfriend, D. R., Blaine, J., Connolly, M. B., & Gladis, M. (2002). Retention in psychosocial treatment of cocaine dependence: Predictors and impact on outcome. *American Journal on Addictions, 11*(1), 24–40.

Sirin, S. R., & Rogers-Sirin, L. (2004). Exploring school engagement of middle-class African American adolescents. *Youth & Society, 35*(3), 323–340.

Skinner, B. F. (1948). Superstition in the pigeon. *Journal of Experimental Psychology, 38*, 168–172.

Skinner, B. F. (1953). *Science and human behavior.* New York: Macmillan.

Skinner, B. F. (1961). Diagramming schedules of reinforcement. *Journal of the Experimental Analysis of Behavior, 1*, 67–68.

Skinner, B. F. (1992). "Superstition" in the pigeon. *Journal of Experimental Psychology: General, 121*(3), 273–274.

Sluzenski, J., Newcombe, N., & Ottinger, W. (2004). Changes in reality monitoring and episodic memory in early childhood. *Developmental Science, 7*(2), 225–245.

Smart, D. W., & Smart, J. F. (1997). DSM-IV and culturally sensitive diagnosis: Some observations for counselors. *Journal of Counseling and Development, 75*, 392–398.

Smith, C., & Smith, D. (2003). Ingestion of ethanol just prior to sleep onset impairs memory for procedural but not declarative tasks. *Sleep, 26*(2), 185–191.

Smith, E. E. (1995). *Concepts and categorization.* In E. E. Smith & D. N. Osherson (Eds.), *Thinking: An invitation to cognitive science* (2nd ed., Vol. 3, pp. 3–33). Cambridge: MIT Press.

Smith, L. J., Nowakowski, S., Soeffing, J. P., Orff, H. J., & Perlis, M. L. (2003). The measurement of sleep. In L. Michael & K. L. Lichstein (Eds.), *Treating sleep disorders: Principles and practice of behavioral sleep medicine.* New York: Wiley.

Smith, T. W., Orleans, C. T., & Jenkins, C. D. (2004). Prevention and health promotion: Decades of progress, new challenges, and an emerging agenda. *Health Psychology, 23*(2), 126–131.

Snarey, J. R. (1985). Cross-cultural universality of social-moral development: A critical review of Kohlbergian research. *Psychological Bulletin, 97*, 202–233.

Snarey, J. R. (1995). In communitarian voice: The sociological expansion of Kohlbergian theory, research, and practice. In W. M. Kurtines & J. L. Gerwirtz (Eds.), *Moral development: An introduction* (pp. 109–134). Boston: Allyn & Bacon.

Snowden, L. R., & Yamada, A-M. (2005). Cultural differences in access to care. *Annual Review of Clinical Psychology, 1*, 143–166.

Snyder, C. R. (2003). "Me conform? No way": Classroom demonstrations for sensitizing students to their conformity. *Teaching of Psychology, 30*(1), 59–61.

Sokol, R. J. Jr., Delaney-Black, V., & Nordstrom, B. (2003). Fetal alcohol spectrum disorder. *JAMA: Journal of the American Medical Association, 290*(22), 2996–2999.

Solan, H. A., & Mozlin, R. (2001). Children in poverty: Impact on health, visual development, and school failure. *Issues in Interdisciplinary Care, 3*(4), 271–288.

Sotres-Bayón, F., & Pellicer, F. (2000). The role of the dopaminergic mesolimbic system in the affective component of chronic pain. *Salud Mental, 23*(1), 23–29.

Soussignan, R. (2002). Duchenne smile, emotional experience, and autonomic reactivity: A test of the facial feedback hypothesis. *Emotion, 2*(1), 52–74.

Southwick, S. M., Vythilingam, M., & Charney, D. S. (2005). The psychobiology of depression and resilience to stress: Implications for prevention and treatment. *Annual Review of Clinical Psychology, 1*, 255–291.

Spearman, C. (1923). *The nature of "intelligence" and the principles of cognition.* London: Macmillan.

Spence, S. A. (2003). Detecting lies and deceit: The psychology of lying and the implications for professional practice. *Cognitive Neuropsychiatry, 8*(1), 76–77.

Sperling, G. (1960). The information available in brief visual presentations. *Psychological Monographs, 74* (Whole No. 498).

Spiegel, D. (1999). An altered state. *Mind/Body Health Newsletter, 8*(1), 3–5.

Spiegel, D., & Maldonado, J. R. (1999). Dissociative disorders. In R. E. Hales, S. C. Yudofsky, & J. C. Talbott (Eds.), *American psychiatric press textbook of psychiatry.* Washington, DC: American Psychiatric Press.

Spitz, R. A., & Wolf, K. M. (1946). The smiling response: A contribution to the ontogenesis of social relations. *Genetic Psychology Monographs, 34*, 57–123.

Spokane, A. R., Meir, E. I., & Catalano, M. (2000). Person-environment congruence and Holland's theory: A review and reconsideration. *Journal of Vocational Behavior, 57*(2), 137–187.

Sprecher, S., & Regan, P. C. (2002). Liking some things (in some people) more than others: Partner preferences in romantic relationships and friendships. *Journal of Social & Personal Relationships, 19*(4), 463–481.

Squier, L. H., & Domhoff, G. W. (1998). The presentation of dreaming and dreams in introductory psychology textbooks: A critical examination. *Dreaming, 10*, 21–26.

Squire, L. R., Schmolck, H., & Buffalo, E. A. (2001). Memory distortions develop over time: A reply to Horn. *Psychological Science, 121*(2), 182.

Stafford, J., & Lynn, S. J. (2002). Cultural scripts, memories of childhood abuse, and multiple identities: A study of role-played enactments. *International Journal of Clinical & Experimental Hypnosis, 50*(1), 67–85.

Stanton, M. D. (2004). Getting reluctant substance abusers to engage in treatment/self-help: A review of outcomes and clinical options. *Journal of Marital & Family Therapy, 30*(2), 165–182.

Steadman, H. J. (1993). *Reforming the insanity defense: An evaluation of pre- and post-Hinckley reforms.* New York: Guilford Press.

Steele, C. M. (2003). Through the back door to theory. *Psychological Inquiry, 14*(3–4), 314–317.

Steele, C. M., & Aronson, J. (1995). Stereotype threat and the intellectual test performance of African Americans. *Journal of Personality and Social Psychology, 69*, 797–811.

Steele, J., James, J. B., & Barnett, R. C. (2002). Learning in a man's world: Examining the perceptions of undergraduate women in male-dominated academic areas. *Psychology of Women Quarterly, 26*(1), 46–50.

Steffens, D. C., McQuoid, D. R., Welsh-Bohmer, K. A., & Krishnan, K. R. R. (2003). Left orbital frontal cortex volume and performance on the Benton Visual Retention Test in older depressives and controls. *Neuropsychopharmacology, 28*(12), 2179–2183.

Steiger, H., Israël, M., Gauvin, L., Kin, N. M. K. N. Y., & Young, S. N. (2003). Implications of compulsive and impulsive traits for serotonin status in women with bulimia nervosa. *Psychiatry Research, 120*(3), 219–229.

Steinhart, P. (1986, March). Personal boundaries. *Audubon,* pp. 8–11.

Stelmack, R. M., Knott, V., & Beauchamp, C. M. (2003). Intelligence and neural transmission time: A brain stem auditory evoked potential analysis. *Personality & Individual Differences, 34*(1), 97–107.

Sternberg, R. J. (1985). *Beyond IQ: A triarchic theory of human intelligence.* New York: Cambridge University Press.

Sternberg, R. J. (1998). Principles of teaching for successful intelligence. *Educational Psychologist, 33,* 65–72.

Sternberg, R. J. (1999). The theory of successful intelligence. *Review of General Psychology, 3,* 292–316.

Sternberg, R. J. (2005). The importance of converging operations in the study of human intelligence. *Cortex. 41*(2), 243–244.

Sternberg, R. J., & Hedlund, J. (2002). Practical intelligence, g, and work psychology. *Human Performance, 15*(1–2), 143–160.

Sternberg, R. J., & Lubart, T. I. (1992). Buy low and sell high: An investment approach to creativity. *Current Directions in Psychological Science, 1*(1), 1–5.

Sternberg, R. J., & Lubart, T. I. (1996). Investing in creativity. *American Psychologist, 51*(7), 677–688.

Stetter, F., & Kupper, S. (2002). Autogenic training: A meta-analysis of clinical outcome studies. *Applied Psychophysiology & Biofeedback, 27*(1), 45–98.

Stompe, T. G., Ortwein-Swoboda, K., Ritter, K., & Schanda, H. (2003). Old wine in new bottles? Stability and plasticity of the contents of schizophrenic delusions. *Psychopathology, 36*(1), 6–12.

Stoner, J. A. (1961). *A comparison of individual and group decisions involving risk.* Unpublished master's thesis, School of Industrial Management, MIT, Cambridge, MA.

Stormshak, E. A., Bierman, K. L., McMahon, R. J., & Lengua, L. J., Conduct Problems Prevention Research Group. (2000). Parenting practices and child disruptive behavior problems in early elementary school. *Journal of Clinical Child Psychology, 29*(1), 17–29.

Storosum, J. G., Elferink, A. J. A., van Zwieten, B. J., van den Brink, W., & Huyser, J. (2004). Natural course and placebo response in short-term, placebo-controlled studies in major depression: A meta-analysis of published and non-published studies. *Pharmacopsychiatry, 37*(1), 32–36.

Strack, F., Martin, L. L., & Stepper, S. (1988). Inhibiting and facilitating conditions of the human smile: A nonobstrusive test of the facial feedback hypothesis. *Journal of Personality and Social Psychology, 54,* 768–777.

Stratton, G. M. (1896). Some preliminary experiments on vision without inversion of the retinal image. *Psychological Review 3,* 611–617.

Streri, A., Gentaz, E., Spelke, E., & Van de Walle, G. (2004). Infants' haptic perception of object unity in rotating displays. *Quarterly Journal of Experimental Psychology: Human Experimental Psychology, 57A*(3), 523–538.

Subramanian, S., & Vollmer, R. R. (2002). Sympathetic activation fenfluramine depletes brown adipose tissue norepinephrine content in rats. *Pharmacology, Biochemistry & Behavior, 73*(3), 639–646.

Sue, D.W., & Sue, D. (2002). *Counseling the culturally diverse: Theory and practice.* New York: Wiley.

Suhr, J. A. (2002). Malingering, coaching, and the serial position effect. *Archives of Clinical Neuropsychology, 17*(1), 69–77.

Sullivan, M. J. L., Tripp, D. A., & Santor, D. (1998). *Gender differences in pain and pain behavior: The role of catastrophizing.* Paper presented at the annual meeting of the American Psychological Association, San Francisco.

Sullivan, P. F., Kendler, K. S., & Neale, M. C. (2003). Schizophrenia as a complex trait: Evidence from a meta-analysis of twin studies. *Archives of General Psychiatry, 60*(12), 1187–1192.

Surawy, C., Roberts, J., & Silver, A. (2005). The effect of mindfulness training on mood and measures of fatigue, activity, and quality of life in patients with chronic fatigue syndrome on a hospital waiting list: A series of exploratory studies. *Behavioural & Cognitive Psychotherapy. 33*(1), 103–109.

Suzuki, L., & Aronson, J. (2005). The cultural malleability of intelligence and its impact on the racial/ethnic hierarchy. *Psychology, Public Policy, and Law, 11*(2), 320–327.

Suzuki, W. A., & Amaral, D. G. (2004). Functional neuroanatomy of the medial temporal lobe memory system. *Cortex, 40*(1), 220–222.

Swanson, J. W., Swartz, M. S., & Elbogen, E. B. (2004). Effectiveness of atypical antipsychotic medications in reducing violent behavior among persons with schizophrenia in community-based treatment. *Schizophrenia Bulletin, 30*(1), 3–20.

Swartz, K. L. & Margolis, S. (2004). *Depression and anxiety.* Johns Hopkins White Papers. Baltimore: Johns Hopkins Medical Institutions.

Sylvester, A. P., Mulsant, B. H., Chengappa, K. N. R., Sandman, A. R., & Haskett, R. F. (2000). Use of electroconvulsive therapy in a state hospital: A 10-year review. *Journal of Clinical Psychiatry, 61,* 534–544.

Szasz, T. (1960). The myth of mental illness. *American Psychologist, 15,* 113–118.

Szasz, T. (2000). Second commentary on "Aristotle's function argument." *Philosophy, Psychiatry, & Psychology, 7,* 3–16.

Szasz, T. (2004). The psychiatric protection order for the "battered mental patient." *British Medical Journal, 327*(7429), 1449–1451.

Tait, R. J., & Hulse, G. K. (2003). A systematic review of the effectiveness of brief interventions with substance using adolescents by type of drug. *Drug & Alcohol Review, 22*(3), 337–346.

Talbot, N. L., Duberstein, P. R., King, D. A., Cox, C., & Giles, D. E. (2000). Personality traits of women with a history of childhood sexual abuse. *Comprehensive Psychiatry, 41,* 130–136.

Tanaka, T., Yoshida, M., Yokoo, H., Tomita, M., & Tanaka, M. (1998). Expression of aggression attenuates both stress-induced gastric ulcer formation and increases in noradrenaline release in the rat amygdala assessed intracerebral microdialysis. *Pharmacology, Biochemistry & Behavior, 59*(1), 27–31.

Tanda, G., & Goldberg, S. R. (2003). Cannabinoids: Reward, dependence, and underlying neurochemical mechanisms-a review of recent preclinical data. *Psychopharmacology, 169*(2), 115–134.

Tang, Y.-P., Wang, H., Feng, R., Kyin, M., & Tsien, J. Z. (2001). Differential effects of enrichment on learning and memory function in NR2B transgenic mice. *Neuropharmacology, 41*(6), 779–790.

Tanner-Halverson, P., Burden, T., & Sabers, D. (1993). WISC-III normative data for Tohono O'odham Native-American children. In B. A. Bracken & R. S. McCallum (Eds.), *Journal of Psychoeducational Assessment Monograph Series, Advances in Psychoeducational Assessment: Wechsler Intelligence Scale for Children–Third Edition* (pp. 125–133). Germantown, TN: Psychoeducational Corporation.

Task Force of the National Advisory Council on Alcohol Abuse and Alcoholism, National Institute on Alcohol Abuse and Alcoholism (2002). *A Call to Action: Changing the Culture of Drinking at U.S. Colleges.* Washington, DC: National Institutes of Health. [On-line]. Available: http://www.collegedrinkingprevention.gov/NIAAACollegeMaterials/TaskForce/TaskForce_TOC.aspx.

Tatrow, K., Blanchard, E. B., & Silverman, D. J. (2003). Posttraumatic headache: An exploratory treatment study. *Applied Psychophysiology & Biofeedback, 28*(4), 267–279.

Taub, E. (2004). Harnessing brain plasticity through behavioral techniques to produce new treatments in neurorehabilitation. *American Psychologist, 59*(8), 692–704.

Taub, E., Gitendra, U., & Thomas, E. (2002). New treatments in neurorehabilitation founded on basic research. Nature Reviews *Neuroscience, 3*, 228–236.

Taub, E., & Morris, D. (2002). Constraint-induced movement therapy to enhance recovery after stroke. *Current Artherosclerosis Reports, 3,* 279–86.

Taub, E., Ramey, S. L., DeLuca, S., & Echols, K. (2004). Efficacy of constraint-induced movement therapy for children with cerebral palsy with asymmetric motor impairment. *Pediatrics, 113,* 305–312.

Taylor, D. J., Lichstein, K. L., & Durrence, H. H. (2003). Insomnia as a health risk factor. *Behavioral Sleep Medicine, 1*(4), 227–247.

Taylor, R. C., Harris, N. A., Singleton, E. G., Moolchan, E. T., & Heishman, S. J. (2000). Tobacco craving: Intensity-related effects of imagery scripts in drug abusers. *Experimental and Clinical Psychopharmacology, 8*(1), 75–87.

Taylor, S. E., & Armor, D. A. (1996). Positive illusions and coping with adversity. *Journal of Personality, 64,* 873–898.

Tebartz van Elst, L, Hesslinger, B., Thiel, T., Geiger, E., Haegele, K., Lemieux, L., Lieb, K., Bohus, M., Hennig, J., & Ebert, D. (2003). Frontolimbic brain abnormalities in patients with borderline personality disorder: A volumetric magnetic resonance imaging study. *Biological Psychiatry, 54*(2), 163–171.

Teesson, M., Hodder, T., & Buhrich, N. (2004). Psychiatric disorders in homeless men and women in inner Sydney. *Australian & New Zealand Journal of Psychiatry, 38*(3),162–168.

Terman, L. M. (1916). *The measurement of intelligence.* Boston: Houghton Mifflin.

Terman, L. M. (1954). Scientists and nonscientists in a group of 800 gifted men. *Psychological Monographs, 68*(7), 1–44.

Terrace, H. S. (1979, November). How Nim Chimpsky changed my mind. *Psychology Today,* pp. 65–76.

Terry, S. W. (2003). *Learning and memory: Basic principles, processes, and procedures* (2nd ed.). Boston: Allyn & Bacon.

Tett, R. P., & Murphy, P. J. (2002). Personality and situations in co-worker preference: Similarity and complementarity in worker compatibility. *Journal of Business & Psychology, 17*(2), 223–243.

Teusch, L., Boehme, H., Finke, J., & Gastpar, K. (2001). Effects of client-centered psychotherapy for personality disorders alone and in combination with psychopharmacological treatment. *Psychotherapy & Psychosomatics, 70*(6), 328–336.

Thapar, A., Fowler, T., Rice, F., Scourfield, J., van den Bree, M., Thomas, H., Harold, G., & Hay, D. (2003). Maternal smoking during pregnancy and attention deficit hyperactivity disorder symptoms in offspring. *American Journal of Psychiatry, 160*(11), 1985–1989.

Theoharides, T. C., & Cochrane, D. E. (2004). Critical role of mast cells in inflammatory diseases and the effect of acute stress. *Journal of Neuroimmunology, 146*(1–2), 1–12.

Thøgersen, J. (2004). A cognitive dissonance interpretation of consistencies and inconsistencies in environmentally responsible behavior. *Journal of Environmental Psychology, 24*(1), 93–103.

Thomas, B. L., & Ayres, J. J. B. (2004). Use of the ABA fear renewal paradigm to assess the effects of extinction with co-present fear inhibitors or excitors: Implications for theories of extinction and for treating human fears and phobias. *Learning & Motivation, 35*(1), 22–52.

Thompson, R. F. (2005). In search of memory traces. *Annual Review of Clinical Psychology, 56,* 1–23.

Thorndike, E. L. (1898). Animal intelligence. *Psychological Review Monograph, 2*(8).

Thorndike, E. L. (1911). *Animal intelligence.* New York: Macmillan.

Thornhill, R., Gangestad, S. W., Miller, R., Scheyd, G., McCollough, J. K., & Franklin, M. (2003). Major histocompatibility complex genes, symmetry, and body scent attractiveness in men and women. *Behavioral Ecology, 14*(5), 668–678.

Thurstone, L. L. (1938). *Primary mental abilities.* Chicago: University of Chicago Press.

Tiihonen, J., Isohanni, M., Rasanen, P., Koiranen, M., & Moring, J. (1997). Specific major mental disorders and criminality: A 26-year prospective study of the 1966 northern Finland birth cohort. *American Journal of Psychiatry, 154,* 840–845.

Tirodkar, M. A., & Jain, A. (2003). Food messages on African American television shows. *American Journal of Public Health, 93*(3), 439–441.

Todorov, A., & Bargh, J. A. (2002). Automatic sources of aggression. *Aggression & Violent Behavior, 7*(1), 53–68.

Tolman, E. C., & Honzik, C. H. (1930). Introduction and removal of reward and maze performance in rats. *University of California Publications in Psychology, 4,* 257–275.

Torrey, E. F., & Yolken, R. H. (1998). Is household crowding a risk factor for schizophrenia and bipolar disorder? *Schizophrenia Bulletin, 24,* 321–324.

Trainor, B. C., Bird, I. M., & Marler, C. A. (2004). Opposing hormonal mechanisms of aggression revealed through short-lived testosterone manipulations and multiple winning experiences. *Hormones & Behavior, 45*(2), 115–121.

Trappey, C. (1996). A meta-analysis of consumer choice and subliminal advertising. *Psychology and Marketing, 13,* 517–530.

Triandis, H. C. (2001). Individualism-collectivism and personality. *Journal of Personality, 69*(6), 907–924.

Trull, T., Sher, K. J., Minks-Brown, C., Durbin, J., & Burr, R. (2000). Borderline personality disorder and substance use disorders: A review and integration. *Clinical Psychology Review, 20,* 235–253.

Tseng, W. (2004). Culture and psychotherapy: Asian perspectives. *Journal of Mental Health, 13*(2), 151–161.

Tsien, J. Z. (2000, April). Building a brainier mouse. *Scientific American,* pp. 62–68.

Tulving, E. (2000). Concepts of memory. In E. Tulving & F. I. M. Craik (Eds.), *The Oxford handbook of memory* (pp. 33–44). New York: Oxford University Press.

Tulving, E., & Thompson, D. M. (1973). Encoding specificity and retrieval processes in episodic memory. *Psychological Review, 80,* 352–373.

Turkington, D., Dudley, R., Warman, D. M., & Beck, A. T. (2004). Cognitive-behavioral therapy for schizophrenia: A review. *Journal of Psychiatric Practice, 10*(1), 5–16.

Turnipseed, D. L. (2003). Hardy personality: A potential link with organizational citizenship behavior. *Psychological Reports, 93*(2), 529–543.

Tversky, A., & Kahneman, D. (1974). Judgment under uncertainty: Heuristics and biases. *Science, 185,* 1124–1131.

Tversky, A., Kahneman, D. (1993). Probabilistic reasoning. In A. I. Goldman (Ed.), *Readings in philosophy and cognitive science.* Cambridge, MA: The MIT Press.

Twenge, J. M., Baumeister, R. F., Tice, D. M., & Stucke, T. S. (2001). If you can't join them, beat them: Effects of social exclusion on aggressive behavior. *Journal of Personality and Social Psychology, 81*(6), 1058–1069.

Uehara, I. (2000). Differences in episodic memory between four- and five-year-olds: False information versus real experiences. *Psychological Reports, 86*(3,Pt1), 745–755.

Uhlmann, E., & Swanson, J. (2004). Exposure to violent video games increases automatic aggressiveness. *Journal of Adolescence, 27*(1), 41–52.

Ulrich, R. E., Stachnik, T. J., & Stainton, N. R. (1963). Student acceptance of generalized personality interpretations. *Psychological Reports, 13*, 831–834.

Underwood, J., & Pezdek, K. (1998). Memory suggestibility as an example of the sleeper effect. *Psychonomic Bulletin & Review, 5*, 449–453.

Unger, J. B., Chou, C-P., Palmer, P. H., Ritt-Olson, A., Gallaher, P., Cen, S., Lichtman, K., Azen, S., & Johnson, C. A. (2004). Project FLAVOR: 1-Year outcomes of a multicultural, school-based smoking prevention curriculum for adolescents. *American Journal of Public Health, 94*(2), 263–265.

Valdez, P., Ramírez, C., & García, A. (2003). Adjustment of the sleep-wake cycle to small (1–2 h) changes in schedule. *Biological Rhythm Research, 34*(2), 145–155.

Van de Carr, F. R., & Lehrer, M. (1997). *While you are expecting: Your own prenatal classroom.* New York: Humanics Publishing.

Van Hulle, C. A., Goldsmith, H. H., & Lemery, K. S. (2004). Genetic, environmental, and gender effects on individual differences in toddler expressive language. *Journal of Speech, Language, & Hearing Research, 47*(4), 904–912.

van IJzendoorn, M. H., & De Wolff, M. S. (1997). In search of the absent father: Meta-analyses of infant-father attachment: A rejoinder to our discussants. *Child Development, 68*, 604–609.

Van Overwalle, F., & Jordens, K. (2002). *An adaptive connectionist model of cognitive dissonance.*

Vanderwerker, L. C., & Prigerson, H. G. (2004). Social support and technological connectedness as protective factors in bereavement. *Journal of Loss & Trauma, 9*(1), 45–57.

Vaughn, D. (1996). *The Challenger launch decision: Risky technology, culture, and deviance at NASA.* Chicago: University of Chicago Press.

Venkatesh, V., Morris, M. G., Sykes, T. A., & Ackerman, P. L. (2004). Individual reactions to new technologies in the workplace: The role of gender as a psychological construct. *Journal of Applied Social Psychology, 34*(3), 445–467.

Vertosick, F. T. (2000). *Why we hurt: The natural history of pain.* New York: Harcourt.

Vickers, J. C., Dickson, T.C., Adlard, P.A., Saunders, H.L., King, C.E., & McCormack, G. (2000). The cause of neuronal degeneration in Alzheimer's disease. *Progress in Neurobiology, 60*(2), 139–165.

Vidmar, N. (1997). Generic prejudice and the presumption of guilt in sex abuse trials. *Law and Human Behavior, 21*(1), 5–25.

Villarreal, D. M., Do, V., Haddad, E., & Derrick, B. E. (2002). MDA receptor antagonists sustain LTP and spatial memory: Active processes mediate LTP decay. *Nature Neuroscience, 5*(1), 48–52.

Vocks, S., Ockenfels, M., Jürgensen, R., Mussgay, L., & Rüddel, H. (2004). Blood pressure reactivity can be reduced by a cognitive behavioral stress management program. *International Journal of Behavioral Medicine, 11*(2), 63–70.

Wagner, U., Hallschmid, M., Verleger, R., & Born, J. (2003). Signs of REM sleep dependent enhancement of implicit face memory: A repetition priming study. *Biological Psychology, 62*(3), 197–210.

Wahl, H., Becker, S., Burmedi, D., & Schilling, O. (2004). The role of primary and secondary control in adaptation to age-related vision loss: A study of older adults with macular degeneration. *Psychology & Aging, 19*(1), 235–239.

Wakfield, M., Flay, B., Nichter, M., & Giovino, G. (2003). Role of the media in influencing trajectories of youth smoking. *Addiction, 98*(Suppl1), 79–103.

Wakimoto, S., & Fujihara, T. (2004). The correlation between intimacy and objective similarity in interpersonal relationships. *Social Behavior & Personality, 32*(1), 95–102.

Walker, E., Kestler, L., Bollini, A., & Hochman, K. M. Schizophrenia: Etiology and course. (2004). *Annual Review of Psychology, 55*, 401–430.

Walker, I., & Crogan, M. (1998). Academic performance, prejudice, and the jigsaw classroom: New pieces to the puzzle. *Journal of Community and Applied Social Psychology, 8*, 381–393.

Wallace, D. C. (1997, August). Mitochondrial DNA in aging and disease. *Scientific American,* 40–47.

Waller, M. R., & McLanahan, S. S. (2005). "His" and "her" marriage expectations: Determinants and consequences. *Journal of Marriage & Family, 67*(1), 53–67.

Waller, N. G., & Ross, C. A. (1997). The prevalence and biometric structure of pathological dissociation in the general population: Taxometric and behavior genetics findings. *Journal of Abnormal Psychology, 106*, 499–510.

Walter, T. & Siebert, A. (1990). *Student success* (5th ed.). Fort Worth, TX: Holt, Rinehart & Winston.

Warner, M. (2004, May 11). *Heart of darkness.* PBS Online NewsHour. [On-line]. Available: http://www.pbs.org/newshour/bb/middle_east/jan-june04/prisoners_5–11.html.

Warr, P., Butcher, V., & Robertson, I. (2004). Activity and psychological well-being in older people. *Aging & Mental Health, 8*(2), 172–183.

Wason, P. C. (1968). Reasoning about a rule. *Quarterly Journal of Experimental Psychology, 20*(3), 273–281.

Wasserman, D., & Wachbroit, R. S. (2000). *Genetics and criminal behavior.* Cambridge, MA: Cambridge University Press.

Watkins, L. M., & Johnston, L. (2000). Screening job applicants: The impact of physical attractiveness and application quality. *International Journal of Selection & Assessment, 8*(2), 76–84.

Watkins, L. R., & Maier, S. F. (2000). The pain of being sick: Implications of immune-to-brain communication for understanding pain. *Annual Review of Psychology, 51*, 29–57.

Watson, J. (1913). Psychology as the behaviorist views it. *Psychological Review, 20*, 158–177.

Watson, J.B., Mednick, S.A., Huttunen, M. & Wang, X. (1999). Prenatal teratogens and the development of adult mental illness. *Development & Psychopathology, 11*(3), 457–466.

Watson, J. B., & Rayner, R. (1920). Conditioned emotional reactions. *Journal of Experimental Psychology, 3*, 1–14.

Watson, J. B., & Rayner, R. (2000). Conditioned emotional reactions. *American Psychologist, 55*(3), 313–317.

Waye, K. P., Bengtsson, J., Rylander, R., Hucklebridge, F., Evans, P., & Clow, A. (2002). Low frequency noise enhances cortisol among noise sensitive subjects during work performance. *Life Sciences, 70*(7), 745–758.

Waynforth, D. (2001). Mate choice trade-offs and women's preference for physically attractive men. *Human Nature, 12*(3), 207–219.

Wechsler, D. (1944). *The measurement of adult intelligence* (3rd ed.). Baltimore: Williams & Wilkins.

Wechsler, D. (1977). *Manual for the Wechsler Intelligence Scale for Children* (Rev.). New York: Psychological Corporation.

Wechsler, H., Lee, J. E., Kuo, M., Seibring, M., Nelson, T. F., & Lee, H. (2002). Trends in college binge drinking during a period of increased prevention efforts: Findings from 4 Harvard School of Public Health College Alcohol Study Surveys, 1993–2001. *Journal of American College Health, 50*, 203–217.

Weinberger, J., & Westen, D. (2001). Science and psychodynamics: From arguments about Freud to data. *Psychological Inquiry, 12*(3), 129–132.

Weiner, B. (1972). *Theories of motivation.* Chicago: Rand-McNally.

Weiner, B. (1982). The emotional consequences of causal attributions. In M. S. Clark & S. T. Fiske (Eds.), *Affect and cognition.* Hillsdale, NJ: Erlbaum.

Weitlauf, J. C., Cervone, D., Smith, R. E., & Wright, P. M. (2001). Assessing generalization in perceived self-efficacy: Multidomain and global assessments of the effects of self-defense training for women. *Personality & Social Psychology Bulletin, 27*(12), 1683–1691.

Werner, J. S., & Wooten, B. R. (1979). Human infant color vision and color perception. *Infant Behavior and Development, 2*(3), 241–273.

Werth, J. L. Jr., Wright, K. S., Archambault, R. J., & Bardash, R. (2003). When does the "duty to protect" apply with a client who has anorexia nervosa? *Counseling Psychologist. 31*(4), 427–450.

Westen, D. (1998). Unconscious thought, feeling, and motivation: The end of a century-long debate. In R. F. Bornstein & J. M. Masling (Eds.), *Empirical perspectives on the psychoanalytic unconscious.* Washington, DC: American Psychological Association.

Whitbourne, S. K. (2005). *Adult development and aging: Biopsychosocial perspectives.* Hoboken, NJ: Wiley.

White, J. R., & Freeman, A. S. (2000). *Cognitive-behavioral group therapy for specific problems and populations.* Washington, DC: American Psychological Association.

White, L. K., & Rogers, S. J. (1997). Strong support but uneasy relationships: Coresidence and adult children's relationships with their parents. *Journal of Marriage and the Family, 59,* 62–76.

White, T., Andreasen, N. C., & Nopoulos, P. (2002). Brain volumes and surface morphology in monozygotic twins. *Cerebral Cortex, 12*(5), 486–493.

Whitney, D. K., Kusznir, A., & Dixie, A. (2002). Women with depression: The importance of social, psychological and occupational factors in illness and recovery. *Journal of Occupational Science, 9*(1), 20–27.

Whorf, B. L. (1956). Science and linguistics. In J. B. Carroll (Ed.), *Language, thought and reality.* Cambridge, MA: MIT Press.

Whyte, M. K., (1992, March-April). Choosing mates—the American way. *Society,* 71–77.

Wickelgren, I. (2002). *Animal studies raise hopes for spinal cord repair.* [On-line]. Available: http://www.sciencemag.org/cgi/content/full/297/5579/178.

Widiger, T. A., & Sankis, L. M. (2000). Adult psychopathology. *Annual Review of Psychology, 51,* 377–404.

Wieseler-Frank, J., Maier, S. F., & Watkins, L. R. (2005). Immune-to-brain communication dynamically modulates pain: Physiological and pathological consequences. *Brain, Behavior, & Immunity, 19*(2), 104–111.

Wigfield, A., & Eccles, J. S. (2000). Expectancy-value of achievement motivation. *Contemporary Educational Psychology, 25*(1), 68–81.

Wild, B., Rodden, F. A., Grodd, W., & Ruch, W. (2003). Neural correlates of laughter and humour. *Brain, 126*(10), 2121–2138.

Williams, A. L., Haber, D., Weaver, G. D., & Freeman, J. L. (1998). Altruistic activity: Does it make a difference in the senior center? *Activities, Adaptation and Aging, 22*(4), 31–39.

Williams, G., Cai, X. J., Elliott, J. C., & Harrold, J. A. (2004). Anabolic neuropeptides. *Physiology & Behavior, 81*(2), 211–222.

Williams, J. E., & Best, D. L. (1990). *Sex and psyche: Gender and self viewed cross-culturally.* Newbury Park, CA: Sage.

Williams, S. S. (2001). Sexual lying among college students in close and casual relationships. *Journal of Applied Social Psychology, 31*(11), 2322–2338.

Williamson, M. (1997). Circumcision anesthesia: A study of nursing implication for dorsal penile nerve block. *Pediatric Nursing, 23,* 59–63.

Willingham, D.B. (2001). *Cognition: The thinking animal.* Upper Saddle River, NJ: Prentice Hall.

Wilson, E. O. (1975). *Sociobiology: The new synthesis.* Cambridge, MA: Harvard University Press.

Wilson, E. O. (1978). *On human nature.* Cambridge, MA: Harvard University Press.

Wilson, R. S., Gilley, D. W., Bennett, D. A., Beckett, L. A., & Evans, D. A. (2000). Person-specific paths of cognitive decline in Alzheimer's disease and their relation to age. *Psychology & Aging, 15*(1), 18–28.

Wilson, T. G. (2005). Psychological treatment of eating disorders. *Annual Review of Clinical Psychology, 1,* 439–465.

Winterich, J. A. (2003). Sex, menopause, and culture: Sexual orientation and the meaning of menopause for women's sex lives. *Gender & Society, 17*(4), 627–642.

Witelson, S. F., Kigar, D. L., & Harvey, T. (1999). The exceptional brain of Albert Einstein. *The Lancet, 353,* 2149–2153.

Wixted, J. T. (2004). The psychology and neuroscience of forgetting. *Annual Review of Psychology, 55,* 235–269.

Wixted, J. T. (2005). A theory about why we forget what we once knew. Current Directions in Psychological Science, 14(1), 6–9.

Wolpe, J. & Plaud, J.J. (1997). Pavlov's contributions to behavior therapy. *American Psychologist, 52*(9), 966–972.

Wood, J. M., Lilienfeld, S. O., Nezworski, M. T., & Garb, H. N. (2001). Coming to grips with negative evidence for the comprehensive system for the Rorschach: A comment on Gacono, Loving, and Bodholdt, Ganellen, and Bornstein. *Journal of Personality Assessment, 77*(1), 48–70.

Wood, W., Christensen, P. N., Hebl, M. R., & Rothgerber, H. (1997). Conformity to sex-typed norms, affect, and the self-concept. *Journal of Personality and Social Psychology, 73*(3), 523–535.

Woods, S. C., Schwartz, M. W., Baskin, D. G., & Seeley, R. J. (2000). Food intake and the regulation of body weight. *Annual Review of Psychology, 51,* 255–277.

World Health Organization. (1992). *ICD-10: International statistical classification of diseases and related health problems* (10th rev. ed.). Arlington, VA: American Psychiatric Publishing.

Wright, J. H., & Beck, A. T. (1999). Cognitive therapies. In R. E. Hales, S. C. Yudofsky, & J. A. Talbott (Eds.), *American Psychiatric Press textbook of psychiatry.* Washington, DC: American Psychiatric Press.

Wright, K. (2003). Relationships with death: The terminally ill talk about dying. *Journal of Marital & Family Therapy, 29*(4), 439–454.

Wright, W. (1998). *Born that way: Genes, behavior, personality.* New York: Knopf.

Wüst, S., Entringer, S., Federenko, I. S., Schlotz, W., & Hellhammer, D. H. (2005). Birth weight is associated with salivary cortisol responses to psychosocial stress in adult life. *Psychoneuroendocrinology. 30*(6), 591–598.

Wyatt, R. C. (2004). Thomas Szasz: Liberty and the practice of psychotherapy. *Journal of Humanistic Psychology, 44*(1), 71–85.

Yacoubian, G. S. Jr., Green, M. K., & Peters, R. J. (2003). Identifying the prevalence and correlates of ecstasy and other club drug (EOCD) use among high school seniors. *Journal of Ethnicity in Substance Abuse, 2*(2), 53–66.

Yamada, H. (1997). *Different games, different rules: Why Americans and Japanese misunderstand each other.* London: Oxford University Press.

Yarmey, A. D. (2004). Eyewitness recall and photo identification: A field experiment. *Psychology, Crime & Law, 10*(1), 53–68.

Yee, A. H., Fairchild, H. H., Weizmann, F., & Wyatt, G. E. (1993). Addressing psychology's problem with race. *American Psychologist, 48,* 1132–1140.

Yeh, S. J., & Lo, S. K. (2004). Living alone, social support, and feeling lonely among the elderly. *Social Behavior & Personality, 32*(2), 129–138.

Yesavage, J. A., Friedman, L., Kraemer, H., Tinklenberg, J. R., Salehi, A., Noda, A., Taylor, J. L., O'Hara, R., & Murphy, G. (2004). Sleep/wake disruption in Alzheimer's disease: APOE status and longitudinal course. *Journal of Geriatric Psychiatry & Neurology, 17*(1), 20–24.

Young, J.D., & Taylor, E. (1998). Meditation as a voluntary hypometabolic state of biological estivation. *News in Physiological Science, 13,* 149–153.

Young, T. (1802). *Color vision.* Philosophical Transactions of the Royal Society, p. 12.

Young, T., Skatrud, J., & Peppard, P. E. (2004). Risk factors for obstructive sleep apnea in adults. *JAMA: Journal of the American Medical Association, 291*(16), 2013–2016.

Zagnoni, P. G., & Albano, C. (2002). Psychostimulants and epilepsy. *Epilepsia, 43,* 28–31.

Zeanah, C. H. (2000). Disturbances of attachment in young children adopted from institutions. *Journal of Developmental & Behavioral Pediatrics, 21*(3), 230–236.

Zebrowitz, L. A., Hall, J. A., Murphy, N. A., & Rhodes, G. (2002). Looking smart and looking good: Facial cues to intelligence and their origins. *Personality & Social Psychology Bulletin, 28*(2), 238–249.

Zhukov, D. A., & Vinogradova, K. P. (2002). Learned helplessness or learned inactivity after inescapable stress? Interpretation depends on coping styles. *Integrative Physiological & Behavioral Science, 37*(1), 35–43.

Ziegert, Jonathan C., & Hanges, Paul J. (2005). Employment Discrimination: The Role of Implicit Attitudes, Motivation, and a Climate for Racial Bias. *Journal of Applied Psychology, 90*(3), 553–562.

Zimbardo, P. G. (1993). *Stanford prison experiment: A 20-year retrospective.* Invited presentation at the meeting of the Western Psychological Association, Phoenix, AZ.

Zimbardo, P. G. (2004). A situationist perspective on the psychology of evil: Understanding how good people are transformed into perpetrators. In A. G. Miller (Ed.), *The social psychology of good and evil* (pp. 21–50). New York: Guilford Press.

Zimbardo, P. G., Ebbeson, E. B., & Maslach, C. (1977). *Influencing attitudes and changing behavior.* Reading, MA: Addison-Wesley.

Zuckerman, L., &, Weiner, I. (2005). Maternal immune activation leads to behavioral and pharmacological changes in the adult offspring. *Journal of Psychiatric Research, 39*(3), 311–323.

Zuckerman, M. (1978, February). The search for high sensation. *Psychology Today,* 38–46.

Zuckerman, M. (1979). *Sensation seeking: Beyond the optimal level of arousal.* Hillsdale, NJ: Erlbaum.

Zuckerman, M. (1994). *Behavioral expressions and biosocial bases of sensation seeking.* New York: Cambridge University Press.

Zuckerman, M. (2004). The shaping of personality: Genes, environments, and chance encounters. *Journal of Personality Assessment, 82*(1), 11–22.

PHOTO CREDITS

Chapter 1

Pages 2–3: Bill Hatcher/NG Image Collection; page 2: (bottom inset) Michael Nichols/NG Image Collection; page 2: (center inset) Martin Harvey/Getty Images; page 5: (top) Henry Groskinsky/Time Life Pictures/Getty Images; page 5: (center) All Points Bulletin/Getty Images; page 7: Bettmann/Corbis Images; page 8: Nina Leen/Time Life Pictures/Getty Images; page 10: Tim Mantoani/Masterfile; page 11: (left) Psychology Archives—The University of Akron; page 11: (right) Deborah Feingold/Archive Photos/Getty Images; page 14: Photodisc/Getty Images; page 18: (top left) Photodisc Red/Getty Images; page 18: (bottom left) Drawing by David Sipress/The Cartoon Bank, Inc.; page 18: (bottom right) Jeffrey Greenberg/Photo Researchers; page 19: (left) Age Fotostock/Superstock; page 19: (right) From Damasio H, Grabowski T, Frank R, Galaburda AM, Damasio AR: The return of Phineas Gage: Clues about the brain from a famous patient. Science, 264:1102–1105, 1994. Department of Neurology & Image Analysis Facility, University of Iowa.; page 21: (left) Chris Carroll/Corbis Images; page 21: (right) Digital Vision/Getty Images; page 22: (top) Science Pictures Ltd./Photo Researchers; page 22: (bottom) Alexander Tsiaras/Stock Boston; page 23: (top) Mehau Kulyk/Photo Researchers; page 23: (center) N.I. H/Photo Researchers; page 23: (bottom) Scott Camazine/Photo Researchers; page 25: Pixtal/Age Fotostock; page 27: (bottom) Jonathan Selig/Getty Images; page 28: (top left) All Points Bulletin/Getty Images; page 28: (top right) Nina Leen/Time Life Pictures/Getty Images; page 29: (top) Photodisc Red/Getty Images; page 29: (bottom) Pixtal/Age Fotostock.

Chapter 2

Pages 32–33: Joel Saget/AFP/Getty Images; page 35: (top) EPA/STR/Landov; page 35: (bottom) Age Fotostock/Superstock; page 36: David Young-Wolff/Stone/Getty Images, Inc; page 38: Courtesy E.R. Lewis, Berkeley; page 41: Reuters New Media Inc./Corbis Images; page 44: Science Pictures Limited/Photo Researchers; page 45: Courtesy Taub Therapy Clinc/UAB Media Relations; page 47: Joel Sartore/NG Image Collection; page 48: (left) Mike Powell/The Image Bank/Getty Images, Inc; page 48: (right) NG Image Collection; page 49: (top) Rob McEwan/Everett Collection, Inc.; page 49: (bottom) Rick Rusing/Stone/Getty Images; page 50: Getty Images News and Sport Services; page 52: (left) Adrian Weinbrecht/Stone/Getty Images; page 52: (center) Corbis Images; page 52: (far left) PictureQuest; page 52: (right) Studio MPM/The Image Bank/Getty Images, Inc; page 54: ©AP/Wide World Photos; page 57: Courtesy Karen Huffman.

Chapter 3

Pages 62–63: ©AP/Wide World Photos; page 62: (bottom) Ellis Years/Taxi/Getty Images, Inc; page 64: Alan Powdrill/Taxi/Getty Images; page 65: Sarah Leen/NG Image Collection; page 66: Photofest Inc.; page 68: Photodisc/Getty; page 69: Bonnier Alba; page 70: Gusto/Photo Researchers; page 71: Stuart Pearce/Age Fotostock; page 72: ©AP/Wide World Photos; page 73: ©AP/Wide World Photos; page 74: Colin Cuthbert/Photo Researchers; page 75: (top) Michael Siluk/The Image Works; page 75: (bottom) Michael Siluk/Index Stock; page 76: Simon McComb Photography/Photonica/Getty Images; page 77: (top) Dennis MacDonald/PhotoEdit; page 77: (bottom) BananaStock/Age Fotostock ; page 78: James Schnepf; page 79: ©AP/Wide World Photos; page 80: Comstock Images/Age Fotostock; page 81: Michael Newman/PhotoEdit; page 82: James Schnepf; page 85: James Schnepf.

Chapter 4

Pages 86–87: Yohio Tomii/Superstock; page 89: (top left) Karen Kasmauski/NG Image Collection; page 89: (bottom) Phanie/Photo Researchers, Inc.; page 90: Terje Rakke/Riser/Getty Images, Inc; page 91: Stephen St. John/NG Image Collection; page 95: ThinkStock/Superstock; page 98: (left) Roy Toft/NG Image Collection; page 98: (center) John Dominis/Time Life Pictures/Getty Images; page 98: (right) ©AP/Wide World Photos; page 100: (top) Ryan McVay/Photodisc Green/Getty; page 100: (center) Stephen Dalton/Photo Researchers, Inc.; page 100: (bottom) VEER Stephen Simpson/Photonica/Getty Images; page 102: (top) Rykoff Collection/Corbis; page 102: (bottom) ©Cordon Art b.v.; page 104: (top) Medford Taylor/NG Image Collection; page 104: (bottom) Baaron Wolman/Woodfin Camp; page 105: Ed George/NG Image Collection; page 107: ©AP/Wide World Photos; page 109: Lindsey Hebberd/Corbis.

Chapter 5

Pages 114–115: George Logan/Taxi/Getty Images; page 117: (top) Patrick Ramsey/Image State/Jupiter Images; page 117: (just below top, left side) Walter Hodges/Stone/Getty Images; page 117: (just below top, right side) Richard Hutchings/Photo Researchers; page 117: (center) Zigy Kaluzny/Getty Images; page 117: (bottom) David Madison/Duomo Photography, Inc; page 120: (top) Hank Morgan/Rainbow; page 120: (bottom) Davis Meltzer/NG Image Collection; page 121: Davis Meltzer/NG Image Collection; page 123: Louis Psihoyos; page 124: (left) ©2006 Nucleus Medical Art. All rights reserved. www.nucleusinc.com; page 124: (center) Courtesy Stephen David Smith, D.M.D.; page 124: (right) Photo Researchers, Inc.; page 125: (top) Zigy Kaluzny/Stone/Getty Images; page 126: Roberto Benzi/Age Fotostock; page 128: (top) John E. Kelly/Stone/Getty Images; page 128: (just below top) Joe Pellegrini/Foodpix/ PictureArts Corp.; page 128: (center) Adam Hart-Davis/Photo Researchers; page 128: (just below center) Uwe Schmid/OKAPIA/Photo Researchers; page 128: (bottom) ©AP/Wide World Photos; page 130: (left) Blair Seitz/Photo Researchers; page 130: (right) Jim Varney/Photo Researchers; page 131: (top left) Courtesy DEA; page 131: (center left) Sinclair Stammers/Photo Researchers, Inc.; page 131: (bottom left) AFP/Getty Images; page 131: (top right) Evan Roberts/Aurora Photos; page 131: (center right) Tom & DeeAnn McCarthy/Corbis; page 131: (bottom right) Matt Cardy/Getty Images; page 133: (left) Emme Lee/Photographer's Choice RF/Getty Images; page 133: (right) Terry Vine/Blend/Getty Images; page 134: (top) Dan Dalton/PhotoDisc Red/Getty; page 135: (left) Bill Bridges/Getty Images; page 135: (right) Michael Newman/PhotoEdit; page 136: (bottom left) Johner Images/Getty Images, Inc; page 136: (bottom right) J. P. Thomas/Jacana/Photo Researchers; page 137: Zigy Kaluzny/Stone/Getty Images.

Chapter 6

Page 142–143: ©AP/Wide World Photos; page 142: (inset) Randy Olsen/NG Image Collection; page 144: Gary Connor/Index Stock; page 146: Courtesy Benjamin Harris; page 147: ©The New Yorker Collection 1998 Sam Gross from the cartoonbank.com. All Rights Reserved.; page 148: Courtesy Benjamin Harris; page 150: Gary Connor/Index Stock; page 152: ©AP/Wide World Photos; page 153: Kaku Kurita; page 155: (top) Cynthia Dopkin/Photo Researchers; page 155: (bottom) Kaku Kurita; page 156: (left) Joe Raedle/Getty Images News and Sport Services; page 156: (right) Courtesy American Philosophical Society; page 157: Joel Sartore/NG Image Collection; page 158: Courtesy Albert Bandura;

TEXT AND ILLUSTRATION CREDITS

Chapter 2

Figures 2.3 and What is happening in this picture: From Kimura, D.; *Sex Differences in the Brain*, Scientific American, Sept. 1992, p. 120 © 1992 Jared Schneidman Designs.

Chapter 3

Table 3.1: From Holmes & Rahe, *The social readjustment rating scale,* © 1967 Journal of Psychosomatic Research, 11 pp. 213–218. Figure 3.4: From Selye; *Stress Without Distress,* © 1974 Harper & Row Publishers. "What a Scientist Sees," pg. 71: From *"Hostility, the metabolic syndrome, and incident coronary heart disease,"* R. Niaura, Ph.D., published in *Health Psychology*, 2002. Table 3.2: From Pomponio, © 2002, John Wiley & Sons.

Chapter 5

Figure 5.3B text adapted from Besnon, Herbert, "Systematic hypertension and the relaxation response". *New England Journal of Medicine*, 296, 1977. Reprinted with permission.

Chapter 7

Pg. 187, top, From Bahrick, Bahrick, and Wittlinger; *Those Unforgettable High School Days*. Reprinted with permission from Psychology Today Magazine, © 1978 Sussex Publishers, Inc.

Chapter 8

Table 8.5: From Sternberg; *Beyond IQ,* © 1985 Cambridge University Press.

Chapter 9

Figure 9.2: From Schaie; *The Life Course of Adult Intellectual Abilities,* American Psychologist, 49, pg. 304–313, © 1994 The American Psychological Association. Reprinted with permission. Figure 9.6: From Frankenburg & Dodds, Denver II Training Manual © 1991 *Denver Developmental Materials.* Pg. 236, "Milestones in motor development." Source: WK Frankenburg, J. Dobbs, P. Archer, H. Shapiro and B. Bresnick, *The Denver II: a major revision and restandardization of the Denver Developmental Screening Test.* Pediatrics, Jan 1992; 89: 91–97. Figure 9.9: From Tanner J. M., Whitehouse, R. N., and Takaislu, M., "Male/female growth spurt." *Archives of Diseases in Childhood*, 41, 454–471, 1966. Reprinted with permission. Figure 9.12: From Chart of elderly achievers. In *The brain: A user's manual.* Copyright © 1982 by Diagram Visual Information.

Chapter 10

Table 10.1: Adapted from *Handbook of socialization theory and research* (1969) D. A. Goslin (ed.). Reprinted by permission of Houghton Mifflin. Figure 10.5: From Papalia, Olds & Feldman; *Human Development* 83, © 2001 McGraw-Hill. Reprinted with permission. Figure 10.6: From Miracle, Miracle, & Baumeister: *Human Sexuality: Meeting Your Basic Needs*, 2nd edition. © 2006. Reprinted by permission of Pearson Education Inc., Upper Saddle River,

NJ. Table 10.4: Are you in the right job? Source: Reproduced by special permission of the Publisher, Psychological Assessment Resources, Inc. from *Making Vocational Choices*, 3rd ed., by John L. Holland, PhD. Copyright 1973, 1985, 1992, 1997 by PAR, Inc. Further reproduction is prohibited without permission from PAR, Inc.

Chapter 11

With Your Own Eyes, "Sensation Seeking," p. 416–417: From Zuckerman; *"The Search for High Sensation"*, Psychology Today, p.38–46. Copyright © 1978 by the American Psychological Association. Reprinted by permission. Figure 11.4: From Maslow, *"A Theory of Motivation,"* in Motivation and Personality, 3e © 1997. Reprinted by permission of Pearson Education, Inc., Upper Saddle River, NJ. Figure 11.7: From Lepper, Greene, and Nisbett; *Undermining Children's Intrinsic Interest In Extrinsic Rewards*, Journal of Personality and Social Psychology, 8e, pp. 129–137, © 1973 American Psychological Association. Figure 11.12: From Hebb, D. O., Organization of Behavior, John Wiley and Sons, Inc. Copyright © 1949. Reprinted by permission of D. O. Hebb and J. Nichols Hebb Paul, sole executor and daughter. Figure 11.21: From Plutchik; *Emotion: A Psychoevolutionary Synthesis,* © 1980 Published by Allyn and Bacon, Boston, MA. Copyright © 1980 by Pearson Education. Adapted by permission of the publisher.

Chapter 12

With Your Own Eyes, Love and the "Big Five": From Cattell: *"Personality Pinned Down,"* Psychology Today, pg. 44, Reprinted With Permission from Psychology Today Magazine, © 1974 Sussex Publishers, Inc. Bottom left: From Buss et al., *"International Preferences in Selecting Mates,"* Journal of Cross-Cultural Psychology, 21, pp. 5–47, © 1990 Sage Publications, Inc.

Chapter 13

Figure 13.3: From Diagnostic and Statistical Manual of Mental Disorder; fourth edition text revision, Washington, DC, © 2000 American Psychiatric Association. Table 13.1: From Diagnostic and Statistical Manual of Mental Disorders, fourth edition text revision, Washington, DC, © 2000 American Psychiatric Association. Figure 13.9: From Gottesman, *Schizophrenia Genesis,* © 1991 W.H. Freeman.

Chapter 14

Figure 14.6: Worksheet printed with permission of Judith S. Beck, Ph.D., © 1993, from *Cognitive Therapy: Basics and Beyond* (1995), Guilford Press.

Chapter 15

Table 15.1: From Rubin, *"Measurement of Romantic Love,"* Journal of Personality and Social Psychology, 16, pp. 265–273, © 1970 Z. Rubin. Figure 15.14: From Janis, Irving L. *Groupthink: Psychological Studies of Policy Decisions and Fiascoes,* 2e, © 1982 Houghton Mifflin. Adapted with permission.

NAME INDEX

Brown, Paul, 106
Brown, R. J., 136, 183, 191
Brown, R. P., 266
Browne, K. O., 364f
Bruck, C. S., 71
Bruggeman, E. L., 263
Brummett, B. H., 70
Brusasca, C., 134
Buchanan, W. T., 183
Buck, R., 304t
Buffalo, E. A., 183
Bugental, B. B., 265
Buhrich, N., 359
Bukowski, W. M., 265
Bulik, C. M., 292
Bull, L., 234
Bunce, D., 80
Burchfield, C., 130, 132
Burden, T., 219
Burgdorf, J., 157
Burger, J. M., 424
Burns, N. R., 211
Burns, R. A., 188
Burns, R. G., 5
Burton, L. A., 180
Bush, George W., 183
Bushman, B. J., 167, 420
Buss, David M., 9t, 316, 408
Butcher, J. N., 333
Butcher, V., 274
Butler, K., 130
Butler, R. A., 285
Butterweck, V., 390
Buunk, B. P., 408
Byrd, James, 142, 164, 166

C

Cacioppo, J. T., 66
Cadoret, R. J., 268
Calkins, Mary, 11, 28
Caltran, G., 248
Camarena, B., 291
Camodeca, M., 267
Campbell, L., 410
Campbell, L. C., 77
Cannella, C., 289
Cannon, T. D., 356
Cannon, Walter B., 300, 301f, 302
Cantor, J. M., 294
Carducci, B. J., 329
Carey, G., 419
Carless, S. A., 274
Carlsmith, J. Merrill, 404
Carmaciu, C. D., 365f
Carnagey, N. L., 167
Carnelley, K. B., 259
Carney, S., 388
Carpenter, S., 407
Carrera, P., 402

Carrere, S., 381
Carskadon, M. A., 118
Carstensen, L. L., 275
Cassidy, J., 258
Casti, J. L., 208f
Catalano, M., 274
Catlett, B. S., 76
Cattell, Raymond B., 211, 314
Cavacuiti, C. A., 382
Ceci, S. J., 188
Cervone, D., 328
Cesaro, P., 90
Ceschi, G., 301
Chabrol, H., 362
Chaikin, N. D., 393
Challem, J., 290
Chan, W. T., 80
Chang, G., 45
Chapelot, D., 289
Charles, S., 275
Charney, D. S., 81, 353
Chartrand, T., 424
Chase-Lansdale, P. L., 247
Chen, F., 409
Cherney, I. D., 269
Cheung, C., 287, 328f
Cheung, M. S., 353, 363
Chi, I., 81
Chiavegatto, S., 268
Chilcoat, H. D., 132
Cho, K., 118
Choi, I., 404
Choi, N., 269
Chomsky, Noam, 208
Chou, K., 81
Chow, Amy, 98f
Christ, A. E., 276
Christ, G. H., 276
Christensen, D., 124
Christensen, P. N., 270f
Chuang, D., 353
Chun, M. M., 178
Cialdini, R. B., 420
Cicchetti, Dante, 317
Cillesen, A. H. N., 267
Clark, Kenneth B., 11, 28, 163
Clark, Mamie P., 11, 163
Clements, C. M., 155
Clinton, S. M., 52
Cloud, J., 388
Clowes-Hollins, V., 350
Coan, J., 381
Coccaro, E. F., 268, 419
Cochrane, D. E., 66
Cohen, A., 364
Cohen, A. B., 97
Cohen, D., 76
Cohen, R. A., 389
Cohen, S., 66, 81

Coley, R. L., 247
Collins, N. L., 259
Collins, P. F., 330
Connolly, S., 130
Connor, J., 116
Connor, K. M., 390
Conway, M. A., 187f
Cook, D. A., 365
Cook, M., 350
Cook, P. F., 422
Coombs, R. H., 129, 130, 291, 292
Cooper, C. J., 211
Cooper, M. L., 259
Cooper, R., 346
Cooper, W. E., Jr., 97, 161, 162
Copeland, P., 294
Corballis, M. C., 201
Corcoran, C., 358
Corey, G., 14, 392
Corkin, S., 172
Costa, J. L., 291, 317f
Costa, P. T., Jr., 314
Courage, M. L., 237
Covinton, M. V., 296
Cowell, Simon, 355
Cowen, P. J., 66
Coyne, S. M., 167
Crabbe, J., 330
Craik, Fergus, 175
Crair, M. C., 100
Crandall, C. S., 292
Craske, M. G., 348
Crews, F. T., 76
Crim, J. L., 132
Cristofalo, V. J., 241
Crogan, M., 422
Cronbach, L., 335
Crone, C. C., 390
Crooks, R., 268t
Cropsey, K. L., 286
Crosby, B., 378
Crow, T. J., 356
Crowe, Russell, 358f, 395
Csernansky, J. G., 419
Cullen, D., 288
Cummings, E., 274
Cunningham, C. L., 14
Curran, D., 268
Currie, S., 77
Curtiss, G., 274
Curtiss, S., 229

D

Dabbs, J. M., 419
Dabbs, M. G., 419
Dackis, C. A., 52
D'Alessio, D., 422
Dallos, R., 272

Dalman, C., 356
Dalton, P., 96
Damak, S., 97
Dana, R. H., 346
Dane, F. C., 191
Daniels, J. R., 269
Dantzer, R., 161
Darley, John, 420, 421
Darmani, N. A., 132
Darwin, Charles, 7, 9t, 37, 302, 305
Dasen, P. R., 393
Daum, I., 200
Davalos, D. B., 356
David, D., 70, 136, 378
Davidson, J. R., 382, 390, 419
Davies, 105, 192
Davis, C., 292
Davis, M., 163
Davison, G. C., 340, 361, 365f, 381
Dawis, R. V., 211
Dawson, K. A., 118
De Charms, Richard, 293f
De Coteau, T. J., 64
De Geus, E. J. C., 217
De Rosnay, M., 258
De Vicente, F., 107
De Wolff, M. S., 257, 258
Deary, I. J., 211, 217
Debaere, F., 160
DeCasper, A. J., 237
Deci, E. L., 296
Deckers, L., 330
DeClue, G., 306
Deffenbacher, K., 191
Deiss, C., 151
Delaney-Black, V., 234f
Delgado-Gaitan, C., 231
Delgado, J. M. R., 419
Delgado, P. L., 353
DeLisi, L. E., 357
Delville, Y., 419
Dement, William C., 116, 118, 121, 123, 123f
Den Boer, J. A., 349
DeNeve, K. M., 419
Denmark, F. L., 270
DePaulo, J. F., 352
Depue, R. A., 330
Der-Karabetian, A., 422
Dermer, M., 409f
Derrick, B. E., 187f
DeStefano, D., 178
DeValois, R. L., 106
Devine, P. G., 406
Devlin, M. J., 291
Devor, E., 268
Diamond, L. M., 259

Greenberg, J., 80, 263
Greenberg, R. P., 122
Greene, D., 297
Greene, K., 247
Greenfield, Patricia M., 9t, 248
Greenwald, A. G., 406
Grekin, E. R., 76
Griffin, D., 402
Grigorenko, E. L., 212
Grilly, D., 76, 130, 234
Grilo, C. M., 291, 292
Grondin, S., 177
Grossmann, K., 257
Grove, W. M., 335
Gruber, U., 45
Grunberg, L., 64
Guarnaccia, P. J., 342, 365f
Gueler, F., 119
Guilarte, T. R., 160
Guilford, J. P., 211
Guinard, J. X., 97
Guinote, A., 407
Gunzerath, L., 20
Gurney, J. G., 79
Gustavson, C. R., 161
Guthrie, E., 376

H
Haddad, E., 187f
Haier, R. J., 217
Haith, M. M., 237
Halberstadt, J., 409
Haller, S., 20
Halperin, J. M., 419
Halpern, D. F., 266, 312
Hamer, D. H., 294
Hamid, P. N., 80
Hamilton, M., 406
Hamm, J. P., 201
Hammack, P. L., 378
Hammen, C., 353
Hammond, D. C., 78, 81, 165
Haney, C., 416
Hanges, P. J., 406
Hanke, M., 76
Hanley, S. J., 288
Hanna, S. M., 382f
Hanson, D. R., 229
Harandi, A. A., 136
Harker, L., 273
Harlow, Harry F., 257, 285
Harlow, J. M., 19f
Harlow, M. K., 257, 285
Harper, F. D., 288
Harper, J. A., 288
Harris, A. C., 270
Harris, P. L., 258
Harrison, E., 134
Hart, A. D., 78

Hart, K. J., 263
Harvey, M. G., 420
Harvey, T., 218
Hatfield, E., 410
Hattie, J. A., 273
Haugen, R., 287
Hauser, R. M., 219
Havighurst, R. J., 275
Hawkins, R. S., 78
Hawkley, L. C., 66
Hay, D. F., 420
Hayashi, T., 330
Hayes, R., 248
Hayflick, Leonard, 241
Hayne, H., 179
Hays, W. S. T., 97
Hazan, Cindy, 259
Healy, A. F., 175
Heaton, T. B., 273
Hedges, D., 130, 132
Hedlund, J., 211, 212, 219
Heffelfinger, A. K., 183
Heider, F., 402
Heimann, M., 248f
Heine, S. J., 403
Helgeland, M. I., 362
Helmholtz, Herman von, 105
Helms, J. E., 365
Henderlong, J., 296
Henke, P. G., 73
Hennen, J., 353
Hennessey, B. A., 296
Henning, K., 267
Henry, W. E., 274
Hergovich, A., 418
Hering, Ewald, 105–106
Heriot, S. A., 384, 386
Herman, C. J., 240
Herman, L. M., 210
Hermann, W. M., 116
Hermelin, B., 216
Hernandez-Reif, M., 257
Hernstein, Richard J., 219
Herrera, E. A., 386
Herron, W. G., 167
Hervé, H., 361
Heslegrave, R., 362
Hettema, J., 330
Hewitt, J. K., 291
Heyder, K., 200
Heyman, G. D., 269
Higgins, N. C., 331, 402
Higher Education Research Institute, 274
Hilgard, E. R., 135, 136
Hill, E. L., 54
Hiller, J. B., 358, 382
Hilsenroth, K. J., 376
Hinton, E. C., 52

Hirao-Try, Y., 70
Hirosumi, J., 289
Hirshkowitz, M., 122
Hittner, J. B., 269
Hobfoll, S. E., 81
Hobson, Alan, 121, 122, 123, 125
Hochgeschwender, U., 291
Hock, H. W., 292
Hodder, T., 359
Hoffman, M. L., 263, 420
Hofmann, V., 258
Hogan, T. P., 215, 334
Holland, John, 274
Holm, R. E., 79
Holmes, Thomas, 64
Holt, N. L., 64f
Holtzworth-Munroe, A., 268, 393
Hong, S., 80
Honts, C. R., 306
Honzik, C. H., 157
Hooley, J. M., 358, 382
Hope, D. A., 64
Hopkins, Anthony, 372
Hopson, D. S., 163
Horiuchi, Y., 353
Horn, J. M., 218
Horney, Karen, 9t, 323, 323f
Horowitz, L. M., 409
Hovland, C. I., 146
Howe, M. L., 186, 187f
Hrebickova, M., 314, 317f
Huang, L. N., 270
Huang, M., 219
Hubel, David, 9t
Huffman, C. J., 177f
Hugdahl, K., 183
Hull, Clark, 9t, 285
Hulse, G. K., 75
Hulshoff Pol, H. E., 356, 357
Huntington, R., 277
Hur, Y., 218
Hvas, L., 239
Hyde, J. S., 263
Hyman, I. E., Jr., 188
Hyman, R., 108, 312
Hypericum Depression Trial Study, 390

I
Iacono, W. G., 306
Ibañez, A., 330
Ice, G. H., 116
Ikemi, A., 393
Ikemoto, K., 357
Ingvar, M., 51
Ippolito, D. L., 289
Irons, C., 353, 363
Irwin, M., 69
Itakura, S., 209

Ivanovic, D. M., 218
Iversen, L., 132
Iyer, P., 118
Izard, Carroll, 304t

J
Jackendoff, R., 210
Jacob, S., 97
Jacobi, C., 291, 292
Jacobsen, T., 258
Jaffee, S., 263
Jain, A., 164
James, J. B., 220
James, William, 7, 11, 28, 116, 300, 301f, 302
Jamison, K. R., 352
Jang, K. L., 361
Janis, I. L., 418
Jarvin, L., 212
Jasperse, M., 240
Javo, C., 154, 155
Jeffries, M., 130
Jeffs, D., 274
Jellison, W. A., 295
Jenkins, C. D., 291
Jennings, L., 392
Jensen, Arthur, 219
Jensen, M. P., 78
Jiang, Y., 178
Joel, J. C., 130
John, U., 76
Johnson, B., 155, 201
Johnson, T. J., 286
Johnson, Virginia E., 268t, 294, 294f
Johnson, W., 211, 330
Johnston, C. A., 386
Johnston, L., 408
Johnstone, L., 389
Jolliffe, C. D., 90
Jones, D. G., 46
Jordens, K., 404
Josephs, R. A., 219, 266
Jourdan, A., 381
Jung, Carl, 9t, 322, 323

K
Kagan, J., 330
Kahneman, Daniel, 202, 204, 402
Kalidindi, S., 356
Kamiya, A., 356
Kane, T. A., 292
Kaneko, T., 183
Kapner, D. A., 77
Kareev, Y., 177
Karmiloff, K., 237
Karmiloff-Smith, A., 237
Kastenbaum, R., 276
Katovich, M. A., 5

SUBJECT INDEX

convolutions in, 53
emotional experience and, 298
Challenger explosion, 194
Change, 228–229
as goal of psychology, 5
Chemical senses, 96–97, 110
Children
aggression in, 419
animistic thinking in, 245. *See also* Cognitive development
awareness of gender roles in, 268
communication and, 207–208
effect of early malnutrition on brain development in, 218
psychosocial development in, 264*f*
sample environmental conditions that endanger, 234*t*
sleep disturbance and, 123
stages of language acquisition in, 208*t*–209*t*
traits in abused and neglected, 317
Chromosomes, 34, 34*f*, 58
Chronic pain, 77–78. *See also* Pain
biofeedback use for, 78*f*
Chronic stressors, 64–65
HPA Axis and, 67*f*
Chunking, 177
in chess, 177
Cingulotomy, 389
Circadian rhythms, 116
disruptions in, 116
physiological functions following, 137
regulation of, 118*f*
Clairvoyance, 108
Classical conditioning, 144
differences between operant and, 149
example of Pavlov's and modern-day, 145*f*
fine-tuning, 146–148
as marketing tool, 164
from prejudice to phobias, 163
relation to learning of, 146
six principles of, 147*f*
techniques of behavior therapists, 383–384
Client-centered therapy, 380–381
Club drugs, 130
effects of, 132
risky behaviors and, 132
Cocaine, 126*f*
Cochlea, 94*f*
Cochlear fluid, 94*f*
Cochlear implant, 95
Coding, 88, 109
Cognition, 198
building blocks, 200–201
facets of, 200
faulty, 350
Cognitive behavioral theory, 156

Cognitive-behavior therapy, 378
"thought journals" of, 379*f*
Cognitive development, 242–249, 243*f*, 251
concrete operational stage, 245
contributions to psychology of, 249
criticisms of theory of, 248–249
formal operational stage, 247–248
Piaget's four stages of, 243*f*
preoperational stage, 244
sensorimotor stage, 244
social learning and, 269
Cognitive dissonance, 404, 410
changing attitudes using principles of, 422–423
creating, 423*f*
culture and, 425
theory, 404*f*
Cognitive expectancies, 329
Cognitive map, 157
Cognitive perspective, 8
Cognitive restructuring, 377
Cognitive retraining, 422
Cognitive-social learning, 166–167
definition of, 156
Cognitive-social theory, 156–158
definition of, 156
Cognitive theories, of motivation, 287
Cognitive therapy, 376–378
Cognitive view, 123
Cohort effects, 230
Collective unconscious, 322
archetypes in, 323*f*
Collectivistic cultures, 270–271, 271*t*
Color
aftereffects, 105, 106
-deficient vision, 106, 107*f*
perception, 105–106
Color blindness, 107*f*
Coma, 116
Committed relationships
factors common to happy, 273
overcoming unrealistic expectations in, 272
Communication. *See also* Language
babies and, 207
by blinking left eye, 114
children and, 207–208
computer-aided, 210*f*
emotional, 299
nonverbal, 207, 208*t*–209*t*, 221
Community mental health (CMH) centers, 394*f*
Comorbidity, 359–360
Companionate love, 410
Competition
genetic transmission of, 284*f*
reducing, 422
Conception, 232*f*

Concepts
artificial, 201
basic-level, 201
formation of, 201
Concrete operational stage, 243*f*, 245
Conditioned emotional responses, 146, 146*f*
Conditioned response, 144
Conditioned stimulus, 144, 148*f*
food or drink as, 161
Conditioning, 7, 144
and learning in everyday life, 163–167
classical, 144
higher-order, 148*f*
operant, 149, 164–166
Conduction deafness, 94–95
Cones, 93*f*, 110
Confidentiality, 14
Confirmation bias, 202
Conflict(s)
approach–approach, 66
approach–avoidance, 66
avoidance–avoidance, 66
interpersonal, 71
stress and, 66
types of, 82
Conformity, 411
advantages of obedience and, 413*f*
factors driving, 412, 426
Congruence, mental health, and self-esteem, 325, 326
Connectionist model, 174
Conscientiousness, 315, 317*f*
Conscious, 318, 336
Consciousness, 114
altering, through meditation and hypnosis, 133–136
control of, 117*f*
drug use and alterations of, 130
William James's description of, 116
levels of, 117*f*
Conservation, 245
experiments used for testing, 246
test for, 245*f*
Consistency
attunement to, 104
need for, 190
Contact comfort, 257, 278
Continuity model, 228
Continuous reinforcement, 150
Control, 80
Control group, 17*f*
Controlled processes, 117*f*
Conventional level, of moral development, 261, 262*t*
Convergence, 102–103
Cooperation, encouraging, 422
Coping
with crisis, tips for, 73*f*
definition of, 78

psychosocial factors related to, 290
role of hypothalamus in, 289–290
stomach and, 289
Hyperopia, 92
Hypnosis, 135–136
features connected to, 136
medical uses of, 136
myths and facts concerning, 135*f*
Hypothalamus, 52, 59
during meditation, 134
role in hunger and eating of, 289–290
suprachiasmatic nucleus (SCN), 118*f*
Hypotheses, 203*f*

I

Iconic memory, 176*f*
Id, 319, 374
Ideality, 333
Identical twins, 35*f*
aggression studies on, 268
intelligence research and, 218
personality studies of, 330
sexual orientation and, 295
Identity crisis, 265*f*, 278
Illness, stress and, 69–73. *See also* Disease
Illusion(s), 99
Ames room, 104
Muller-Lyer, 99
Imaginary audience, 247–248
Imitation
learning language and, 208
phobias resulting from, 350
of role models, smoking and, 74
Immune system
cancer cells and, 69, 69*f*
compromising of, stress and, 69*f*
exercise and, 80
impairment of, 66
Implicit bias, 406
Implicit/nondeclarative memory, 180
Impossible figures, 102, 102*f*
Imprinting, 256
Incentive theory, 287
Incus (anvil), 94*f*
Independent variable, 16, 29
Individualistic cultures, 270, 271*t*
Individual psychology, 322
Infants. *See also* Babies
assimilation and accommodation of, 242–243
attachment and, 256, 257. *See also* Attachment
brain development of, 235
expressions of emotion in, 305
measuring perceptual abilities and preferences of, 237
motor development of, 236
research on imitation of facial expressions of, 248*f*

schemas of, 242
sensory and perceptual development of, 236
stimuli preferred by, 285
vision in, 236–237
Inferiority complex, 322
Information
and elaborative rehearsal, 181*f*
encoding, for LTM, 180
integration into long-term memory, 179
processing, 175
aging and, 241
dreams as, 123
remembering, mood congruence and, 182
in sensory memory, 176
in short-term memory, 177, 178
Information processing approach, 8
Information processing model, 174
Informed consent, 14, 28
Infrequency, statistical, 342
Ingroup, 406
Ingroup favoritism, 407
Inheritance
Alzheimer's disease and, 185
genetic, 34–37
research of human, 35
Injury
memory loss and, 184–185
traumatic brain, 184
Inner ear, 94*f*, 110
Inner-ear deafness, 95
Insanity, 344*f*
Insanity plea, 344*f*
Insight, 157
learning through sudden, 168
Insight learning, 157
Insight therapies, 374–382
Insomnia, 123
Instinctive drift, 162
Instincts, 284–285, 308
and aggression, 419
repressed, 319
Institutionalization, 394–395
alternatives to, 394*f*
Intellectualization, 320*t*
Intelligence
debate over definition of, 211
ethnicity and, 219
extremes in, 216–217
formal definition of, 211
genetic and environmental influences on, 218
influence of brain on, 217–218
measuring, 213, 214*f*, 215
ongoing debate over nature and measurement of, 220
research on, 211–212
unusual form of, 216*f*
Intelligence quotient (IQ), 213. *See also* IQ tests
Intelligence tests, 213, 214*f*. *See also* IQ tests

assessing mental ability at the extremes, 216–217
reliability of, 215
standardization of, 215
stereotype threat and group differences in, 220
validity of, 215
Interactional perspective, 229
Interference theory, 187*f*
Intermediate cultures, 271*t*
Intermittent schedule of reinforcement, 151
Internal locus of control, 80
Interneurons, 47*f*
Interpersonal attraction, 408–409
Interposition, 105*f*
Interpretation, 375–376
Interviews, 19
for personality assessment, 332
strengths and challenges of, 334
Intrinsic motivation, 296, 297, 308
Introversion, 330, 331
Investment theory, 204–205
IQ tests, 213–215. *See also specific tests*; Intelligence tests
questions of cultural bias related to, 219
Iris, 93*f*, 110

J

James-Lange theory, 300, 301*f*
drawbacks of, 302
Jerusalem syndrome, 355
Jet lag, 118
Just noticeable difference (JND), 89*f*, 109

K

Ketamine (Special K), 132
Kinesthesia, 97, 98, 98*f*
Kinesthetic sense, 98*f*

L

Language. *See also* Communication
abilities of chimpanzees and gorillas, probing, 209–210
building blocks of, 207*f*
development, 207–208
stages of, 208*t*–209*t*
necessities for producing, 206
proponents of animal, 210
thought and, 206
Language acquisition device (LAD), 208
Latent content, 375
Latent learning, 157
Lateral hypothalamus (LH), 289–290
Lateralization, 54
Law of effect, 149–150, 168
Learned helplessness
punishment and, 155*f*
theory, 353

Monozygotic twins, 35*f*
    intelligence research and, 218
Mood congruence, 182
Mood disorders, 352
    causes of, 353
Mood stabilizer drugs, 388
Moral development, 260–263, 278
    assessing Kohlberg's theory of, 263
    conventional level, 261
    Kohlberg's stages of, 260, 262*t*, 261–263
    postconventional level, 261
    preconventional level, 260
Morality, 260
Morality principle, 319
Motion parallax, 103, 105
Motivated forgetting theory, 187*f*
Motivation
    and behavior, 289–297
    biologically oriented theories of, 284–285
    biopsychosocial theories of, 287–288
    effect of attributions on, 287
    expectancies and, 287
    intrinsic versus extrinsic, 296, 297
    psychosocial theories of, 287
    research on, 282
    sensation seeking and, 286
    theories and concepts of, 284–288
    traits distinguishing high achievement, 293
Motor cortex, 53
    body representation of, 55*f*
Motor development, 236
    and culture, 236
Motor neurons, 47*f*
Motor reproduction, 159*f*
MRI (magnetic resonance imaging) scan, 23*t*
Müller-Lyer illusion, 99
Multiphasic inventories, 332–333
Multiple intelligences, theory of, 212
Multiple personality disorder, 354
Multitrait inventories, 332–333
Mutations, genetic, 37
Myelination, 235*f*
Myelin sheath, 38
Myopia, 92

N
Naikan therapy, 393
Narcolepsy, 123, 123*f*
Narcotics. *See also* Drugs
    effects of, 128*f*
    pain relief and, 130
Naturalistic observation, 18–19
Natural killer cells, 69*f*
Natural selection, 37
Nature-nurture controversy, 5, 28, 228. *See also*
    Behavioral genetics
    attachment behavior and, 256
    language acquisition and, 208*f*
    and theories of schizophrenia, 356–357

Nearsightedness, 92
Need for achievement, 293
Negative punishment, 153
Negative reinforcement, 150, 150*t*
Neo-Freudians, 322–323
Nerve deafness, 95
Nerves, 37
    parasympathetic, 49
Nervous system
    branches of, 45*f*
    influence of psychoactive drugs on, 126
    organization, 44–48
    tools for studying, 20
Neurochemistry, as major contributor to
    personality, 330–331
Neurogenesis, 44–46, 58
Neuroleptics, 388
Neurons, 37, 58
    basic features of, 37–38
    communication
        and, 38
        within, 39*f*
        between, 40*f*
    sensory and motor, 47*f*
    structure of, 38*f*
Neuroplasticity, 44, 58
Neuroscience, 34, 58
    breakthrough in, 45*f*
    learning and, 160–161
Neuroscientific/biopsychological perspective, 8
Neurosis, 344, 345
Neuroticism, 315*f*, 317*f*
Neurotransmitters, 38, 41*f*, 58, 75
    effect of cocaine on, 126*f*
    in genetically engineered mice, 183
    imbalances of, 353
    known or suspected effects of, 42*t*
    release of, 70
    schizophrenia and, 357
    substance P in, 90
Neutral stimulus, 144, 148*f*
Nicotine, 75
    effects of, 128*f*, 130
    as teratogen, 234
Nightmares, 124, 125*f*
Night terrors, 124, 125
NMDA (N-methyl-D-aspartate), 183
Nodes, 38
Non-rapid-eye-movement (NREM) sleep. *See*
    NREM sleep
Noradrenaline, 42*t*
Norepinephrine, 42*t*, 43. *See also* Neurotrans-
    mitters
    imbalances of, 353
    release of, 75
Norms
    definition of, 412
    for personal space, 412
    violation of, 342

NREM sleep, 121
    sleepwalking during, 125

O
Obedience, 413
    advantages of conformity and, 413*f*
    factors influencing, 414, 415
    Milgram's shock generator and, 414–415
    overcoming destructive, 423–424
    relaxed moral guard and, 424
    situational factors and, 424
Obesity
    epidemic proportions of, 290
    as risk factor for heart disease, 70
Objective personality tests, 332–333
    strengths and challenges of, 334–335
Object permanence, 244*f*
Observational learning, 158, 386
    and destructive behaviors, 166–167
    and gendered behavior, 268–269
    processes, 158, 159*f*
    techniques, 384
Observations
    for personality assessment, 332
    strengths and challenges of, 334
Obsessive-compulsive disorder (OCD), 345,
    349
Occipital lobes, 54
Olfaction, 96–97, 96*f*, 110
Olfactory bulb, 96*f*, 110
Olfactory epithelium, 96*f*
Openness, 315*f*, 317*f*
Operant conditioning, 149, 164–166
    beginnings of, 149–150
    differences between classical and, 149
    in everyday life, 165*f*
    techniques, 384
Operations, 244
Opiates. *See also* Drugs
    effects of, 128*f*
    medical uses of, 130
Opponent-process theory, 105–106
    example of, 106
Originality, 205
Outer ear, 94*f*, 110
Outgroup, 406
Outgroup homogeneity effect, 407
Oval window, 94*f*
Overgeneralization, 378
Overlearning, 27

P
Pain. *See also* Chronic pain
    exacerbation of, 77
    functions of, 77
    perception, 90
    phantom limb, 90
    relief with narcotics, 130

cognitive behavioral, 156
cognitive dissonance, 404*f*
cognitive-social, 156–158
damage, 241
decay, 187*f*
disengagement, 275, 279
encoding failure, 187*f*
evolutionary/circadian, 122
frequency, 94
gate-control, 90, 109
humanistic, 325–327
interference, 187*f*
investment, 204–205
motivated forgetting, 187*f*
multiple intelligences, 212
opponent-process, 105–106
personality-job fit, 274
of personality, psychoanalytic/psychody-
     namic, 318–324
place, 94
programmed, 241
repair/restoration, 122
retrieval failure, 187*f*
social-cognitive, 328–329
socioemotional selectivity, 275, 279
trait, 314–317
triarchic, of successful intelligence, 212
trichromatic, 105, 106
Theory of multiple intelligences, 212
     and possible careers, 212*t*
Therapists
     eclectic approach of, 392
     finding, 392
     need for gender sensitivity of, 394*f*
Therapy(ies)
     behavior, 383–386
     biomedical, 386–389
     cognitive, 376–378
     cultural issues in, 392–393
     five most common goals of, 391*f*
     goals and effectiveness, 391–392
     group, family, and marital, 381–382
     humanistic, 378, 380–381
     insight, 374–382
     psychoanalysis/psychodynamic, 374–376
     woman and, 393–394
Thermogenesis, 289, 291
Thinking, 200, 221. *See also* Language
     formal operational, 247
Thorndike box, 149*f*
Time management strategies, 26
Tip-of-the-tongue phenomenon, 187*f*
T-lymphocyte, 69*f*
Tobacco use, dangers of, 74–76. *See also*
     Smoking
Toddlers, using modified sign language with,
     207

Tolerance, 129, 359
Top-down processing, 108, 108*f*, 111
Traditional three-stage memory model,
     175–180, 175*f*, 182
     long-term memory, 179–182
     sensory memory, 176–177
     short-term memory, 177–178
Trait disorders, 346
Traits, 34, 314. *See also* Trait theories
     in abused and neglected children, 317
     Big Five, 315, 316
     heritability of, 36
     important in the 1800s, 333
     polygenic, 35
Trait theories. *See also* Traits
     evaluating, 317
     five-factor model (FFM), 314–315
Trait theorists
     areas of study for, 314
     criticism of, 317
     early, 314
Transduction, 88, 109
Transference, 375
Trauma. *See also* Post-traumatic stress disorder
     (PTSD)
     to brain, 184
     repression and, 192
     tips for coping with, 73*f*
Traumatic brain injury, 184
Triarchic theory of successful intelligence,
     212
Trichromatic theory, 105, 106
Trichromats, 106
Twins, 35–36
     fraternal, 35*f*
     identical, 35*f*
     Minnesota Study, 218
Tympanic membrane, 94*f*
Type A personality, 70
     hostility and, 71
Type B personality, 70

U

Ulcers, 72–73
     psychosomatic explanation for, 73
Unconditional positive regard, 326, 336, 381
Unconditioned response, 144
Unconditioned stimulus, 144
Unconscious, 318, 336
     unlocking secrets of the, 374–376
Unconscious mind, 117*f*
Unconsciousness, 116, 117*f*

V

Variable interval, 151*t*
Variable ratio, 151*t*

Ventromedial hypothalamus (VMH),
     289–290
Vestibular sense, 97
     confusion of, 98*f*
     function of, 98
Virtual reality therapy, 384
Visible light, 91*f*
Vision
     abnormalities of, 92
     after middle age, 240
     color, 105
     color-deficient, 106, 107*f*
     within first two years of life, 236–237
     properties of, 92*f*
     stereoscopic, 102
Visual cliff, 102, 103*f*
Visual learning, 24
Visual memory, 176*f*

W

Wavelength, 92
Websites, 5, 6
Wechsler Adult Intelligence Scale for
     Children (WISC-III) (WAIS), 214*f*
Wechsler Adult Intelligence Scale (WAIS),
     214*f*
Wechsler Preschool and Primary Scale of
     Intelligence (WPPSI), 214*f*
Wernicke's aphasia, 54
Wernicke's area, 54
Wiitiko psychosis, 362–363
Windigo psychosis, 362–363
Withdrawal, 129, 130, 132, 359
Women
     and mate selection, 316
     anxiety disorders in, 348
     behavior often shown by, 268*t*
     eating disorders in, 291
     effect of stereotype threat on, 220
     physical changes in, 238–239, 239–240, 266
     in psychology, 11
     stereotypical media portrayals of, 166
     and therapy, 393–394
     top four traits for, 316
Work, 274
Working memory
     as central executive, 178*f*
     short-term memory as, 178

Z

Zygote, 232